ISBN 978-1-331-42357-7
PIBN 10188146

1 MONTH OF
FREE
READING

at
www.ForgottenBooks.com

By purchasing this book you are eligible for one month membership to ForgottenBooks.com, giving you unlimited access to our entire collection of over 700,000 titles via our web site and mobile apps.

To claim your free month visit:
www.forgottenbooks.com/free188146

English
Français
Deutsche
Italiano
Español
Português

www.forgottenbooks.com

Mythology Photography **Fiction**
Fishing Christianity **Art** Cooking
Essays Buddhism Freemasonry
Medicine **Biology** Music **Ancient
Egypt** Evolution Carpentry Physics
Dance Geology **Mathematics** Fitness
Shakespeare **Folklore** Yoga Marketing
Confidence Immortality Biographies
Poetry **Psychology** Witchcraft
Electronics Chemistry History **Law**
Accounting **Philosophy** Anthropology
Alchemy Drama Quantum Mechanics
Atheism Sexual Health **Ancient History**
Entrepreneurship Languages Sport
Paleontology Needlework Islam
Metaphysics Investment Archaeology
Parenting Statistics Criminology
Motivational

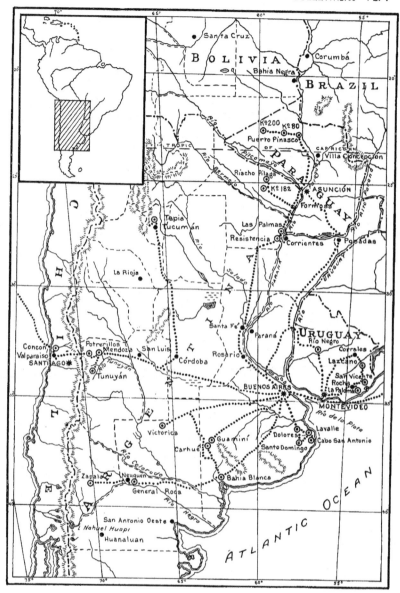

MAP OF SOUTHERN SOUTH AMERICA, TO SHOW COLLECTING LOCALITIES
MENTIONED IN THE PRESENT REPORT. ROUTE OF TRAVEL IS INDICATED
BY HEAVY DOTTED LINE

SMITHSONIAN INSTITUTION
UNITED STATES NATIONAL MUSEUM
Bulletin 133

OBSERVATIONS ON THE BIRDS OF ARGENTINA, PARAGUAY, URUGUAY, AND CHILE

BY

ALEXANDER WETMORE

Assistant Secretary, Smithsonian Institution

WASHINGTON
GOVERNMENT PRINTING OFFICE
1926

ADVERTISEMENT

The scientific publications of the National Museum include two series, known, respectively, as *Proceedings* and *Bulletin*.

The *Proceedings*, begun in 1878, is intended primarily as a medium for the publication of original papers, based on the collections of the National Museum, that set forth newly acquired facts in biology, anthropology, and geology, with descriptions of new forms and revisions of limited groups. Copies of each paper, in pamphlet form, are distributed as published to libraries and scientific organizations and to specialists and others interested in the different subjects. The dates at which these separate papers are published are recorded in the table of contents of each of the volumes.

The *Bulletin*, the first of which was issued in 1875, consists of a series of separate publications comprising monographs of large zoological groups and other general systematic treatises (occasionally in several volumes), faunal works, reports of expeditions, catalogues of type-specimens, special collections, and other material of similar nature. The majority of the volumes are octavo in size, but a quarto size has been adopted in a few instances in which large plates were regarded as indispensable. In the *Bulletin* series appear volumes under the heading *Contributions from the United States National Herbarium*, in octavo form, published by the National Museum since 1902, which contain papers relating to the botanical collections of the Museum.

The present work forms No. 133 of the *Bulletin* series.

Alexander Wetmore,
Assistant Secretary, Smithsonian Institution.

Washington, D. C., *December 3, 1925.*

TABLE OF CONTENTS

OBSERVATIONS ON THE BIRDS OF ARGENTINA, PARAGUAY, URUGUAY, AND CHILE

By Alexander Wetmore

Assistant Secretary, Smithsonian Institution[1]

INTRODUCTION

The successful operation of the Federal Migratory Bird Treaty in according protection, hitherto uncertain, to migratory game and insectivorous birds, with resultant increase in many species, led naturally to inquiries regarding the present status of birds that migrate in winter beyond our borders. Particularly was this the case with those species, mainly shore birds, that pass south into the southern portion of the South American Continent. To gather first hand information on the questions involved the Biological Survey, United States Department of Agriculture, in May, 1920, dispatched the writer to Argentina with instructions to carry on the desired observations. During the extended period of field work incident to such a task there was abundant opportunity to make representative collections of native birds and to record many points of interest concerning their distribution and habits.

Since observation of shore birds was the main object in mind, travel was restricted largely to the level sections where suitable shores and marshes were available, so that collections were made in the main in the lowlands. The area covered extended from northern Paraguay south to northern Patagonia, and from the eastern border of Uruguay west to the foothills of the Andes in Mendoza, and included a limited section near Valparaiso, Chile. Points for work were chosen carefully to allow comprehensive survey of as large an area as practicable. Studies of the specimens secured together with field observations, where pertinent, are presented herewith in as much detail as is warranted.

For assistance while in South America, thanks are due especially to Dr. Roberto Dabbene of the Museo Nacional in Buenos Aires,

---

[1] The investigations covered in the following pages were made when the author was on the staff of the Bureau of Biological Survey, U. S. Department of Agriculture. The report has been brought to completion since he became Assistant Secretary of the Smithsonian Institution.

our foremost authority on Argentine birds, for information and assistance in organizing field work in Argentina, and many valuable details regarding the country. Dr. Roberto Sundberg, of the Defensa Agricola, among others, was instrumental in securing permits necessary for work in Uruguay, and Señor Juan Tremoleras, of Montevideo, an experienced naturalist, gave valuable information regarding his country. In Valparaiso Dr. Edwyn Reed was most cordial in arranging for work in the vicinity, and later forwarded a number of valuable specimens from Juan Fernandez Island. Acknowledgment is due also to many friends for courtesies in connection with the prosecution of field work, for living quarters in remote sections, and for transportation where travel was difficult.

In subsequent work on the collections secured, loan of material necessary for comparison has been obtained through the friendly cooperation of the authorities of the Museum of Comparative Zoölogy, the Field Museum of Natural History, the Carnegie Museum, the American Museum of Natural History, and the Museum of Vertebrate Zoology, while visits made to the four institutions first named have permitted more comprehensive studies. Finally acknowledgment is made of the constant advice of Dr. C. W. Richmond, Associate Curator of Birds in the United States National Museum, in particular with regard to obscure and difficult points in nomenclature.

ITINERARY

In the following paragraphs is given an itinerary of travel performed while engaged in the duties outlined in the preceding paragraphs, with brief descriptions of the localities where specimens were secured. These have been located as definitely as possible since many of them are not shown in current atlases. They may be found also on the accompanying map (pl. 1), where the route followed is shown by a dotted line, and collecting localities are indicated by a line drawn below the name of the place.

On June 21, 1920, after a 24-day journey from New York, I arrived in Buenos Aires, where Dr. Roberto Dabbene, at the Museo Nacional de Historia Natural, received me with the greatest cordiality and on this and many subsequent occasions accorded me the freedom of the museum collections, gave me letters of introduction to naturalists throughout Argentina, and aided in many other ways. Several days were occupied in necessary preliminaries and in securing needed information. On June 29, during a day afield near Berazategui, Province of Buenos Aires, distant 27 kilometers southeast of Buenos Aires, I secured my first specimens of Argentine birds. The region was one of level fields, with bordering lines of willows and eucalypts, that gave way to muddy shores and rush or brush grown marshes at the border of the Rio de la Plata.

On the evening of July 3 I left Buenos Aires for the north by rail, and arrived the following noon at Santa Fe, where I had opportunity for a trip afield in the afternoon in level country near an open lagoon. As this was the hunting season many men were out in pursuit of birds of all kinds, as was the case at Berazategui. That same evening I continued north by rail to Resistencia, capital of the Territory of Chaco, where I arrived the following night. At Resistencia I was fortunate in meeting Señor Enrique Lynch Arribalzaga, with whom I was associated for several days. Field work was carried on here in a limited area of marsh, small lagoons, brush-grown fields and pastures, thickets, and small woodland near the Rio Negro, a small stream north of town. On July 11 I continued to Barranqueros, on the banks of the Rio Parana, 8 kilometers from Resistencia, and crossed by steamer to Corrientes. The following morning at daybreak I embarked on a small steamer for the port of Las Palmas. Travel on the great inland river systems at this time was uncertain, due to a strike among sailors who manned the steamers, and most of the boats normally available were not running. Government police boats were pressed into service for transportation of mails, and it was one of these that afforded communication at the time between Corrientes and Formosa. The river at this point varied from 200 to 400 meters in width, with its swift, turbid current enclosed between cut banks 2 or 3 meters high. Above Corrientes the shores were wooded or open by turns, with scant sign of habitation. At one point a hill 10 meters high made a marked eminence in an otherwise level landscape.

The port for Las Palmas, Chaco, is located on the western bank of the Rio Paraguay, a short distance above the confluence of that stream with the Parana. The steamer cast anchor, swung in to the shore, a plank was thrust out to the bank, and the few passengers and luggage for this point disembarked. A small narrow-gauge railroad led inland for 9 kilometers to the little village of Las Palmas, headquarters for a large estancia that covered 60 leagues of land. The manager, R. A. Young, to whom I presented a letter from Señor Enrique Cáceres, governor of the Territory of Chaco, received me here and granted permission to carry on work on the lands under his charge. Quarters were obtained at a little fonda in the village. A strike among workmen employed in the quebracho and sugar mill was in progress, and at times the factory, guarded by militia imported for the purpose from Buenos Aires, was virtually in a state of siege. The Rio Quia, known familiarly as the " Riacho," passed the northern border of town, with numerous small lagoons and old river channels, or esteros, adjacent. The Rio de Oro, a large stream, drained an area farther north. Tracts of woodland, partly open and partly dense jungle, bordered streams and channels, with broad savannas on either side, through which were scattered groves of trees, many 18 or 20 meters tall. Slight depressions in the prairies were filled with water and many low tracts were grown with tall stands of saw grass, known as *paja brava*. Lagoons were bordered by rushes and covered with floating masses of vegetation. Suitable tracts in the higher savannas were under cultivation, and grazing cattle had opened trails through forest that otherwise would have been impassable. (Pls. 2, 3, and 4.)

Work was completed here on August 2, when I went down again to the port to board another police boat bound up the Paraguay for Formosa. The current was swift, rendering progress slow. The banks were wooded, with game or cattle trails leading to water at intervals, or with occasional clearings in the vicinity of the few small towns. (Pl. 4.)

Formosa, the capital of the Territory of Formosa, located on the west bank of the Rio Paraguay, was reached early on August 3. The land on the river bank near the town is comparatively high, but inland and to the north becomes low and swampy. A line of railroad built by the National Government to promote development of the country extended northwest from Formosa for a distance of 297 kilometers, on a line midway between the Bermejo and Pilcomayo rivers. It was contemplated to extend it finally to Embarcacion, where it would connect with other lines from the south. Stations on this road at this time bore numbers corresponding to their distance in kilometers from Formosa. On August 5 I took the biweekly train to Kilometer 182, a point that had been recommended by Mayordomo, Cacique of the Tobas, whom I had met at Las Palmas. As the railroad leaves Formosa it enters the Chaco, a broad nearly level area of alternate forest and marshy savanna, cut by several large streams, that extends west of the Rio Paraguay from northern Santa Fe north through Chaco, Formosa, and western Paraguay into Bolivia. For miles our train traversed a roadbed built through an interminable estero, with broad swamps and prairies on either hand, dotted with slender trunked palms interspersed with stands of saw-edged grass and rushes, and bordered by bands of low-growing hardwoods, prominent among which was the quebracho, valuable for its dye product. Hundreds of acres were covered with ant hills built up 3 or 4 feet above the surrounding level to raise them above inundations caused by the summer rains. At intervals we crept out to higher ground and stopped at some little station, with a cluster of low houses or grass-thatched huts about it. Elsewhere no signs of man were visible; bands of rheas,

BORDER OF LOW WOODLAND IN FORMOSAN CHACO
Riacho Pilaga, Formosa, August 8, 1920

SMALL LAGOON IN CHACO, PARTLY COVERED WITH FLOATING PLANT
KNOWN AS LLANTEL DE CABALLO (PISTIA OCCIDENTALIS)
La Sabana, Chaco, July 5, 1920

PARTLY DRY LAGOON IN CHACO, BORDERED BY DENSE GROWTH OF SAWGRASS

Resistencia, Chaco, July 8, 1920

TERMITE HILL OPENED BY FLICKERS (COLAPTES CAMPESTROIDES)

Las Palmas, Chaco, July 30, 1920

A STAGNANT ESTERO, OR ANCIENT RIVER CHANNEL, IN CHACO, BORDERED
BY FOREST AND COVERED WITH GROWTH OF WATER HYACINTH; HAUNT
OF JACANAS (JACANA JACANA)

Las Palmas ,Chaco ,July 30, 1920

MOUTH OF RIO DE ORO WHERE IT DEBOUCHES INTO THE RIO PARAGUAY.
THE LOW WOODED BANKS ARE TYPICAL OF THE CHACO

Below Puerto Bermejo, Chaco, August 2, 1920

SLUGGISH STREAM DRAINING A BROAD PALMAR, OR MARSH GROWN WITH
SLENDER PALMS

Near Formosa, Formosa, August 23, 1920

OPEN TRACT WEST OF PUERTO PINASCO, PARAGUAY.   DRY GRASS,
GROWING IN SANDY LOAM, HAS BEEN BURNED BY INDIANS TO DRIVE OUT
GAME

Riacho Salado ,170 kilometers west of Rio Paraguay, September 24, 1920 flocks of maguari storks, courlans, and other strange birds were numerous. In mid-afternoon I reached my destination, Kilometer 182 (known locally as Fontana), and there left the railroad at the hospitable invitation of Don Pedro Upitz to continue by oxcart northwest for 15 kilometers to the estancia Linda Vista on the Riacho Pilaga. Señor Upitz had come in here as a colonist four years before, and had established himself beyond the limit of scant settlement in open country ranged by the Tobas. For several miles on either side of the railroad the forest had been cut away, but at the Riacho Pilaga tree growth was in its original condition. Open savannas, often of a marshy nature, mingled with scattered groves, while near the small sluggish streams, known as riachos, were extensive forests with a jungle undergrowth that, as it was not grazed, required a machete to penetrate. Several lagoons, some covered with matted vegetation that drifted about with the wind, offered attraction to water birds. The savannas were grown with bunch grass that seldom attained great height as it was burned yearly by the Indians to drive out concealed game. An extensive forest, known as the Monte Ingles, lay near a little frequented stream, the Riacho Ingles. The country as a whole was higher than that immediately west of Formosa and was now comparatively dry. It is inundated extensively during the summer rains. Frost was frequent; the first intimation of spring came toward the close of my stay with the blossoming of the tree known as lapacho (*Tecoma obtusata*). On August 21 I returned to Formosa for further work for a few days. (Pls. 2 and 5.)

On August 26 I passed my equipment through the Argentine customs in Formosa and crossed by rowboat to Alberdi, Paraguay, a little town on the opposite side of the Rio Paraguay, where passage was secured by steamer for Asuncion. The following morning I had a view of the winding outlet of the Rio Pilcomayo, and a short time later landed in Asuncion, the capital of Paraguay. Through the kindness of officials of the International Products Co., I received permission to visit their extensive land holdings in Alto Paraguay, and on August 28 set out up river again for Puerto Pinasco.

The Rio Paraguay this season was higher than normal by several feet, the water was tinged a dull olive, though with little sediment, and the current ran swiftly. On the west the shores were uniformly low, but low hills appeared at intervals on the eastern bank. At long intervals we stopped at small towns, and once or twice remained for several hours to take on wood used as fuel. The boat arrived at Puerto Pinasco, marked on older maps as Puerto Stanley, at daybreak on August 31. At this point the river is deflected to the west by a long hill projecting from higher country behind, and flows in general in an east and west direction for several kilometers before turning again to the south. At Puerto Pinasco the International Products Co. maintained a quebracho mill and headquarters for their cattle ranches, that in the aggregate covered several hundred square leagues of land. The Americans stationed here received me with greatest hospitality, and I owe much to their friendly assistance, in particular to Frank Branson, in charge of the large ranch at Kilometer 80, to Carl Hettman, his assistant, and to Fred Hettman, engineer for the company at Puerto Pinasco. On September 1 I visited a low hill, covered with dense forest located 35 kilometers west of the port. This hill or cerro, noted as being the only elevation of the sort in this part of the Chaco, was formed by an outcrop of what appeared to be quartzite, porphyritic in spots, overlaid with a deposit of limestone in which were traces of molluscan fossils. It rose 15 to 18 meters above the surrounding level. Apparently it is an outlier of the higher land that here forms the eastern bank of the stream. On September 3 I worked near the river at Puerto Pinasco and on September 4 proceeded inland to the ranch at Kilometer 80, located 80 kilometers west of the port. A narrow-gauge railroad used in transporting quebracho logs and supplies ran out for 56 kilometers; the rest of the journey was performed on horseback. The region showed the diversity usual in the Chaco. Broad savannas were broken by belts of low woodland, with dense undergrowth of spiny plants, or had scattered bushes and trees over their surface. Abundant growth of grasses furnished almost limitless feed for cattle. Lagoons, usually U shaped, and often a kilometer or two long, were numerous, and harbored many water birds. A meandering stream, the Riacho Jacare, wound across the country which was divided by fences into huge pastures 5 kilometers square each, thus embracing a league of land. Broad areas were covered with open stands of tall slender palms. A large open lagoon at the ranch house furnished an attractive point at which to observe shore birds passing abundantly in migration. It was much warmer here, there was no frost, and the discomforts of cold quarters in the Chaco of Argentina were soon forgotten. Warm, dry winds from the north prevailed.

On September 23, in company with Carl Hettman, I made a brief trip into the unexplored interior in a motor car. We continued west to a puesto, or outlying shelter hut at Kilometer 110, where we remained for the afternoon and night, collecting about a lagoon. The following morning we passed out through the last fences and continued west over Indian footpaths. The country was level and, as in the Chaco in general, the alternate belts of savanna and woodland ran east and west. Large areas grown with scattered palms were evidently inundated by summer rains; the forest growth became lower as we progressed. The dry season was near its close, and lagoons and streams were disappearing rapidly where they were not already dry. We crossed the stream bed of the Riacho Salado several times, now dry except for occasional alkaline pools, and passed two fair-sized lakes, Laguna Lata and Laguna Perdido. At Kilometer 200 we camped at a lake named Laguna Wall. Here we were beyond the limits of the Anguete Indians and encountered the first Lenguas. Seventy-five kilometers beyond was Laguna Verde, and still farther we heard there was a large Indian village. Considerable areas of slightly rolling country, with loose sandy loam, were traversed, and extensive thickets of a heavy thorned shrub, known as *vinal*, were encountered. There was evident approach to a more arid section, different from that found nearer the river. (Pl. 5.) We saw one jaguar, greatly astonished at the apparition of our rapidly moving car, encountered two otters traveling in search of permanent water, and startled occasional small deer, or rheas. Birds were numerous. On September 26 we returned to the home ranch, and on the 28th I arrived again in Puerto Pinasco. On September 30 an Indian took me across to the eastern bank of the Rio Paraguay, where I spent the day on the long hill already mentioned, the Cerro Lorito, of limestone formation, which rises 100 meters or more above the stream. Tall forest growth came to the water's edge and harbored species of birds not seen in the Chaco. Broad stretches of quiet water on either side of the river were covered with masses of floating water hyacinth and other growth, known collectively as *camalote*.

On October 2 I took the steamer to Asuncion, where I arrived on October 3, and continued on the 7th by rail to Buenos Aires, reaching that city on the 9th. Various matters of business consumed the period until October 19 when I proceeded to Dolores, in the eastern part of the Province of Buenos Aires, by rail, and then on October 22 continued east to Lavalle, traveling by motor as far as Conessa and by horse-drawn vehicle for the remainder of the distance. This region is a vast plain, elevated only slightly above sea level, with winding channels or cañadones bordered by rush-grown marshes at frequent intervals. Land was divided into extensive estancias given over mainly to grazing, so that rural population was limited. Lavalle (formerly called Ajo) is a straggling village on the banks of a small tidal stream known as the Rio Ajo. The land here is lower than at Dolores, so that exceptional tides force water up into some of the streets of the village; 10 kilometers below Lavalle the Ajo flows into the Bay of Samborombon. For this distance the stream is bordered by marshes and alkaline barrens, grown with *Salicornia peruviana*, with occasional little elevated spots that support a few low trees or bushes. Tidal channels with soft clay bottoms, difficult and often dangerous to cross, wind about through the marshy flats. Across the level pampa inland a few slight elevations of a meter or so are grown with groves of native trees that form veritable islands in an apparently limitless level plain, otherwise broken only by widely scattered estancias with their plantations of eucalypts. While at Lavalle it was my good fortune to spend a number of days at the Estancia Los Yngleses, the home of the late Ernest Gibson, an ornithologist well known for his careful and painstaking observations on the birds of this region. The Gibson estate is located about 6 kilometers south of Lavalle, and is surrounded by well-established groves of eucalyptus in addition to the lower tala, ombú, and coronillo trees native to the pampa. For work quarters it was my privilege to occupy a little building erected by Mr. Gibson for a study and museum. From this hospitable point I crossed on November 3 to the coast where camp was made in a little hut, 25 kilometers south of the northern point of Cabo San Antonio, on property belonging to the Estancia Tuyú. Here a broad sand beach extended north and south as far as the eye could reach, bordered inland by a stretch of shifting sand dunes 400 meters wide, with a marshy swale intervening between the dunes and the more elevated grazing lands beyond. Like most of this coastal region, this tract was visited only by occasional herdsmen or by parties from one of the estancias. Shore birds were encountered in migration from the north, with a great flight of long-tailed and parasitic jaegers, while large bands of pintails and other ducks came up from the south. A tremendous storm that endured for two days interfered somewhat with field work. On November 8 I returned to Los Yngleses, and November 11 continued to Lavalle for a few more days at the mouth of the Ajo and the vicinity. On November 16 I crossed by stage coach to Santo Domingo on the railroad, a distance of 18 leagues across the green plains, with only an occasional grove or an estancia to break the line of the horizon. For the first half of the distance marshes were frequent, but beyond the land became higher. The great storm of 10 days before was reported to have killed 300,000 sheep in the Province of Buenos Aires alone, and in many places we passed piles of their bodies. (Pls. 6 and 7.)

On November 17 I returned by rail to Buenos Aires, and on the 20th left again for the south. On the following morning the train passed through the barren hills of the Sierra de la Ventana and arrived in Bahia Blanca, where the route turned west. After leaving the level flats near the sea the railroad traversed an arid section slightly elevated and rolling, covered with low scrub and occasional tracts of scanty grass. At Rio Colorado descent was made to the stream valley of that name, and continued along it to Fortin Uno, where we crossed another elevated region to the valley of the Rio

Negro. At night on November 21 I reached the station of Rio Negro and obtained quarters in the village of General Roca, 2 kilometers distant. The valley of the Rio Negro here was about 6 kilometers wide, with a line of low rolling hills of sand and water-worn gravel at the north, cut by winding valleys that became steep-sided barrancas where first they opened on the flats below, and then disappeared. The region was arid and had vegetation of the usual desert types. Thorny mesquites (*Prosopis strombulifera*) were common, mingled with a yellow-flowered shrub (*Caesalpina praecox*), and creosote bush (*Covillea divaricata* and *C. nitida*). An opuntia (*Opuntia hickeni*) and a turkey-head cactus (*Echinopsis leucantha*) was fairly common. The valley floor, sloping gently to the Rio Negro, 5 kilometers from Roca, was covered with a scrub of atriplex (*Atriplex lampa* and *A. crenatifolia*) and creosote bush. Considerable areas were cleared, and, under irrigation, yielded abundant crops. The actual flood plain of the stream was of sandy loam, interspersed with much gravel. Here were thickets of willows, some attaining the size of trees, and baccharis (*Baccharis dracunifolia*), with a varied flora of herbs. Cottonwoods and tamarisk (*Tamarix gallica*) have been planted along irrigation ditches. The Rio Negro is a broad, swiftly-running stream, rather heavy with grayish white sediment. Its course was broken by low islands bordered by small channels, and little lagoons of quiet water were common. On the opposite shore a rock escarpment, with steep talus-strewn slopes at the base, rose to an elevation of 100 to 125 meters. The soil in general in this area was strongly alkaline. The crested tinamou, small flycatchers, finches, and odd tracheophones were common, while water birds abounded along the river. The region supported an avifauna far different from that of country covered previously. (Pl. 16.)

On the evening of December 5 I continued west by rail and on December 6 reached Zapala, in the Gobernacion de Neuquen, a town of 30 or 40 houses, at that time the terminus of the railroad, located on a broad flat on the watershed between the Limay and Neuquen Rivers, in sight of the distant snow-capped Cordillera. Here the land was thrown into broad ridges, with shallow depressions between that led down into a broad valley draining to the eastward. The region was arid, but supported various shrubs and a certain amount of grass. Elevation was about 900 meters, and the region lay in a higher life zone than Roca, except for certain hot north-facing valleys. Violent winds were frequent. Small seed snipe were here on their breeding grounds. Work was continued here until December 11. (Pl. 17.)

On December 12 I arrived in Bahia Blanca, Buenos Aires, and on the following day visited the flats about the bay at Ingeniero White, the port for the city. Here were broad stretches of alkaline barrens stretching inland from muddy bays, where shore birds were common. Woody vegetation was confined to low shrubs (mainly *Grahamia bracteata*), except where willows or eucalypts grew about scattered houses. Broad areas were covered with *Salicornia*. December 14 I moved north by rail to Carhue, Buenos Aires, a small town just beyond the Sierra de la Ventana, with a large lake of strongly saline water known as Lago Epiquen near by. The country here was rolling and was divided, as usual, into large estates given to grazing or the cultivation of wheat. Hollows on the pampa were occupied by little ponds or marshes, and extensive uplands were grown with bunch grass. Thousands of grebes were present on the large lake, attracted perhaps by myriads of brine shrimp, but were preparing to move to their breeding grounds elsewhere. After a few days spent in observation and collecting in this somewhat high pampa, on December 21 I continued by rail by way of Alta Vista, Darragueira, and Pico to Victorica, in the Gobernacion de Pampa, where I arrived December 22. Rolling sand dunes mingled with more level areas at this point and there were extensive tracts of open forest of caldén (*Prosopis nigra*), a thick-trunked, short-limbed tree, with lower growth of smaller trees, spiny shrubs, and stalked cacti. The region had been reported as one of many lagoons, but the present season had been extremely dry so that little water remained, and water-loving birds had perforce departed for other regions. The drought broke after Christmas, and two tremendous downpours filled all hollows with water. Small brush and tree haunting birds were very abundant here, but their tenure is limited, as land is being steadily cleared for wood or for cultivation. The belt of forest (pl. 9) was reported to extend for many kilometers north and south. On December 31 I departed by rail for Buenos Aires, where I arrived the following day.

January 7, 1921, I embarked on a steamer and early the following morning arrived in Montevideo, Uruguay, where I was occupied in various official matters for nearly two weeks, with time only for excursions along the coast to Carrasco, a summer resort, or to some of the numerous parks. On January 22 I proceeded by train to San Carlos and crossed by motor to Rocha in eastern Uruguay. The following morning I went for the day by train to the port of Rocha, known as La Paloma, where I found an extensive sand beach bordered inland by rolling pampa, cut by steep-walled gullies that sheltered dense thickets. Returning to Rocha the following morning I continued by motor from Rocha to San Vicente de Castillos, called locally Castillos, and shown on most maps as San Vicente. Inland, rounded hills, with frequent exposures of granite, rose with slopes grown with thickets and low trees. The Cerro Navarro, northeast of town, was especially promi-

VIEW ACROSS LEVEL PAMPAS OF EASTERN BUENOS AIRES

From lookout station at Estancia Los Yngleses, LaValle, Buenos Aires, October 30, 1920

PLANTATION OF TREES ON PAMPA, ATTRACTIVE TO SMALL BRUSH INHABIT-
ING BIRDS. AN OMBÚ TREE AT LEFT

Estancia Los Yngleses, LaValle, Buenos Aires, November 10, 1920

BROAD SALINE FLATS GROWN WITH SALICORNIA AND PAMPAS GRASS

Mouth of Rio Ajo, below LaValle, Buenos Aires, October 25, 1920

MARSH, OR CAÑADON, GROWN WITH RUSHES, TYPICAL OF PAMPAS REGION

Estancia Los Yngleses, LaValle, Buenos Aires, October 29, 1920

nent. The Laguna Castillos, Laguna Negro, and other lakes, swamps, and cienagas, covered broad areas on the coastal plain, interspersed with great forests of palms and open prairies. The palm groves were of a different type than those of the Chaco, as the trees had thick, heavy trunks and long fronds, whose heavy bases, when dry, furnished a valuable source of firewood. These groves covered large tracts extending toward the Brazilian frontier, a few miles distant. (Pls. 11 and 12.) Rains were frequent throughout my stay. It was said in San Vicente that I was the first North American to visit that section, and I was received with every courtesy. On February 2 with all my equipment I continued north and east of north in a two-wheeled cart drawn by horses over a little-traveled road heavily washed and gullied by rains. We crossed a range of rolling hills and then descended into a broad valley drained by the Arroyo Sarandi. The country was sparsely populated, rheas were abundant, and many other birds were seen. At the Paso Alamo on the Sarandi, 30 kilometers north of San Vicente, I made a camp for a few hours and collected a number of birds in low thickets and prairies. That night I slept at a *boliche* near the level marsh known as the Bañado de la India Muerta, and on the following morning did some collecting in the vicinity. On February 3 I reached Lazcano, 20 kilometers north of the marshes just mentioned, and there remained until February 9. Low rocky hills here bordered a broad valley drained by the Rio Cebollati. The stream itself was bordered by dense thickets and low trees, forming a band nowhere wide but still of fair extent, considering the type of country. On either hand were broad saw-grass swamps and meandering channels in fording which my horses frequently sank until head and neck alone projected above the water. Water birds abounded, thicket-haunting species were found along the stream, and prairie-inhabiting forms were encountered on the bare uplands. (Pl. 13.)

February 9 I left Lazcano by coach, crossed the river on a ferry or balsa at the Paso del Santafecino to enter a more populous region that continued to Corrales in the Department of Corrales, where I arrived that night. On February 10 I returned to Montevideo by railroad, and on February 13 continued by rail to Rio Negro, Department of Rio Negro, in northwestern Uruguay. Here a high, rolling plain was cut by a broad swift stream, the Rio Negro, bordered by low thickets with lagoons and marshes of small size in its flood plain. Birds were abundant but were in molt, and so were quiet. Rains were frequent and the weather enervating because of humidity and intense heat. Completing field work here on February 22, I went by train to Salto on the Rio Uruguay, and crossed that broad stream by ferry to Concordia in Argentina, where, after the usual customs examination, I secured quarters until there was a train for Buenos Aires, where I arrived again on February 24. On February 25 the day was spent in examining collections in the museum at La Plata, in company with Dr. Carlos Bruch.

March 2 I went by rail to Guamini, in the southwestern part of the Province of Buenos Aires, not far from Carhue, where observations had been made in December. Broad, open pampa (pl. 8), slightly rolling, extended for an apparently limitless distance, with occasional slight depressions occupied by lakes of more or less alkaline water that served to support fish or water stock. Guamini was built on the shore of the Laguna del Monte that had risen recently and flooded some of the lower streets of the town. Broad barrens, covered in part with alkaline efflorescences, stretched on either hand, with salicornia, chenopodaceous plants, and other salt loving herbs in abundance. Fall was at hand, the pampan vegetation had turned brown, the sky was often overcast and the wind cold. Colder weather in Patagonia was driving shore birds north, and the lake shore furnished attractive resting places where they flocked by hundreds.

On March 8 I left for Buenos Aires, where I arrived on the following morning. Here I was fortunate in meeting James L. Peters, who was traveling for the Museum of Comparative Zoölogy at Harvard University. As we were both bound for western Argentina we joined forces and traveled in company for a period. After a farewell visit to Doctor Dabbene at the Museo Nacional, we left by train on March 11, and on the following evening arrived in Mendoza, Province of Mendoza (altitude approximately 750 meters). On March 13 we collected across rough arid flats cut by many dry washes west of the city. Thorny brush of a desert type was scattered over sandy, gravelly slopes with abrupt hills in the background. In lower sections broad areas produced abundant crops through irrigation. On advice of Dr. Carlos S. Reed, then in charge of the Educational Museum in Mendoza, we proceeded on March 15 to Potrerillos, Mendoza, on the line of the transandean railroad, within the Andean foothills above the junction of the Rio Blanco and Rio Mendoza. The country was rough and broken, with bowlder-strewn winding valleys leading between steeply sloping hills; inland rose the snow-covered ridges of the Sierra del Plata. The altitude at the railroad was given as 1,370 meters. Our collecting was carried on mainly at about 1,500 meters. The region was only slightly less arid than the open plains below and supported the usual desert types of cacti, thorny shrubs, and bunch grass. The nights were cold and sharp and westerly winds from the higher slopes carried the chill of snow. On March 19 we rode inland to an old estancia, known as El Salto at an elevation of 1,800 meters, where a number of species of higher zone affinities not found below were taken. (Pls. 17 and 18.)

At Potrerillos we met W. B. Alexander, engaged at the time in studying parasites of cactus for introduction into Australia, who returned with us to Mendoza, March 21, and continued in our company on the following day, when we went by train to Tunuyan, Mendoza, 81 kilometers south. At this point a broad cultivated valley of fertile black loam bordered a small stream, known as the Rio Tunuyan. Waste land along the river and extensive marshy cienagas furnished suitable places for water birds (pls. 18 and 19), and broad fields where hemp was harvested attracted seed eaters. On March 24 we drove west by motor for 50 kilometers to the arid slopes below the mountain foothills, where we found the usual desert shrubs. On the eastern shore of the Rio Tunuyan was a range of rolling sand hills from 20 to 60 meters in elevation, with many dry washes and arroyos covered with thorny shrubs. Weather in general had become colder.

March 29 we returned to Mendoza and left by train the following day, Mr. Peters for Buenos Aires, and I for Tucuman by way of Villa Mercedes, Rio Quarto, and Cordoba. Mr. Alexander, who had journeyed ahead, joined me at San Luis. We arrived April 1 in Tucuman, where Mr. Peters joined me again on April 5. We met Dr. Miguel Lillo and examined his excellent collections, and also made the acquaintance of Señores L. Dinelli and E. Budín. Peters and I had planned to penetrate here into the higher mountains on the west, but found that the rainy season, which normally terminated in March, was still in progress, making mountain trails uncertain and in places impassable. As the next alternative we went on April 6 to Tapia, Tucuman, a well-known collecting spot, which though only 30 kilometers north of Tucuman, is in the edge of a more arid belt of lessened rainfall. Tapia was merely a station on the railroad with a few small houses and no regular accommodation for travelers. Through courtesy of the station agent, Señor Maximo Kreutzer, we were allowed to use a corner of the depot baggage room for work and sleeping quarters, and remained here until April 14. The region was one of small knolls and long hills that rose in places into small cumbres, the whole covered with a low scrub forest in which occasional clearings had been made. (Pl. 19.)

Large barrancas and scattered cattle trails made convenient passageways through the thickets, though ordinarily the growth was not sufficiently dense to impede passage. The altitude was approximately 700 meters. Bird life was abundant and of great variety.

The season was fall; many birds apparently had come down from the mountains and migratory movement was still in progress. At the same time we were far enough north to escape rigorous cold, so that insect feeding species were present in numbers. A red-flowered epiphyte (*Psittacanthus*) that formed brilliant patches of color, visible in the trees for long distances, drew many hummingbirds, among them a beautiful species with long tail (*Sappho sapho*).

The night of April 14 we returned to Tafi Viejo, Tucuman, and on the 16th visited Señor Budín in Tucuman. On April 17 we climbed the Sierra San Xavier above Tafi Viejo, a mountain rising to an altitude of 2,300 meters. The town lies at about 600 meters, with small cultivated fields or chacras extending up a gradual slope to the base of the hills at about 1,300 meters. At this point we entered a steep-sided valley and traversed a trail that zigzagged up the slopes through a heavy rain forest dense with creepers, ferns, and parasitic plants, and with an undergrowth of huge nettles, other soft-stemmed plants, and low shrubs. At about 1,800 meters on the trail this forest terminated, though on southeast exposures it ran up 250 meters farther. Beyond were openings, with grass waist high, and groves of tree alder and other strange trees that formed forest of another type in certain areas. At 2,100 meters tree growth, except in sheltered gulches, gave way to rounded slopes covered with bunch grass. Among such diverse habitats we obtained a number of birds not seen before and regretted that our departure was imperative on the following day.

In Tucuman, on April 18, we parted company and I returned to Mendoza, where I arrived April 20. At 5 the following morning I passed my baggage through the Argentine customs, and shortly after left on the trans-Andean railroad for Valparaiso, Chile, where I arrived at midnight. Through Dr. Edwyn Reed, to whom I was indebted for many courtesies, I removed on April 23 to Concon, a tiny settlement where the Rio Aconcagua enters the sea, going by rail to Viña del Mar and by motor car to a little road house at Concon, where I arrived at 9 in the evening in a drenching rain. At Concon the Aconcagua meandered through a level, fertile valley with rounded hills grown with brush on either hand. A broad gravel or sand beach lay on the ocean front, with rocky cliffs to the south. The weather was cool but pleasant, and with the general aspect of the country gave a strong reminder of California. On April 29 I returned to Valparaiso, and on the following morning embarked on the Grace Line steamer *Santa Elisa* for the States. Stops were made at Antofagasta, Chile, May 2; Iquique, Chile, May 3; Mollendo, Peru, May 4; and Callao, Peru, May 6. On May 11 we passed through the Panama Canal, and May 18 arrived at New York. On the following morning I again reached Washington.

LIFE ZONES

After some hesitation on the part of the author there has been included in this report a brief sketch of the life zones of the region traversed, a treatment that is necessarily tentative, since it is based on an amount of work in the field wholly inadequate when the great extent of territory included is considered. The limits assigned to the various divisions are thus merely suggestive. Attempt is made only to call attention to major zonal divisions as they appear to an eye trained to such observations in North America. Definite limits and characteristics may be given only with extensive data that may change some of the inferences presented at this place.

Doctor Dabbene in his Ornitologia Argentina (pp. 169–182) for the whole of Argentina has outlined five major faunas of somewhat different significance than the zones here outlined. Mr. Peters in his recent paper on the summer birds of northern Patagonia (pp. 281–283) has found three life zones indicated in the Territory of Rio Negro, which, beginning with the lowest, he numbers Zones 1, 2, and 3.

In the present instance four zones only are considered, the Tropical, Lower, and Upper Austral, and Temperate. Though the term Austral as applied to a life zone was originated to designate a region in the Northern Hemisphere it may without violence be utilized for the corresponding zone south of the Equator, since in reality it signifies an area adjacent to the Tropics. It is considered preferable to use an established name rather than coin a new one. The zone above the two divisions of the Austral is termed the Temperate in accordance with established usage of Goldman, Chapman, Todd, and others in regions near the Equator.

TROPICAL ZONE

If, as seems logical from experience in other parts of the New World, we adopt the occurrence of frost as marking the southern limit of the Tropical Zone, then the southern Chaco, north to the Rio Pilcomayo in Chaco and Formosa is not tropical, since heavy frosts are of regular occurrence there. In passing up the Rio Paraguay from Asuncion a large-leaved *Cecropia*, a tropical tree, was first recorded near the little village of Curuzu-Chica, while the mango tree was first observed a short distance above, at Antioquiera. This appeared to be the limit of dilute Tropical Zone along the Paraguay, though even at this point bananas appeared to have been slightly touched by frost. The winter of 1920, however, had been unusual for severity of cold. Puerto Pinasco and the Chaco behind it appeared to be within the lower limit of the Tropical Zone, though even here it did not seem to me that the typical tropics had been reached. From this point the line limiting the Tropical Zone may swing down in the west into the lowlands of Salta and Jujuy, and in the east to include part of the Territory of Misiones. As the land through the interior is comparatively level, transition between Tropical and the succeeding zone is very gradual. The following forms of birds taken at Puerto Pinasco were not secured farther south. Some of them, however, are recorded for eastern Salta, Jujuy, or for Misiones:

Heterospizias meridionalis meridionalis.
Pyrrhura frontalis chiripepe.
Trogonurus variegatus behni.
Dendrocolaptes picumnus.
Lepidocolaptes angustirostris certhiolus.

Synallaxis albilora.
Troglodytes musculus musculus.
Turdus albicollis.
Basileuterus hypoleucus.
Basileuterus flaveolus.
Myospiza humeralis humeralis.

LOWER AUSTRAL ZONE

A moderate climate, one where frost may occur regularly but snow only casually, characterizes the greater part of the level sections of eastern and northern Argentina, extending south to the valley of the Rio Negro and on the north including Uruguay and a portion of southern Paraguay. As the zone that succeeds the Tropical belt it may be called the Lower Austral Zone. Though varied in its characteristics, it is readily divisible into at least two sections—one arid and the other humid. The level eastern pampas lie within the humid section of this zone, which includes also the Argentine Chaco. Toward the interior there is a gradual decrease in amount of annual rainfall, with a corresponding transition to a condition of aridity characterized by broad, dry plains grown with scattered scrub of caldén (*Prosopis nigra*), piquillin (*Condalia lineata*), or perhaps broad areas covered with creosote bush (*Covillea divaricata*), atriplex (*Atriplex lampa* and *A. crenatifolia*), and others. This zone covers the broad flats of eastern Mendoza to the base of the mountains, where it penetrates among the winding valleys into the foothills to about 1,200 meters altitude (in that latitude) and extends in the interior from Santiago del Estero and La Rioja south to the valleys of the Rio Colorado and Rio Negro in northern Patagonia. It corresponds to Peters' Zone 1 in Patagonia, which, according to him, prevails in the Territory of Rio Negro below "1,000 to 1,500 feet," and extends up the valley of the Rio Limay to a point between Paso Limay and Senillosa. It is also the zone of central Chile.

The Lower Austral Zone is characterized by the following breeding birds:

Rhea americana albescens.
Calopezus elegans morenoi.
Nothura maculosa nigroguttata.
Nothura darwini mendozensis.
Nothoprocta perdicaria perdicaria.
Rhynchotus rufescens pallescens.
Spiziapteryx circumcinctus.
Larus cirrocephalus.
Sterna trudeaui.
Leptotila ochroptera chlorauchenia.
Picazuros picazuros reichenbachi.
Notioenas maculosa fallax.
Amoropsitta aymara.
Myiopsitta monachus monachus.
Dyctiopicus mixtus malleator.
Chrysoptilus melanolaimus perplexus.
Lepidocolaptes angustirostris angustirostris.
Lepidocolaptes angustirostris praedatus.

Drymornis bridgesii.
Furnarius rufus rufus.
Leptasthenura fuliginiceps paraensis.
Siptornis patagonica.
Stigmatura budytoides flavocinerea.
Rhinocrypta lanceolata.
Teledromas fuscus.
Spizitornis parulus curatus.
Spizitornis parulus patagonicus.
Troglodytes musculus bonariae.
Mimus triurus.
Mimus patagonicus tricosus.
Trupialis defilippii.
Phrygilus carbonarius.
Diuca minor.
Brachyspiza capensis argentina.
Brachyspiza capensis choraules.
Embernagra olivascens gossei.

UPPER AUSTRAL ZONE

Beyond the valley of the Rio Negro in Patagonia climatic conditions become more austere and snow and ice are regular features of a prolonged winter. This zone, which may be called the Upper Austral, since it corresponds to that zone in the north, runs southward apparently into Santa Cruz, perhaps almost to the Straits of Magellan and extends to the base of the Andes in the west. It was found at Zapala, Neuquen, though a narrow tongue of Lower Austral came along the floor of a deep valley almost to Zapala, and from there runs northward along the arid mountain slopes, being found above 1,500 meters in Mendoza. It corresponds to Zone 2 in Peters' statement of the life zones of Patagonia. Its occurrence in Chile is uncertain, but it should be found along the northern border of the southern forest region. In the south this zone is wholly arid and covers an area of rolling plateaus, broken by rocky hills and rough valleys, covered with low, thorny bushes or mats of spiny, stiffstemmed plants that persist throughout the year. Various flowering annuals appear with a somewhat rigorous spring and persist for a short period. There are numerous lakes, particularly in the west (many of them alkaline), with occasional patches of permanent green vegetation in springy localities. The region is one of high winds, that blowing from the west, sweep with them air currents cold from Andean snows.

The following birds are characteristic of this area:

Pterocnemia pennata.
Tinamotis ingoufi.
Calopezus elegans elegans.
Thinocorus orbignyianus.
Metriopelia m. melanoptera.
Geositta rufipennis.

Enicornis phoenicurus.
Muscisaxicola capistrata.
Muscisaxicola maculirostris.
Sicalis lebruni.
Phrygilus aldunatei.

Others may be added as the southern distribution of birds in Patagonia becomes better known. In this zone *Vultur gryphus* finds its lowest point of regular occurrence. A considerable number of species range in both Upper Austral and Transition Zones.

TEMPERATE ZONE

As field work did not carry me above the Upper Austral Zone, I am dependent on the accounts of other travelers for information on this higher zone. A statement regarding the Temperate Zone as it is found in western Patagonia is based mainly on Peters' remarks concerning his Zone 3, which he says covers the east Andean slopes in western Rio Negro and is characterized by a temperate forest with normal rainfall. Part, at least, of the forest of southern Chile, where precipitation is heavy, must belong here, but the southern limit of this zone is uncertain. In passing north along the Andes, tree growth becomes scanty a short distance north of Lago Nahuel Huapi, and in western Neuquen entirely disappears. Beyond, through the length of Argentina, the higher mountain slopes are arid and bare with scant vegetation, which is restricted mainly to the valley floors and the gentler inclines above. The bolder mountain masses show bare rock exposures, in the main too young to have weathered into permanent soil. Zonal delimitation under these conditions is difficult. In Rio Negro Peters placed the upper limit of his Zone 3 at about 1,000 meters. The dry, arid slopes to the northward have forced zonal lines rather abruptly upward as in crossing on the trans-Andean railroad from Mendoza it appeared to me that Temperate Zone began at the Rio Tupungato at about 2,800 meters' altitude.

The following birds include mainly species of the southern forested region. The list may be expanded by including some of the water birds peculiar to the Magellanic region.

Attagis gayi gayi.
Chloroenas araucana.
Microsittace ferruginea.
Strix rufipes.
Ipocrantor magellanicus.
Cinclodes patagonicus rupestris.
Sylviorthorhynchus desmurii.
Pygarrhichas albo-gularis.

Scytalopus magellanicus.
Scelorchilus rubecula.
Pteroptochos tarnii.
Agriornis livida fortis.
Lichenops perspicillata andina.
Melanodera melanodera.
Melanodera xanthogramma.

Above the Temperate Zone is a great Paramo Zone extending to the line of perpetual snows on the mountains, a cold, bleak region with little bird life, at present insufficiently known, that will be dismissed at this place with bare mention.

NOTES ON MIGRATION

Though ordinarily we may think of extensive and widespread migratory movements among birds as something more or less peculiar to the Northern Hemisphere, yet on investigation we find pronounced migrations among the birds of South America, especially in the southern part of the continent. The migratory flight in the latter region may be considered as of two kinds, first, that of birds come from North America for the period of the northern winter, and, second, that of species that pass south to breed, and with the close of the period of reproduction withdraw again toward the Tropics.

Though a number of passerine and other birds from North America come commonly to the northern part of South America, comparatively few of these species pass as far south as the section covered by Paraguay, Uruguay, Argentina, and Chile. Among the few of the smaller land species that perform this extended flight, the barn swallow and the bobolink are worthy of mention, especially the latter, as though the barn swallow occurs during the northern winter months from the West Indies southward, the bobolink withdraws wholly into the Chaco. The yellow-billed cuckoo, cliff swallow, olive-backed thrush, nighthawk, and Swainson's hawk are of more or less common occurrence in the northern half of the region in question, but are not found in abundance. In addition to these may be mentioned the parasitic and long-tailed jaegers (that have been recorded casually, but that occasionally at least, occur in great abundance along the coast of Buenos Aires), Cabot's, royal, and arctic terns, and the red and northern phalaropes. The great body of North American migrants, however, are shore birds, some of which as the two yellowlegs, the sanderling, and the spotted sandpiper have extended winter ranges, while others as the Hudsonian godwit, the upland plover, the buff-breasted, pectoral, Baird's, and white-rumped sandpipers find in the pampas and in Patagonia their winter metropolis. With these last may be mentioned the Eskimo curlew now nearly, if not actually, extinct.

A few individuals of these northern species arrive in the south in July and August, but their main southward flight occurs from September to November. In other words, they pass south with the coming of fall in the Northern Hemisphere, and below the Equator follow the advance of southern spring to their winter home, remain during the southern summer, and with the coming of colder weather in February and March withdraw northward until they cross the Equator and follow the northern spring in its advance to their breeding grounds in the northern United States, Canada, and Arctic

America. Their itinerary thus takes advantage of the shifting seasons in both continents.

Because some of these species now or formerly occurred in the Argentine in great abundance, it has been held by some that there are in these species two groups of individuals, a northern body that breeds in North America and migrates south to clement regions in Mexico or Central America, and a southern group, that occupies a breeding ground in Patagonia, the islands of Antarctic seas, or even the great Antarctic Continent, that comes north to winter in the Argentine. This belief was based in part upon the seemingly irregular occurrence of some of these migratory birds, with records (scattered and few) of certain species that were found on the pampas during the northern breeding season, and in part upon disbelief in the powers of flight in creatures apparently small and weak. There are certain species, such as the pied-billed grebe, cinnamon teal, fulvous tree duck, and others that have a breeding range in both North and South America. In some of these individuals from the two colonies appear indistinguishable; in others the two groups may differ slightly in minor characters. There has never been any certain indication, however, of the breeding south of the Equator of such species as the golden plover, Hudsonian godwit, the yellowlegs, and other species considered as migrants from the north. The scattered individuals that remain in Argentina during the northern summer are wounded, sterile, or otherwise diseased individuals that have been unable to perform the long flight northward, or that have lacked the physiological incentive to do so. The few supposed occurrences of their nesting have, on investigation, proven erroneous, and the migration and seasonal movements of these species is so well understood that there is no question that they nest in the north and pass south of the Equator only in migration. Data from birds banded in the north eventually will authenticate these facts.

In their movements after reaching the northern coast of South America these northern species have three main routes, one north and south along the Atlantic coast, one that passes along the Pacific coast line, and a third that follows the great interior north and south river system of the Paraguay and Parana. Some of the birds that follow this last route on their southern journey apparently drive straight south across the pampas until they strike the southern coast of Buenos Aires, and then swing around to follow up to some wintering ground in the eastern pampas, or near the mouth of the Rio de la Plata. In November, on the eastern coast of Buenos Aires, I witnessed a curious phenomenon where one line of northern migrants came driving south down the coast, and a second, traveling in the opposite direction, came sweeping up from the south. My only supposition was that birds on the latter course had come south by the interior route perhaps to reach the coast near Bahia Blanca, and were now turning to seek their winter homes.

With regard to the migratory movement of native birds, particularly in Argentina, many instances are noted in the writings of Dabbene, Gibson, Hudson, and others, while Peters has given an account of the arrival of a number of species in northern Patagonia. In general, such migratory movements are as readily evident to the field observer as in northern regions. Large numbers of ducks of various kinds, seed snipe. small ground-haunting flycatchers (*Lessonia r. rufa*), a subspecies of house wren, and other birds appear in the Province of Buenos Aires from more southern regions at the commencement of winter, and wholly or in part withdraw again as summer approaches. Other species, as *Thermochalcis longirostris*, are regular birds of passage from Brazil to Patagonia. The migratory flights of the fork-tailed flycatcher are as evident as those of the northern kingbird (*Tyrannus tyrannus*), for at the end of January these birds gather in flocks and begin a northward movement that carries all to Brazil during the following month. The jacana, the sulphur-bellied flycatcher (*Myiodynastes solitarius*), two species of martins (*Progne elegans* and *Phaeoprogne t. tapera*), and a small swallow (*Iridoprocne albiventris*) are summer visitants near Buenos Aires that retire at the approach of cold, as do the greater part of other small species, some of whose individuals are hardy enough to remain.

Even in the Paraguayan Chaco, in the edge of the Tropics, the spring migration was easily evident, as with the approach of warmer days in September *Podager nacunda* passed in small numbers to the south (making as regular a flight as the North American nighthawk), a kingbird (*Tyrannus m. melancholicus*), and another flycatcher (*Myiodynastes solitarius*) appeared, and a little goatsucker (*Setopagis parvulus*), hitherto absent, began its tremulous calls at evening.

The low woodland of the level reaches of the Chaco, with its dense jungle impervious to cold winds, and its tangled openings, where the sun may be warm even on sharp frosty mornings, harbors many winter visitants from the more open country to the south, or from the mountain slopes to the west. Here many small flycatchers, warblers, and other insect-eating birds rest in comfort and security, remaining quiet during brief spells of cold and becoming active when the sun appears. The woodlands to the westward that cover the low hills in northern Tucuman are also attractive at this season, and at times small birds are so abundant there that they fairly swarm.

Altitudinal migrations were easily evident in the Andean foothills in Mendoza, as flycatchers and others came working down the mountain slopes in little bands, traveling down toward the plains where they passed on northward. With heavy storms in the higher reaches, these movements become more pronounced and at times include hill inhabiting species that temporarily pass down to the warmer lowlands until the stress of weather has passed.

It was interesting also to observe the migrational movements of a form of the monarch butterfly (*Anosia erippus*) that wintered in numbers in the Chaco, and in spring flew southward to spread over the pampas.

ANNOTATED LIST OF BIRDS

The following account consists of an annotated list of the species of birds collected, with observations on a few of which no specimens were taken. Measurements, made in millimeters, have been taken by the method usual among present day American ornithologists. The wing measurement is the chord of the distance from the bend of the wing (the metacarpal or wrist joint) to the tip of the longest primary taken with dividers except in large birds where the measurement is made with a straight rule, but without flattening the wing. The length of tail, measured with dividers, is the distance from the base of the median rectrices on the uropygium to the tips of the longest tail feathers. The culmen has been measured from the base in all cases except where (as in parrots) it is specified as taken from the cere, etc. The distance, measured with dividers, is taken in a straight line from the basal point to the extreme tip. The tarsal length, likewise taken with dividers, is secured by placing one point of the instrument at the upper end of the tarsus on its posterior side, and the other at the end of the middle trochlea of the metatarsus.

With shifts in generic names, so common in our modern nomenclature, at times a change is required in the current designation for a family. Current family names have been derived from the appellation of some genus considered as typical of the group concerned, in the main from the oldest genus name in the family. Where this genus name has been changed through the application of the law of priority in publication, type fixation, or other cause, change in the family name necessarily follows. Certain ornithologists (mainly in England) are advocating derivation of a new family name from the next oldest generic appellation. This, however, may cause confusion since it may result in shifting the type genus from a group that has been held typical of the family concerned, to one that is aberrant or possibly even to a genus of doubtful allocation. The confusion that may arise is easily evident. It seems preferable to allow the same generic group to remain as typical of the family regardless of change in its appellation; in other words, to allow the family name to change with shifts in the name of the type genus. Such changes though lamentable are less confusing than shifts in the facies of a family complex, such as might result if the other course that has been outlined is adopted.

Such a course is implied in the International Code of Nomen- clature (art. 5), which specifies that "the name of a family or sub- family is to be changed when its type genus is changed."

Order RHEIFORMES

Family RHEIDAE

RHEA AMERICANA (Linnaeus)

Struthio americanus LINNAEUS, Syst. Nat., ed. 10, vol. 1, 1758, p. 155. (Sergipe, Brazil.)

In spite of continued pursuit by Indian and white hunters the rhea still remains in fair abundance in the wilder sections of the Chaco, while on many of the extensive estancias in the pampas of Argentina and Uruguay the birds are preserved in bands that in many instances include a large number of individuals. In settled districts, where land has been divided into small holdings, the great birds have been largely exterminated, a fate that will befall the majority as rural population increases. At the Riacho Pilaga, For- mosa, in August, 1920, Indians brought in bundles of rhea plumes for trade, to be sold later in Buenos Aires where they were made into feather dusters. Near the railroad at this same locality rheas still occurred in some of the open camps, but were more abundant farther inland toward the Rio Pilcomayo. Occasional bands were observed from the train in traversing the railroad line leading northwest into the interior from the town of Formosa.

In the Paraguayan Chaco west of Puerto Pinasco rheas were common. In 1920 fences on the holdings of the International Prod- ucts Co. had been extended westward to a point 120 kilometers from the Rio Paraguay. Outside this boundary rheas were encountered frequently but were wild and wary, as they were subject to pursuit by Indians who frequently offered bundles of plumes or sections of skin for sale. Small bands were to be seen within the fences in some of the league square *potreros*, where open savannahs offered suitable range, and near the ranch at Kilometer 80, west of Puerto Pinasco, rheas were observed frequently, especially in the region along the Riacho Jacare. On my arrival in that region on Sep- tember 6 I was told that a rhea's nest containing 43 eggs had been found a week previous, and during the period of my stay male rheas were heard booming during the morning hours. On one oc- casion (September 12) in company with Carl Hettman I heard this deep boom from a depression bordering a stream, and soon after saw a rhea running away through the acacias. The sound is deceptive, as it frequently seemed to come from a great distance when, as in the present instance, the bird was quite near. In fact the rhea sounds nearly as loud when far distant as when close at hand. Though the birds frequented the open camps, they followed narrow trails through bands of forest leading from one open tract to another. When not alarmed they walked slowly along, feeding from the ground. When approached they took sudden alarm and ran away with long strides, often with spread wings, covering the ground rapidly.

Rheas were to be stalked only with great care. In hunting them the Tobas cut small, leafy limbs from shrubs that did not wither quickly, and tied these on their bodies until they resembled bushes. In this disguise, one by the way that was most effective, they worked slowly down on the unsuspecting birds, advancing when the rheas were feeding with heads down, and remaining motionless when the rheas raised their heads to observe the country. Advance was made until within a few meters when the birds were killed with bow and arrow, or by a discharge from a single-barreled shotgun loaded with slugs. The Tobas and Pilagas in Formosa claimed that these birds possessed a keen sense of smell and were always careful to hunt them up wind. Whether there is truth in the assertion is uncertain, but it may be remarked that many Indians were readily detected even where the olfactory sense in the observer was only moderately developed.

On September 23, at Kilometer 110, I purchased two young rheas only 3 or 4 days old, for a yard of light-weight canvas from Capita-í, an Angueté. These young had a mournful little whistle, repeated constantly, that carried for some distance. They were interesting little birds, erect in carriage, with a preternaturally old appearance that was betrayed at once by their stumbling over slight obstacles as they walked or ran. In resting they frequently leaned against some object instead of lying prone, as do the young of many other long-legged birds.

One young rhea from this same brood was kept alive. It proved to be tame and unsuspicious, and, in fact, sought human company. At freedom in the patio at the ranch house it responded readily to an imitation of its note, and spent many hours reclining against my feet and ankles as I worked on notes and specimens. It was especially prone to do this toward evening when it became tired, and apparently in its eyes long-legged humans filled the place normally occupied by a long-legged father rhea. Young birds were common in a domestic state in many of the regions visited. When small they form odd and amusing pets, fearless and friendly in every way;

like some other pets, however, they have a surprising rapidity in growth, and soon develop to a point where they become a nuisance through an appetite that is satisfied omnivorously with whatever may offer that is small enough to swallow from the vegetables prepared for dinner, seized instantly when the cook's attention is attracted elsewhere, to the watch or shaving soap of the unfortunate owner.

An egg secured on September 23, 120 kilometers west of Puerto Pinasco, is between pale olive-buff and olive-buff in color, and has the shell roughened by fine short corrugations that at short intervals form slit-like pores several times longer than wide, with their axes in general longitudinal to the axis of the egg. This egg measures 135.2 by 96.5 mm. Rhea eggs, made into a batter with flour and fried, were excellent eating, and were sought after during the early breeding season. A single egg, thus prepared, was sufficient for three persons.

I am indebted to Carl Hettman for the following note on this species, based mainly on observations made on the upper Rio Pilcomayo. In that region the rhea nest mainly in September and October. The male is said to select a nest site on loose sandy soil, among tall grass in some secluded corner near forest or perhaps in a small, well-screened opening in the monte. A hole more or less circular a meter across is scratched out to a depth of from 100 to 150 mm. The females deposit their eggs in this. Frequently, in fact, nearly always, single eggs known as *guacho* (stray) eggs are found near by. It is supposed that they are deposited by females who visit the nest to find it occupied by some other member of the harem of their polygamous mate. Should the nest be found, the male is encountered near at hand. To attract attention from his treasures he dashes about with spread wings, but makes no effort (in the wild bird) to attack.

The breeding season varied, I found, with the locality. Mention has been made of the period in the Paraguayan Chaco. On December 8, in the hills back of Zapala, in the Territory of Neuquen, I encountered a male that had either eggs or small young concealed in a broad hollow, though search failed to reveal them. On February 2, north of San Vicente, Department of Rocha, Uruguay, I noted a male with chicks a week old, and were told that others were breeding.

Rhea flesh is eaten, the wings forming the portion most highly prized, and, in addition, parts of the bird figure as remedies in the country medicine chest. From the upper part of the stomach, preserved in a dried form, portions, cut up as needed, are boiled to make a tea said to be a specific for indigestion, a curious use for the powerful digestive agents found in the stomach of this bird. An oil found in the hollow tibiotarsal bone, when applied externally, is claimed to be an excellent remedy in cases of rheumatism.

The rhea is known as *avestruz*, or more commonly in the north as *ñandu*, a guaraní term also used to designate a spider. Occasionally when there was danger of confusion the rhea was indicated as *ñandu guaçu*, or large *ñandu*. The bird was also called *suré*, while to the Angueté Indians it was known as *pil-ya-pin*. The booming of the male was spoken of as *bureado ñandu*.

On large estancias where rheas are not molested they increase rapidly in number, and many landowners complained that the great birds were expensive, as they consumed much feed otherwise available for stock. Some said that their daily consumption of food equaled that of a sheep; others placed it as equivalent to that of a steer. As there was little return from sale of feathers, sentiment in many quarters is arising against them.

The only specimens secured were two chicks (mentioned above), taken September 23, 1920, at Kilometer 110 west of Puerto Pinasco, Paraguay. These were apparently about 3 days old. Both are females, and though of the same sex show considerable difference in tone of color, one being browner than the other.

Three subspecies of *Rhea* have been recognized, the typical *americana* from North Brazil, *intermedia* Rothschild and Chubb[2] from South Brazil and Uruguay (type locality Barra San Juan, Colonia, Uruguay), and *albescens* Lynch Arribalzaga and Holmberg[3] from Argentina (type locality Carhue, Province of Buenos Aires). The bird from Argentina was separated by Brabourne and Chubb[4] under the subspecific name *rothschildi* on the basis of specimens from the Estancia Los Yngleses, near Lavalle (formerly Ajo), Province of Buenos Aires. The name *Rhea albescens*, proposed for a supposed distinct species, the white rhea, though based on albinistic specimens, is obviously applicable to the present form, since Carhue is far to the northward of the known range of Darwin's rhea. In passing it may be noted that *Rhea americana*, var. *albinea* of Doering[5] is simply a new name for *albescens* of Lynch Arribalzaga and Holmberg.

The status of rheas from Paraguay is uncertain. Two suppositions are open, either that they represent an undescribed form or that they are representative of *intermedia* known from South Brazil. The two juvenile specimens from Puerto Pinasco differ from a newly hatched specimen of *R. a. albescens*, taken near Bahia Blanca, Argen-

[2] Nov. Zool., vol. 21, 1914, p. 223.
[3] El Naturalista Argentino, vol. 1, pt. 4a, April, 1878, p. 101.
[4] Ann. Mag. Nat. Hist., 1911, p. 273.
[5] Exped. al Rio Negro, Zool., 1881, p. 58.

tina (from University of Kansas Museum, collected in 1903), in having the wings more heavily marked with dark bands and the neck grayer.[6]

PTEROCNEMIA PENNATA (d'Orbigny)

Rhea pennata d'ORBIGNY, Voy. Amer. Merid., Itin., vol. 2, 1834, p. 67. (Bahia San Blas, southern Buenos Aires.)

On December 6, 1920, at Zapala, in western Neuquen, I examined an adult female of Darwin's rhea secured by a police officer, who had killed the bird about 50 kilometers southwest of town by a fortunate shot with a revolver. As the rhea had been skinned roughly to prepare it for the table, I was able only to note the curious arrangement of the feathering on the front of the tarsus and to secure the skull. On December 10 I saw a living young bird, recently hatched, of the same species that had been captured between a point below San Martin de Los Andes and Zapala. The call of this chick was lower and somewhat harsher than that of *Rhea americana*, but had a similar mournful, whistled inflection.

The skull secured has the dorsal elongation of the lachrymal bone, short and triangular, extending back only to a point well anterior to the rearward extension of the nasals, with the orbital margin of the postorbital process smooth. In *Rhea americana* the lachrymal is produced as an elongate spine that ends at the level of the posterior end of the nasals, while there is a distinct notch where the anterior margin of the postorbital process joins the margin of the orbit. I am not able to distinguish the difference in the form of the temporal fossa in the two species described by Pycraft.[7]

A second skull taken from the mummied body of a bird only half grown found on a butte south of Zapala December 9, exhibits the same distinguishing characters in the lachrymal as the adult.

Order TINAMIFORMES

Family TINAMIDAE

CALOPEZUS ELEGANS ELEGANS (Is. Geoff. Saint-Hilaire)

Eudromia elegans "D'Orb. et Is. Geoff.," Is. GEOFF. SAINT-HILAIRE, Mag. Zool., 1832, cl. 2, pl. 1. (Mouth of Rio Negro.[8])

On December 15, 1920, while working through the rolling pampa south of the shore of Lago Epiquen, near Carhue, Buenos Aires,

---

[6] For a description of the habits, economic value, hunting, and domestication of the rhea see La Cultura Argentina, Muñiz, F. J., Escritos Científicos, 1916, pp. 83–218, an acccount reprinted from old numbers of La Gaceta Mercantil.

[7] On the Morphology and Phylogeny of the Palaeognathae, Trans. Zool. Soc. London, vol. 15, December, 1900, p. 270.

[8] Designated by Peters, Bull. Mus. Comp. Zoöl., vol. 65, May, 1923, p. 287.

I was astonished to hear the low whistle of the martineta and to catch sight of an occasional crested bird as it ran aside through the weeds. Others were noted in this same region on December 18. It was said that the martineta was encroaching slowly on the range of *Rhynchotus rufescens* in southern Buenos Aires, and that as *Calopezus* came in it drove out and replaced the rufous-winged bird. Barrows* in 1881 recorded *Calopezus* only from the neighborhood of Bahia Blanca, though he covered the region to the northward as far as Carhue and Puan, so that there may be something in the belief that the species is extending its range. Carhue is situated in west central Buenos Aires, a point within the more watered section of the eastern pampas. No specimens of crested tinamou were secured here, so that these notes are placed questionably under the subspecies *elegans*. The form known as *morenoi* which occurs in western Pampa was found in more arid country, though the eastern limit of its range is not known.

On December 15 a *copetón*, as the birds were known locally, flushed with a startled note direct from a nest containing three beautiful eggs. The nest was a slight hollow scratched out under the lee of a low hillock of earth in ground partly bare of vegetation, though a fringe of grasses partly overhung the nest cavity. A few bits of grass stems carelessly arranged formed an attempt at nest lining, but lay at one side where they were no protection to the eggs. The whole formed as crude and carelessly constructed a nest as I have seen, save among such groups as shore birds and goatsuckers. The eggs have the usual shining glasslike surface and vary in color from cosse to calliste green. One has the side discolored to a light yellowish olive. They measure as follows: 51.9 by 40.5, 51.9 by 38.9, 51.7 by 38.5 mm. (Pl. 8.)

Many *Calopezus* were offered for sale in the markets of Buenos Aires during winter. Those examined were in part at least of the typical subspecies. It is probable, however, that the forms marketed there include the western and northern subspecies as well.

CALOPEZUS ELEGANS MORENOI Chubb

Calopezus elegans morenoi CHUBB, Bull. Brit. Orn. Club, vol. 38, Dec. 12, 1917, p. 31. (Neuquen, Argentina.)

Adult females of the crested tinamou were taken at General Roca, Rio Negro, Argentina, on November 25 and 26, 1920, and males on December 2 and 3. A chick not more than 2 days old was collected on December 3. An immature female about half grown was taken December 27 near Victorica, Territory of Pampa, and an adult female with a nearly grown male at Tunuyan, Province of Mendoza, on

---

* Auk, vol. 1, 1884, p. 318.

PLUMES OF THE PAMPAS GRASS

Guamini, Buenos Aires, March 3, 1921

NEST AND EGGS OF CRESTED TINAMOU (CALOPEZUS E. ELEGANS)

Carhue Buenos Aires, December 15, 1920

OPEN FOREST OF CALDÉN (PROSOPIS NIGRA). NESTING GROUND OF BANDED FALCON (SPIZIAPTERYX CIRCUMCINCTUS) AND BOYERO (TAENIOPTERA IRUPERO)

Near Victorica, Pampa, December 23, 1920

THE SOMBRE TODO (IODINA RHOMBIFOLIA), A COMMON, SPINY-LEAVED TREE OF THE PAMPAS

Victorica, Pampa, December 28, 1920

March 27. The small series secured at General Roca came from a point about 80 kilometers east of the town of Neuquen, the type locality of the subspecies described as *morenoi* by Chubb. These differ from birds from the mouth of the Rio Negro (designated by Peters, Bull. Mus. Comp. Zool., vol. 65, May, 1923, p. 287, as type locality of *C. e. elegans*), in grayer coloration and in lighter, less heavy barring of the underparts that tend to become immaculate on the abdomen. The two birds from Tunuyan, Mendoza, differ from those from General Roca in slightly browner coloration, with the light spots and bars on the upper surface larger, giving a distinctly more speckled appearance to the back. This same tendency is exhibited in the juvenile specimen from Victorica, Pampa. Birds from the three localities, however, may be allocated to *morenoi* without violence, giving this form a range extending from the Rio Negro, in Neuquen and western Rio Negro (probably from the southern side of the watershed of this stream), north through the plains and lower Andean foothills to central Mendoza, and east through the western Pampas to extreme north central Pampa (probably through San Luis). In San Juan *morenoi* is replaced by the peculiar pale *Calopezus e. albidus* Wetmore,[10] while to the northward are found *Calopezus e. formosus* Lillo in eastern Tucuman and northwestern Santiago del Estero and *C. e. intermedius* Dabbene and Lillo [11] in the Andean valleys of western Tucuman and La Rioja. *Calopezus elegans elegans* is thus confined to eastern and southern Patagonia and southern Buenos Aires.

An adult male of *morenoi* from General Roca taken December 2 is molting and has new feathers of the body plumage appearing on the back. These new plumes are considerably darker in ground color than the old feathers, while the light markings are suffused with a deeper shade of buff, indicating that the dry arid climatic conditions found in the haunt of this bird induce considerable fading of the plumage.

The chick (U. S. Nat. Mus. No. 283661) has the ground color of the head buffy brown, with a line of dull white extending from the base of the nasal groove backward on either side of the crown down over the back of the neck. This line has the brown feathers on either margin tipped with points of black that form a broken border for it. A well-developed straight crest of 8 or 10 filamentous plumes extends from the back of the crown; this is buffy brown in color with the feathers marked with black below the extremity. Lores buffy brown extending as a narrow line almost to edge of eyelid; super-

---

[10] *Calopezus elegans albidus* Wetmore, Journ. Washington Acad. Sci., 1921, p. 437 (San Juan).

[11] An. Mus. Nac. Hist. Nat. Buenos Aires, vol. 24, July 22, 1913, p. 104.

ciliary, beginning behind lores and extending down over sides of neck white, interrupted by a black bar above center of eye, bordered by a broken black line above, with a similar line beginning behind eye; auricular region deep mouse gray; stripe below and behind eye to auricular region buffy brown; supramalar streak white, extending from base of bill as a narrow line across lower loral region, broadening below eye and extending over side of neck; malar streak buffy brown with a narrow line of black on either side; hind neck and back buffy brown, the back with many slender white plumes interspersed among the brown forming white lines, and the brown feathers tipped with prominent spots of black; this same coloration extends over base of wings and flanks where the spots become smaller; wings vinaceous buff mixed with white; throat, lower breast, and abdomen dull white, a faintly indicated line of blackish and buffy-brown spots leading down from lower margin of ramus; a poorly defined band of vinaceous buff mixed with neutral gray across breast.

This chick of *morenoi* from General Roca is markedly paler than a chick four days or more older of *C. e. elegans* from Bahia Blanca. The latter has the ground color of the down on the crown tawny-olive, while on the back it is slightly duller than tawny-olive. The young *elegans* is more heavily banded across the breast and has underparts decidedly browner in color. The distinction between these two young is more decided than in the few adults examined.

The martineta or crested tinamou was observed from a train near the town of Rio Colorado, Rio Negro, on November 21, while at General Roca, Rio Negro, the birds were common from November 25 to December 3. Small bands containing from three to six or eight were encountered among the arid hills lying north of town, often near the mouths of little valleys that opened out on barren flats. The birds ranged back and forth in the open thorny scrub, from the bottoms of the draws to tops of low hills, passing out onto the flats below or penetrating (pl. 16) inland among the hills. The flocks were composed mainly of males that were not breeding, some in a condition of partial molt of the body plumage. The presence of flocks was betrayed by their curious, three-toed tracks in the sand, though the birds themselves usually hid. As I ranged back and forth over the low slopes in search, the tinamou finally took alarm, usually when I had returned the second or third time over ground where I suspected that they were concealed, and burst out with a roar of wings like a pheasant, to pass out of sight over the slopes. Occasionally one, more wary than the others, came out from 40 to 100 meters behind me and dropped at once over the crest of a hill without offering a shot. The birds rise swiftly from 3 to 6 meters in the air and then go straight away, perhaps climbing gradually to pass some low ridge, or if flushed on a high slope, drop to disappear behind some shoulder. The wings beat rapidly for 15 or 20 strokes and then are set for a short sail, to be followed by another series of wing beats. The flight is swift but the birds are easily killed. They usually offer quartering shots, and a slight wound at long range is sufficient to bring them to ground when they may run a short distance or may crouch with eyes half closed to await whatever fate may overtake them. Occasionally one or two ran rapidly away under cover of bushes, head and neck erect, and tail drooped so that in form they resembled guinea fowl. Such birds frequently gave a low whistled call *cheef*, heard occasionally as they took wing. During the warmer part of the day they spent much time in dust baths in the shade of low bushes, presumably to rid themselves of vermin. On two that I shot fresh from such baths, rows of mallophaga occupied the slender feathers of the crest, apparently a refuge from the dangers of asphyxia as the insects crawled down immediately into the head feathers of the dead bird.

The call of this species, a low mournful whistle given slowly, that may be represented as *wheet whee whee* was on the order of the more musical note of the rufous-winged tinamou (*Rhynchotus*), but with far less carrying power.

On the level flats above the stream bed of the Rio Negro, crested tinamou were breeding, though those in the hills a few miles away did not seem to be in pairs. In traversing the broad flats I saw their tracks or occasionally had a glimpse of a gray form running rapidly through the brush, but the cover of *Atriplex* and creosote bush was dense, and it was seldom that the birds flushed. By careful stalking it was possible to work close to whistling males, but I seldom saw them. On December 3 I found a nest containing broken eggs and later surprised an adult bird with several chicks. The parent flapped away on its breast with beating wings to attract my attention, while the young disappeared instantly. After careful search I located two and was able to capture one though the other escaped. The grayish color of the down on these tiny birds simulated that of the earth on which they crouched so that it was difficult to single them out. They lay motionless with head outstretched, but unlike young gallinaceous birds slipped away to one side, when opportunity offered, to a new hiding place. They remained under cover of the thorny bushes where it was difficult to get at them and with crouching step ran from cover to cover. On stopping they suddenly changed direction and ran a few inches to one side, a maneuver executed so quickly that it frequently eluded the eye.

The dung of the adult birds is greenish in color and soft in consistency with a very offensive odor. The crop and alimentary tract of the chick taken was filled with what was unmistakably the ordure of the parent—certainly a curious circumstance—taken either to supply partly digested vegetable food, or water, in a region where succulent vegetation was scarce and moisture absent.

On December 6 I noted many martinetas, as these tinamou are called, in traveling by train to Zapala in western Neuquen. Scattered individuals were common, while it was not rare to see 30 or 40, or even 100, all adults, banded together. These frequently exhibited little alarm, appearing graceful in their attitudes in contrast to their stiff, stilted motions when startled. At times one ran out and bowed abruptly, throwing the head down almost between the feet. Near Zapala on December 7 I found where some predatory animal had eaten a martineta, but noted no further sign of them there.

Near Victorica, Pampa, these tinamou were fairly common and an immature bird was taken. Males were heard whistling on December 29. On March 27, on the flats bordering the Rio Tunuyan, a short distance south of Tunuyan, Mendoza, half a dozen were found in company. On the low brush-covered sand hills east of the river, the birds were abundant and tracks were seen in the sand in many places. An adult female secured on this day was about to lay, so that the breeding season seems to vary considerably with the locality.

The hugely developed caeca found in the intestine of this bird, differing greatly from those in any other known species, have been described and figured by Beddard.[12] They are thin-walled sacs with the external surface divided into many lobular projections well marked toward the base, and tending to disappear at the free end. The size is immense in proportion to the bulk of the bird. In one specimen that I examined they measured roughly 130 mm. long by 25 mm. in diameter, in another 125 mm. long by 22 mm. wide. The distal end becomes smooth and more attenuate than the base. In discussing a specimen of *Calopezus e. formosus* (female), collected by R. Kemp, at Laguna Alsina, Bonifacio de Cordoba, C. Chubb[13] gives a figure, taken from a sketch by the collector on the original label of the bird, where the caeca are shown as elongate cylindrical organs, somewhat swollen at intervals. The figure, however, does not agree with the field notes, given immediately above it, as there Mr. Kemp states, " Caeca—100 and 140 mm. Large, conical and sacculated." It must be presumed that there has been some error in attributing the sketch to the present bird as from the delineation, no one would describe the caeca as large, conical, or sacculated. Personally I examined the caeca in about a dozen specimens of *Calopezus elegans*, including birds of both sexes and in all found them of the conical lobulated type figured and described by Beddard, though

---

[12] Ibis, 1890, pp. 61–66.   [13] Ibis, 1919, p. 14.

varying somewhat in the smoothness and elongation of the free end. They constitute an odd development in an interesting bird, one connected without doubt with bodily function, perhaps of aid in some way in the conservation of water in a species specialized for life in arid regions.

NOTHURA MACULOSA NIGROGUTTATA Salvadori

Nothura nigroguttata SALVADORI, Cat. Birds Brit. Mus., vol. 27, 1895, p. 560. ("Central Pampas," Argentina.)

From specimens examined in the collection of the United States National Museum, this form of the spotted tinamou seems to range through the pampas of the Province of Buenos Aires north into western Uruguay. (One specimen, seen, collected by Capt. T. J. Page, in August, 1860, marked "Uruguay" without more definite locality is similar to birds from northern Buenos Aires.) *Nothura m. nigroguttata* is similar to *N. m. maculosa* but is paler in general coloration (more buffy, less rufescent) with the markings of the under surface bolder, darker, and better defined. The northward range in Argentina is at present uncertain. Hartert and Venturi [14] record a specimen from Mocovi, Santa Fe, as *nigroguttata*, but an old skin from Corrientes, taken by Page in November, 1859, may represent a distinct form ranging between *nigroguttata* of the South and true *maculosa* of Paraguay: Above this bird resembles *nigroguttata* in type of marking, but the general tone of the upper parts is distinctly browner with less black, and the markings of the underparts are restricted to scanty narrow bars on the sides and flanks, and to narrow streaks on the breast and throat. In form of markings this bird thus suggests *Nothura m. savannarum* Wetmore, but has the bold black color of that form replaced by browns.

On October 21, 1920, near Dolores, Province of Buenos Aires, I flushed several of these tinamous in low ground near a marsh, while on the following day a dozen or more were noted in crossing from Dolores to Lavalle. Near Lavalle the species was common from October 23 to November 17. It was noted subsequently near Carhue from December 15 to 17, and at Guamini, from March 3 to 8. The tinamou recorded from Rio Negro, western Uruguay, from February 15 to 19, was supposed to be the present form, but no specimens were taken.

The spotted tinamou, a bird of the open country, thrives especially in the pampas, but ranges into more wooded country where open savannahs or prairies cut through the groves and forests. In closely grazed pastures it is often found in open tracts where the only cover is a few weed stalks, or a clump or two of dead grass left standing from the previous year. In such regions it is encountered

---

[14] Nov. Zool., vol. 16, 1909, p. 266.

many times in low, marshy ground where cover is more abundant than on the uplands. The birds seem more at ease and more common, however, in regions where low grasses or similar vegetation offer shelter.

Males were heard giving their low piping whistle, a single note repeated with increasing rapidity until it terminated in a trill, during the whole year, though more commonly in spring. They run rapidly with head and neck erect, and sloping back so that in attitude they suggest guinea fowl. In feeding they walk rapidly with nodding heads, pecking at the tender herbage. When frightened the birds crouch and remain motionless, or run quickly aside to flush with a thunder of wings when closely pressed.

Spotted tinamou rise with a rush, throwing the feet back under the tail when under way; after a series of rapid strokes of the rounded wings they scale for a short distance and then stroke again to avoid losing momentum, continuing to fly and sail alternately until safe, when they scale to the ground. Flight is accompanied by a strange whirring whistle that might be considered vocal, but is evidently mechanical in its origin, as the sound is heard only when the wings are beating swiftly, and ceases when the birds are scaling with motionless pinions. The flight is well controlled, as at times I saw them rush away down wind at a tremendous speed. In alighting, however, the birds often seem awkward, as they scale down to within 2 or 3 feet of the ground and then throw the wings up and drop heavily to earth, when two or three hops or a little run are taken to stop their momentum; often the bird may stumble and fall forward in its breast. The flight is direct and swift, and the roar made in rising disconcerting but the birds are easily killed once one has learned to gauge their speed.

Near Carhue on December 16 I flushed one direct from a nest placed in a little hollow under a clump of grass growing on a low hillock. The nest hollow was lined with grasses and contained four eggs, with incubation begun. The eggs are dusky drab in color with highly polished surface that, though apparently smooth, contains many small rounded pits. These eggs measure 40.4 by 31.1, 41.1 by 31.9, 41.8 by 31.3, 42.7 by 32.0 mm. On December 17 in this same region I flushed several young birds ranging in size from the bulk of a zenaida dove to nearly grown, and found that they flew as readily as adults. Adults and young fly with the neck curved so that the head is held slightly erect. The species is known universally as *perdiz*.

Spotted tinamous are exposed for sale in large numbers in the markets of the city of Buenos Aires, and form one of the staple game birds offered in hotels, restaurants, or occasionally as a substitute for the chicken served unfailingly at meals on railroad dining cars. Many thousands are killed each year by sportsmen, and the hunting of this species is a favorite pastime with those addicted to such sport. Like the bobwhite of North America, the spotted tinamou seems to have adapted its habits to changes brought about by man in its haunts, so that when it receives the slightest encouragement it remains common in spite of persecution. The meat is white and palatable and the bird larger in bulk than a quail. It is hunted with dogs, and though it has a tendency to run before them, makes a very satisfactory game bird that might thrive if introduced in the more temperate portions of the United States

NOTHURA MACULOSA SAVANNARUM Wetmore

Nothura maculosa savannarum WETMORE, Journ. Washington Acad. Sci., vol. 11, Nov. 4, 1921, p. 435. (San Vicente, Department of Rocha, Uruguay.)

The type specimen of this well-marked subspecies, an adult female, was taken at San Vicente, Department of Rocha, Uruguay, on January 27, 1921. As pointed out in the original description, the bird differs from *Nothura m. nigroguttata* in much bolder, heavier black markings on the dorsal surface, paler, more finely streaked hind neck, and more restricted, darker markings on the breast. In addition, the lateral bars on the sides and flanks are heavier and do not extend as far out on the abdomen and upper breast. The same characters set it off from true *Nothura m. maculosa*, while *savannarum* in addition is paler, more buffy, less rufescent above and below. Specimens of *Nothura m. minor* (Spix) are not available, but from Hellmayr's observations,[15] this form, described from Diamantina (formerly called Tejuco), Minas Geraes, Brazil, resembles *savannarum* in paler, more buffy coloration and restricted ventral markings, but is distinctly smaller. The wing in the type of *savannarum* measures 139.5 mm., while the same measurement in a series of five *minor*, according to Hellmayr, varies from 111 mm. to 116 mm. The subspecies described as *savannarum* is supposed to range through eastern Uruguay into Rio Grande do Sul, Brazil, and to meet *minor* somewhere to the northward in southern Brazil. Names that have been applied to spotted tinamous seem to refer entirely to other subspecies than the present one. *Tinamus major* (Spix)[16] is said by Hellmayr[15] to be a synonym of *N. m. maculosa*. *Tinamus medius* Spix,[17] also a synonym of true *maculosa*, is said to be based on an immature bird. There may be confusion in regard to these names, as *T. medius* and *T. minor* were described from Tejuco, now called Diamantina, Minas Geraes, while *T. major* is given as from " Campis

---

[15] Abband. Kön. Bayerischen Akad. Wiss., vol. 22, 1906, pp. 707–708.
[16] Av. spec. nov. Brasiliam, vol. 2, 1825, p. 64, pl. 80.
[17] Av. spec. nov. Brasiliam, vol. 2, 1825, p. 65, pl. 81.

Minas Geraes" prope pagos Tejuco et Contendas." However, the types of *T. major* and *T. medius* are said by Hellmayr to have the rufescent cast of the dorsal surface found in *N. m. maculosa*, while *N. m. minor* is a small form so that none of these names can apply to the bird I have described as *savannarum*. The *Nothura media* of Salvadori[18] is a synonym of *N. m. minor* (Spix). While *Nothura assimilis* G. R. Gray[19] described from "South America" is also identical with *N. m. minor*.[20]

Near La Paloma, port of the town of Rocha, the spotted tinamou was seen on January 23, 1921, while from January 25 to February 2 it was common near San Vicente, where the birds were especially abundant in grassy fields near the Laguna Castillos. The female described as the type of the present form, taken here on January 27, contained a well-formed egg almost ready to be laid. The birds were heard whistling in all directions, and half-grown young were seen on January 30. From February 3 to 9 they were noted in numbers near Lazcano.

NOTHURA MACULOSA BOLIVIANA Salvadori

Nothura boliviana SALVADORI Cat. Birds Brit. Mus., vol. 27, 1895, p. 561. (Bolivia.)

Specimens of the spotted tinamou taken at Las Palmas, Chaco, Kilometer 182 (Riacho Pilaga), Formosa, and Kilometer 80, Puerto Pinasco, Paraguay, are representative of the Bolivian form, though with adequate material there is no question but that they will be found to constitute one or more distant races allied to that bird. In general they are characterized by a strong grayish cast above and below, with the markings of the underparts restricted, and on the breast formed into lines. All are sharply cut off from the brighter colored *Nothura m. maculosa* and *N. m. nigroguttata* that range east and south of the Chaco. (*Nigroguttata* is said to occur at Mocovi, Santa Fe, in the southern end of the Chaco region.) Three distinct types of coloration are represented by the three localities from which specimens called here *boliviana* are available. A female taken at Las Palmas, Chaco, on July 16, 1920, differs from others here described under the name *N. m. boliviana*, in being more deeply buff in coloration, especially on the wings and sides of the neck, and in having bolder, heavier black markings on the hind neck. Two males from the Riacho Pilaga, 10 miles northwest of Kilometer 182, Formosa, taken August 15 and 18, are much grayer than this Las Palmas bird, and have the hind neck grizzled with

---

[18] Cat. Birds Brit. Mus., vol. 27, 1895, p. 563.
[19] List Birds Brit. Mus., pt. 5, Gallinae, 1867, p. 105.
[20] See Salvadori, Cat. Birds, Brit. Mus., vol. 27, 1895, p. 564, and Hellmayr, Abhand. Kön. Bayerischen Akad. Wiss., vol. 22, 1906, p. 707.

dull black and olive-buff, with a slight buffy tinge. A female from Kilometer 80, west of Puerto Pinasco, Paraguay, is grayer throughout than any of the others, while the markings on the hind neck and on the underparts are greatly restricted.

Spotted tinamous are birds of sedentary habit that have been divided into a number of subspecies, even under the more or less cursory examination that has been granted them by ornithologists up to the present time. When series of specimens are available from their entire range it will be found that a number of geographic races have been overlooked, as it is probable that every extensive river system may have a distinct form ranging through the plains of its drainage basin. In the material at hand in the United States National Museum three types of coloration are readily distinguished among the spotted tinamous; *Nothura m. maculosa, N. m. nigroguttata,* and *N. m. savannarum* (probably *N. m. minor,* which I have not seen) a group of subspecies characterized by more or less intense buffy coloration and bold markings of the dorsal surface, ranging from the well-watered pampas north to the Chaco and through eastern Paraguay, Uruguay, and southern Brazil; *Nothura m. boliviana, N. m. peruviana,* and *N. m. agassizi* of grayish color, strongly marked above and streaked on the breast below, covering eastern Peru, Bolivia, and the Chaco in Bolivia, Paraguay, and Argentina;[21] and *Nothura d. darwini, N. d. mendozensis,* and *N. d. salvadorii,* of grayish cast, with fine, vermiculated lining above and diffuse markings below, from Patagonia, and the arid regions of western Argentina, north into Salta. *Nothura maculosa* and *N. darwini* are at present recognized as distinct species, while the group characterized by *boliviana* (including the forms given above) would also seem distinct, specially from *N. maculosa* in the characters that have been enumerated. It is significant that a form identical with or close to *boliviana* was taken at Las Palmas on the west bank of the Rio Paraguay, while a specimen in the United States National Museum, from Corrientes a few miles below on the eastern shore, just below the confluence of the Parana and Paraguay, has the buffy coloration and bold markings of the true *maculosa* group. The relationships of these birds are points to be settled only when additional series are available.

Nothura m. boliviana does not seem to have been recorded previously from Argentina.

One who has garnered from desultory reading on South American natural history that the spotted tinamou is a bird of weak, uncertain

---

[21] *Tinamus boraquira* Spix (Av. spec. nov. Braziliam, vol. 2, 1825, p. 63, pl. 79) if correctly delineated in the original plate should be placed in the genus *Nothoprocta,* a group that differs from *Nothura* in having the posterior face of the tarsus covered with small reticulate, hexagonal scales instead of with two rows of large scutes, the outer of which is much broader and more distinct than the inner.

flight must perforce abandon this idea at the first encounter in the field. The birds rise with a disconcerting whistle and roar of wings that startled me into missing my first bird clean, but on my second encounter I retaliated by making a double, though one bird was lost in the grass, as I failed to mark it properly. At Las Palmas, in the Territory of Chaco, *Nothura m. boliviana* was common from July 14 to July 28. The birds were encountered in savannas and prairies bordered by heavy groves, where growths of grasses a foot or so high afforded cover. Occasionally one ran out with neck extended to the utmost to watch me, but more frequently they lay close and were overlooked in the heavy cover. At the Riacho Pilaga, Formosa, from August 8 to 20, tinamou were common in the drier areas. Though it was winter and the air often sharp and frosty, I heard the plaintive piping whistle of males frequently as I worked at specimens or walked through the open savannas in early morning. Cover was so heavy here that the birds were seen less easily than in the case of the form inhabiting the open pampa, and though heard constantly they were seldom flushed. On one occasion I saw one walking across an open space where there was little cover, but a minute later, when I returned with a gun, the tinamou had hidden and could not be found, though it was only a few feet away. When flushed they rose from 2 to 5 meters from the ground and darted swiftly away. It was difficult to make them rise a second time.

One was noted in the outskirts of the town of Formosa on the Paraguay River on August 23, and several were seen near Puerto Pinasco on September 3. At Kilometer 80 they were abundant from September 6 to 26, and were seen in the open camps through the Chaco for a distance of 200 kilometers west of the river. A female killed on September 6 had the ovaries developing, while males were calling constantly. The call consists of a repetition of one note that begins slowly, becomes louder and somewhat more rapid, and then dies away, a pleasant and agreeable sound. Though brush-grown areas were common here, the birds ranged entirely in open country. On September 25 at Laguna Wall a set of five fresh eggs was secured from a Lengua Indian who had taken them that morning. These eggs are slightly paler than dusky drab (being lighter than the set of *N. m. nigroguttata* described from Carhue) and measure (in millimeters) as follows: 42.3 by 31.2, 43.1 by 31.6, 44.2 by 31.5, 44.9 by 32.2, 45.1 by 31.5.

The Angueté Indians knew this species as *seh' en likh'*.

The soft parts of a bird secured at Las Palmas were colored as follows: Maxilla fuscous; sides of maxilla, and mandible cartridge buff; iris apricot orange; tarsus and toes vinaceous buff.

NOTHURA DARWINI MENDOZENSIS Chubb

Nothura darwini mendozensis C. CHUBB, Bull. Brit. Orn. Club, vol. 38, Dec. 12, 1917, p. 31. (Mendoza, Argentina.)

No specimens of this tinamou were preserved so that the allocation of field notes under this heading is provisional. Spotted tinamous of this type were recorded at General Roca, Territory of Río Negro, from November 23 to December 3, 1920, where the birds frequented alfalfa fields and other green growth near the river. At Victorica, Pampa, on December 28 I shot an immature bird, but was forced to kill it at too close range because of bushes among which it was found so that it was not possible to preserve it. It was light in color and resembled *mendozensis*. Others were noted here between December 26 and 29. The bird at times frequents open brushy areas as well as fields and prairies, in this differing from the preceding spotted tinamous. At Tunuyan, Mendoza, tinamou were fairly common from March 23 to 28; on the latter date I killed one but lost it in high weeds.

There is one specimen in the United States National Museum from Cordoba (taken July 8, 1913, by Renato Sanzin) that differs from specimens from Mendoza, the type locality, in more buffy coloration and in possessing bolder markings above.

In notes and general habits *N. darwini* is similar to *N. maculosa*.

NOTHOPROCTA CINERASCENS (Burmeister)

Nothura cinerascens BURMEISTER, Journ. für Orn., vol. 8, 1860, p. 259. (Tucuman.)

An adult female about to lay was killed near Tapia in northern Tucuman on April 10, 1921. These tinamou were fairly common here in areas where thickets of small thorny shrubs grew in dense clumps, interspersed with irregular openings covered with weeds growing from 1 to 3 feet high. The call of males was a whistle suggesting that of *Calopezus elegans* but far less musical. Though heard frequently from April 6 to 14, they lay so close that only three or four were seen during this period. They rise with a startling roar of wings almost at one's feet and dash swiftly away, frequently giving a loud clucking call, dodging almost at once behind some clump of brush that offers protection. I was forced to kill the one taken at too close range, as it was about to disappear behind a bush, and tore it badly.

Tinamou heard whistling above the city of Mendoza on March 13 may have been the present species, as also one that flushed with an excited note from a rocky hill slope near Potrerillos, Mendoza, on March 19.

Nothoprocta cinerascens is distinct from *N. perdicaria* in having coarser, larger reticulations on the posterior face of the tarsus.

NOTHOPROCTA PERDICARIA PERDICARIA (Kittlitz)

Crypturus perdicarius KITTLITZ, Mem. Acad. Imp. Sci. St. Petersbourg, Divers Savans, vol. 1, Livr. 2, 1830, p. 193, pl. 12.   (Valparaiso, Chile.[22])

A female taken at Concon north of Valparaiso, Chile, on April 27, 1921, was prepared as a skin, while another, secured at the same time, was preserved as a skeleton.   Conover [23] has shown that the tinamou of southern Chile differs from that of more northern localities in darker coloration and more brownish upper parts, with undersurface clay color instead of gray, and has named it *N. p. sanborni* (type locality Mafil, Valdivia).

Conover considers *Nothoprocta coquimbica* Salvadori,[24] named from a bird taken at Coquimbo by Doctor Coppinger, in the month of June, indistinguishable from true *perdicaria*.   According to the ranges assigned by Salvadori *perdicaria* is found in northern and central Chile, while *coquimbica* occurs in South Chile.[25]   Manifestly the ranges as given are interchanged as Coquimbo lies 350 miles north of Valparaiso.   The bird from Concon, while coming from within a few miles of the type locality of *perdicaria*, has the breast decidedly grayer than a small series of old skins in the United States National Museum, from near Santiago, Chile, in this resembling the description of *coquimbica*, but, on the other hand it is blacker above, with paler, browner markings than is shown in the plate of *coquimbica* given by Salvadori.

On April 27 I encountered several of these tinamou in a steep-sided brush-clothed gulch in the rolling hills, south of the mouth of the Rio Aconcagua at Concon.   Some ran aside, as I approached, to hide in the brush, while others rose with excited whistling calls and dashed away behind cover of trees.   Others were noted on April 28.   In the female bird noted above, the maxilla, save on the posterior cutting edge, was fuscous-black; remainder of maxilla and mandible drab-gray, with the tip of the mandible shaded with fuscous; iris Rood's brown; tarsus and toes slightly duller than chamois; nails fuscous.

Tinamou were offered for sale in the markets of Valparaiso in considerable numbers, and were sold in the streets in pairs by itinerant vendors.

RHYNCHOTUS RUFESCENS PALLESCENS Kothe

Rhynchotus pallescens KOTHE, Journ. für Ornith., January, 1907, p. 164. (Tornquist, Buenos Aires.)

The southern, gray race of the rufous-winged tinamou, distinguished by its grayer coloration and larger size from the typical

---

[22] According to Chrostowski (Ann. Zool. Mus. Pol. Hist. Nat., vol. I, no. 1, Sept. 30, 1921, p. 18), Kittlitz' type specimen was killed near Valparaiso on Apr. 3, 1827.

[23] Auk, 1924, p. 334.

[24] Cat. Birds Brit. Mus., vol. 27, 1895, p. 554, p. 15.

[25] See Brabourne and Chubb, Birds of South America, 1912, p. 6.

form, must bear the subspecific name *pallescens*. Though Kothe recorded a specimen from Tornquist in the Province of Buenos Aires as *Rhynchotus rufescens catingae* Reiser, a form that ranges far to the north in Piauhy, Brazil, at the same time he proposed for it the designation *pallescens*, a name that has been usually overlooked, as it has not been listed in the Zoological Record. I collected a pair of these birds in the sand-dune region 15 miles south of Cape San Antonio, Province of Buenos Aires, on November 4, 1920, and two others (one of which was preserved as a skeleton) on November 6. The skull of a third specimen was secured on the same day. An adult male taken on the Riacho Pilaga, Formosa, on August 18 seems to represent a northern race, ranging between *pallescens* of the pampas, and *R. r. alleni* of Matto Grosso. It has the bold black markings and general gray cast of *pallescens*, but differs in having the foreneck, hindneck, and upper breast washed distinctly with brown, and the rictal stripe much heavier. A bird in the collection of the National Museum from Cordoba is somewhat intermediate between the birds of northern and central Argentina, as it has a slight buffy wash on the neck. I have considered it inadvisable to describe the specimen from the Formosan Chaco until further material is available.

The rufous-winged tinamou, though common in many localities, is so shy that in spite of its size it is difficult to see and still more difficult to collect. The call of the male, a musical, slowly given whistle that bears a strong resemblance to the song of the Baltimore oriole, may be heard frequently, but it requires careful stalking to obtain sight of the bird. In fact, for several months this note was a puzzle to me. I heard it first at Las Palmas, Chaco, and again at the Riacho Pilaga, Kilometer 182, Formosa, coming from the long grass of the savannas. On many occasions I followed the clear whistled call out across the open without catching sight of the elusive musician, and, until I traced the note to its proper source, I was inclined to attribute it to a blackbird, *Gnorimopsar chopi*, a species almost ubiquitous in the Chaco that frequently flushed from the spot from which the call seemed to come. At Las Palmas in July I had a glimpse of one as it ran swiftly through the grass at the border of a wood, but did not secure a specimen until August 18, when at the Riacho Pilaga, one burst out at my feet with a thunder of wings and rushed away 2 meters above the ground, to be dropped at 40 meters with a charge of number eight shot. In the Chaco, Indians were said to hunt the large tinamou as they did the rhea, disguised by branches from a thick-leaved bush, so that they resembled dense shrubs. Tinamou were lured out into the open by an imitation of their whistle, and killed with bow and arrow or shotgun by the hunter invested in his blind.

Near Lavalle, in eastern Buenos Aires, the species had been exterminated on the fertile, pastured uplands, but was common in a desert stretch of sand dunes lying parallel to the beach below Cabo San Antonio. From November 3 to 8, while camped in a small hut in this region, I found the birds in comparative abundance. For two days during a tremendous storm the tinamou, discouraged by the downpour of rain, were entirely silent, but later when the weather cleared after the *temporal* they sang from all sides. By following the note it was usually simple to locate a pair and by startling them to force them to wing. The flight is swift but heavy and direct, so that they are easily killed. At this season they were breeding, as females taken contained eggs ready to lay.

While crossing on a train near Sierra de la Ventana on November 21 I noticed a number of pairs walking quietly about in the bunch grass that covered the pastures. On December 17 near Carhue I heard several calling from growths of thistles below the crest of a hill, where the birds were sheltered from wind, but though the musical, somewhat labored calls came from near at hand, the tinamou retreated through the dense growth as I advanced, and I did not catch sight of them. On March 3 and 4, 1921, the species was heard calling at long intervals near Guamini. It is said that these birds can not compete with the crested tinamou, *Calopezus elegans*, so that when the latter invades a region the rufous-winged bird disappears.

In Uruguay the note of the rufous-winged tinamou was heard on February 2, 1921, at the Bañado de la India Muerta, south of Lazcano, but none were seen.

This tinamou is hunted with dogs, and it is claimed that after the bird makes two or three flights it is exhausted and may be taken by hand. Be this as it may, I can testify that the initial flight is vigorous. The flesh of the rufous-winged tinamou is white in color and delicious in flavor, far exceeding in taste that of the other species that I encountered. The bird is so heavy that the tender muscles of the breast are frequently split as it falls to the ground when shot. The caeca of this species are long, slender, and cylindrical in form, entirely different from those of *Calopezus*.

Order SPHENISCIFORMES

Family SPHENISCIDAE

SPHENISCUS MAGELLANICUS (Forster)

Aptenodytes magellanica FORSTER, Comm. Soc. Reg. Scient. Gottingensis, vol. 3, 1781, p. 143, pl. 5. (Staten Island, Tierra del Fuego, and the Falkland Islands.)

On January 23, 1921, I found over 100 dried bodies of penguins cast up on the beach at La Paloma, Uruguay, and carried away one or two complete and a series of skulls. Fishermen there told me that it was frequent to find these birds along the beaches in winter and apparently there is heavy mortality among them. The *pajaro niño*, as the penguin is called, was also reported as frequent on the eastern coast of the Province of Buenos Aires.

Order COLYMBIFORMES

Family COLYMBIDAE

COLYMBUS DOMINICUS BRACHYRHYNCHUS Chapman

Colymbus dominicus brachyrhynchus CHAPMAN, Bull. Amer. Mus. Nat. Hist., vol. 12, Dec. 23, 1899, p. 255. (Chapada, Matto Grosso, Brazil.)

On August 9, 1920, near the Riacho Pilaga, 10 miles northwest of Kilometer 182, Formosa, Argentina, I found two of these grebes swimming slowly across the open water of a rush-bordered lagoon. At the time I was navigating a crude balsa made of a bundle of cattails bound together on which I knelt and paddled with a small pole. With this unwieldy craft I managed to approach near enough to secure one of the grebes, but the other dived and was lost in the rushes.

In the specimen taken the wing measures 102 mm. and the culmen 21.5 mm., so that in size this bird agrees with the measurements given by Chapman in his original diagnosis of the subspecies *brachyrhynchus*. It is also slightly darker below than *C. d. brachypterus* Chapman. The present specimen, when compared with a considerable series of *brachypterus*, has the sides of the breast grayer, and the band across the upper breast lighter in color.

Colymbus dominicus differs from *C. chilensis*, the other common small grebe of this region, in having the outer webs of the inner primaries and the secondaries margined, at least near the tip, with dull gray, while the scutes on the tarsus and middle toe are broader and less numerous. In *dominicus* the large scutes on the front of the tarsus number from 12 to 14, and those on the basal joint of the middle toes from 11 to 12, a total of from 23 to 26 in the combined spaces. In *C. chilensis* there are from 15 to 17 scutes on the front of the tarsus, and 14 to 17 on the basal segment of the middle toe, a total of from 30 to 34.

Should the small grebe from southern South America prove separable from northern examples, the name *Podiceps speciosus* Felix Lynch Arribalzaga,[26] based on a specimen in winter plumage taken in May, 1873, on Baradero Island, Province of Buenos Aires, is available. La Ley, a large folio sheet, was a daily paper, published for a short period only, under the editorship of Señor Enrique

---

[26] La Ley, Buenos Aires, July 2, 1877, p. 1.

Lynch Arribalzaga and his brother. Because of the rarity of this publication the pertinent part of the description of *P. speciosus* is transcribed here from notes made from a partial set of La Ley in the Bibliotheca Nacional in Buenos Aires.

LA LEY DIARIO DE INTERESES GENERAL

Administración y Dirección, Maipu 211.   Buenos Aires, Julio 2 de 1877   Redacción anónima

[Page 1.] Descripcion de una especie del genero Podiceps, por Felix Lynch. [An introductory statement gives notes on the grebes found in Argentina, and states that the new species, taken in May, 1873, at the Isla de Baradero, was supposed to be a migrant from Entre Rios.]

PODICEPS SPECIOSUS (Nobis)

La parte superior de la cabeza es parda oscura en la frente y vertex; sus plumas son bastante largas y forman un copetillo cuya punta se dirije hacia atras. Cada pluma lleva un pequeña borde de color castano. Cuando el ave se asusta eleva algo las plumas de la cabeza.

El occipucio es blanco, pero sus plumas tienen el extremo pardo oscuro. La parte inferior de la cabeza, sus costados, region parotida y algo del cuello de color blanco sucio, pero la cara y las plumas que cubren los oidos son jaspeados de oscuro; la mancha blanca de los costados de la cabeza adquiere gradualmente un tinto acanelado claro hacia su borde posterior y en lo alto de cuello, este último es castaño claro en lo anterior y costados, y pardo oscuro en lo posterior. La base de el es algo mas oscura by sus plumas se asemejan a pelos. El dorso del ave y las coberteras de las alas pardo oscuras con jaspe castaño claro á causa de que las plumas oscuras llevan un ribeto de aquel color.

El lomo hasta la rabadilla, negro. La cola, blanca acanelada con algunas plumas negras. El pecho, vientre y costados acanelados claros, con debil bañlo vinoso y cierto reflejo plateado sobre todo en los dos primeros.

Nueve de las remeras primarias son oscuras por encima con sus barbas externas rojizas y por de bajo son prises plateadas; la decima remera es blanca, pero el bordo externo, el mastil y gran parto del extremo son de color negruzco, las tres primeras remeras secundarias son blancas pero manchadas de negruzco.como la décima primaria; el color negruzco desminuye gradualmente de una á otra en intensidad y estension hasta que, a contar de la 3a secundaria las demás son blancas puras con mastil del mismo color.

Las tapadas, blancas con algun baño acanelado. Iris rojo carmin. Pico negruzco en la mandibula superior, azulado sucio en la enferior; el estremo de esta última del color de la superior. Pies aplomados oscuros—Longitud total, 23 cet.—Pico desde su ángulo hasta la punta, 0.02—Tarso 6,94.—Dedo medio, 0.05 ;[27] interno 9,035 ; externo 9,05, y plugar 0,01.

This paper was noticed in El Naturalista Argentino,[28] where it is said to have been accompanied by a plate. No plate was given in the original publication, so that it was probably issued as a separate sheet and so lost.

---

[27] Probably an error for 9,05.
[28] Vol. 1, pt. 1, Jan. 1, 1878, p. 32.

With regard to the genus name *Colymbus*, I have followed American custom in applying it to a group of grebes, though there is uncertainty as to whether it belongs to that family or to the loons. Dr. Witmer Stone in the Auk, 1923 (pp. 147–148), has reviewed the case briefly and is inclined to consider that *Colymbus* should go to the loons. His argument on the matter is readily accessible and need not be quoted here. It may be noted, however, that though Gray in 1840 and 1841 cites *C. glacialis* Linnaeus as type of *Colymbus* without comment, in an appendix issued in 1842 containing revisions to his second edition he remarks (p. 15), " *Colymbus*, after L. add (1735)," thus indicating that here, as in 1855, he had the edition of Linnaeus for 1735 in mind. The earliest definite fixation of type for *Colymbus* Linnaeus 1758 is apparently that of Baird, Brewer, and Ridgway, Water Birds (vol. 2, 1884, p. 425), where it is cited as *Colymbus cristatus* Linnaeus. For much of the matter given above I am indebted to Dr. C. W. Richmond.

COLYMBUS CHILENSIS (Lesson)

Podiceps chilensis " Garnot " Lesson, Man. d'Orn., vol. 2, June, 1828, p. 358. (Concepción Bay, Chile.)

This species was first described by Lesson in his Manuel d'Ornithologie under the names of *Podiceps chilensis* and *P. americanus*, of which *P. chilensis* has anteriority, as *P. americanus* is given lower down on the same page. The species has been commonly accepted under the name *americanus*, dating from *Podiceps americanus* Lesson and Garnot,[29] which, however, is preoccupied by the name used above. The designation *Podiceps chiliensis* occurs in the work last cited, but on page 601.

On November 2, 1920, at the Estancia Los Yngleses, near Lavalle, Province of Buenos Aires, I secured an adult male of this species in full plumage. While watching a small pool surrounded by rushes, I had a glimpse of the neck of one of these birds projecting above the surface of the water, but it disappeared at once. I remained hidden for several minutes, making a variety of cooing and grunting calls, until suddenly, without a ripple on the water or a sound, the bird appeared in the center of the pool directly in front of me, where by a quick shot, it was secured. The bill of this individual was black; iris slightly lighter than carmine; tarsus and toes dark neutral gray, blotched with deep to light olive gray, with the under surface of the webs blachish slate.

At General Roca, Rio Negro, on December 3, these small grebes were common along quiet channels bordered with rushes and

---

[29] Voyage Autour du Monde, Coquille, Zoologie, vol. 1, November, 1829, p. 599.

willow thickets near the Rio Negro. Here the birds floated about with back and rump feathers expanded to receive the warm rays of the sun, or swam with slender necks erect in ordinary grebe attitude. At times one rose, and with body erect and neck extended forward, fluttered along the surface for a few feet, an action probably reminiscent of mating displays given in spring. At such occasions the white secondaries showed prominently. At times the grebes gave a whistled call, often in an explosive tone, that at first I supposed was the song of some passerine unknown to me. One female taken was in molt and had cast all of the feathers in the primary and secondary series in the wing; new feathers growing in had just burst the sheaths. The birds were observed preening their plumage and eating discarded feathers. An individual that floated near shore was so fearless that I turned it about with my hand while attempting to get it in proper position for a photograph.

At Carhue, in the Province of Buenos Aires, on December 16 a single bird of this species was observed among great flocks of *C. occipitalis* on the alkaline waters of Lago Epiquen. Near Guamini the species was noted in small numbers in open water of a large lake on March 3, 4, and 6, 1921.

In the small series at hand are specimens of this grebe secured in localities ranging from the Straits of Magellan to Bolivia and central Chile. I am, however, unable to differentiate those from the various areas as subspecies. The grebe described from the Falkland Islands as *rollandi* by Quoy and Gaimard is a large representative of the present species, distinguished by much larger size and darker coloration. It is sufficiently distinct to be recognized as a separate species on the basis of material available at present, though formerly the name *rollandi* was used for all grebes of this type in South America.

COLYMBUS OCCIPITALIS OCCIPITALIS (Garnot)

Podiceps occipitalis GARNOT, Ann. Sci. Nat., vol. 7, January, 1826, p. 50. (Falkland Islands.)

The name in common use for this bird, taken from *Podiceps calipareus*,[30] is antedated by *Podiceps occipitalis* Garnot, as cited above. It may be noted that the description by Lesson and Garnot just cited is also antedated by plate 45 of the same work, published with livraison 5 in October, 1827, where the bird is figured under the caption *Podiceps kalipareus*.[31] Chapman [32] has confirmed the validity of *C. o. juninensis* (Berlepsch and Stolzmann)[33] described from

---

[30] Lesson and Garnot, Voyage autour du Monde, Coquille, Zoologie, vol. 1, May, 1830, p. 727.

[31] See Matthews, Austral Avian Record, vol. 2, October 23, 1913, p. 53.

[32] U. S. Nat. Mus., Bull. 117, 1921, p. 49.

[33] Ibis, 1894, p. 112.

Lake Junin, Peru. Comparison of a skin from Santiago, Chile, with two from Argentina shows no differences that may not be ascribed to individual variation. Specimens from the Falkland Islands, the type locality, are not available to me at present.

On December 15, 1920, as I came down from Carhue, Province of Buenos Aires, to the shore of Lago Epiquen, I made out the forms of thousands of small white-breasted birds resting on the surface of the water. On closer approach I found that they were grebes of the present species swimming, preening, resting, or feeding in loose flocks and bands that extended down the lake until lost to view in a shimmering heat haze that danced over the water. From where I rested on a small hillock overlooking the barren shores of the saline lake (similar in formation and salinity of water to Great Salt or Owens Lakes in the United States) fully 10,000 grebes were in sight, while the number on the entire expanse of the lake, a body 35 kilometers long by 20 kilometers wide, must have been immense. All were in full plumage and at a short distance appeared entirely white. Though the majority were quiet, mating activities were carried on in a few areas. Pairs or occasional parties of five or six individuals partly rose on the surface, and with sides touching, dashed off across the water for 10 or 12 meters. Usually as they stopped one, or more in case of a small flock, rose, and with extended neck and fluttering wings, splattered off for a short distance alone. Pairs approached one another with the posterior portion of the body lowered and breast raised, frequently to remain with breasts opposed, as they turned and pressed against one another, for a minute or more. At such times the breast feathers were expanded laterally, so that the birds appeared large. A call note resembling *tick tick*, given in an excited tone, was heard constantly. On the whole, the actions of the birds reminded me of the American eared grebe (*Colymbus nigricollis californicus*), but were more subdued. It is possible that they became more active a little later in the season, as not more than 5 per cent as yet felt the mating impulse. It is presumed that the species may nest in lakes in the mountains. E. Budín, of Tucuman, according to Hartert and Venturi,[34] found the species nesting in a lake in the Cumbres de Calchaqui at an altitude of 4,300 meters. The birds were in all probability attracted to Lago Epiquen by the abundant food available in the form of brine shrimp (*Artemia*, species), but would not remain since the heavy, saline water and lack of aquatic vegetation were not suitable for breeding colonies.

In early morning flocks swam up into the mouth of an arroyo where fresh water entered the lake, and here on December 15 I secured two and on December 16 another. These birds were easily

---

[34] Nov. Zool., vol. 16, December, 1909, p. 256.

killed, as they seemed less agile in rapid diving than other species of grebes that I have taken.

An adult male had the bill dull black, more grayish at base; iris scarlet, with a narrow circle of baryta yellow around the margin of the pupillar opening; outer face of tarsus, fourth toe, and under surface of all the toes, dusky neutral gray; inner faces of first and second toes washed with vinaceous buff; rest of tarsus and toes neutral gray.

AECHMOPHORUS MAJOR (Boddaert)

Colymbus major BODDAERT, Tabl. Planch. Enl., 1783, p. 24. ("Cayenne.")

This large grebe, when in full plumage, bears a striking resemblance to Holboell's grebe in pattern of markings and color. The bill in general form agrees with *Aechmophorus occidentalis* (the type of the genus), but is slightly heavier, while the base of the bill is somewhat more heavily striated, and the nostril, bordered above by a stronger membrane, is less elongate in form. In addition the feathers on the sides of the mandibular rami form a more obtuse angle and in most specimens do not extend as far forward, while the streaked plumage of juvenile birds is strikingly different from that of the plain gray of the chick of *occidentalis*.

The species was first observed on the Parana River between Holt and Zarate on October 9, 1920. On October 25 six or eight were seen on salt water in the mouth of the Rio Ajo, below Lavalle, Buenos Aires, and three, all in immature dress, were taken. Two of these, preserved as skins, have indistinct dark streaks on the side of the head, and lack the full bright color of adults. As my boat approached in the narrow river, these birds worked to one side and finally made a long dive to allow me to pass. Three were observed sleeping as they floated on the water, with the neck drawn back so that the bill rested on the shoulder at the side of the neck with the point ahead. As this threw the rounded head in the middle of the back it produced a curious outline. From this attitude the birds dived with no loss of time in swinging the bill to the front, an evident advantage of this attitude over that assumed under similar circumstances by the pied-billed grebe in which the bill is turned and inserted among the feathers of the back. One of the grebes taken here had the end of one toe and part of the web on another bitten out as though by a turtle or a small shark. The species seems to range along the seashore, as on November 6 when I was on the coast below Cabo San Antonio one washed ashore after a heavy storm, while on the following day two more, one an adult male, were secured under similar conditions. These had been dead for two days at least, and were preserved either as skulls or skeletons.

Near General Roca, Rio Negro, on November 27, one was seen breasting the swift current of the Rio Negro in main stream. On January 9, 1921, in the mouth of the Arroyo Carrasco, below Montevideo, I found two pairs apparently on their breeding grounds. When I first arrived only two grebes were visible, but these were joined at once by two others, one of which came sliding out from the reeds on the opposite bank. The birds, handsome in their full plumage, swam back and forth with arching necks and raised crests, finally diving as I came out from cover of the bushes. Two others were seen offshore in salt water opposite Carrasco on January 16, while near Lazcano, Department of Rocha, Uruguay, two in full plumage were noted February 7. On March 4 one was recorded on the Laguna del Monte at Guamini, Province of Buenos Aires, Argentina.

It is probable that birds from Patagonia and Chile may be separated from those from Buenos Aires, and farther north by smaller size, smaller, more slender bills, and in adult plumage, by grayer hind neck. The series at hand, however, is insufficient to make these points certain.

PODILYMBUS PODICEPS ANTARCTICUS (Lesson)

Podiceps antarcticus LESSON, Rev. Zool., July, 1842, p. 209. (Valparaiso.)

The pied-billed grebe of southern South America may be segregated from typical *podiceps* of North America as a subspecies under the name *antarcticus* of Lesson, on the basis of average grayer coloration above, including sides of neck, and more brownish hind neck. It is a large robust form with strong, heavy bill. Specimens have been examined from Buenos Aires, Rio Negro, and Chubut, Argentina, and Santiago, Chile, but whether this form covers the whole of South America remains to be determined. Females of this species in breeding dress may be distinguished from males by the greater extent of the dusky markings on the feathers of the undersurface, especially on the lower breast and abdomen, so that they appear much darker, a point that needs attention in making comparisons. Following are measurements, in millimeters, of three specimens taken near General Roca, in the Territory of Rio Negro, Argentina: Two males, wing 136.5, 145.0; culmen, 25.6, 25.0; tarsus, 45.5, 46.0; one female, wing, 133.8; culmen, 22.5; tarsus 42.5.

Near Lavalle, Province of Buenos Aires, from October 30 to November 9, pied-billed grebes were seen at intervals in pools in.the marshes. They were very shy and seemed to be on their breeding grounds. At General Roca, in Rio Negro, they were found in the quiet water of narrow lagoons, bordering the present channel of the river. On November 30 I killed three here, a mated pair and a fully grown male still in immature plumage. The adult male taken called at intervals with the usual sonorous notes of this species. The birds were observed here from November 27 to December 3. Near Lazcano, Department of Rocha, Uruguay, one was seen February 7, 1921, and near Concon, Chile, several were noted on a small slough on April 25.

One harboring the old delusion that grebes possess no tails has only to watch the present species during the breeding season to be undeceived, as at that time males frequently swim about truculently with the short tail held erect, so that it is very prominent. In diving, the birds, if not frightened, often lower the fore part of the body and then sink slowly beneath the surface, turning the head about for a last view, before they finally disappear without leaving a ripple on the water.

Order PROCELLARIIFORMES

Family DIOMEDEIDAE

DIOMEDEA MELANOPHRIS Temminck

Diomedea melanophris "Boie," TEMMINCK, Nouv. Rec. Pl. Col. d'Ois., livr. 77, April, 1828, pl. 456. (Cape of Good Hope.)

The black-browed albatross, common off the coast of Brazil below latitude 22° 37′ S. from June 15 to 19, 1920, was observed in small numbers in the great mouth of the Rio de la Plata on June 20, below Montevideo. On January 23, 1921, I found a skull of this species cast up on the beach near La Paloma, Department of Rocha, Uruguay.

Family HYDROBATIDAE

MACRONECTES GIGANTEUS (Gmelin)

Procellaria gigantea GMELIN, Syst. Nat., vol. 1, pt. 2, 1789, p. 563. (Staaten Land.[35])

On May 2, 1921, I collected the skull of a giant fulmar on the beach at Antofagasta, Chile. Like many of the other birds cast up by the waves at this place, the bird was covered with crude oil, suggesting that it had been killed through becoming saturated with oil floating on the sea.

PUFFINUS PUFFINUS PUFFINUS (Brünnich)

Procellaria puffinus BRÜNNICH, Orn. Bor., 1764, p. 29. (Faroes and Norway.)

On November 7, 1920, following a heavy storm that had endured for three days, a Manx shearwater washed in on the beach 25 kilometers south of Cabo San Antonio, Province of Buenos Aires. The

---

[35] See Mathews, Birds of Australia, vol. 2, pt. 2, July 31, 1912, pp. 184, 186.

bird was so emaciated that the pectoral muscles were reduced to thin bands overlying the sternum, a condition due apparently to lack of food, as there was no indication of disease. The measurements of this specimen, in millimeters, are as follows: Wing, 227.5; tail, 71.5; culmen, 35.6; tarsus, 42.6. The feathers of wings and back are somewhat worn, while the rectrices seem to have been renewed recently. The specimen is of the type with blackish upper surface (differentiating it from the grayer *P. p. yelkouan* and *P. p. mauretanicus*). A shearwater of this species (subspecific form unknown) has been recorded from Iguape, on the coast of Sao Paulo, Brazil,[36] but there seems to be no previous note of occurrence for Argentina.

PUFFINUS CREATOPUS Coues

Puffinus creatopus COUES, Proc. Acad. Nat. Sci. Philadelphia, 1864, p. 131. (San Nicholas Island, California.)

Skulls of this shearwater were preserved from two mummied specimens found on the beach at Antofagasta, Chile, on May 2, 1921.

PUFFINUS GRAVIS (O'Reilly)

Procellaria Gravis O'REILLY, Greenland, Adj. Seas. etc., 1818, p. 140, pl. 12, fig. 1. (Cape Farewell and Staten Hook to Newfoundland.)

On January 23, 1921, I found the dried body of a great shearwater on the beach at La Paloma, Department of Rocha, Uruguay, and secured the skull. There seems to be no other record for the occurrence of the species in Uruguay, though it is known from the Falkland Islands and Tierra del Fuego north into North Atlantic seas.

PUFFINUS GRISEUS (Gmelin)

Procellaria grisea GMELIN, Syst. Nat., vol. 1, pt. 2, 1789, p. 564. (New Zealand.)

A number of mummified shearwaters of this species were found on the beach at Antofagasta, Chile, on May 2, 1921, and three skulls were collected. About 9 in the evening on May 5, while the steamer was passing 12 kilometers west of the Balliesta Islands, Peru, several came on board attracted by the lights, and near the same hour on May 7, when 16 kilometers west of Lobos Afuera Island, we encountered large numbers. On this last occasion 40 or 50 blundered aboard, and several hundred in the water were observed scurrying aside as the ship passed by the lights from the promenade deck. Those that came aboard fell sprawling on deck and then scuttled along, half erect, with rapid awkward steps. The obliquely placed feet, with the comparatively slight flexure of which they were susceptible, made their stride short and stilted. Some

---

[36] Von Ihering, Aves do Brazil, 1907, p. 37.

hurried into dark corners; others attempted to clamber into the lighted passageways, pulling themselves over the raised thresholds of the doorways by aid of their bills. Occasionally one gave a raucous call, with widely opened mouth. All resented handling by biting savagely. When thrown overboard they fluttered heavily down to the water or turned to swing in again toward the ship. One that I skinned proved to be an immature female, and all of those handled appeared to be young birds. During the daytime they were observed resting on the water or scaling over the waves, when they were recognized by the light under wing coverts that showed in contrast with their otherwise somber coloration.

PROCELLARIA AEQUINOCTIALIS AEQUINOCTIALIS Linnaeus

Procellaria aequinoctialis LINNAEUS, Syst. Nat., ed. 10, vol. 1, 1758, p. 132. (Cape of Good Hope.)

This species, noted commonly at sea from June 15 to 20, 1920, from latitude 22° 30′ S. to the mouth of the Rio de la Plata, frequently came close to the stern of the steamer in following over the wake. The birds circled with set wings, scaling swiftly along, often tilting sideways at such an angle that they seemed about to overturn. As their momentum slackened they rose again with a few quick wing strokes, and then swinging in a short circle scaled away once more. Occasionally they alighted on the water, where they floated high like gulls. They were characterized by sooty black plumage, rounded tail, and a bill marked with yellow and slaty black. The silence of these great sea birds, in time, impresses one as uncanny.

On November 3 and 4 many were circling just outside the breakers below Cabo San Antonio, on the eastern coast of Buenos Aires, during a heavy gale. At times they swung in to within 100 meters of the shore. On a dead bird picked up on the beach at La Paloma, Rocha, Uruguay, on January 23, 1921, the culmen measured 50.5 mm., while there was a small white interramal chin spot. The head of this specimen was preserved.

OCEANODROMA TETHYS (Bonaparte)

Thalassidroma tethys BONAPARTE, Journ. für Ornith., 1853, p. 47. (Galapagos Islands.)

An immature female Galapagos petrel came aboard ship on the evening of May 9, 1921, when we were about 10 miles west of La Plata Island, on the coast of Ecuador. As Loomis [37] has indicated, the tail in this species is slightly forked, the incision in the present specimen amounting to 5 mm., so that *tethys* seems to belong in the

---

[37] Proc. California Acad. Sci., ser. 4, vol. 2, pt. 2, no. 12, Apr. 22 1918, p. 153.

genus *Oceanodroma* rather than in *Procellaria*, where in *P. pelagica*, the type species, the tail is slightly rounded.

OCEANITES GRACILIS GRACILIS (Elliot)

Thalassidroma gracilis ELLIOT, Ibis, 1859, p. 391.   (Chile.)

On the evening of May 7, 1921, 15 kilometers west of Lobos de Afuera Island, Peru, four graceful petrels, attracted by the lights, were captured on board ship. On deck they were helpless and even by aid of their wings were barely able to walk. When handled they gave a low chirping call. All were males, in which the outermost primaries had been molted recently, so that one or two of the outer ones were still inclosed in sheaths. They agree in color and characters with the type of this species preserved in the National Museum, save that the wing is shorter, due to the molting primaries. In most of a series studied by Loomis [38] the molt came from late November to early January, though one June specimen had recently shed the primaries. My skins vary in amount of white on the abdomen from a diffuse wash on the tips of the feathers in one to another in which the white forms a solid well-defined patch. The wing in the type specimen of the species measures 131.2 mm. Measurements for two of my specimens are 118.3 mm. in each case, but these birds, as stated above, have just completed a molt and may have the primaries not quite fully grown. They seem, however, to be smaller than the form described as *Oceanites g. galapagoensis* Lowe [39] from the Galapagos Islands.

Mathews [40] has noted that in six specimens of the graceful petrel five had the tarsus booted, while the other showed indistinct signs of scutellation. In the six birds that are before me the tarsus is booted with faint scutes indicated for a short distance at either end.

Order CICONIIFORMES

Family PHALACROCORACIDAE

PHALACROCORAX VIGUA VIGUA (Vieillot)

Hydrocorax vigua VIEILLOT, Nouv. Dict. Hist. Nat., vol. 8, 1817, p. 90. (Paraguay.)

The common cormorant, known universally as viguá from its appellation in Guaraní, was observed in many localities. The species was fairly common on the Rio Paraguay, from Corrientes as far as Asuncion and increased in abundance from that point to Puerto Pinasco. Scattered individuals were observed at Las Palmas, Chaco,

---

[38] Proc. California. Acad. Sci., ser. 4, vol. 2, pt. 2, Apr. 22, 1918, p. 181.
[39] Bull. Brit. Orn. Club, vol. 41, June 8, 1921, p. 140.
[40] Birds of Australia, vol. 2, pt. 1, May 30, 1912, p. 9.

on July 30, and on August 16 at the Riacho Pilaga, Formosa, one was shot as it circled past a large lagoon. At Puerto Pinasco the species was common and passed constantly along the river. Inland in the Chaco it was less numerous but was noted occasionally in lagoons. At Kilometer 80 one was seen September 15, and at Kilometer 110 one was noted on September 24.

In the Province of Buenos Aires cormorants were seen in small numbers near Dolores on October 21 and 22, while near Lavalle they were common from October 23 to November 15, both in channels that traversed the marshes wherever there was sufficient water and along the tidal reaches of the Rio Ajo. A male was taken here on October 8. At General Roca, Rio Negro, single birds were observed occasionally from November 24 to December 3. Near Montevideo, Uruguay, on January 9 and 16, 1921, cormorants were common along the beaches as far as Carrasco and were observed fishing in salt water, flying along parallel to the coast or resting in flocks in close formation on sandy beaches. One was observed at La Paloma, Rocha, on January 23. Near San Vicente, Rocha, the species was fairly common on January 31 at the Laguna Castillos. On February 2 one was noted inland at the Paso Alamo on the small Arroyo Sarandí. Near Guamini, Buenos Aires, they were recorded from March 3 to 8 in flocks that contained as many as 300 adults and young. The species seemed to be distributed universally wherever deep water or quiet lagoons offered suitable feeding grounds, though most common near large streams. On the Paraguay River it increased in abundance to the northward.

The two specimens that I secured were both immature individuals.

Family ANHINGIDAE

ANHINGA ANHINGA (Linnaeus)

Plotus anhinga LINNAEUS, Syst. Nat., ed. 12, vol. 1, 1766, p. 218. (Brazil.)

A snakebird was observed along the Riacho Quia near Las Palmas, Chaco, on July 30 and 31, 1920.

Family ARDEIDAE

NYCTICORAX NYCTICORAX NAEVIUS (Boddaert)

Ardea naevia BODDAERT, Tabl. Planch. Enl., 1783, p. 56. (Cayenne.)

The only specimen taken is a female, secured October 31, 1920, near Lavalle, Buenos Aires, in very worn plumage that seems to be that of the second year. The bird is a peculiar shade of grayish brown above with slightly indicated streaks of whitish on the crown, neck, and lesser wing coverts. Below it is whitish with the sides of the head, neck, and breast streaked with grayish brown. The abdomen is white. This specimen measures as follows (in millimeters) : wing, 289; tail, 105.5; exposed culmen, 65.7; tarsus, 80. The status of the South American night herons is at present obscure, but so far as I can determine from available material there is no distinction between the lighter colored *Nycticorax* from Argentina north into northern South America and that of North America. Doctor Chapman [41] recently has recognized *N. n. tayazu-guira* (Vieillot) as a valid race, while Hartert [42] has considered it a synonym of *naevius*. The latter course is the one here followed.

The night heron, known as the *zorro de agua* (water fox), had the habits usual to the species in other regions. On the pampas, where growths of rushes formed extensive cover in lagoons and swamps, they were fairly common. In Uruguay they were observed in wooded swamps. None were seen in the Chaco. The species was recorded as follows: Lavalle, Buenos Aires, October 31 and November 9, 1920; General Roca, Rio Negro, December 3 (one very light and one very dark bird observed) ; Carhue, Buenos Aires, December 15 to 18; San Vicente, Uruguay, January 31, 1921; Lazcano, Uruguay, February 7; Rio Negro, Uruguay, February 16 to 18; Guamini, Buenos Aires, March 3; Tunuyan, Mendoza, March 26 and 28.

BUTORIDES STRIATUS CYANURUS (Vieillot)

Ardea cyanura VIEILLOT, Nouv. Dict. Hist. Nat., vol. 14, 1817, p. 421. (Paraguay.)

Though Todd [43] considers variation in this species individual and recognizes no subspecies, the adult green heron of southern South America in the series that I have seen may be distinguished from that of the northern portion of the continent (including Venezuela, Colombia, and the Guianas) by paler, less grayish abdomen. Immature birds have the streaks on the foreneck heavier and the throat more heavily spotted with black in the median line than those from northern localities. Vieillot's name, *Ardea cyanura*, based on Azara's account of this heron in Paraguay, is available for this southern subspecies, of which I have seen specimens from northern Argentina, Uruguay, and Paraguay.

Cancroma grisea of Boddaert,[44] referring to the Crabier de Cayenne of Daubenton, which has Surinam as its type locality, must be considered a synonym of *striatus*. *Ardea noevia* J. F. Miller [45] and *Ardea naevia* Shaw [46] seem to represent a North Ameri-

---

[41] U. S. Nat. Mus., Bull. 117, 1921, pp. 51–54.
[42] Bull. Brit. Orn. Club, vol. 35, Oct. 14, 1914, p. 15.
[43] Ann. Carnegie Mus., vol. 14, 1922, p. 136.
[44] Tabl. Planch. Enl. Hist. Nat., 1783, p. 54.
[45] Var. Subj. Nat. Hist., no. 6, 1782, pl. 35.
[46] In J. F. Miller. Clm. Phys., 1796, p. 70 (pl. 35).

can green heron, as the sides of the head are figured and described as distinctly brown or ferruginous. In the reference mentioned to the Cimelia Physica, Sherborn and Iredale [47] have called attention to the transposition of the text and plates for *Ardea naevia* and *A. torquata*, in which the plate of *naevia* is accompanied by text headed *torquata* and relating to that species, which follows. The confusion is easily cleared, however, by careful reading of the descriptive portion.

Green herons were not widespread in abundance in the localities that I visited, but are reputed to be more common in more northern and eastern regions. Their habits are those common to green herons the world over.

At Puerto Pinasco I noted one or two near the Rio Paraguay during the first week in September, 1920, but saw none in the interior Chaco. In the swamps and lowland lagoons of eastern Uruguay the birds were common. I found them first at San Vicente on January 31, and again on February 2 at the Arroyo Sarandi, to the northward. At Lazcano they were common near the Rio Cebollati from February 5 to 9 and were recorded at Rio Negro, Uruguay, from February 14 to 19. A female was secured at San Vicente January 31 and a male at Lazcano February 6. The birds fed in open, marshy swamps or more frequently along shallow pools surrounded by thickets of water-loving shrubs. During the heat of the day they retired to shaded perches in trees or thickets near water. Intruders were greeted with complaining squawks, reiterated as the birds flew to more distant perches, and perhaps continued after a point of safety had been reached. One was observed as, in a crouching attitude, it crept slowly, with tail twitching nervously, toward the margin of a channel, intent on reaching striking distance of a school of minnows that played in the shallows.

SYRIGMA SIBILATRIX (Temminck)

Ardea sibilatrix TEMMINCK, Nouv. Rec. Planch. Col. Ois., livr. 46, May, 1824, pl. 271. (Paraguay and Brazil.)

Five specimens of the whistling heron were taken, one of which was preserved as a skeleton. The first one shot, an adult male, killed at Las Palmas, Chaco, July 22, 1920, had the soft parts tinted as follows: Distal third of bill black; remainder light grayish vinaceous tinged with a bluish shade at the base of the mandibular rami; bare skin on side of head light squill blue, shading to deep dull violaceous blue on base of bill, with a narrow band of this color extending across base of culmen; iris olive-buff; tarsus and toes black. Adult males were taken near the Riacho Pilaga at Kilo-

---

[47] Ibis, 1921, p. 306.

meter 182, Formosa, on August 9 and 14, and an adult pair at Kilometer 80 on September 10.

These herons were noted in small numbers in the Chaco, where they frequented open marshy lagoons and the borders of esteros. Frequently they walked about on masses of floating vegetation where the water was a meter or so deep, or stalked slowly about pools in wet meadows. They were wary and, as they fed in the open, difficult to approach. They fly with a peculiar short stroke of the wing that is highly characteristic, and in flight appear dull gray, with light tail and duller forepart of body, so that when a fallen bird is retrieved the beautiful, blended colors of the plumage come as a sharp surprise. Their alarm note is a harsh *quah-h-h quah-h-h*, resembling that of other herons, but in addition they give a shrill whistled note that is repeated frequently. The latter call I heard only from birds that were flying, and noted that, as it was given, the neck was outstretched, to be retracted as the whistling was finished. A pair observed in display about a small pool, on September 24, flew swiftly back and forth and then set the wings to sail rapidly in short circles while they turned first one side and then the other to show alternately the dark back and the light breast. The performance was executed with a dash and speed that would have done credit to a duck and reminded me in a way of the darting maneuvers executed at times by shore birds.

In the Chaco the whistling heron was observed as far west as Laguna Wall, 200 kilometers west of Puerto Pinasco, on the Paraguay River. Occasional individuals were seen from January 24 to February 2 at San Vicente, in the Department of Rocha, eastern Uruguay, where I found them at times walking about in dry fields in search of the abundant grasshoppers. Others were noted at Lazcano, February 6, 8, and 9, and one was seen at Rio Negro, Uruguay, February 16.

The species is known as *garza chifflon* or simply as *chifflon*. The Toba Indians called it *pilh' la tse de*, while the Angueté knew it by the impressive cognomen of *pat gwa zhi gwa mokh*. Indians occasionally offered the plumes of the wing coverts and nape for barter.

CASMERODIUS ALBUS EGRETTA (Gmelin)

Ardea egretta GMELIN, Syst. Nat., vol. 1, pl. 2, 1789, p. 629. (Cayenne.)

Two adult egrets were seen near Kilometer 80, west of Puerto Pinasco, Paraguay, on September 13, 1920, and four were observed at Carrasco, near Montevideo, Uruguay, on January 16, 1921, a sad commentary on the present status of a bird once found in abundance throughout southern South America. Breeding colonies were reported on the Rio Pilcomayo and near the Rio Cebollati in eastern Uruguay. In Argentina herons are protected and traffic in their plumes is prohibited.   In the Chaco, however, where the aborigines live almost entirely by the chase, Indians are permitted to kill herons and other birds for food at any season, a concession that has led to organized hunting of herons for plumes.   In August, 1920, when I was at the Riacho Pilaga in Formosa, the Tobas were preparing for an extended plume hunt in heron rookeries located somewhere near the Rio Pilcomayo and wished me to accompany them.   When I inquired concerning the condition of the plumes at that season I found the Tobas well versed in the matter, as they remarked without hesitation that eggs should now be hatching in the nests so that in a few days the plumes would be ripe.   I was informed that during the previous year Cacique Mayordomo, chief of the Tobas of that section, had organized plume hunting on a cooperative scale, and had secured 78 kilograms of plumes.   These had been sold to traveling merchants for between 8,000 and 9,000 pesos (at normal exchange 9,000 paper pesos is equivalent to about $3,965) for shipment to Buenos Aires.

ARDEA COCOI Linnaeus

Ardea cocoi Linnaeus, Syst. Nat., ed. 12, vol. 1, 1766, p. 237.   (Cayenne.)

This heron was observed in fair numbers, but no specimens were secured.   The cocoi heron, when seen in the field, resembles the great blue heron of North America in haunt and habits, as it does in call notes and general appearance.   It was observed solitary on the shores of large lagoons and rivers, and was more common in the northern portion of the section traversed than in the Pampas.   It was recorded in small numbers along the Parana and Paraguay Rivers from Corrientes to Puerto Pinasco, from July to September, 1920.   About the 1st of October adult birds suddenly increased in number along the Rio Paraguay, and many were observed in passing by steamer from Puerto Pinasco to Villa Concepcion on October 2.   It is probable that at this time they had come out from the drying Chaco to feed and secure food for young.   At Lavalle, Buenos Aires, the species was recorded from November 8 to 16, and at General Roca, Rio Negro, on November 27.   Young birds of the year were observed at San Vicente, Rocha, on January 31 and February 2, 1921, and near Lazcano, Rocha, from February 5 to 9.   Cocoi herons were noted at Guamini, Buenos Aires, on March 6 and 7, and in the vicinity of Tunuyan, Mendoza, on March 23.

TIGRISOMA MARMORATUM (Vieillot)

Ardea marmorata Vieillot, Nouv. Dict. Hist. Nat., vol. 14, 1817, p. 415.   (Paraguay.)

An adult female, taken at Kilometer 80, west of Puerto Pinasco, Paraguay, on September 17, 1920, measures (in millimeters) as follows: Wing, 330; tail, 127.5; exposed culmen, 100; tarsus, 107.   A male shot at the same locality on September 20 has the following measurements: Wing, 358; tail, 127.5; exposed culmen, 106; tarsus, 112. Both of these birds are in full plumage and are similar save that the female, in addition to being slightly smaller, is paler throughout, and has the lower neck barred with black on the sides and behind. A third bird, a female, shot 110 kilometers west of Puerto Pinasco on September 23 is identified as the present species with some reservation. It is juvenile, molting from juvenal plumage, and may represent *bolivianum* instead of *marmoratum*. In color it is buff barred with black, as usual in young tiger bitterns, save that the black bars are more restricted in width than in other specimens examined. The crown and hind neck vary from mikado brown to cinnamon, barred narrowly with black. New feathers that are appearing on the upper back are dull black, barred narrowly with wavy, irregular bars of cinnamon; others on the sides of the foreneck are russet margined tipped and barred with black. This bird measures as follows: Wing, 316; tail, 115.5; exposed culmen, 99; tarsus, 108 mm.

It is probable that *Tigrisoma marmoratum* will prove to be a subspecies of *T. lineatum*, a species of northern range, from which it differs mainly in larger size as far as may be judged from available descriptions.

The tiger bittern is a species that frequents open shores of marshy lagoons, often among growths of cattails or other rushes. Two ranged about a lagoon at the Riacho Pilaga in Formosa during August, 1920, but were wild and wary. On one occasion I knocked one over with a broken wing and waded for it in water reaching to my armpits, but was so impeded by mud and aquatic growth that the heron swam to a mass of floating vegetation and was lost before I reached it. Near Kilometer 80, west of Puerto Pinasco, Paraguay, three were observed in Laguna Palmas, where they walked about on floating plants that choked the water. When they flushed, one alighted on a shaded perch in a tree, where I stalked it and shot it. This bird, an adult female, had the bill black, with the lower half of the mandible pale olive-buff, a color that extended along the gonys to the tip; bare skin from above eye to base of bill, wax yellow; a line from anterior canthus of eye to bill, and another above commissure to below eye, deep neutral gray; rest of bare skin on side of head and over ramus of lower jaw, citron yellow; iris antimony yellow; front of tarsus, and toes shading from chaetura black to chaetura drab; back of tarsus for upper half varying from tea green to vetiver green.

The flight of these herons is slow and direct, accomplished with slowly flapping wings, and in the air they appear as large as a great blue heron. They were seen in flight to various feeding grounds at dusk. At a large lagoon at Kilometer 110 they were fairly common and were encountered among rushes, from which they flushed with a low note that resembled *wok wok.* In the Guaraní language the species was known as *hocó.* The adult was called *nhe ha na* by the Anguetés, while the bird in barred immature plumage was known as *ca pi a tik.*

IXOBRYCHUS INVOLUCRIS (Vieillot)

Ardea involucris VIEILLOT, Encyc. Meth., vol. 3, 1823, p. 1127. (Paraguay.)

Azara's least bittern is similar in haunt and habit to the least bittern of the West Indies and the United States, so that it may be more common than would appear from the few occasions on which I encountered it. On October 31, 1920, at the Estancia Los Yngleses, near Lavalle, Buenos Aires, while wading about in water nearly to my knees, I saw three among rushes in a cañadon. The birds were wild, flushed at a distance of 60 meters, and flew rapidly away above the marsh vegetation, to drop down again when at a considerable distance away. A male was killed at long range on this date and one was taken on November 2. Another was seen November 9. The birds appear light in color when flying and are of surprisingly rapid flight for a bird of this group. At Carhue, Buenos Aires, one flushed from a clump of cattails in an arroyo on December 15 and one was seen the following day. One was recorded at the Laguna Castillos, south of San Vicente, Uruguay, on January 31, 1921.

Family CICONIIDAE

JABIRU MYCTERIA (Lichtenstein)

Ciconia mycteria LICHTENSTEIN, Abhandl. Kön. Akad., Wiss. Berlin (Phys. Klass.) for 1816–17, 1819, p. 163. (Brazil.)

At the Riacho Pilaga, Formosa, two jabirus were seen on August 16, 1920, as they soared at least 1,000 meters above the earth. The birds, in appearance snow white with dark heads, turned in short circles with set wings, and finally sailed away out of sight. At Puerto Pinasco, Paraguay, on September 3 one passed high overhead. I had no other opportunity to observe this species.

MYCTERIA AMERICANA Linnaeus

Mycteria americana LINNAEUS, Syst. Nat., ed. 10, vol. 1, 1758, p. 140. (Brazil.)

On November 16, 1920, a flock of 12 wood ibis was noted between Lavalle and Santo Domingo, Buenos Aires; the species was seen nowhere else in Argentina. At the Laguna Castillos, below San

Vicente, Uruguay, wood ibis were fairly common on January 31, 1921, and an immature male was taken. Here they ranged in bands of 10 to 20 individuals, often accompanied by a roseate spoonbill or two, that rested in open grassy areas adjoining shallow rush-grown lagoons. They were wary, but under cover of rushes I crept within 40 meters of one flock and lay for a time watching them as they rested motionless or preened their feathers. At a sudden alarm they rose in confusion, and I killed one that I had singled out in advance. When flocks were flushed they usually sailed and flapped for a time in narrow circles a hundred meters above the earth and then flew on to some safer resting place.

On February 2 a flock was observed in a bañado near the Arroyo Sarandi; 30 miles northwest of San Vicente, and from February 6 to 9 scattered individuals were seen near Lazcano. The wood ibis was known locally as *fraile*.

The specimen secured still shows traces of the nestling down on the nape and the back of the neck. In this species neck feathers of the juvenal plumage often burst the sheaths near the base, while the tip is still inclosed in a corneous case so that the tips of the feathers appear as though waxed.

EUXENURA GALATEA (Molina)

Ardea Galatea MOLINA, Sagg. Stor. Nat. Chili, 1782, p. 235. (Chile.)

In Molina's account of the birds of Chile there are two composite names that refer in part to the present species and in part apparently to the egret (*Casmerodius*). The first of these, *Ardea galatea* (p. 235), is described as " di color di latte col becco giallo lungo quattro pollici, e le gambe crem fine; queste gambe, come pure il collo, hanno due piedi, e sette pollici di altezza," while the Latin diagnosis, in a footnote, says "Ardea occipite subcristato, corpore lacteolo, rostro luteo, pedibus coccineis." The length of bill cited is too short for *Euxenura* and nothing is said of black in the wings, but in general color of feathers and legs and in size this can fit only the stork, and the name is here taken for that species. *Tantalus pillus* of the same work (p. 243) is also a composite, but here again general size and color of body are those of *Euxenura*, while length of bill and color of legs are those of *Casmerodius*. Both names seem more applicable to *Euxenura* and are here accepted for that bird, a course that obviates necessity for change in the name of the egret.

The Maguari stork was common through the Chaco as far north as the Territory of Formosa, but was not observed in Paraguay. From Garabate, Santa Fe, northward to Charadai, Chaco, many were seen from the train on July 5, 1920. Single individuals were recorded at Las Palmas, Chaco, on July 27 and 31, while near the Riacho Pilaga, in east central Formosa, the species was common. On August 21 I examined one that had been killed near the station known as Fontana at Kilometer 182 on the Government railroad. One was observed near the town of Formosa on August 24, and a short distance inland the birds were common. In eastern Buenos Aires scattered individuals were seen east of Dolores on October 22, while from October 27 to November 16 the species was common in the marshes in the vicinity of Lavalle and extended west as far as Santo Domingo. None were observed in western Buenos Aires. In Uruguay I saw this species in small numbers at the Laguna Castillos, near San Vicente, on January 31, and farther north near the Paso Alamo on the Arroyo Sarandi on February 2. Scattered individuals were noted near Lazcano on February 6 and 7.

This handsome bird is an inhabitant of wet, open savannas where woodland does not encroach too closely, or of extensive marshes and wet meadows on the pampas. Its large size and contrasted colors render it conspicuous, and it is a species that will become rarer as its range is invaded more extensively by man. In the wilder districts Maguari storks were wary, as is any large bird that is hunted constantly, but on some of the extensive estancias in eastern Buenos Aires, particularly at Los Yngleses, the great birds were seldom molested, so they had become accustomed to herdsmen and others passing through their haunts and paid little attention to men. In the air these storks fly with neck outstretched and legs extended, beating the broad wings strongly to gain momentum for a glide or sail that may carry them for a long distance. At times they circle with outspread wings, frequently rising a hundred meters or more in the air. They evidence considerable interest in intrusions in their haunts and swing back and forth overhead, turning the head curiously to eye the intruder below. Where not molested they may pass at 50 or 60 meters, but usually are more wary. In flight the peculiar fork of the short, black-colored tail is readily seen through the mesh of the white under tail coverts that project beyond the ends of the longest rectrices.

It was not unusual in favorable situations in the Chaco to find 30 or 40 of these storks gathered in a scattered band, though elsewhere they were less gregarious. It is possible that these congregations represented migratory bands come up for winter from the south. These storks were silent so far as my observation extended, save that a bird with a broken wing clattered its bill loudly, but made no attempt to strike at me. The species is known in Guaraní as *tuyuyú* and in Spanish as *cigüena*. A male and a female that I killed on October 28 near Lavalle, Buenos Aires, were not breeding, though in fully adult plumage. The female showed the following colors: Tip of bill blackish, bordered with a wash varying from pompeiian red to brick red; base of bill puritan gray, shaded anteriorly into tea green; iris cream color; bare carunculated skin around eye, and between rami of mandibles between nopal red and brazil red; bare skin on sides of throat primuline yellow, bare area on breast duller than brazil red, shaded to primuline yellow laterally and anteriorly; tarsus and crus a peculiar purplish red between pomegranate purple and bordeaux; nails black. The wing is diastataxic.

Family THRESKIORNITHIDAE

MOLYBDOPHANES CAERULESCENS (Vieillot)

Ibis caerulescens VIEILLOT, Nouv. Dict. Hist. Nat., vol. 16, 1817, p. 18. (Paraguay.)

In the Chaco, west of Puerto Pinasco, Paraguay, this ibis was fairly common from September 5 to 25, 1920, but was not observed elsewhere. A female was taken September 7 and two males on September 12 (one preserved as a skeleton), while a female killed on September 13 was so badly shot that I saved only the skull. In an adult female the bill was black; bare throat deep Quaker drab; bare loral region dark neutral gray; iris orange chrome, somewhat paler on inner margin where it bordered the pupil; tarsus and toes testaceous; nails black. In another female the bill and bare skin on the head were black; lower eyelid pale vinaceous lilac; iris mikado orange.

These striking birds were found on marshy ground or about such small pools of water as remained in nearly dry lagoons. Frequently they were seen at rest in the tops of dead trees, where they had a commanding outlook, but always over or near water. The flight is direct, accomplished by steady flapping, and the passage of the bird is often announced by loud trumpet calls, *kree kree kree*, unlike any other bird note that is familiar to me. They fly with neck and legs outstretched in usual ibis fashion and are strong and muscular of body, so that they are hard to kill and difficult to skin when finally secured. The body gives off an unpleasant musty odor similar to that of the glossy ibis. While calling they frequently sail with motionless decurved wings. They stalk slowly about probing in mud and water often to the full extent of their long curved bills, or rest quietly in the sun and preen their feathers. At a distance they are to be distinguished from *Theristicus caudatus* by the head, which appears thickened and heavy because of the bushy nuchal crest. The species was commonly known as *bandurria*, or in Guaraní as *curucau moroti*. (The last word, signifying "light," serves to distinguish the present species from *T. caudatus*, which is characterized as "blue.") The Angueté Indians call this ibis *tay tit*.

THERISTICUS CAUDATUS (Boddaert)

Scolopax caudatus Boddaert, Tabl. Planch. Enl. Hist. Nat., 1783, p. 57. (Cayenne.)

This large ibis was common in the Paraguayan Chaco, west of Puerto Pinasco, from September 7 to 26, 1920. The birds ranged in small flocks of 9 or 10 individuals in openings among palms near the borders of lagoons, where at times they were accompanied by one or two crested ibis (*Molybdophanes caerulescens*). It was usual to find them walking about in search for food in short grass beneath the palms or at rest on fallen palm trunks or on the ground. Their call note was a loud *kree kree kree-ee*, a trumpetlike call similar to that of the crested ibis, but higher in pitch. A wounded bird emitted a grunting note like *kwah-ah*. In flight they travel directly, with head and feet extended, and steady, regular wing beat. In silhouette the head appears more slender than in the crested ibis, so that they may be distinguished in situations where colors are not readily visible. In Guaraní the species is known as *Curucau yobí*, and otherwise as *bandurria*.

In a male and two females killed on September 13 the colors of the soft parts were similar in the two sexes. In a female they were as follows: Bill and bare skin on head dull black, save for lower eyelid, which is pale vinaceous lilac; iris nopal red; tarsus eugenia red, shading on toes to old rose; nails dull black.

Salvadori [48] has given an excellent account of the South American ibises of the genus *Theristicus*, to which there is little that may be added. It may be noted that *caudatus*, in which the undersurface below the neck is more or less uniform, has the transverse grayish band, prominent on the upper breast in *melanopis* and *branickii*, indicated by a distinct grayness of the feathers of that area, while in addition in one female *caudatus* secured the feathers of the breast below this band are washed distinctly with a rusty color, an indication of the light breast patch prominently developed in the two other species. The three known species of this genus are complementary in their ranges and differ from one another in a series of characters in such regular and progressive manner as to indicate close affinity; indeed, it is not impossible that eventually intergrades may be secured that will link the three as geographic forms of one wide ranging species. The following key (adapted in large part from Count Salvadori's paper) may serve to identify them:

a^1. Breast and abdomen black or blackish (save for faintly indicated grayish band) ; greater wing-coverts white; culmen 145 mm. or more in length.
    caudatus.
a^2. Breast, below transverse gray band, white with more or less rufescent wash ; greater wing coverts not clear white; culmen 135 mm. or less in length.

[48] Ibis, 1900, pp. 501–517.

b^1. Abdomen black; breast more or less rufescent; greater wing coverts grayish white; culmen longer, 128 mm. or more in length___ melanopis.
b^2. Upper portion of abdomen white; breast white with very faint rufescent tinge; greater wing coverts gray; culmen shorter, less than 125 mm. in length_____ branickii.

PLEGADIS GUARAUNA (Linnaeus)

Scolopax guarauna LINNAEUS, Syst. Nat., ed. 12, vol. 1, 1766, p. 242. (Brazil.)

The white-faced glossy ibis was irregularly distributed throughout regions of open marshes. Near Berazatequi, Buenos Aires, June 29, 1920, of two flocks seen, one contained about 40 birds. An immature individual that had been killed by a hunter was examined. One was seen near Las Palmas, Chaco, on July 31, and at Formosa, Formosa, on August 23, a flock of 40 passed south above the Rio Paraguay. Near Lavalle, Buenos Aires, on October 22, six were noted in a pasture among sheep. A female taken November 2, though in immature plumage, showed some development of the ovaries. At Santo Domingo, Buenos Aires, November 16, nearly 1,000 were scattered through flooded fields. One was observed near Car-- hue in western Buenos Aires on December 21. At the Laguna Castillos below San Vicente, Uruguay, the birds were fairly common on January 31, but were not observed elsewhere in that country. On March 2, I noted many from the train between the stations of Canuelas and 25 de Mayo, Province of Buenos Aires, and on March 4 saw 50 near Guamini. A small flock fed in a marshy meadow near Tunuyan, Mendoza, on March 27.

Scanty material available seems to indicate that birds from southern South America may average smaller than those from western United States. The female secured near Lavalle, Buenos Aires, measures as follows: Wing, 235; tail, 83; culmen 105.5; tarsus, 79 mm. It would seem that guarauna is entitled to specific rank as distinguished from the true glossy ibis. In a fair series from the western United States and southward all adults possess a distinct white line on the forehead and at times on the sides of the head behind the bill, a mark that is definitely lacking in the other species, both from the New World and from the Eastern Hemisphere.

Family PLATALEIDAE

AJAIA AJAJA (Linnaeus)

Platalea ajaja LINNAEUS, Syst. Nat., ed. 10, vol. 1, 1758, p. 140. (Brazil.)

A roseate spoonbill was seen at intervals about a lagoon at the Kilometer 80 ranch west of Puerto Pinasco, Paraguay, during September, 1920, but was so shy that it flew away to other lakes whenever men appeared. On the Estancia Los Yngleses near Lavalle, Buenos Aires, a flock that at times numbered 15 was observed regularly from October 22 to November 2. Others were seen in crossing from Lavalle to Santo Domingo on November 16. Near San Vicente, Uruguay, a few were observed in marshes in company with wood ibis on January 31, 1921, and one was seen not far from Lazcano, also in the Department of Rocha, on February 6.

Near Lavalle the birds were found about shallow, open pools at the borders of rush-grown cañadones, or were observed as they crossed the marshes or circled in a close flock above the rushes. In flight, neck and legs were fully extended, and the broad wings beat slowly and steadily. Occasionally they uttered low grunting or croaking calls. Whether in the air or walking along the border of some pool the bright colors of their plumage showed clearly in beautiful contrast with a background of blue sky or green rushes. Locally the birds were known as *rosado* or as *cucharón*.

A male and a female about ready to breed were shot on November 2. The soft parts in the male were colored as follows: Maxilla mineral gray, with margin and irregular spots over surface dark neutral gray, and scales at base mineral gray; mandible pale ecru drab, with a wash toward center of mineral gray, margined and blotched with dark neutral gray, with scales near base mineral gray; bare skin on sides of head to behind eye, and a transverse line behind base of maxilla pale zinc orange; crown and sides of head above ears pale turtle green; gular skin for an inch behind symphysis of mandible glaucous, rest of pouch light ochraceous buff; skin through ears and across back of head black; iris scarlet red; tarsus and crus old rose, tarsus more or less clouded with fuscous; toes fuscous. This male, though otherwise in full plumage, had the center of the crown and the nape still covered with feathers.

On examining these two birds while they were fresh I found that the gular area contained a distensible air sac that apparently was maintained partly inflated, and was connected with extensive passages of the cervical air sac along the sides of the neck. This gular sac formed a large oval chamber lying beneath the tongue, constricted behind to a small orifice leading into the cervical series of air cells mentioned above. When fully inflated it forced the thin skin forming the floor of the mouth in front of the small tongue upward until the bladderlike distension was raised against the partly elevated or opened upper mandible so that it gave a most curious appearance. This development was present in both sexes.

These two spoonbills from Lavalle appear to be somewhat larger than birds from the southern United States and seem to represent a form that it may be possible to recognize, though at present I do not care to consider the matter definitely, as I do not have other specimens from South America available. The two birds taken have the following measurements: Male, wing, 370; tail, 105; culmen, 172; tarsus, 112; female, wing, 360; tail, 97; clumen, 163, tarsus, 105 mm.

Family PHOENICOPTERIDAE

PHOENICOPTERUS CHILENSIS Molina

Phaenicopterus Chilensis MOLINA, Sagg. Stor. Nat. Chili, 1782, p. 242. (Chile.)

In suitable localities in the Province of Buenos Aires the flamingo is fairly common, incongruous, though beautiful, figures in the landscape to one accustomed to think of them as birds of more tropical regions. At the mouth of the Rio Ajo near Lavalle on November 15, 1920, a dozen beat heavily past me in a strong wind, barely out of gun range, while on the following day 50 in company were seen feeding in a canal. Near Carhue, in southwestern Buenos Aires, flamingos were observed frequently from December 15 to 18. A form of brine shrimp (genus *Artemia*) swarmed in the heavily alkaline waters of Lake Epiquen and may have formed the food of the great birds, as the water was too heavily impregnated with salts to permit the occurrence of mollusks. At any rate flamingos occurred there in lines or loose flocks, never separated far from one another, often in water nearly to their bodies. While some fed by immersing their heads, others rested quietly or preened their feathers with bills awkwardly developed for such a use. As I approached they walked slowly a little way and then extended their wings, raised the long legs suddenly, and started away, striking the water with alternating strokes of their webbed feet in a clearly audible patter until they had gained sufficient momentum to rise in the air. Occasionally they gave low, honking calls.

On January 31, 1921, several bands were seen at the Laguna Castillos near San Vicente, Rocha, Uruguay, where they ranged with coscorobas and black-necked swans, a beautiful trio whose pleasantly contrasting colors, visible for a long distances, linger clearly in memory.

Family ANHIMIDAE

CHAUNA TORQUATA (Oken)

Chaja torquata OKEN, Lehrb. Naturg., Th. 3, Zool., Abth. 2, 1816, p. 639. (Paraguay and La Plata.[49])

Brabourne and Chubb, when they proposed[50] the name *Chauna salvadorii* for the crested screamer, to replace *Palamedea cristata*

---

[49] In the original the type locality is given as " Paragui, um La Plata," based apparently on Azara, who says " habita no solo el Paraguay, sino tambien las dos bandas del río de la Plata." Since La Plata was separated from Paraguay as early as 1620 it must be supposed that Oken meant Paraguay and La Plata, not Paraguay near La Plata as a literal translation of the German would read.

[50] Birds of South America, vol. 1, December, 1912, p. 53.

Swainson (1837), antedated by *Palamedea cristata* Linnaeus (1766), a name that refers to the cariama, overlooked a note by Dr. C. W. Richmond to the effect that the proper name for the bird in question was found in *Chaja torquata* Oken (1816).

The screamer was found in the Chaco in remote regions where settlements were few, and was common in the pampas on large estancias, where the birds were given more or less protection. Though formerly distributed throughout this entire region they have been killed or driven away throughout extensive areas. Occasional screamers were noted from the train in crossing the marshy region in northern Santa Fe on July 5. At the Riacho Pilaga, in the interior of Formosa, single birds were observed about lagoons from August 10 to 21, and on August 16 an adult male was taken. Near Puerto Pinasco, Paraguay, they were common from September 6 to 30, and were found in the interior Chaco to the westward as far as I penetrated (to Kilometer 200). One was observed on the Rio Paraguay itself on September 30.

Screamers ranged usually in pairs, but at times congregated in some numbers. On one occasion I saw 14 in a flock, circling in the air like vultures, 18 gathered in a band at the border of a lagoon and others scattered about near by, until in all I had 40 of the great birds under observation at one time. They were found ordinarily on floating masses of vegetation over deep water or in damp meadows where marshy growth was not too luxuriant. When alarmed or suspicious they flew up to perch in the low tops of near-by trees, where they were able to view the country. On alighting on the *camalote*, as the masses of water hyacinth and other vegetation that formed floating mats in the water were called, they frequently extended the wings for a few seconds, until they had tested the footing, but their long toes enabled them to walk over these insecure masses without trouble. The approach of any suspicious object was the occasion of loud trumpeting calls, rather gooselike in nature, that resembled the syllables *chah hah*, given slowly and with equal emphasis. These calls were loud, so that they carried for long distances, and had a certain stirring quality that was more or less pleasing, but were repeated so incessantly that in time they tended to become irritating, particularly when more desirable game was put on the alert by the alarms sounded by these efficient sentinels. These loud calls were often followed by a curious rattling, rumbling sound, audible only for a short distance, that resembled the noise produced by rubbing and compressing a dried, distended bladder. This sound was wholly internal and seemed to be produced when air was forced from the large air sacs into the smaller cells that lie between the skin and the body. At times the forepart of the body was slightly elevated as it was produced.

On September 25 I examined a young screamer about half grown, but still rather helpless, in the possession of an Indian boy. The bird gave a low whistling, piping call.

In the vicinity of Lavalle, Buenos Aires, screamers were common from October 27 to November 9, and after my experience with them in the Chaco it seemed strange to find them walking about in marshy spots among scattered bands of sheep. The flight of screamers is strong, and they rise heavily with loud swishing wings. I saw them occasionally soaring in circles high in the air. At the Estancia Los Yngleses I was told that 50 had gathered to feed in a small tract of alfalfa and that it had been necessary to drive them away to prevent damage. On November 6, after a severe storm, an immature bird washed ashore on the beach below Cape San Antonio. I supposed that it had been blown out to sea during a heavy gale and drowned.

· In Uruguay screamers were seen at the Laguna Castillos, near San Vicente, on January 31, and the Arroyo Sarandi (Paso Alamo) February 2. A few were noted near Lazcano on February 6 and 8. The birds were very wary here and were much hunted. Their flesh is dark and coarse fibered, but I found it palatable. The species is known universally as *Chaja*, a name given in imitation of the common call.

The adult male collected in Formosa on August 16 weighed 6.7 pounds. The soft parts were colored as follows: Maxilla and tip of mandible blackish brown No. 3; rest of mandible olive gray, shading to pale olive gray at base; bare space about eye vandyke red, shading to dull Indian purple on chin, rami of mandible and space behind nostrils; iris orange cinnamon; tarsus and toes alizarine pink, slightly darker toward crus; nails black. At Kilometer 80, west of Puerto Pinasco, Paraguay, I killed a female on September 13 and a pair on September 17. These were all in partial molt. In one, new growth covering the larger spur on the wing had pushed the older covering away so that on one wing the sheath slipped off as I handled it. The wing spurs vary in development and may be more perfect in young individuals than in older ones, in which they may have suffered more or less injury.

Order ANSERIFORMES

Family ANATIDAE

CYGNUS MELANCORIPHUS (Molina)

Anas Melancoripha MOLINA, Sagg. Stor. Nat. Chili, 1782, p. 234. (Chile.)

The beautiful black-necked swan was recorded October 28, 1920, and again on November 16, near Lavalle, Buenos Aires. At General Roca three were observed resting in backwater from an eddy on the

Rio Negro. On January 31, 1921, when I visited the Laguna Castillos, near San Vicente, Rocha, the great birds were common but
were too wary to permit approach, so that in the end no specimens
were obtained. On the wing the birds form a beautiful picture
from the contrast in color between the black neck and the snow white
body. As they pass they may utter low honking calls suggesting
those of geese.

CAIRINA MOSCHATA (Linnaeus)

Anas moschata LINNAEUS, Syst. Nat., ed. 10, 1758, p. 124. ("Brasilia.")[51]

The Muscovy duck was fairly common in the wilder sections of the
Chaco where lagoons offered permanent water during the dry season
of winter. The first noted were seen on August 13, 1920, near the
Riacho Pilaga, a few leagues south of the Rio Pilcomayo in the
Territory of Formosa, Argentina. As two rose from a marshy
lagoon and passed me, beating heavily against a strong wind, I shot
and wounded one. The birds made a short circle and alighted on a
large horizontal limb of a quebracho tree growing in the open, where
they rested for some time. On this occasion I had the misfortune
to lose the crippled bird in a dense tract of *monte*. A single one of
these ducks was observed at frequent intervals in open water on a
large lagoon, and when alarmed swam out into growths of cat-tails
and hid. It was shot on August 17 and proved to be an immature
male. Another dead bird was examined in possession of some
Indians at this same point.

At Kilometer 80, west of Puerto Pinasco, Paraguay, from September 7 to 17 these great ducks were common in flocks of three to
a dozen about lagoons densely grown with sedges and rushes, where
pools of open water were small and infrequent. Conditions were
somewhat similar to those frequented many times by mallards,
though in this case the locality may have been chosen through necessity, as it was near the end of the dry season and all lagoons were
greatly reduced in area. When alarmed the Muscovies rose readily
in spite of their weight and flew off low over the tops of the dense
groves of palms that surrounded the marshes. Often instead of
continuing to other lagoons, after a flight of a few yards the birds
alighted on the larger limbs of some dead deciduous tree standing
among the palms, where they rested in company. It was a source of
continual surprise to me to flush them from such locations. I found
that their claws were curved and sharp pointed as an aid in a firm
grasp on the limbs.

The flight of the Muscovy duck is heavy and rather slow. At each
stroke of the wings the white shoulders of adult birds flash promi-

---

[51] Linnaeus gives the type locality of the present species as "India." Berlepsch
and Hartert (Nov. Zool., vol. 9, April, 1902, p. 131) have corrected this to "Brasilia"
as the species in a wild state is known only from the New World.

LARGE DOMED NESTS OF STICKS CONSTRUCTED BY PSEUDOSEISURA LOPHOTES
Near Rio Negro, Uruguay, February 17, 1921

ABANDONED NEST OF LEÑATERO (ANUMBIUS ANNUMBI) OCCUPIED BY
DIUCA FINCHES (DIUCA MINOR)
Near Victorica, Pampa, December 26, 1920

LOWLAND THICKETS NEAR LAGUNA CASTILLOS, HAUNT OF BRUSH INHABIT-
ING BIRDS THAT HERE HAVE OUTPOST COLONIES IN THE PAMPAS

Near San Vicente, Uruguay .January 31, 1921

SUMMIT OF THE CERRO NAVARRO, A GRANITE OUTLIER OF THE HILL
FORMATION OF NORTHERN URUGUAY

San Vicente, Uruguay, January 28, 1921

nently, making a conspicuous field mark. At no time were they observed flying at an altitude of more than 80 meters in the air, while usually they passed barely high enough to clear low trees, from 15 to 30 meters above the ground. In alighting they flapped heavily to break their momentum as they came down into the grass. Their tree-perching habit may be the outcome of life in a region where during the rainy season there is nowhere else to rest save in the water.

An adult male taken September 7 was completing a molt of the body plumage and had the sexual organs dormant. No indication of breeding was noted among them.

To the Angueté Indian the Muscovy duck was known as *meh dik tee.*

The immature male (fully grown) secured in Formosa shows patches of old brown feathers among glossy black plumes that recently had been renewed. It does not have the broad white shoulder of the adult; there are scattered black feathers over the loral region, and the skin behind the eye is closely feathered. The caruncles of the adult are barely indicated. The adult taken in Paraguay in life had the soft parts colored as follows: Nail on both mandible and maxilla dark neutral gray; remainder of tip of bill pale drab gray, washed with livid brown on margin; spot behind nostrils, line of culmen between nostrils, as well as central portion of mandibular rami pale drab gray; band across bill in front of nostrils extending around on mandible, and base of bill, including bare skin on side of head, black; caruncles black at base, elsewhere purplish vinaceous; iris cream buff; tarsus and toes black.

DENDROCYGNA BICOLOR BICOLOR (Vieillot)

Anas bicolor VIEILLOT, Nouv. Dict. Hist. Nat., vol. 5, 1816, p. 136. (Paraguay.)

Near Lavalle, Province of Buenos Aires, the fulvous tree duck was common among the cañadones from October 28 to November 9, 1920. The birds ranged in flocks, frequently 30 or 40 together, that were found in open ponds where the water was a meter deep. They were frequently active at dusk. When flushed they rose with the whistled wheezy calls that gave them their local name of *pato sifflón* and passed on, often flying rather high, to more distant resting places. In the air they seldom show color, appearing simply as silhouettes of black against the sky. The birds on the wing differ in appearance from other ducks and offer a remarkable resemblance to ibises as they pass with rather slow wing beat and long necks outstretched, a similarity engendered by the long, bluntly pointed wing. The flight is only moderately fast. A female taken on

October 28 and a male on November 10 showed no indication of breeding.

At Tucuman, Argentina, on April 3 and 5, 1921, tree ducks, supposed from their notes to be the present species, passed overhead during the evening toward some feeding ground south of town.

The typical subspecies may be distinguished from *Dendrocygna bicolor helva* Wetmore and Peters from North America, by its slightly duller coloration and heavier, broader bill.

DENDROCYGNA VIDUATA (Linnaeus)

Anas viduata LINNAEUS., Syst. Nat., ed. 12, vol. 1, 1766, p. 205. (Lake Cartagena.)

The white-faced tree duck was seen only in the Paraguayan Chaco west of Puerto Pinasco. From September 6 to 8, 1920, the species was fairly common near the ranch at Kilometer 80, and a male was taken here on September 6. The birds rested in flocks on open shores or mud bars at the borders of lagoons, frequently 100 or even more together. They were most active toward night, and at dusk, or even after dark, passed overhead to distant feeding grounds, their approach heralded by their strange, sibilant, whistled calls *swee ree ree, swee ree ree*. After September 8 these flocks disappeared, but on September 24 and 25 the species was again encountered farther west at Laguna Wall, approximately 200 kilometers west of Puerto Pinasco. Flocks seen here ranged in size from 40 to 100. In evening the birds circled or passed high overhead, frequently flying in wavy lines like pintails, calling constantly. During early morning and in evening they were observed feeding in shallow ponds, where they waded along, working eagerly in the mud and water like huge teal. As the birds advanced the head was swung from side to side to cover the feeding ground thoroughly, while the silt collected was sifted rapidly through the bill. At this season they were not breeding. In Guaraní the species is known as *suiriri*, a good imitation of the call, while the Angueté Indian distinguishes it as *kwah te gwi jah*.

The male secured on September 6 had the soft parts colored as follows: Band across tip of bill, nostrils, and extreme base of culmen puritan gray; rest of bill black; iris Rood's brown; tarsus and toes clear green-blue gray; nails black.

After careful comparison of a fair series of white-faced tree ducks from both African and South American localities, I am unable to find sufficient grounds to warrant separating birds from the two continents as subspecies. African birds have the upper back somewhat more finely vermiculated, the black markings on the sides narrow, and the brown of the breast somewhat duller. These differences, however, are so slight as to seem almost intangible, though it seems probable that with extensive material two or more forms may be separated.

COSCOROBA COSCOROBA (Molina)

Anas Coscoroba MOLINA, Sagg. Stor. Nat. Chili, 1782, p. 234. (Chile.)

The coscoroba was first observed on November 15, 1920, at the mouth of the Rio Ajo, Lavalle, Buenos Aires, when four passed in company with flamingos. Other white birds that I took to be this species were noted occasionally flying across the marshes, but none came within gun range. On January 31, 1921, at the Laguna Castillos, near San Vicente, Uruguay, I found a considerable number gathered on open shores with *Cygnus melancoriphus*. The white coloration of these fine birds, especially when massed in flocks, made them conspicuous at long distances, and they were correspondingly wary. Before I was able to creep up within 100 meters the flocks flew out a meter into the lagoon, and then swam away over the high waves out of range, to return when I had passed. At rest they appear entirely white, but as the wings are extended the black at the tips of the primaries is revealed. The bill appears to be light reddish in color. No specimens were taken.

DAFILA SPINICAUDA (Vieillot)

Anas spinicauda Vieillot, Nouv. Dict. Hist. Nat., vol. 5, 1816, p. 135. (Buenos Aires.)

Six adults of the southern pintail were collected near Lavalle in Buenos Aires as follows: One female on November 2, 1920, another on November 3, and three males and one female on November 6. An adult male taken November 2 was preserved as a skeleton. In addition to the usually recognized characters of longer tail and greenish black speculum used to distinguish males from females, it may be noted that in females the throat is so nearly immaculate as to appear almost white, while in males it is strongly spotted. In a small series of these birds from Chile, Argentina, Paraguay, and Peru, I am unable to detect differences in size or coloration that may be correlated with geographic range.

An adult secured at Lavalle had the soft parts colored as follows: Stripe down culmen nearly to nail, nail and margin of maxilla adjacent, serrate margin of bill, and tip of mandible including nail, black; space behind nail on maxilla and mandible gray number 7; rest of bill mustard yellow; iris Vandyke brown; tarsus and toes olive gray, clouded on joints with neutral gray; webs slate color.

The coloration of young in the down, less than a week old, taken from one of four specimens secured at General Roca, Territory of Rio Negro, on November 27 is as follows (Cat. No. 283,675, U.S.N.M. male); forepart of crown buffy brown, becoming clove brown, mixed with buffy brown, on hind crown, and hind neck; loral region and stripe extending under eye buffy brown, with a faintly indicated whitish spot at base of bill on loral region; streak behind eye clove brown; superciliary streak, lower eyelid, and poorly defined auricular streak whitish; lower hind neck, upper back, and wings with down clove brown basally, buffy brown distally, the lighter color predominating and extending down on sides of breast; rest of upper parts deeper than clove brown; streak extending across posterior side of forearm to base of wing, and from there in a somewhat irregular line on either side of back to line of thighs whitish; underparts whitish with a slight buffy tinge; and indistinct collar of buffy brown across upper breast. The specimens taken are very uniform in color.

The southern pintail was common in the fresh water of marshy pools and channels near Lavalle in the Province of Buenos Aires at the end of October, 1920. On November 6, following a severe storm of wind and rain that flooded large areas and killed many thousand sheep, a great migration of these birds came in from the south, the flight continuing during morning and evening for a period of three days. At the time I was in camp in the sand-dune area on the east coast of the Province of Buenos Aires, about 24 kilometers south of Cape San Antonio.

In flocks and pairs pintails came swinging in to feed on areas of flooded land adjacent to the dunes. The birds showed little fear, and if I merely crouched on the ground had no hesitancy in passing within 60 meters, so that I killed four without difficulty. On the evening of November 7 the flight from the south was greatly increased, the birds passing in flocks of half a dozen to one hundred. The larger flocks traveled in irregular lines, the birds more or less abreast, passing steadily to the northward from 30 to 60 meters from the ground, while occasional bands swung around to drop in on some suitable feeding ground. At intervals I noted pairs of ducks, male and female flying together, alone or in company with flocks, but during the entire movement certainly more than 95 per cent of the pintails seen were males. The following morning the flight began again at daybreak and continued until about 10 in the morning. Flocks scattered out to feed over the pampa covered with shallow water from the rains. The total number of birds seen on this day between 5 o'clock in the morning and noon was estimated at between 15,000 and 20,000 individuals. As before, more than 95 per cent of those observed near at hand were males, and it was assumed that this proportion held among those noted at a greater distance. Male birds that I shot were still in full breeding condition. From the make-up of the flocks it was my belief that in *Dafila spinicauda*, as in *Dafila acuta* in the United States, males desert their mates as soon as the eggs are deposited, and then band together to spend the remainder of the summer in company. Pintails nested commonly through the pampas in this immediate region and were breeding at this time. It was my opinion, however, that many of the birds observed, patently in migration, had come from more southern regions in Patagonia, where the species nests commonly. The extreme eastern part of the Province of Buenos Aires, behind the Bay of Samborombon and Cape San Antonio, is divided into great estancias, with small rural population. Broad marshes, swamps, and wet meadows, known as cañadones and bañadones extend for miles and furnish feeding and loafing grounds suitable for these birds where they may pass the hot weather and molt in lazy idleness. After November 9 these ducks remained abundant, but, although they roamed over the country in search of feeding grounds, there was no concerted movement among them, as the great migration from the south seemed at an end. On November 15 I found many (nearly all males) along the Rio Ajo below Lavalle, where few had been observed on October 25.

The breeding season, as in many other birds of this region, may be irregular. A paired female taken November 4 was not yet in condition to lay. On November 16 in traveling from Lavalle to Santo Domingo, Buenos Aires, I observed what seemed to be a mating flight of this bird. As a pair circled over a cañadon, high in air, the male at short intervals swung under and slightly in front of the female, while she at each approach swerved to one side or the other leaving him again behind.

At General Roca, in the Territory of Rio Negro, from November 23 to December 3, pintails were common along the Rio Negro. A female, in company with eight young 3 or 4 days old, was found on November 27, and four of the young birds were taken. On December 3 a female was seen with a brood of immature birds at least three-quarters grown. In both instances the females (who were not accompanied by males) were very solicitous and thrashed about in the water to attract my attention or flew back and forth overhead. A single bird was seen on a salt lagoon at Ingeniero White, the port of Bahia Blanca, on December 13, while from December 15 to 18 the species was common at Carhue, in western Buenos Aires. The fresh skull of an adult male was secured from the camp of a hunter. One was seen at a small pool near Victorica, Pampa, December 29.

In Uruguay two were noted January 9, 1921, on an arroyo below Carrasco, a bathing resort near Montevideo, while on January 31 a few were found on the Laguna Castillos near San Vicente, in the

Department of Rocha. Others were recorded at Lazcano, in the same general region, on February 7.

Around the Laguna del Monte at Guamini, Province of Buenos Aires, Argentina, the species was common from March 3 to 8. I was surprised to note occasional true pairs of these birds here, though the breeding season was past, as fall was well advanced. I judged that these were pairs that for some reason had been unsuccessful in nesting during the summer and that had not as yet parted company. Near Tunuyan, Mendoza, a flock of 25 pintails was seen on March 25, and others the following day. Early on the morning of March 28 a flock of 100, and later two smaller groups, passed due north, flying high in air. These last seemed to be flight birds in fall migration.

In general appearance and habits *Dafila spinicauda* is similar to *Dafila acuta*. The birds frequent the open water of lagoons or rest on bars, muddy shores, or projecting points where they have open outlook. They impress one as alert and intelligent, eminently able to care for themselves. In wilder sections, where not molested, they exhibit little fear, but when hunted it was many times almost impossible to come within range of them, especially on open pampa, where there was little or no opportunity for concealment. The birds often feed by immersing the head and neck as they paddle across shallow pools or bays, or in deeper water tip in order to reach the bottom. Where heavy winds or rising waters flood areas of muddy flat the pintails follow the creeping advance of the water line to feed eagerly in the windrow of seeds and dead or drowning insects that it carries with it. Recently flooded areas of shallow water are always attractive. The flight is swift and direct. On the wing the birds resemble *D. acuta*, but appear heavier in the neck. Though females resemble males, they may be distinguished sometimes when in the air by the shorter, less-pointed tail, especially when flocks swerve in passing overhead. The call of the male is a mellow, trilled whistle, a purling sound pleasing to the ear, resembling that of the northern pintail. It is given frequently as parties of males pass on the wing. The note of the female is a low *ka-ack* or *qua-ack*, slightly lower in tone than that of our pintail.

The species is one of the abundant ducks of the pampas and was common among birds offered for sale during winter in the great markets of the city of Buenos Aires.

PAECILONITTA BAHAMENSIS RUBRIROSTRIS (Vieillot)

Anas rubrirostris VIEILLOT, Nouv. Dict. Hist. Nat., vol. 5, 1816, p. 108. (Buenos Aires.)

Near Carhue, in western Buenos Aires, the southern Bahama duck was common from December 15 to 18, 1920. The birds ranged in small flocks, at times as many as 50 together, near the mouths of fresh-water arroyos draining into the strongly saline Lago Epiquen, or in small ponds common here in slight depressions through the undulating pampa. The birds rose with a high-pitched call, and on the wing in flight and form resembled *Dafila spinicauda*, a species from which they were easily distinguished by the buffy-brown tail (in color distinctly lighter than the back) and by the sharply defined lines of their bicolored heads.

As no specimens were taken these notes are allocated under the subspecies *rubrirostris* on the basis of Bangs's recent review of the group.[52]

NETTION FLAVIROSTRE (Vieillot)

Anas flavirostris Vieillot, Nouv. Dict. Hist. Nat., vol. 5, 1816, p. 107. (Buenos Aires.)

The curious nesting habits of the tree or yellow-billed teal have been well described by the late Ernest Gibson in his notes on birds of the Cape San Antonio region, Province of Buenos Aires.[53] While working at Mr. Gibson's estancia, Los Yngleses, near Lavalle, from October 30 to November 9, 1920, I found the birds fairly common. The breeding season had begun and the teal were nesting in huge stick nests of the monk parrakeet (*Myiopsitta monachus*) placed in the tall eucalyptus trees lining the driveways near the estancia house. The birds themselves spent much of their time resting 40 or 50 feet from the ground on open horizontal limbs in the eucalyptus, where they stood on one leg asleep with the bill in the feathers of the back as calmly as though they rested on some mud bar in a lagoon. Though six or eight frequently congregated in these situations, when flushed the birds separated in pairs that circled swiftly over the open fields to return to some safer haven among the trees. The males gave a low whistle and the females a high-pitched *kack kack ka-ack*, notes that in both cases resembled those of the similar sex in the green-winged teal (*Nettion carolinense*). In fact, the resemblance to the call of the northern bird was so close that I never overcame a feeling of surprise when I heard the present species call from the treetops.

After heavy rains the tree teal descended to shallow pools in the grassy fields near at hand, but at other times flew out to feed in the marshes and swamps in company with other ducks. Males taken were in full breeding condition. On October 31 I observed a male on the wing in pursuit of a female, giving his musical whistled note. The two circled and swung swiftly through the tops of the trees in a

---

[52] Proc. New England Zool. Club, vol. 6, Oct. 31, 1918, p. 89.
[53] Ibis, 1919, pp. 20–21.

mating flight similar to that of *Nettion carolinense* or *N. crecca*. A brood of newly-hatched young was reported but I did not see them. As the parrot nests occupied by the teal are frequently from 15 to 25 meters from the earth there was considerable speculation as to how the ducklings reached the ground. It may be supposed, however, that they merely tumbled out, their slight weight and resilient bodies being sufficient guarantee against injury from the fall to the grass-padded carpet below.

Along the Rio Negro south of General Roca, Territory of Rio Negro, occasional birds were seen from November 27 to December 3. They were still breeding here and were found in pairs in quiet side channels bordered with heavy growths of willows. A single bird was noted in company with the southern pintail (*Dafila spinicauda*) near Carrasco, below Montevideo, Uruguay, on January 9, 1921. On March 8 six were observed resting in shallow water near Guamini, Province of Buenos Aires. Near Tunuyan, Mendoza, two were seen March 25, and on March 28 one, apparently a flight bird from the south, passed in company with pintails.

An adult male was taken October 30 and another November 9 at Los Yngleses near Lavalle. Both birds were in full breeding plumage. Specimens from Chile and Argentina do not seem to differ appreciably in size or coloration.

NETTION LEUCOPHRYS (Vieillot) [54]

Anas leucophrys VIEILLOT, Nouv. Dict. Hist. Nat., vol. 5, 1816, p. 156. Paraguay.)

Though it is probable that teal seen about open lagoons near Las Palmas, Territory of Chaco, Argentina, at the end of July, 1920, were the present species, the ring-necked. teal was taken only in the Paraguayan Chaco near Puerto Pinasco. On September 8 near the ranch at Kilometer 80 west of Puerto Pinasco a flock of a dozen passed me swiftly to alight in a small channel that had been filled by heavy rains a few hours before. Two that I secured were females, both immature birds that had just attained full growth. On September 24 and 25 several mated pairs of these small teal were observed at Laguna Wall, 200 kilometers west of the Rio Paraguay, beyond the locality given above. All seen here were mated, and an adult male taken September 25 was in breeding condition.

In habits the ring-necked teal is similar to related ducks. When startled the birds spring into the air and dart away with swift direct flight. On the wing the forepart of the head appears very light while as the birds pass the flash of the white patch on the greater coverts on the otherwise dark wing makes a good field mark. At

---

[54] See Oberholser, Proc. Biol. Soc. Washington, vol. 30, Mar. 31, 1917, p. 75, for change in name for this species from the current *N. torquatum*.

Laguna Wall the birds frequented shallow, open pools in marshy areas, and when flushed circled swiftly away low over the marsh vegetation. The call of the female was a high-pitched, somewhat varied note, that may be represented as *qua-a*, *qua-er* or *qua-ack*. Those taken were very fat.

The Angueté Indian called the present species *peh ro a pah*, while to my Lengua boy at Laguna Wall it was known as *pil wa pah*.

NETTION BRASILIENSE (Gmelin)

Anas brasiliensis GMELIN, Syst. Nat., vol. 1, pt. 2, 1789, p. 517. (Brazil.)

At Kilometer 80, west of Puerto Pinasco, Paraguay, the Brazilian teal was noted in small numbers in company with other ducks on the bare, open shore of a large lagoon. The male of a mated pair was crippled one morning, but was not actually recovered until several days later on September 18, 1920, when I came across him again. To my surprise the female bird still accompanied him, though the male was unable to fly. I recorded the note of the male as a high, whistled call not so clear as that of the green-winged teal. The call of the female was a loud *qua-ack*.

Near Rio Negro, in west central Uruguay, a few of these ducks were seen about a small rush-grown lagoon near the shore of the Rio Negro. Two males killed here on February 18, 1921, had completed the wing molt and were able to fly, though the new primaries were not quite fully grown. New feathers were growing in over the breast and back on these birds, but there is no indication of an eclipse plumage, as old and new feathers are similar in color. These two birds were past breeding, as the intromittent organs were shrunken and small, though in one the testes were still 28 mm. long. (In the other the testes were greatly reduced.) On this same occasion, however, I noted several mated pairs, while males frequently joined in little flocks so that sometimes four or five were found together.

These ducks fed in the shallows of swampy lagoons or swam about, threading their way through the floating surface plants that in many places covered the water in a mat. On the wing the black shoulder with the white bar on the tips of the secondaries showed prominently, while with binoculars it was possible to see the elongate patch of the white axillars alternately hidden and displayed with the movement of the wings. The call note of the female, as noted here, was a high pitched *kack kack*, while the males gave a high *swees swees swee* that suggested the call of a wigeon. Males had the habit, common among teal, of bowing to one another or to their mates.

The colors of the soft parts in life in the male taken west of Puerto Pinasco were as follows: Upper mandible between dark vinaceous purple and Indian purple; lower mandible etruscan red; iris bone brown; tarsus and toes coral red; nails deep mouse gray.

MARECA SIBILATRIX (Pöppig)

Anas sibilatrix Pöppig, Froriep's Notizen, vol. 25, no. 529, July 1829, p. 10. (Chile.)

The strikingly marked Chiloë wigeon was first observed on November 6, 1920, on the Estancia Tuyu, south of Cape San Antonio, Province of Buenos Aires, when a few came to ponds behind the dune region bordering the beach. On November 9, in company with B. S. Donaldson, at the Estancia Los Yngleses I killed a fine male, a bird in full plumage, very heavy and fat but not in breeding condition. Along the Rio Negro below General Roca, Territory of Rio Negro, the wigeon was common from November 27 to December 3. All those observed were males, gathered in small flocks that rested on sand or gravel bars bordering the swiftly flowing main channels of the river, or that frequented the quieter waters of the lagoons bordering the stream on either side. The birds examined closely were all adult, evidently past breeding. One taken was beginning to molt. At Carhue, Buenos Aires, wigeon were observed from December 15 to 18 in small numbers. A few observed on December 18, when about 30 were seen, were in pairs, but the majority of those noted were males. On the wing the white shoulders of this species form a prominent field mark. Males have a low whistled note resembling the syllables *wheur, wheur,* accompanied at times by low chattering calls. The birds were wild and alert, so that it was difficult to collect them.

Birds of this species in fresh, unworn plumage have the light margins on the feathers of the dorsal surface broad and conspicuous so that they appear much paler and more conspicuously streaked than others that are somewhat worn.

QUERQUEDULA CYANOPTERA (Vieillot)

Anas cyanoptera Vieillot, Nouv. Dict. Hist. Nat., vol. 5, 1816, p. 104. (La Plata Region and Buenos Aires.)

On November 16, 1920, between Lavalle and Santo Domingo, in the Province of Buenos Aires, I noted six pairs of cinnamon teal and occasional additional males through the extensive swamps of this region in open marshes and pools. Near General Roca, Territory of Rio Negro, Argentina, occasional pairs were seen from November 23 to 30 on quiet channels of the Rio Negro, and on December 3 I observed a flock of 30 males, all in full plumage, evidently birds that had bred and were preparing to molt. On this same day one was seen in company with a flock of wigeon. Near Carhue, western Buenos Aires, a male was recorded December 15, and a female with six newly hatched young on December 16. The head of an adult male, discarded by some hunter, was found on the day following. At Guamini, in this same region, I noted ten or a dozen cinnamon teal on March 3, 1921, among them two males in full brown plumage. Others were noted on March 4. On March 28 a female was observed near Tunuyan, Mendoza.

QUERQUEDULA VERSICOLOR VERSICOLOR (Vieillot)

Anas versicolor VIEILLOT, Nouv. Dict. Hist. Nat., vol. 5, 1816, p. 109. (Paraguay.)

The gray teal was noted in numbers near Lavalle, Province of Buenos Aires, Argentina, from October 28 to November 9, 1920, but was not observed elsewhere in the regions visited. At Los Yngleses the birds were common in the cañadones and were found in pairs or in small flocks. They frequented shallow, open pools in the marshes, and when flushed flew with swift darting flight rather low over the rushes. Many times they were observed in passage over the marsh, but never traveled at high altitudes in the air. On November 9 I encountered several flocks, each containing 8 or 10 males, that apparently were banded together after having bred. These remained separate from the mated birds, frequenting the water of open pools, or standing on the shore of some pond or open marsh. A female was taken October 28 and three others, a male and two females, on October 31.

From examination of a small series of these teal in the United States National Museum, it appears that birds from the Straits of Magellan may be separated as *Querquedula versicolor fretensis* (King).[55] A bird sexed as a female but probably a male examined from Gregory Bay differs from northern birds in larger size and in bolder, heavier markings on the underparts, especially on the abdomen. The bill in particular is long and heavy. The measurements, in millimeters, of this specimen are as follows: Wing, 203.0; tail, 75.5; culmen, 47.0; tarsus, 36.5. Specimens that represent true *versicolor* have been examined from Paraguay, Province of Buenos Aires, Argentina, and central Chile. The bill in these ranges from 38.2 to 43 mm., the wing from 180.9 to 192.5 mm. The northward range of the subspecies *fretensis* is at present uncertain.

SPATULA PLATALEA (Vieillot)

Anas platalea VIEILLOT, Nouv. Dict. Hist. Nat., vol. 5, 1816, p. 157. (Paraguay.)

A few males were noted November 8, 1920, near Lavalle, Buenos Aires, in company with southern pintails recently arrived in migra-

---

[55] *Anas fretensis* King, Proc. Zool. Soc. London, pt. 1, Jan. 6, 1831, p. 15. (Straits of Magellan.)

tion. A number were observed exposed for sale in the markets of Buenos Aires at the end of June. No specimens were secured.

METOPIANA PEPOSACA (Vieillot)

Anas peposaca VIEILLOT, Nouv. Dict. Hist. Nat., vol. 5, 1816, p. 132. (Paraguay and Buenos Aires.)

The rosy-billed duck was one of the common species found in the region surrounding Lavalle, Province of Buenos Aires, from October 23 to November 9, 1920. The birds frequented open pools in the marshes where the water stood from a few centimeters to a meter or so deep, and though not averse to frequenting small ponds surrounded by high vegetation did not penetrate among the rushes. In form and habits the species is closely similar to the American redhead (*Marila americana*), a species from which it is not dissimilar in color pattern aside from the prominent rosy-colored knob developed on the bill in males. Females, when on the wing or when resting on the water, resembled female redheads closely, but were marked by the sharply outlined white under tail coverts that made a prominent field mark. The flight was swift and direct, and birds showed entire lack of fear of any object not wholly visible to them, so that to secure a shot it was often only necessary to crouch in the grass or rushes when they were circling on the wing.

At this season rosy-billed ducks were found in pairs or were engaged in mating. Frequently four or five males swam in pursuit of one female, who remained in the lead while in turn her suitors rose to flutter along for several meters with the rear portion of the body dragging on the surface of the water. The note of the male is a purring *kah-h-h*, a low call that carries for only a short distance.

On December 3 several were found on quiet channels and lagoons near the Rio Negro below General Roca, Territory of Rio Negro. Others were noted at Carhue, Buenos Aires, December 17, and on February 6, 1921, the species was seen in a large marsh near Lazcano, in the Department of Rocha, Uruguay. On February 8, at Lazcano, I examined a young bird three-quarters grown that had been killed recently by some gunners. This bird displayed dull markings slightly darker than those of the female, especially on the breast. In form it showed the strong, heavy leg muscles characteristic of the young of deep-water ducks, while the muscles of the breast were thin and undeveloped, though the wing quills were half grown. Two adult males were observed at Rio Negro, in west-central Uruguay, on February 16. The species is commonly known as *pato picaso*.

HETERONETTA ATRICAPILLA (Merrem)

Anas atricapilla MERREM, in Ersch u. Gruber, Allg. Encyc., sec. 1, vol. 35, 1841, p. 26. (Buenos Aires.)

The curious black-headed duck was encountered only in the cañadones on the Estancia Los Yngleses, south of Lavalle, Province of Buenos Aires. The birds frequented pools surrounded by rushes where the water was from 2 to 3 feet deep, and were shy and retiring, so that it was difficult to observe them. On my first encounter I found two pairs in a small pond in company with coots and other ducks. To my astonishment the black-headed ducks dived when startled and disappeared like so many grebes, evidently seeking the shelter of the rushes. Another pair was observed on November 9, 1920, when I was fortunate in securing the male, though both birds dived instantly at the flash of the gun. The female disappeared and was not seen again. When in the water the birds suggest ruddy ducks, though the tail is not held at an angle as in the genus *Erismatura.*

Black-headed ducks were evidently breeding during the first week in November. Females noted swam about with heads erect, behaving like other ducks. Males followed them or faced them with neck drawn in and throat puffed out, at intervals raising the point of the bill and giving a low note *quah quah,* barely audible at 45 meters. It is possible that in diving quickly the birds use wings as well as feet, but on this point I was not certain. The species is widely distributed but is not common as it was not encountered elsewhere.

The colors of the soft parts in the male that I secured were as follows: Top of bill behind nostrils, line of culmen, nail and space behind it black; base of maxilla shading from Rocellin purple anteriorly to vinaceous buff toward feathers; rest of maxilla deep Dutch blue; mandible tilleul buff, becoming deep Dutch blue at base; iris bone brown; front of tarsus and toes deep olive buff, becoming neutral gray on sides, joints, and webs.

The specimen taken was an adult bird in breeding condition. At the back of the mouth on either side was a vertical slit 12 mm. long, forming the entrance of a thin-walled cheek pouch that extended backward, and to a slight degree downward, for about 25 mm., above and external to the hyoidean muscles. This sac was evidently capable of considerable inflation. For at least the anterior half it was overlaid by a thin fascia of muscle, probably a portion of the *cucullaris.* In addition, the upper end of the esophagus is full and large, with thin walls that are pouched outward, evidently capable of expansion. Midway of the neck the walls of the esophagus became normal. There was no tracheal air sac.

The trachea, proceeding from above, was soft for the upper half, below which it was hardened and firm for the space of 15 mm.; apparently the rings tended here to ossify. At this point the trachea broadened gradually, and became again thin walled, until it reached a diameter of 12 mm., when it contracted slowly until at the level of the shoulders it was once more of normal breadth. There was no bulla ossea.

On examining the freshly killed bird, I found that the skin of the throat hung in a loose fold that began below the line of the eye. In skinning this specimen it was noted that the neck was large, so that the skin passed readily over the head.

The black-headed duck is a species of somewhat uncertain affinities that requires more detailed anatomical study before its position may be definitely known. The lack of a broad lobe on the hind toe, the somewhat weakened form of the bill, and small feet are characters that assign it to its present position in the subfamily Anatinae near the freckled duck *Stictonetta naevosa*. In the full, loose skin of the neck, development of special, distensible sacs about the head in the male, small wings, glossy, shining plumage, and lack of a bulla ossea it suggests the Erismaturinae, a group from which *Heteronetta* differs, however, in lack of a lobed hind toe, small feet, presence of an enlargement in the center of the trachea, and elongated upper tail coverts. It is possible that the characters that ally it with the latter group are due to convergent evolution, as the duck in habits is similar to the ruddies. For the time it may be left in its present position.

ERISMATURA VITTATA Philippi

Erismatura vittata PHILIPPI, Wiegmann's Arch. für Naturg., 1860, pt. 1, p. 26. (Chile.)

The small southern ruddy duck, though widely distributed, seems to be rather rare in occurrence, as it was seen only in northern Patagonia near General Roca, in the Territory of Rio Negro. The birds were found in channels and long lagoons bordering the main stream of the Rio Negro, where the water, though often deep, had slight current, and was in places bordered by clumps of cattails or overhung by low willows. A pair seen on November 27, 1920, were evidently nesting, but the majority of the birds observed were males that seemed to have completed breeding for this season. Frequently three or four were found in company swimming with heads drawn in and spread tails held up at an angle. They were undemonstrative save when occasionally one swam out to jerk the head up and down two or three times, a custom common among males of many ducks. A breeding male secured on November 27 was in full plumage, though the rectrices showed wear and in part were faded in color. The soft parts in this bird were colored as follows: Maxilla and base of mandible Eton blue; nail and anterior margin of maxilla marked with fuscous; mandible, except at base, pale brownish vinaceous; iris Rood's brown; tarsus and toes storm gray, becoming dark neutral gray on joints and webs. A male shot on December 3 did not have the sexual organs developed.

On seizing my first specimen of this ruddy duck, I was pleased to detect in its neck an air sac that, to the sense of touch, appeared similar to one that I had described in the North American *Erismatura jamaicensis*.[56]

The body of this bird was preserved carefully in alcohol for subsequent examination. On dissecting it I find the arrangement on the throat is considerably different from what it is in the North American bird. In *Erismatura vittata* the larynx is less highly specialized than in *E. jamaicensis*. The fold of connective tissue, attached to the thyroid cartilage, that I have called the "ligula laryngis" is reduced to a slight ridge. The larynx in general has the cavity anteriorly compressed from side to side, and posteriorly expanded. There is no marked division or vestibule in its lower portion, and the lateral pads found in *jamaicensis* that I termed the "pulvini laryngis" are wanting. On the dorsal surface of the trachea just behind the larynx is a transverse slit that interrupts or divides the upper tracheal ring. The succeeding ring is broadly notched for half its width, but is not cut entirely through. This slit forms the opening from the trachea into the tracheal air sac that extends down between trachea and esophagus.

The neck of this tracheal air sac is long and narrow, while the elongate sac itself is but little enlarged or swollen. In its total length from opening to free lower end the sac measures 65 mm., while at its greatest distension it is apparently not more than 10 mm. in diameter. It is thin walled and transparent in texture, ends below at the level of the shoulders, and has no connection, save through the trachea, with the series of pulmonary air sacs. The sternotrachealis muscle expands somewhat over the sac, but is developed merely as a broad, thin fascia of little muscular strength.

The weak development of the sac, with the small tracheal slit and lack of specialization of the larynx, are notable when compared with this structure in *Erismatura jamaicensis*. The esophagus of *Erismatura vittata*, however, is remarkable. The pharyngeal end has the surface rugose as usual; immediately below the tube swells in an elliptical expansion that contracts again to normal size at the level of the shoulder, so that it occupies the same position

---

[56] Proc. U. S. Nat. Mus., vol. 52, Feb. 8, 1917, pp. 479–482; Condor, vol. 20, January, 1918, pp. 19–20.

as the tracheal air sac. The dorsal wall of the esophagus bordering the line of the muscular neck is thickened, while the distended anterior saclike portion is much thinner. This portion of the esophagus is obviously capable of considerable inflation, and remains broad and full even in the alcoholic specimen. Below the distension the esophageal walls are thickened as is normal.

The lingual muscles covering the hyoids are unusually heavy and well developed, as are the muscular slings that support the hyoidean apparatus below the head. Muscular attachments are evident on the lateral walls of the upper end of the esophagus, but in the specimen available I am unable to make out their arrangement.

It is evident that the atrophy of the tracheal air sac has been replaced by this curious esophageal expansion, a structure entirely absent in *E. jamaicensis*. The syrinx has no lateral bulb, agreeing thus with *jamaicensis*. Female specimens of *vittata* were not examined, but it seems probable that the structures described above are of a sexual nature and confined to the male.

Order FALCONIFORMES

Family CATHARTIDAE

CATHARTES URUBITINGA Pelzeln

Cathartes Urubutinga. PELZELN, Sitz. Kais. Akad. Wiss., vol. 44, 1861, p. 7. (Sapitiba, Irisanga, and Fort San Joaquim, Brazil.)

In his account of the Birds of British Guiana, 1916 (p. 211), Chubb takes *Cathartes ruficollis* Spix as the name for the yellow-headed vulture, a usage that more recently has been followed by Swann.[57] On examining the original account of *Cathartes ruficollis*[58] it is found that it is described as having the head red, the wing coverts brownish, and the shafts of the primaries dark, characters that indicate that this name refers to *Cathartes aura*, so that *ruficollis* is not available for the yellow-headed bird, which must be known as *urubitinga* Pelzeln as above noted. In his original description Pelzeln described *urubitinga* on the basis of specimens collected by Natterer in Brazil without definite citation of locality. Later[59] he cites nine specimens collected by Natterer at Sapitiba, in the district of Rio Janeiro, Irisanga, near the Rio Mogyguassu in northern São Paulo, and Fort San Joaquim, on the Rio Branco in extreme northern Brazil, not far from the frontier of British Guiana. The name *urubitinga*, therefore, is based on birds from these three localities.

[57] Syn. Accipitres, ed. 2, pt. 1, Sept. 28, 1921, p. 4.
[58] Spix, Avium Spec. Nov. Brasiliam, vol. 1, 1824, p. 2.
[59] Ornith. Brasiliens, 1871, p. 1.

The yellow-headed vulture was observed first at Resistencia, Chaco, on July 8 and 10, 1920. From July 16 to August 1 the species was recorded at Las Palmas, Chaco, and from August 9 to 19 at the Riacho Pilaga, Formosa. In the locality last named it was more common than the red-headed turkey vulture. At Puerto Pinasco, Paraguay, the yellow-headed vulture was seen on September 1, and at the ranch at Kilometer 80, west of that point, it was recorded on September 8, 9, and 15. One was observed at Lazcano, Uruguay, on February 8, 1921. The bird was found in the same territory as the turkey vulture, often in company with that species, of which it is a counterpart in general appearance, actions, and habits. On cold winter mornings the birds remained in their roosts until the sun had warmed the air, and on cold, cloudy days remained wholly inactive; but in sunny weather I seldom sat down to care for birds that I had killed without one or more of these vultures swinging overhead to observe what I was about. Like the turkey vulture, they have a graceful flight as they quarter tirelessly back and forth in search for food, or soar in great circles high in air. At times they rest in dead trees, or may alight in heavy woods if attracted by any movement or activity that seems to promise food. They appear slight in body for their wing expanse, and are tough and hard to kill. In a wounded individual I noted that the sides of the concave horny tongue tip were capable of being appressed to a considerable degree. The species was readily distinguished in life from the turkey vulture by the distinctly yellow head. A male killed near Las Palmas, Chaco, on July 20 had the head colored in a striking manner. The bill was cream buff, shading to vinaceous buff on a broad area that extended onto the forehead, behind the nostrils; side of the head in general, including eyelids, deep chrome; center of crown dark Tyrian blue, bordered on either side by a broad band of stone green; skin of throat posteriorly deep chrome, becoming paler forward, to shade into olive buff toward base of bill; space between mandibular rami spotted with dark Tyrian blue; a dull spot of slate blue beneath the nostrils on either side; iris carmine; tarsus cartridge buff, shading to neutral gray on the toes, where the interscutal spaces have a scurfy whitish appearance.

A female taken at Kilometer 182, Formosa, on August 13, 1920, was preserved as a skeleton.

The species does not seem to have been seen previously in Uruguay, but I include the Lazcano record without hesitation, though no specimen was taken, as one bird rested on a fence post while I passed at a distance of 10 meters, so that I had an exceptionally clear view of it.

The red-headed and yellow-headed vultures have been involved in much confusion, as though easily distinguished in life, in the field, in dried skins they appear closely similar. The following may be of aid in separating the two species:

$a.^1$ Head (in life) mainly yellow; general coloration more uniform, blacker, sheen of feathers of dorsal surface with green predominating, purple restricted or nearly absent; wing coverts black without distinct paler edgings_____C. urubitinga.

$a.^2$ Head (in life) mainly red; general coloration less dark, more variegated with brownish edgings to feathers, sheen of feathers of dorsal surface with purple predominating, green restricted or nearly absent; wing coverts distinctly margined with brownish, this color often extensive_____C. aura.

The color of the shafts in the outer primaries is not of definite value, as it is variable in both species. It has been alleged that in *urubitinga* the feathering on the back of the neck extended farther forward. This, however, is merely an age character, as immature birds of either species have the neck more or less feathered to the base of the cranium or even onto the nape, while in adults this area is naked.[60]

That *aura* and *urubitinga* are distinct species there can be no doubt. Though *aura* has a much greater zonal range, in tropical and subtropical regions, *aura* and *urubitinga* are found together throughout extensive areas, while they are sufficiently distinct to controvert any theory that might consider them color phases of one species.

The specimen from Las Palmas has the following measurements: Wing, 514; tail, 217 (culmen defective); tarsus, 60; middle toe with claw, 76 mm. These dimensions are somewhat larger than those of birds from eastern Brazil, British Guiana, and Venezuela, so that it is possible that a race characterized by slightly greater size inhabits the basin of the Rio Paraguay.

CATHARTES AURA RUFICOLLIS Spix

Cathartes ruficollis SPIX, Avium Spec. Nov. Brasiliam, vol. 1, 1824, p. 2. (Interior of Bahia and Piauhy.)

The geographical forms of the turkey vulture and the nomenclature applied to them have been involved in much confusion and uncertainty, a state that has not been remedied by the recent action of Chubb [61] in adopting *ruficollis* as the proper name for the yellowheaded vulture. A review of the entire group has shown that there are apparently five forms of *aura* that may be recognized, two from North America and the West Indies and three from South America. In general it may be said that turkey vultures from North America and the West Indies differ from those of South America in browner coloration and more distinct brown edgings on the wing coverts.

[60] See Bangs and Penard, Bull. Mus. Comp. Zoöl., vol. 62, April, 1918, p. 34.
[61] Birds of British Guiana, vol. 1, 1916, p. 211.

With this in mind, the following brief synopsis of my findings as to the valid forms and their distribution will be clear.

CATHARTES AURA AURA (Linnaeus).

Vultur aura LINNAEUS, Syst. Nat., ed. 10, vol. 1, 1758, p. 86.   (State of Vera Cruz, Mexico.[62])

The typical form of the turkey vulture has the wing coverts distinctly margined with brown like *C. a. septentrionalis*, but is smaller as the measurements of the wing range from 475 to 510 mm. It ranges from Panama (Empire, Fort Lorenzo) northward into Mexico, and in the West Indies. Swann[63] has described the bird from Cozumel Island as *C. a. insularis*. From the description this is apparently similar in size to *aura* (wing 470–505 mm.), but is said to be darker. As Cozumel Island lies only 15 miles from the mainland, I am inclined to consider this as doubtfully distinct (no specimens are at hand), and for the present a synonym of *aura*, especially since there is some variation in degree of blackness in that form. The type of *Cathartes burrovianus* Cassin[64] has been examined by Nelson and pronounced identical with *aura*.

CATHARTES AURA SEPTENTRIONALIS Wied.

Cathartes septentrionalis WIED, Reise Nord-America, vol. 1, 1839, p. 162. (Near New Harmony, Indiana.)

The northern turkey vulture is similar in color to typical *aura*, but is larger, with a wing ranging from 520 to 553 mm. It ranges from the northern part of the Mexican table-land north through the United States into southern Canada. The southern limit attained by this form in Mexico is uncertain.

CATHARTES AURA RUFICOLLIS Spix.

Cathartes ruficollis SPIX, Avium Spec. Nov. Brasiliam, vol. 1, 1824, p. 2. (Interior of Bahia and Piauhy.)

The turkey vulture of eastern and northern South America is similar in size to typical *aura*, as the wing ranges from 495 to 510 mm., but differs in blacker color and in restriction of the brown edgings on the wing coverts. It ranges from Paraguay (probably from northern Argentina and Uruguay) north through Brazil into the Guianas and Venezuela. The applicability of the name *ruficollis* to the red-headed turkey vulture is discussed under the account of the yellow-headed vulture. Hellmayr[65] has stated that Spix's type of *ruficollis* was no longer extant, and referred the name to *aura* with a query. I think there can be no valid reason for not recog-

---

[62] Type locality fixed by Nelson, Proc. Biol. Soc. Washington, vol. 18, Apr. 18, 1905, p. 124.

[63] Syn. Accipitres, ed. 2, pt. 1, Sept. 28, 1921, p. 3.

[64] Proc. Acad. Nat. Sci., Philadelphia, vol. 2, 1845, p. 212.

[65] Abhandl. K. Bayerischen Akad. Wiss., II Klass. vol. 22, pt. 2, 1906, p. 567.

nizing that this name denotes a red-headed turkey vulture, and as such it is the oldest name available for the small vulture of eastern and northern South America. *Oenops pernigra* Sharpe,[66] described from Guiana, Amazonia, and Peru, must be placed as a synonym here.

CATHARTES AURA JOTA (Molina).

Vulcur (sic) *Jota* MOLINA, Sagg. Stor. Nat. Chili, 1782, p. 265.   (Chile.)

The form that must bear this name is similar in color to *ruficollis* Spix, but is larger (wing from 530 to 550 mm.). It ranges from the Straits of Magellan through Chile north through the Andes apparently to Colombia. *C. a. meridionalis* Swann[67] must be considered a synonym of Molina's *jota*.

CATHARTES AURA FALKLANDICA (Sharpe).

Oenops falklandica SHARPE, Cat. Birds Brit. Mus., vol. 1, 1874, p. 27. (Falkland Islands.)

No specimens of the turkey vulture from the Falkland Islands are at hand. From descriptions it is similar in size to *C. a. ruficollis*, but is distinguished by distinct grayish margins on the median wing coverts and secondaries. According to Swann,[68] the Falkland Island vulture ranges from the Falkland Islands north along the coast of southern Chile.

A female turkey vulture secured on September 11, 1920, at Kilometer 80, west of Puerto Pinasco, Paraguay, has a wing measurement of 500 mm. and so is representative of the form here called *ruficollis*. Additional notes assumed to belong under this form which follow are not validated by specimens; Vera, Santa Fe, July 5, 1920; Las Palmas, Chaco, July 14, 17, 21, and 26; Riacho Pilaga, Formosa, August 19; Kilometer 80, west of Puerto Pinasco, Paraguay, September 11, 15, 16, 17, and 20; San Vicente, Uruguay, January 26, 28, and February 2, 1921; Lazcano, Uruguay, February 7 and 9.

In Paraguay this species was known as *urubu capiní*, literally translated as bald-headed buzzard.

The following records may pertain to the present form or may refer to *C. a. jota* (Molina): General Roca, Rio Negro, November 23 to 29, 1920 (fairly common); Zapala, Neuquen, December 7 to 9; Tunuyan, Mendoza, March 27; Tapia, Tucuman, April 12; Tafi Viejo, Tucuman, April 17. The status of the turkey vultures from central Argentina must remain in abeyance until specimens can be measured and examined. It is probable that *jota* comes north into northern Patagonia if not farther, and that it also occurs through

---

[66] Cat. Birds Brit. Mus., vol. 1, 1874, p. 26.
[67] Syn. Accipitres, ed. 2, pt. 1, Sept. 28, 1921, p. 3.
[68] Who, in the reference just cited, p. 4, gives this form as *iota* Molina.

the eastern foothills of the Andes, along the entire western border of the Republic.

CATHARTES AURA JOTA (Molina)

Vulcur Jota MOLINA, Sagg. Stor. Nat. Chili, 1782, p. 265. (Chile.)

On April 25 and 27 turkey vultures were observed near Concon, in the Intendencia of Valparaiso, Chile. No specimens were secured so that these notes are allocated here solely on geographical evidence.

CORAGYPS URUBU FOETENS (Lichtenstein)

Cathartes foetens LICHTENSTEIN, Verz. Ausg. Saüg. und Vög. Zool. Mus. Kön. Univ. Berlin, 1818, p. 30. (Paraguay.)

The black vulture, common in the warmer regions that I visited, had habits identical with those of the species in the southern United States. It was recorded as follows: Resistencia, Chaco, July 8, 1920; Las Palmas, Chaco, July 17 to August 1; Riacho Pilaga, Formosa, August 7 to 19; Formosa, Formosa, August 23 and 24; Puerto Pinasco, Paraguay, September 1 to 23 (seen west to a point 110 kilometers from the Rio Paraguay; Las Flores, Maldonado, Uruguay, January 22, 1921; San Vicente, Uruguay, January 26; Tunuyan, Mendoza, March 22 and 24; Tapia, Tucuman, April 11; Concon, Chile, April 25. About temporary camps of woodcutters in the Chaco these birds gathered in flocks to secure offal from the killing pens where meat was prepared for human consumption. In cattle country carcasses of horses and other animals offered a supply of food. In the town of Rio Negro, Uruguay, on February 18 I saw a tamed bird running about in the streets, with no fear of dogs or pedestrians. It was of interest to note that in the Chaco I saw three species of vultures in view at the same time on several occasions, while a white-breasted bird seen soaring high in air may have been the king vulture, so that it may be possible there to find four forms of this family together.

The matter of subspecies in the black vulture is still open to question, as in a limited series I do not find any sharply trenchant difference between northern and southern birds. Todd [69] in a recent consideration of the bird does not recognize geographic races. Specimens from Florida and Georgia have wing measurements ranging, irrespective of sex, from 420 to 436 mm. In one from Chile the wing is 405 mm., while in an adult female that I killed on April 11, 1921 (skull alone preserved), the wing measured 433 mm. It is possible that there are more than two forms involved. I have followed current usage in recognition of a southern race as the specimens available are not sufficient to enable an independent opinion in the matter. *Cathartes foetens* of Lichtenstein is given in

---

[69] Ann. Carnegie Mus., vol. 14, 1922, p. 142.

a catalogue of specimens with no description, but with a reference to the *iribu* of Azara. On consulting Azara [70] it is found that the *iribu* is the black vulture so that this name will antedate *Vultur brasiliensis* Bonaparte.[71]

The type locality for the southern form, is thus Paraguay. Hellmayr [72] has held incorrectly that *Vultur urubu* of Vieillot [73] is a synonym of *Cathartes aura*. Although in his discussion of the species, Vieillot states that "un rouge sanguin colore la peau de la tete et du cou" his diagnosis and plate can apply only to the black vulture. In the general account of this species Vieillot points out clearly the distinctions between the black and turkey vultures, and it may be that his note on the color of the head and neck under the black vulture refer to the purplish suffusion found on these parts in the adult bird when recently killed.

VULTUR GRYPHUS Linnaeus

Vultur gryphus LINNAEUS, Syst. Nat., ed. 10, vol. 1, 1758, p. 86.   (Chile.)

On March 19, 1921, at El Salto, an estancia above Potrerillos, Mendoza, at an elevation of 2,000 meters, I saw one of these birds sailing above the valley. Three were observed above Uspallata, Mendoza, on April 21. The male of this species is known as *condór*, the female as *buitre*, a distinction that may lead to confusion as it would indicate that two distinct species were intended.

Family FALCONIDAE

MILVAGO CHIMANGO CHIMANGO (Vieillot)

Polyborus chimango VIEILLOT, Nouv. Dict. Hist. Nat., vol. 5, 1816, p. 260. (Paraguay.[74])

The chimango seems more common in the pampas region than elsewhere in its range, as there it is often found in such numbers that it may be said to be abundant. To the northward it is replaced by the allied *Milvago chimachima* throughout the Chaco, save in the extreme southern part. Records for *M. chimango* made during my field work follow: Berazategui, Buenos Aires, June 29, 1920; Vera to Los Amores, Santa Fe, July 5, (the most northern point at which the species was observed); Dolores, Buenos Aires, October 21;

---

[70] Apunt. Hist. Nat. Pax. Paraguay, vol. 1, 1802, p. 19.

[71] Consp. Gen. Av., vol. 1, 1850, p. 9.

[72] Abhandl. Kön. Bayerischen Akad. Wiss., Kl. II, vol. 22, pt. 3, 1906, p. 567; and Nov. Zool., vol. 28, May, 1921, p. 174.

[73] Hist. Nat. Ois. l'Amer. Sept., vol. 1, 1807, p. 23, pl. 2.

[74] The type locality for *chimango* of Vieillot, established by common usage as Paraguay (see Brabourne and Chubb, Birds of South America, vol. 1, December, 1912, p. 63, and Swann, Synoptical List of the Accipitres, ed. 2, pt. 1, Sept. 1921, p. 16), is in a way unfortunate, as Vieillot, translating Azara's comment, says that the chimango is rare in Paraguay but common along the Rio de la Plata, a condition that holds to-day.

Lavalle, Buenos Aires, October 23 to November 15; General Roca, Rio Negro, November 23 to December 3; Zapala, Neuquen, December 7 to 9; Carhue, Buenos Aires, December 15 to 18; Victorica, Pampa, December 26 to 30; Carrasco, Uruguay, January 9 and 16, 1921; La Paloma, Uruguay, January 23; San Vicente, Uruguay, January 26 to February 2; Lazcano, Uruguay, February 3 to 9; Guamini, Buenos Aires, March 3 to 8; (Potrerillos, Mendoza, March 20 a rectrix found); Tunuyan, Mendoza, March 24 to 28; Concon, Chile, April 26 to 28. Adult males taken at Lavalle, November 8, General Roca, November 26, Carhue, December 15, and an immature female shot at Tunuyan, March 24, are all representative of the typical form. *Milvago c. temucoensis* Sclater [75] described from Palal near Temuco, Province of Cautin, Chile, and said to range in the Provinces of Cautin and Valdivia, is represented in the United States National Museum by specimens collected by naturalists from the United States Fish Commission steamer *Albatross*, at Laredo Bay, Straits of Magellan, a considerable extension of range over that previously known for this form. These birds agree with one from Valdivia in darker coloration and more heavy persistent barring below than is found in typical *chimango*. Apparently *temucoensis* ranges throughout the region of heavy rains in southern Chile. Specimens from near Santiago, Chile, Tunuyan, and General Roca show a tendency toward dark coloration, but are so near *chimango* as to be indistinguishable. A skull and a skeleton of adult males were preserved at Victorica, Pampa, on December 28 and 31, and the skull of an immature bird with the cranial bones not yet ankylosed was found near Guamini, Buenos Aires.

The chimango is a common species of the open country, and is seen almost inevitably by the naturalist on every day afield during work within its range. The birds are at their best as scavengers along muddy shores where they feed or forage in little groups, often in company with gulls, or are found beating back and forth over areas where food may be found. In the open country they rest on the ground or on fence posts, or perch in low bushes or trees where such are available. The birds have little fear, as they are despised for their manner of living, and, save where they become too numerous, are seldom molested. It was common to have them come about fearlessly while I examined dead birds in the field, and care was necessary to guard specimens from damage. The flight of the chimango is comparatively weak, though the birds often delight in soaring and sailing during windy weather, particularly at the beginning of the breeding season. When in the air they show a white

---

[75] Bull. Brit. Orn. Club, vol. 38, March 4, 1918, p. 43.

rump and two light patches on either wing. Small birds show no fear of them, and feed or rest with unconcern though chimangos may be near at hand, except when they have nests containing eggs or young, when these little hawks are harried mercilessly by everything from small passerines to the spur-winged lapwings, doubtless for good cause, as the chimango delights in helpless prey. On one occasion I observed one pecking steadily in an attempt to drag out the entrails of a lamb, too helpless from some disease to move other than to flinch at the cruel strokes of the bird's beak. At other times the chimango may be of considerable economic value, as in Uruguay, during a period of invasion by locusts, chimangos were seen in bands that at times numbered 30 or 40 individuals gathered to feed on this food. The hawks walked or ran about on the ground or swooped down at their prey from above, and fed until completely satiated. At Carrasco, Uruguay, on January 16, I observed 16 gathered over an area of sand dunes to feed on a small cicada (*Proarna*, species) abundant at the time. Their feet are too weak to afford firm grasp with the talons, but on the ground they walk with ease and freedom. One that I wounded slightly ran so swiftly that it was captured only after a long chase. The birds are usually more common in the vicinity of water than elsewhere, and drink copiously and frequently even though the water may be quite brackish in taste.

MILVAGO CHIMACHIMA CHIMACHIMA (Vieillot)

Polyborus chimachima VIEILLOT, Nouv. Dict. Hist. Nat., vol. 5, 1816, p. 259. (Paraguay.)[76]

In addition to the color characters assigned by Bangs and Penard[77] to *Milvago chimachima cordata* from Panama, this northern form seems to be slightly smaller, as in the type, a female, the wing is given as 292 mm., and the tail 196 mm., and in a male topotype the wing is 275 mm., and tail 183 mm. In two adult females of the southern form from Las Palmas, Chaco, and Riacho Pilaga, Formosa, the measurements are as follows: Wing, 305, 302 mm.; tail, 197, 197 mm. An adult male from Kilometer 80, Puerto Pinasco, Paraguay, measures, wing 292 and tail 187 mm. There is considerable variation in color of the head and underparts in adults of the typical form, some being much paler than others. I believe that the plumage, as it ages after the molt, is subject to considerable bleaching.

This small carrion hawk, known as the chimango, or, more properly in Guaraní, as *kiriri*, was encountered first at Las Palmas, Chaco. On July 27, 1920, one or two were found in the tops of low

---

[76] No locality is designated in Vieillot's original description but the type locality has been assumed to be Paraguay as the description is taken from Azara.

[77] Bull. Mus. Comp. Zoöl.. vol. 62, April, 1918, p. 35.

trees scattered across a prairie, and at my appearance greeted me with harsh squalls. An adult female was taken. Another was observed on July 30. At the Riacho Pilaga, Formosa, a female in streaked juvenal plumage was killed on August 7, and an adult female on the following day. Others were recorded here until August 19, and one was seen near the town of Formosa on August 24. In the region about Puerto Pinasco, Paraguay, they were fairly common and an adult male was collected September 15.

This form has customs similar to that of *Milvago chimango*, a species that it replaces in the north, but is more of an inhabitant of wooded areas. The partly open, partly forested Chaco seemed fitted especially for its needs, and here it was encountered where scattered trees furnished shelter. The birds are scavengers in habit and feed on any waste animal food that is available. They also bear a bad reputation among the housewives on the estancias for their propensity to filch young chickens and ducks, and, as the kiriri comes familiarly about buildings, its depredations may at times be considerable. My "squeaking" to attract small birds from dense coverts always drew these small hawks when they were about, and their squalling calls were often annoying when I was straining my ears to catch some faint bird note from the surrounding thickets or trees. Carrion hawks often came while I was engaged in cleaning bird skeletons, and walked about on the ground to pick up bits of flesh that I threw out to them, while it was necessary to hang skeletons put out to dry in places where they were secure from the sharp eyes of these prying marauders. The chimango is a bird of weak flight, flapping and sailing rather slowly, and never, so far as I am aware, is it directly aggressive to other birds unless it encounters young or individuals that have been injured in some way. Adults when on the wing appear light in color on the body and tail, with a light bar in either wing. Their call, a harsh squall, is reminiscent of that of *Ibycter ater*.

POLYBORUS PLANCUS PLANCUS (Miller)

Falco Plancus MILLER, Var. Subj. Nat. Hist., 1777, pl. 17. (Tierra del Fuego.)

The carancho was almost universal in occurrence throughout Argentina and Uruguay, as it ranged throughout wooded regions as well as on the open pampas. None were observed at the localities worked in Rio Negro and Neuquen in northern Patagonia, but this was due in all probability to the short time occupied in field work in those regions. The carancho, as the bird is known in the south, is a bird of strong flight, though it does not delight in soaring or circling in the air as is customary in vultures and many hawks. Its vigorous form, with contrasted light and dark colors, is one that constantly attracts attention when traveling through its range, whether the birds are observed at rest in the top of some tree that commands an outlook over open country, or in steady direct flight toward some distant point. In the Chaco the caracará, as the bird is called in Guaraní, is known as a scavenger that is tolerated so that it is tame, and at times almost aggressive in its approach to man. In the pampas these hawks are killed relentlessly because of their depredations in eating out the eyes of newly-born lambs, and in many districts the bird is becoming rare.

At Las Palmas, Chaco, caracaras were common from July 13 to August 1, 1920. When waste from a sugar factory killed many fish in a small stream caracaras gathered in bands to feed on them, and, it may be added, filched a number of mouse traps that I had supposed were securely hidden in the brush along the bank. One was observed eating a cavy, found lying dead, which like all large prey was held firmly under one foot and torn into small bits with the heavy bill. Near the Riacho Pilaga, Formosa, caracaras were recorded from August 7 to 19. One flew down frequently to disturb feeding flocks of monk parrakeets (*Myiopsitta m. cotorra*) in a sweet-potato field in hope that some cripple, wounded by the shooting of Indians, might fall into its clutches. It was not able to seize the uninjured birds. On August 12 one flew to a large stick nest, 9 meters from the ground in a quebracho tree standing in an open savanna, to bring a bit of stick to add to the structure. After this had been arranged satisfactorily the bird settled for an instant in the nest cavity, and then flew to a limb overhead and surveyed the nest carefully. At the town of Formosa, on August 23, caracaras flew back and forth above the strong current of the Paraguay River in search for any carrion that might come downstream.

In the vicinity of Lavalle, Buenos Aires, caracaras were observed frequently from October 27 to November 15. The skull of a male was secured on October 31. The birds were wary here as they were shot relentlessly by the estanceros because of their destructiveness to young stock. During the day carranchos ranged over the open pampa but returned at nightfall to roost in trees in occasional groves of ombú, tala, or eucalyptus. On November 6 I collected a set of two fresh eggs near the coast about 25 kilometers south of Cape San Antonio. The site was a small tree in a little grove planted about a water hole, a spot remote from habitation and the only suitable one available in a radius of several kilometers. The nest, placed about 6 meters from the ground, was an untidy structure, bulky and heavy in appearance, made of dried stems of a sharp-pointed rush, with broken ends stuck out in all directions. The deeply cupped interior was lined in part with a felted mass of pellets ejected by the parents, that formed a soft bed for the two handsomely marked eggs. As I examined these and packed them in my hat to remove them both caracaras hovered with harsh grating calls a few feet above my head. The ground color of these eggs varies from light-pinkish cinnamon to pinkish cinnamon, obscured and in places almost obliterated by a heavy irregular wash that varies from auburn and chestnut to hessian brown and liver brown. About the large end these blotches become heavier and more concentrated, and in places are almost black. These eggs measure 59.8 by 48 mm. and 54.5 by 48.5 mm. A caracara was gathering sticks for a nest on November 15.

A few were observed near San Vicente, Uruguay, from January 26 to February 2, 1921, and one was seen at Lazcano on February 7. North of Cordoba, Argentina, many were noted along the railroad on March 31 and at Tapia, Tucuman, the species was fairly common from April 7 to 13. A few were recorded on the slopes of the Sierra San Xavier, above Tafi Viejo, on April 17.

On my return from Paraguay to the pampas of Buenos Aires I noted that the carrancho of the south seemed larger than that observed a few days previous in the northern Chaco, an impression that has been sustained by a study of specimens. Skins from Chile (one) and Argentina from the Straits of Magellan northward (nine) show a wing measurement that varies from 410 to 442 mm. (average 431 mm.). Skins from Brazil (Pernambuco, and one from Bahia or Rio de Janeiro) and Puerto Pinasco, Paraguay (three in all) range from 365 to 405 mm. (average 387 mm.). (There seems to be no constant difference in size correlated with sex in the caracaras.) The large form apparently ranges throughout Argentina as I killed an adult female at Las Palmas, Chaco, on July 20, 1920, with a wing measurement of 429 mm.) and into Paraguay as a bird from that country without definite locality (taken on the Page expedition in the fifties) has the wing 425 mm. As the type locality of Miller's *Falco plancus* has been cited by Shaw [78] as Tierra del Fuego, the southern form will stand as *Polyborus plancus plancus*.

POLYBORUS PLANCUS BRASILIENSIS (Gmelin)

Falco brasiliensis GMELIN, Syst. Nat., vol. 1, pt. 1, 1788, p. 262. (Brazil.[79])

As has been stated above a male caracara secured at Kilometer 80, west of Puerto Pinasco, Paraguay, on September 15, 1920, has a wing measurement of 405 mm., and so appears to belong to the northern form, for which the name *Falco brasiliensis* of Gmelin is available. This is assumed to be the form that ranges from

---

[78] Cim. Phys., 1796, p. 34.
[79] Type locality hereby fixed as Pernambuco.

northern Paraguay northward through Brazil. The status of the bird from the coastal region of extreme southern Brazil (Santa Catherina and Rio Grande do Sul) is uncertain. From September 6 to 30, 1920, I found caracaras common in the country about Puerto Pinasco. As they were not persecuted in this region they came in numbers about ranch buildings in search of offal from the killing pens. Their habits do not differ from those of the southern form.

HERPETOTHERES CACHINNANS QUERIBUNDUS Bangs and Penard

Herpetotheres cachinnans queribundus BANGS and PENARD, Bull. Mus. Comp. Zoül., vol. 63, June, 1919, p. 23. (Pernambuco, Brazil.)

The southern laughing falcon was seen first on July 21, 1920, near Las Palmas, Chaco, when an adult female was taken as it rested on a dead stub in an opening in the forest. Farther northward at the Riacho Pilaga, in central Formosa, from August 8 to 20 the species was more common, and two additional females were shot on August 14 and 20 (the latter preserved as a skeleton). At Kilometer 80, west of Puerto Pinasco, Paraguay, one was seen September 12, and at Kilometer 200 two were found on September 25. The birds, usually in pairs, inhabit the taller growths of heavy forest where they rest on open perches surrounded by dense heavy growth. Where not molested they come out into more open regions. Though of heavy build, seeming strong and powerful, they are sluggish in ordinary habit and are seldom seen save at rest. On only one occasion did I observe one turning in circles in the air above the trees.

One is not long in the haunts of the laughing falcon without becoming familiar with its strange loud notes, though it may be some time before the bird is seen. The call begins as a single note, given at short intervals, and then changes to a more rapid repetition of varied sounds. After two or three minutes the mate of the performer may join in and the birds call rapidly, first in alternate short notes and then in a strange medley, a weird, unearthly concert, startling indeed to one not familiar with its source, that may be continued without cessation for 10 minutes. These strange duets were especially impressive when heard at dusk. Countrymen related that the birds were very observant and announced by their calls the passage of men through the forest.

The falcons themselves are handsome birds, their heavy white crests and boldly marked black head being no less impressive than their notes. The Toba Indians knew them as *gua kow* in evident imitation of their calls, while in Guaraní they were called Guaycurú, a word signifying a Chaco Indian, usually designating a warlike type.

The three adult females taken had the tips of the rectrices much worn, in one to such an extent that the bare shaft projected for nearly 10 mm. beyond the barbs of the feathers. The bill was black; cere and gape primuline yellow; iris walnut brown; tarsus and toes deep olive buff; claws black.

These southern birds represent the light extreme of the form described as *queribundus*. Two females offer the following measurements: Wing, 290–298; tail, 230–238; culmen from cere, 24–23; tarsus, 57–66.5 mm. (Measurements of the specimen from Formosa given first in each case.) An old specimen in the collection of the United States National Museum, Cat. No. 16526, collected by T. J. Page, recorded by Bangs and Penard in the original description of the subspecies as from Parana, in reality was secured on the Paraguay River in southern Matto Grosso, between Corumba and the Paraguayan border.

MICRASTUR SEMITORQUATUS (Vieillot)

Sparvius semitorquatus VIEILLOT, Nouv. Dict. Hist. Nat., vol. 10, 1817, p. 322. (Paraguay.)

Critical study of names proposed for this hawk indicates that the designation above is the proper one. The *esparvero faxado* of Azara [80] is without question an immature of the present species, and as such furnished the sole basis for Vieillot's epervier a demi-collier roux *Sparvius semitorquatus* from Paraguay. Azara's description of the bird, excellent and unmistakable in its details, states that "baxo de la cabeza es blanco; pero cada pluma tiene a lo largo una tirita obscura." Vieillot in translating this makes an error in ascribing these markings to the crown, as he says " les plumes du dessus de la tête ont un trait transversal noirâtre sur un fond blanc." Had he written *dessous* instead of *dessus* the transcription would have been correct. The remainder of Vieillot's description coincides with that of Azara except that the tail is said to be 241 mm. long instead of 235 mm. Vieillot, in his Encycl. Méth., follows his own statements, translating them into latin. It would appear that the original transcription was erroneous, perhaps a mere typographical error, and that the name must be recognized for the present species, as it has priority of pagination over *Sparvius melanoleucus* (which is described on page 327 of the same work) based on Azara's *esparvero negriblanco*, the adult of the present bird.

The single specimen obtained was shot near Kilometer 80, west of Puerto Pinasco, Paraguay, on September 9, 1920. The bird was attracted by my squeaking to call up smaller species and flew in to perch in the heavy growth a few feet away. The flight was almost

---

[80] Apunt. Hist. Nat. Paxaros Paraguay, vol. 1, 1802, p. 126.

noiseless, so that the downy margins on the flight feathers seem to serve to deaden the sound of the wings. In life the facial ruff was almost as prominent as in the marsh hawk.

Swann[81] has described a subspecies from Sarayacu, Ecuador, as *buckleyi* on basis of small size. Specimens available agree more or less in measurement throughout the entire range of the bird, so that it is possible that this bird from Ecuador, with a wing measurement of only 217 mm., may be an aberrant specimen or may belong elsewhere.

My skin from Paraguay is a male (apparently in its second year) of the extreme of the rufescent phase for this species. The dorsal surface is entirely warm brown in color, with the transverse lighter markings on wings and back more definitely indicated than is usual. A few dusky feathers are in evidence on the crown and back. It measures as follows: Wing, 246; tail, 248; culmen from cere, 18; tarsus, 81.5 mm.

SPIZIAPTERYX CIRCUMCINCTUS (Kaup)

Harpagus circumcinctus KAUP, Proc. Zool. Soc. London, 1851 (publ. Oct. 28, 1852), p. 43. ("Chili"=Argentina.)

This rare hawk was found in small numbers near Victorica, Pampa, on December 23, 27, and 28, where three specimens were secured, a pair on December 23 and an adult male on December 28.

The birds frequented the larger growth of woodland in this region, usually where the forest of heavy-limbed, stocky trees was fairly open. To avoid the intense rays of the sun, they chose shaded perches in such trees as the caldén, where the foliage was confined largely to the tips of the branches and did not obscure the outlook below or at the side. Attention was drawn by the querulous whining calls of these falcons, similar to a note of the brown thrasher, but given in a much louder tone. At times the birds, rather than fly, hopped agilely through the limbs to place a screen of branches between themselves and the observer. Their flight was direct, like that of a small falcon, with the white rump displayed as a prominent identification mark. When they appeared in the open they were pursued hotly by fork-tailed flycatchers and other related species. The birds taken were breeding (though no nests were observed) and were in somewhat worn plumage. (Pl. 9.)

An adult female shot December 23 had the tip of the bill dull black; base of maxilla light grayish olive; base of mandible mignonette green; cere, gape, skin of lores, and bare skin about eye wax yellow; iris light cadmium; tarsus and toes slightly paler than deep colonial buff; nails black.

[81] Syn. Accipitres, ed. 2, pt. 1, Sept. 28, 1921, p. 25.

The species may not have been taken in the Territory of Pampa before, as the " Biga de la Paz, Pampa," of Burmeister may refer to the town known as Paz in southern Santa Fe. At Victorica the species is near its southern range, though it may range south to the limit of the Pampan monte, somewhere northwest of Bahia Blanca. With the destruction of this forest for wood, the bird will, of necessity, become extinct in this area through lack of suitable cover.

In parts of the Province of Cordoba *Spiziapteryx* may be common, as on April 19, 1921, between Quilino and Cordoba, from a train window I noted 8 or 10 at rest in the morning sun, perched like sparrow hawks on dead stubs or telegraph poles. The species has been reported previously from Santa Fe (?), Mendoza, Cordoba, Santiago del Estero, Catamarca, La Rioja, Tucuman, and Salta.

CERCHNEIS SPARVERIA CINNAMOMINA (Swainson)

Falco cinnamominus SWAINSON, Anim. Menag., 1838, p. 281. (Chile.)

Treatment of the sparrow hawks from the southern part of South America, with existing material, is difficult and uncertain. Two forms are currently recognized, *australis* of Ridgway of eastern and northern range, and *cinnamomina* of Swainson, described from Chile. These two differ *inter se* in size, in the marking of the tail and to a slight degree in coloration of the under surface. Material in the United States National Museum representing them is far from satisfactory, but from study of this and from literature it appears that the male of *cinnamomina* differs from *australis* in larger size (wing 187–199, average 193; tail 129–143, average 134 mm.), in narrower subterminal band on the tail (9–16 mm.), in more or less rufous on the tips of the rectrices, and in having the outer rectrix with only one bar (rarely more) and the inner web rufescent. The female has the wing 197–209 mm., and the black bars on the rectrices narrower and less complete. In the male of *australis*, as represented by birds from Brazil, the wing is shorter (175–185, average 181; tail, 122–131; average 127 mm.), subterminal tail band broader (18–22 mm.), tail tipped with white or gray, inner web of outer rectrix white, with three or more black bars, and the underparts whiter. The female has the wing 179–190 mm., the black bars on the tail wider, more complete, and the subterminal band wider.

Skins from Patagonia and the eastern base of the Andes in Argentina agree well with *cinnamomina*. Those from the pampas region northward into Uruguay and Paraguay are more or less intermediate between *cinnamomina* and *australis*. This broad area of intergradation between the two forms, as here considered, is puzzling, but may be explained in a way by considering some of the intermediates taken in the north that most nearly resemble typical *cinnamomina*, as possible winter migrants from more southern breeding grounds. Four specimens that I secured have the following char-acteristics: An adult male shot at Victorica, Pampa, on December 28, 1920, a breeding bird, has the tip of the tail mixed rufous and white, the black subterminal band 12 mm. wide at its widest point, the inner web of the outer rectrix rufescent, unmarked save for the subterminal bar, and the following measurements: Wing, 186; tail, 128 mm. It thus has the color markings of *cinnamomina*, but is smaller than most of that form. A female taken at Guamini, Buenos Aires, March 8, 1921, has the black tail bands interrupted centrally, very slightly restricted, and the following measurements: Wing, 197.5; tail, 130.5 mm. It is within the limit of measurement for *cinnamomina*, but is more boldly marked. An adult male shot at Las Palmas, Chaco, July 27, 1920, has the tips of the tail without rufous, the subterminal band 16–17 mm. wide, the outer rectrix with the inner web partly white, with one black spot in addition to the subterminal band, the wing 190 and the tail 128 mm. The bird thus slightly approaches *australis* in coloration, but is large. An adult breeding male from San Vicente, in eastern Uruguay, taken January 25, 1921, has the tip of the tail rufous, the subterminal band 16 mm. wide, the external rectrix with two bars on the white and rufous inner web, the wing 189 and the tail 123 mm. This bird from its geographic position might be supposed to be near *australis*, but seems as near *cinnamomina* as the others. From the review above it will be seen that these specimens are all more or less inter-mediate, but I have considered them all nearer *cinnamomina* than *australis*.

The sparrow hawk was recorded as follows: Las Palmas, Chaco, July 16 to August 1, 1920; Riacho Pilaga, Formosa, August 21; Formosa, Formosa, August 23 and 24; Puerto Pinasco, Paraguay, September 1 to 20; General Roca, Rio Negro, November 23 and 27; Victorica, Pampa, December 24 to 29; La Paloma, Uruguay, January 23, 1921; San Vicente, Uruguay, January 25 to February 2; Lazcano, Uruguay, February 5 to 9; Rio Negro, Uruguay, February 17 and 18; Guamini, Buenos Aires, March 3 to 8; Mendoza, Mendoza, March 13; Potrerillos, Mendoza, March 16 to 21; Tunuyan, Mendoza, March 23 to 29; Tapia, Tucuman, April 6 to 13.

Wherever found the bird was recorded as a watchful observer from some commanding perch from which it had a clear outlook over open country. In the breeding season it circled about scream-ing *killy killy killy*, but at other seasons it was silent, only taking wing when too closely pressed. Occasionally in the Chaco I saw one stooping swiftly at some inoffensive *Heterospizias* at rest in a tree top that perhaps had roused the ire of the smaller bird through the usurpation of a favored perch. At Kilometer 80, near Puerto Pinasco, sparrow hawks seemed to be nesting on September 7, but in the pampa they did not appear to breed until December. In Uruguay, where the present bird is common, it is an efficient enemy of the locust hordes that devastate the cultivated lands.

FALCO FUSCO-CAERULESCENS FUSCO-CAERULESCENS Vieillot

Falco fusco-caerulescens VIEILLOT, Nouv. Dict. Hist. Nat., vol. 11, 1817, p. 90. (Paraguay.)

Near the Riacho Pilaga, Formosa, and from that point eastward to the Rio Paraguay, aplomado falcons were fairly common during the middle of August, 1920. They frequented open savannas where stubs of dead quebrachos offered lookout stations, or failing these, even rested on the tops of bushes near the ground. Their flight was swift and direct, performed with strong, quick beats of the wings, and in general appearance they suggested small duck hawks. At the Riacho Pilaga the sight of these little falcons brought consternation to the screeching flocks of monk parrakeets that fed in the open in old sweet-potato fields. A male falcon taken on August 12, a bird fully grown but in dark immature dress, had the tip of the bill black, shading posteriorly through gray number 7 to mustard yellow at base, cere and bare skin about eye mustard yellow; iris Rood's brown; tarsus and toes primuline yellow; claws black. The species was not seen again until April 9, 1921, at Tapia, Tucuman, when a female was brought down with a broken wing as it passed me above a wooded slope. This bird ran swiftly on the ground to cover and was captured only after a rapid chase down a brush-grown slope. On April 10 two were seen, evidently hunting, as one dashed down into little openings in the woods and then, disappointed in seeing prey, rose again to continue its direct flight. On April 17 a male was killed from a little tree above a mountain pool at an elevation of 2,300 meters, in the Sierra San Xavier, above Tafi Viejo, Tucuman. This was an immature bird of the year, while the female taken at Tapia is probably in its second year, as it is distinctly gray above.

The subspecies *Falco f. septentrionalis* Todd [82] proposed for the aplomado falcon of North America may be distinguished by slightly larger bill, longer tail, and by greater average size in all measurements. In color northern and southern birds appear identical. The wing measurement in this species seems somewhat variable and in the series at hand is not of definite value in separation, save when used in averages. The bill, however, is slightly larger and longer, and the tail longer in *septentrionalis*. A single female from La Raya, in the Andes of Peru, that greatly exceeds any other specimens in general size (wing, 313 mm.) save that the bill is small, probably

---

[82] Proc. Biol. Soc. Washington, vol. 29, June 6, 1916, p. 98. (Fort Huachuca, Arizona.)

is representative of an unnamed subspecies. Following is a synopsis of measurements of specimens in the collections of the United States National Museum:

FALCO F. FUSCO-CAERULESCENS.

Five males (Santiago, Chile; Conchitas, Buenos Aires; Kilometer 182, Formosa; Tapia, Tucuman; Paraguay) wing, 235–255.5 (245.4); tail, 148.5–165 (156.3); culmen from cere, 14.5–16 (15.3); tarsus 43–45 (43.5 mm.). Three females (Chile; Conchitas, Buenos Aires; Tapia, Tucuman) wing, 280–290 (285); tail, 181.8–183.3 (182.5); culmen from cere, 16.9–18 (17.6); tarsus, 47.8–50.5 (49.1 mm.).

FALCO F. SEPTENTRIONALIS.

Nine males (Mirador, Vera Cruz; Alta Mira, Tampico; Lake Palomas, Chihuahua; Hachita, and Luna County, New Mexico; Fort Huachuca, Arizona), wing, 245–272.5 (259.1); tail, 146.5–187 (172.2); culmen from cere, 15.1–17.7 (16.4); tarsus, 43–47.5 (45.3 mm.).

Four females (Mirador, Vera Cruz; Mazatlan, Sinaloa; Otero County, New Mexico; Fort Huachuca, Arizona), wing, 287–298 (293); tail, 188.3–203.5 (197.1); culmen from cere, 18.5–19.8 (19.1); tarsus, 46.5–53. (49.9 mm.).

GAMPSONYX SWAINSONII SWAINSONII Vigors

Gampsonyx swainsonii VIGORS, Zool. Journ., vol. 2, April, 1825, p. 69.
(Ten leagues W. S. W. of Bay of San Salvador, Bahia, Brazil.)

A male of this beautiful little falcon was taken September 25, 1920, at Kilometer 200, west of Puerto Pinasco, Paraguay. The bird was killed as it rested on a dead limb in the top of a tree that stood in the border of a tract of forest. The call note of this individual was a low *kee kee*. In attitude the bird resembled a sparrow hawk at rest, but was distinguished by its shorter tail.

Compared with two skins from Diamantina, Brazil, this specimen differs in a broader collar of black across the hind neck. One side is nearly immaculate, the other is streaked slightly with rufous. Measurements are as follows: Wing, 154; tail, 92.5; culmen from cere, 13; tarsus, 29.5 mm.

Family ACCIPITRIDAE

ELANUS LEUCURUS LEUCURUS (Vieillot)

Milvus leucurus VIEILLOT, Nouv. Dict. Hist. Nat., vol. 20, 1818, p. 563.
(Paraguay.)

On September 24, 1920, near an alkaline stream known as the Riacho Salado in the Chaco, 170 kilometers west of Puerto Pinasco, Paraguay, I saw one of these kites in a savanna dotted with low trees, and after some difficulty secured it. The bird, an adult male, rested on the top of a low shrub, balancing in the wind, but flew before I came within range to circle and sail gracefully for several minutes before it chose to rest. Finally, blown by a gust of wind, it miscalculated its distance in passing me and fell at a long shot. On the following day two were found resting in the top limbs of low trees in an open marsh grown with saw grass on the border of Laguna Wall, 30 kilometers farther west. When flushed they swung about lightly and gracefully, seldom more than a few yards from the ground. One was taken and preserved as a skeleton. The Lengua Indians called this species *Kabuko*.

At the Estancia Los Yngleses near Lavalle, Buenos Aires, the *alcón blanco*, as this kite is known, was found on October 27, and two males were taken. The first was observed as it hovered in the wind 15 or 18 meters from the ground, stationary above one spot of grass that it watched intently. At a hasty glance its light coloration gave it the semblance of a gull. Another was secured from a perch in the top of an ombú tree where it rested in a part of a grove sheltered from wind. In one of these birds, apparently full adult, the cere and upper mandible were chamois; gape and base of lower mandible slightly grayer than primuline yellow; remainder of bill black; iris orange chrome (verging toward orange rufous); tarsus and toes slightly duller than apricot yellow; claws black. The other male, though fully adult in other respects, retained an indication of the dark spotting at the tips of some of the rectrices that is found in juvenal plumage.

Bangs and Penard [83] have described the white-tailed kite from North America as *Elanus leucurus majusculus* (type-locality, Florida) on the basis of slightly greater size. The difference between birds from North America and those from South America, while slight from the series examined in the U. S. National Museum, seems constant. A small series from Venezuela, Brazil, Paraguay, Argentina, and Chile seem uniform in size and coloration. Measurements of South American specimens that I secured as noted above follows:

No.	Locality	Date	Sex	Wing	Tail	Culmen from cere	Tarsus
				Mm.	*Mm.*	*Mm.*	*Mm.*
283734	Puerto Pinasco, Paraguay	Sept. 24, 1920	Male ad	301	151. 5	18. 0	34. 5
283733	Lavalle, Buenos Aires	Oct. 27, 1920	...do	304	163. 0	17. 0	36. 4
283735	...do	...do	...do	302	174. 0	16. 5	35. 0

[83] Proc. New England Zool. Club, vol. 7, Feb. 19, 1920, p. 46.

ROSTRHAMUS SOCIABILIS SOCIABILIS (Vieillot)

Herpetotheres sociabilis VIEILLOT, Nouv. Dict. Hist. Nat., vol. 18, 1817, p. 318. (Corrientes and Rio de la Plata.)

As the everglade kite from Florida, described by Ridgway[84] as *Rostrhamus sociabilis plumbeus*, may be distinguished from South American specimens by the grayish wash on the upper surface, the typical form will bear the trinomial designation used above.

Though Azara states that he had never seen his *gabilán de estero sociable* in Paraguay, I found it about the lagoon at Kilometer 80, west of Puerto Pinasco, on September 8 and 9, 1920, and noted many little piles of empty snail shells at the bases of palm stubs, where they had been carried and dropped when empty. The hawks were shy, so that I had no shots at them. On the Estancia Los Yngleses, near Lavalle, Buenos Aires, the everglade kite, known as the *caracolero* was fairly common from October 28 to November 16. A female in immature dress was taken October 28, and a male in the same plumage, with one fully adult, on October 31. Apparently the dark adult feathering is not assumed until the third year.

These hawks were found about open marshes often in little flocks, that in one instance numbered as many as 11. When fence posts were not available and there were no trees near at hand, the hawks rested on the ground on very slight elevations or perched in the rushes. At other times they soared in company in short circles, often following this pastime for the space of nearly an hour. The bluntly pointed wing, with its broad expanse and well-rounded outline, and sharply square-cut tail served to identify the species at once, even when far distant. At this season they were somewhat noisy and emitted a rasping chattering call that was audible at no great distance, especially on days when the wind was strong. Near their resting places I found piles of empty snail shells (in this case of *Ampullaria insularum* d'Orbigny) where they had been discarded. None of the shell heaps were extensive at this season and I judged that the hawk was migratory and had only recently returned from the north. In the Paraguayan Chaco, where the snail hawks frequented marshy savannas, I saw them perched frequently in the leafy tops of trees.

Near San Vicente, in eastern Uruguay, several everglade kites were seen on January 31, 1921, near the Laguna Castillos, and one was noted February 2 on a bañado bordering the Arroyo Sarandi near the Paso Alamo. At Lazcano, Uruguay, they were seen over saw grass marshes from February 5 to 7, and an immature female was shot on February 7. At Rio Negro, Uruguay, they were ob-

---

[84] Baird, Brewer, and Ridgway, Hist. North American Birds, Land Birds, vol. 3, 1874, p. 209.

served near marshes on February 17 and 18. On March 8 one was seen from the train near 25 de Mayo, Buenos Aires, and others were noted in the same manner west of the city of Buenos Aires on March 13.

The immature female taken October 28, when fresh, had the bill, anterior to the cere, black; base of bill, including the mandibular rami, the skin back as far as the eye and a narrow external rim on the eyelids zinc orange; iris liver brown; tarsus and toes dull yellow ocher; claws black. The male in adult plumage secured on October 31 had the bill mainly black; cere, bare skin in front of eye, gape, and mandibular rami flame scarlet; iris carmine; tarsus and toes apricot orange; claws black. The adult thus was much brighter in color. The immature female from Uruguay, fully grown, but a bird of the year, has the paler markings in the plumage much darker, more rufescent, than in birds in second-year dress.

Doctor Oberholser [85] proposes to replace *Rostrhamus* of Lesson by *Cymindes* Spix [86] on the ground that *Cymindes* is a new name at the reference cited, a suggestion, however, that is in error. Spix gave diagnoses for all of the genera that he used without citing the name of the founder, so that *Aquila*, *Polyborus*, and *Cathartes* which precede *Cymindes* are characterized in the same manner as the name under discussion. On reference to the index to the first volume of Spix, which immediately precedes the text, the genus in question is given as *Cymindis*, on page 7, as stated, it stands as *Cymindes*, while plate 2 is lettered *Cymindis Leucopygus*. It is obvious, therefore, that *Cymindes* is simply an emendation (apparently unintentional) of *Cymindis* Cuvier and as such has no priority over the generic name *Rostrhamus* for the everglade kite.

GERANOSPIZA GRACILIS (Temminck)

Falco gracilis TEMMINCK, Nouv. Rec. Planch. Col. Ois., livr. 16, April, 1822,[87] pl. 91. (Eastern Brazil.)

Hellmayr [88] considers *gracilis* of Temminck a subspecies of *Geranospiza caerulescens*. In the material available the two appear so different, with no indicated intergrades, that this usage is not justified.

This bird was encountered on two occasions near the ranch at Kilometer 80, west of Puerto Pinasco, Paraguay. On the morning of September 15, 1920, one rested quietly on a post above a pond in one of the corrals, and was killed from the door of the kitchen. On September 20 in a tract of heavy monte one flew into the top of

---

[85] Proc. Biol. Soc. Washington, vol. 35, Mar. 20, 1922, p. 79.
[86] Av. Spec. Nov. Brasiliam, vol. 1, 1824, p. 7.
[87] From Sherborn, Ibis, 1898, p. 488.
[88] Nov. Zool., vol. 28, May, 1921, p. 177.

a tall tree to a perch above the surrounding leaves and peered about. giving a note resembling *whaow* in a drawn-out nasal tone. In Guaraní the species was called *taguatoi*.

Both birds taken are males, one in full plumage, and the other in process of molt from a lighter, immature dress. This second bird has the throat white, and is lighter, less distinctly marked below than the adult. The second specimen, when first killed, had the maxilla and tip of the mandible black; remainder of mandible and a spot on the maxilla below the nostril glaucous-gray; iris marguerite yellow; cere deep neutral gray; tarsus and toes carnelian red; nails black.

CIRCUS CINEREUS Vieillot

Circus cinereus VIEILLOT, Nouv. Dict. Hist. Nat., vol. 4, 1816, p. 454. (Paraguay and Rio de la Plata.)

The small marsh hawk was recorded only near Lavalle, Buenos Aires, where at the Estancia Los Yngleses on October 29, 1920, I secured a male as it came sailing across a marsh with several *Agelaius thilius* in hot protest of its passage. This specimen, hatched apparently the previous summer, is in brown immature plumage save for one or two gray clouded feathers in the dorsal region, and a grayish wash on some of the primaries. In life the tip of the bill was dull black; base of maxilla light Payne's gray; base of mandibular rami, gape, and cere light olive yellow, changing laterally on the cere to asphodel green; iris pale pinard yellow; tarsus primuline yellow; claws black.

Near the coast below Cape San Antonio in this same region I found a pair that evidently nested somewhere near at hand in the rush-covered marshes that here alternated with sand dunes. The male, an adult bird in full plumage, that appeared very light in color on the wing, was taken November 6. This species is similar in appearance and manner of hunting to the North American marsh hawk and has the same light graceful soaring flight that enables it to scan the grass closely in its search for food.

CIRCUS BUFFONI (Gmelin)

Falco buffoni GMELIN, Syst. Nat., vol. 1, pt. 1, 1788, p. 277. (Cayenne.)

Of two specimens (both males) of this marsh hawk taken, one was secured at Las Palmas, Chaco, on July 26, 1920, and the second at the Riacho Pilagu, Formosa, on August 15. The first of these was an adult bird in full dark plumage, with sexual organs one-fourth developed. The breast and neck are entirely black save for an obscure white patch on the chin and partly concealed white markings on the ruff and upper breast, while the abdomen varies from russet to mars brown, and the thighs and flanks are nearly black. Feathers of the lower abdomen and under tail coverts are marked or tipped obscurely with white. The second specimen is an immature bird in process of molt from the first year plumage: On the dorsal surface new, nearly black feathers are appearing. Below, the breast and abdomen are dirty white, with more or less streaking of fuscous. It appears that three years at least may be required to gain the full adult plumage, since the new feathers growing in on the back in this individual are obscurely margined with rusty, a character absent in the fully adult bird. The plumages and plumage change in this hawk are of considerable interest, but can be studied successfully only with a considerable series. Apparently the species is dimorphic since light or dark individuals are found in the young stages.

This beautiful hawk inhabited the open savannas of the Chaco, or the extensive marshes of the pampas. Almost invariably it was seen skimming in true harrier style along the borders of channels or lagoons where it might hope to encounter prey, its large size serving to distinguish it from *C. cinereus* found in the same regions. The dark plumaged adults were especially handsome, their coloration being frequently visible at a considerable distance.

The species was found at a number of points, but was more common in the Chaco than elsewhere, as will be observed in the following records: Las Palmas, Chaco, July 26 and 31, 1920; Riacho Pilaga, Formosa, August 15 and 17; Kilometer 182 to Formosa, Formosa, August 21, several; 200 kilometers west of Puerto Pinasco, Paraguay, September 25, one; Lavalle, Buenos Aires, November 2 and 4; San Vicente, Uruguay, January 31, 1921, and at the Paso Alamo, Arroyo Sarandi, February 2, one; Lazcano, Uruguay, February 3, one.

The adult male taken had the distal half of the bill black; base gray number seven; cere vetiver gray; iris ochraceous tawny; tarsus and toes light orange-yellow; claws black.

URUBITINGA URUBITINGA URUBITINGA (Gmelin)

Falco urubitinga GMELIN, Syst. Nat., vol. 1, pt. 1, 1788, p. 265. (Eastern Brazil.[89])

Near Las Palmas, Chaco, on July 14, 1920, I killed a male in immature plumage from a tree above a pool of water in heavy forest. This bird was only recently from the nest, as, though in complete plumage, down filaments still clung to some of the rectrices and secondaries. Above it is dark brown, with markings of sayal brown

---

[89] Gmelin writes that the bird was found in Brazil, but Berlepsch and Hartert (Nov. Zool., vol. 9, 1902, p. 113) from the original sources of Gmelin's information have given the type locality as " Bras. or.", eastern Brazil.)

on the borders of the feathers, and the head and neck blackish, streaked with white. The wings have obscure mottlings of light-pinkish cinnamon, and the long tail is obscurely banded with narrow alternating bars of blackish and a lighter color that varies from whitish and light-pinkish cinnamon to dull brown. The under surface of the body is whitish, with streaks of blackish. The thighs are obscurely barred with black, ivory yellow, and sayal brown.

At the Riacho Pilaga, Formosa, on August 16, a pair were hunting along the border of a lagoon in search for food. They flew from perch to perch on low rounded masses of reeds that projected from floating vegetation lodged against the rushes, or dropped down to the partly submerged stuff below. As they flew the white band across the tail was so prominent as to attract attention at once. They paid no attention to me as I drifted down on them with the wind in an unwieldy *cachiveo* made from the trunk of a silk-cotton tree, until I came near enough to secure both birds. They were male and female, and though not in breeding condition, I was of the opinion that they were paired. The soft parts, similar in both sexes, were as follows: Bill mainly black; a space on maxilla below nostril and base of mandible gray number 7; cere and gape chamois; iris verona brown; tarsus and toes primuline yellow; claws black.

On August 21, while passing by train from the station at Kilometer 182 to Formosa, I saw several of these hawks flying over open marshes. At Puerto Pinasco, Paraguay, I killed an adult male on September 10. The note emitted by this bird was a shrill *Ker-r-r-re-e-e-e*, with a cadence similar to that of a policeman's whistle. The species was found, in my experience, mainly in open country and appeared to hunt over marshes, where its long legs may have been of service in resting on partly submerged perches or in securing prey from the water.

BUTEO POLYOSOMA (Quoy and Gaimard)

Falco Polyosoma QUOY and GAIMARD, Voy. Uranie Physicienne, Zool., August, 1824, p. 92, pl. 14. (Falkland Islands.)

A handsome adult male of this species was taken at General Roca, Rio Negro, on December 2, 1920, as it rested on a pole and tore at the body of a cavy held in its feet. Others were observed soaring over the dry gravel hills of this region, in appearance and action suggesting red-tailed hawks. On December 6, at Kilometer 1097, between Neuquen and Zapala, I observed a nest of this hawk placed on a telegraph pole, where it was supported by the wires. The owners of the structure rested in the tops of low bushes a few feet away. A hawk of this species was observed near Zapala, Neuquen, on December 8. Buteonine hawks were seen in the foothills of the

Andes in Mendoza, and again in Tucuman, but identity of these birds is often uncertain, even when specimens are available, and I do not care to hazard a guess as to what they may have been. The question of species and subspecies in the South American representatives of the genus *Buteo* is much involved and can be made clear only by collection of extensive series throughout the continent.

According to Stresemann,[90] *Buteo erythronotus* (King)[91] becomes *Buteo polyosoma* (Quoy and Gaimard).

BUTEO SWAINSONI Bonaparte

Buteo swainsoni BONAPARTE, Geog. and Comp. List, 1838, p. 3. (Near the Columbia River.)

On April 17, 1921, a Swainson's hawk was soaring in company with other hawks over the summit of the Sierra San Xavier above Tafi Viejo, Tucuman.

GERANOAËTUS MELANOLEUCUS (Vieillot)

Spizaetus melanoleucus VIEILLOT, Nouv. Dict. Hist. Nat., vol. 32, 1819, p. 57. (Paraguay.)

Near the city of Mendoza this species was seen on March 13, 1921, and above Potrerillos one, that had perhaps been captive at some time as a string dangled from its leg, was observed on March 16. Several were found in company with smaller hawks, soaring over the Cumbre above Tafi Viejo, Tucuman, on April 17. When awing the tail of this bird appears strongly rounded.

RUPORNIS MAGNIROSTRIS PUCHERANI (J. and E. Verreaux)

Asturina Pucherani J. and E. VERREAUX, Rev. Mag. Zool., ser. 2, vol. 7, 1855, p. 350. (Paraguay.[92])

From available material it seems that this subspecies is characterized by the dark, almost black throat (in the adult), very narrow rufescent bars of the undersurface, slight longitudinal stripes on foreneck and upper breast, and by the rufescent tinge of the lighter bars in the tail. Two adult females collected at Las Palmas, Chaco, on July 26 and 28, 1920, agree with specimens examined from Paraguay in these particulars. An adult female from San Vicente, Uruguay, taken on January 31, 1921, is not wholly in agreement with *pucherani*, as it has the heavier barring on the undersurface found in *R. m. nattereri*, but in other respects is more like the present

---

[90] Journ. für Ornith., 1925, pp. 309–319.

[91] *Haliaeetus erythronotus* King, Zool. Journ., vol. 3, 1827, p. 424. (Port Famine, Straits of Magellan.) No locality is cited in connection with the original description, but on p. 426, in closing his account of the hawks secured, King remarks that " all of the above species of Falconidae were collected at Port Famine." See also Swann, Syn. Accipitres, pt. 2, Jan. 3, 1922, p. 85.)

[92] Type locality designated by Brabourne and Chubb, Birds of South America, vol. 1, December, 1912, p. 68.

form with which it is identified. An immature male from Rio Negro, Uruguay, secured February 15, is also slightly intermediate in its characters.

Hellmayr [93] has proposed to replace the subspecific name *pucherani* by *superciliaris* from *Sparvius superciliaris* Vieillot [94] based on the *esparvero pardo ceja blanca* of Azara. Azara, however, says in the beginning of his description that the feathers of the head and nape in his bird were pointed, while the remaining plumes of the dorsal surface had rounded ends, a character not to be found in *Rupornis*, so that he must have had some other hawk in mind. Though the remainder of the description may fit the immature stages of the present bird, this first statement must identify the bird described as one of some other genus. It is curious that Azara's *Esparvero indaye*, which is undoubtedly a *Rupornis*, was not given a name by those who republished his descriptions.

This small hawk was common in open wooded regions of the Chaco in northern Argentina and Paraguay, and was observed in less abundance in Uruguay. It frequented the borders of groves where it might perch in the shade or the open as desired. The *caranchillo*, as the bird was called, was fearless often to a point of stupidity, and was seldom alarmed even by a gunshot fired close at hand. In regions where it was common it came almost invariably when I was "squeaking" to draw warier denizens of the thickets from cover, and perched near at hand with jerking tail while it peered about to locate the sound. Though most other birds were little afraid the squalling calls of the hawk caused some to remain partly concealed, and even frightened shyer ones from appearing at all, so that at times I found *Rupornis* considerable of a nuisance. Toward the end of August the mating season seemed near as the hawks became very noisy, and screaming shrilly, often turned in short circles two hundred meters in the air, while others squealed in answer from the tree tops below.

An adult female taken July 26 had the soft parts colored as follows: Tip of bill dull black, base clear green-blue gray, becoming deep colonial buff on rami of mandible; cere primuline yellow; iris massicot yellow; tarsus and toes honey yellow; claws black.

These hawks were recorded as follows: Resistencia, Chaco, July 9 and 10; Las Palmas, Chaco, July 13 to 31; Riacho Pilaga, Formosa, August 7 to 21; Formosa, Formosa, August 23 and 24; Puerto Pinasco, Paraguay, September 3 to 25 (observed west to a point 200 kilometers from the Rio Paraguay); San Vicente, Uruguay, January 31, 1921 (two seen); Rio Negro, Uruguay, February 15 to 18.

The pelvic powder downs in these hawks are well developed.

---

[93] Nov. Zool., vol. 28, 1921, p. 183.
[94] Nouv Dict. Hist. Nat., vol. 10, 1817, p. 328.

RUPORNIS MAGNIROSTRIS SATURATA (Sclater and Salvin)

Asturina saturata SCLATER and SALVIN, Proc. Zool. Soc. London, 1876, p. 357. (Apolo[95] and Tilotilo, Bolivia.)

A male in fresh fall plumage that agrees with the characters assigned to this form was taken near Tapia, Tucuman, on April 18, 1921. From *R. m. pucherani* of areas farther east it is distinguished by darker coloration throughout, and by the much bolder, heavier markings of the under surface. The bird was killed as it rested in a tree in low scrubby forest. On April 12 another was observed as it turned in short circles far above the earth and gave the shrill squealing calls common in hawks of this species more especially in spring. This subspecies does not seem to have been recorded previously in Argentina.

HETEROSPIZIAS MERIDIONALIS MERIDIONALIS (Latham)

Falco meridionalis LATHAM, Ind. Orn., vol. 1, 1790, p. 36.   (Cayenne.)

An adult female of this hawk taken at Kilometer 80, west of Puerto Pinasco, Paraguay, on September 7, 1920, appears to be a representative of the typical northern form as it has the following measurements: Wing, 412; tail, 214; culmen from cere, 23; and tarsus, 100 mm. The species was fairly common both at Puerto Pinasco and in the Chaco to the westward from September 1 to 23. At this season the birds were found in pairs and the female secured showed some growth in size of the ovary, so that the breeding season seemed near. The birds frequented open savannas dotted with small trees that offered convenient resting places. Though in appearance (save in color) and in action they resemble the red-tailed hawk (*Buteo borealis*) they seem slower and less aggressive. On one occasion I saw one stoop at a guira cuckoo on the ground but miss it, and the cuckoo was then able to escape in spite of its slow, weak flight. On windy days the hawks rested facing the wind with head lowered to a level with the body, and the tail raised so as to offer as little resistance to the gusts as possible. Sparrow hawks drove at them occasionally but the large hawks merely dropped their heads to avoid being struck, and made no attempt to punish their assailant. The call note of this large hawk is a high pitched note resembling *kree-ee-ee-er* that terminates in a drawn out wail. It suggests in a way the squealing call of *Buteo borealis*, but is less forceful and vigorous. Occasionally a pair circled about in the air a hundred meters from the earth and emitted a snarling, grunting *kweh kweh kwuh-h kweh kweh*. To the Angueté Indians they were known as *so bas gookh*, and to the Lenguas as *nata pais shar o*.

---

[96] So spelled in the Century Atlas of the World.

HETEROSPIZIAS MERIDIONALIS AUSTRALIS Swann

Heterospizias meridionalis australis SWANN, Auk, vol. 38, 1921, p. 359.
(Laguna de Malima, Tucuman, Argentina.)

Swann recently has separated the southern hawks of this species as above, on the basis of larger size and darker coloration. From the specimens at hand it appears that this action may be sustained on the basis of size, but that constant color differences correlated with geographic range, are not present. Nine specimens of *meridionalis* from Panama, Brazil (Para, Pernambuco, Diamantina, and Chapada), show the following measurements: Wing, 378–417; tarsus, 90–111.5 mm. The series examined contains three females, one doubtful male, and five birds without indication of sex. Three skins from Argentina (Kilometer 182, Formosa, Corrientes, and Conchitas, Buenos Aires) measure as follows: Wing, 423–450; tarsus, 109–113 mm. Doctor Allen [96] has described the plumage changes with age in this species, findings that are verified in the series here at hand. In general, birds during their first season are very dark brown, almost black, save for more or less white on the under surface and some rufous in the primaries and greater coverts. During the second year the amount of rufous in the wings is increased and invades more or less of the underwing surface as well as the lesser wing coverts. In the third year the under parts and head become rufous, barred below, save on the throat, with blackish, but the back remains fuscous brown. In the fully adult plumage, apparently in the fourth year, the upper back assumes an ashy shade, but otherwise the bird is similar to what it was in the plumage of the preceding year. Apparently the type of Swann's *australis* is a bird in third year plumage. I am unable to detect any difference in color between a bird in third-year stage from Para (*meridionalis*), and one in a similar plumage from near the city of Buenos Aires (*australis*). An adult female (third year) that I secured on August 12, 1920, near the Riacho Pilaga, Formosa (wing 423 mm.), is somewhat intermediate between the northern and southern forms, but has been identified with the southern bird. A specimen in the United States National Museum secured by Capt. T. J. Page at Corrientes (wing 448 mm.) seems to be typical *australis*. As Corrientes is just south of the Paraguayan border it is not improbable that though *meridionalis* was found in the Chaco at Puerto Pinasco, the form of eastern Paraguay is the larger southern bird, in which case Swann's name will become a synonym of *Circus rufulus* Vieillot,[97] based on the *gavilán acanelado* of Azara.

This handsome hawk was first observed near Las Palmas, Chaco, in July, 1920, but specimens were not secured until I entered the

[96] Bull. Amer. Mus. Nat. Hist., vol. 5, 1893, pp. 145–146.
[97] Nouv. Dict. Hist. Nat., vol. 4, 1816, p. 466.

Territory of Formosa, where the bird was more common. As at Puerto Pinasco, this hawk was sluggish in its movements. At this season it was found alone, usually perched in the top of some low tree that gave command of an open space. Occasionally one sailed along across the savannas a few feet from the ground on the lookout for food. Grass fires attracted these birds, and I saw them frequently near areas where fires were about burned out where no doubt small rodents and other similar prey offered desirable food. When on the wing the handsome markings of this beautiful bird are displayed to the fullest advantage so that, save for the dark tipped wings, it appeared wholly rich reddish brown.

The Toba Indians called this species *mi yuh*.

ACCIPITER GUTTIFER Hellmayr

Accipiter guttifer HELLMAYR, Verh. Ornith. Ges. Bayern, vol. 13, September 20, 1917, p. 200. (Bolivia.)

According to Bertoni [98] *Sparvius guttatus* of Vieillot [99] founded on the *Esparvero pardo y goteado* of Azara [1] refers to the immature of *Accipiter pileatus* (Temminck); in accordance with this, Hellmayr has given the present bird, long known as *guttatus*, the name *Accipiter guttifer*.

A female shot at Tapia, Tucuman, on April 10, 1921, dashed into a clump of bushes in front of J. L. Peters and me, in pursuit of a small bird. On seeing us, hardly 3 meters away, it checked its flight abruptly, alighted for an instant on a stump, irresolute as to the best manner of escape and then darted off. When dropped with a broken wing it ran swiftly on the ground.

The bird is an adult female in post nuptial molt, with new plumes appearing in wings, tail, upper breast, crown, and back.

ACCIPITER ERYTHRONEMIUS (Kaup)

Nisus vel. Acc. erythronemius "G. Gray," KAUP, in Jardine, Contr. Ornith., 1850, pt. 3, p. 64. (Bolivia.)

On February 19, 1921, at Rio Negro, Uruguay, one of these small hawks was killed as it flew past bearing something in its talons. Its prey, whatever it may have been, dropped in high grass where it could not be found. Later another was seen on a perch near the ground at the border of a small opening in heavy brush. As it flew it came near me and was secured. The long tail, rounded wings, and the head apparently drawn in on the shoulders give this bird the appearance usual in small Accipiters.

---

[98] An. Soc. Cient. Argentina, vol. 75, February, 1913, p. 79.
[99] Nouv. Dict. Hist. Nat., vol. 10, 1817, p. 327.
[1] Apunt. Hist. Nat. Paxaros Paraguay, vol. 1, 1802, p. 113.

The two taken are males in molt from an immature to adult plumage. Old feathers still appear in wings and tail, but elsewhere have been replaced, though all of the new feathers are not fully grown. The narrow bars on the under surface are mainly dark gray, or grayish brown with little mixture of rufous. These two measure as follows: Wing, 164–171; tail, 117–123.5; culmen from cere, 9.6–10; tarsus, 47–50 mm.

Swann [2] treats *Accipiter salvini* Ridgway from Venezuela as a subspecies of *A. erythronemius*.

Order GALLIFORMES

Family CRACIDAE

ORTALIS CANICOLLIS (Wagler)

Penelope canicollis WAGLER, Isis, 1830, p. 1112. (Paraguay.)

Nine specimens, all adult, secured in the Chaco were taken as follows: Riacho Pilaga, Formosa, August 13, 14 (one skeleton), 18 (one in alcohol), and Kilometer 80, Puerto Pinasco, Paraguay, September 6, 10, and 20 (two). At Puerto Pinasco the birds were found inland to Kilometer 200. Birds from the two localities are similar; males are larger and usually paler on the posterior dorsal surface than females. A female taken August 13 at the Riacho Pilaga had the soft parts as follows: Bill fawn color; soft operculum over nostrils, and space behind hair brown; bare skin on sides of head fawn color, on throat tinged with Pompeian red; iris army brown; tarsus and toes avellaneous; claws fuscous. The skin of the throat was more heavily tinged with red in males than in females.

The charatá as *Ortalis canicollis* is usually known was a common species in the more extensive forests of the wilder, less-frequented portions of the Chaco. It was typically a tree-haunting bird that frequented open tree limbs, the borders of trails or edges of groves where dense cover close at hand furnished shelter at any alarm. They were found in bands that included from four to eight individuals, until in September they separated in pairs for the purpose of breeding. On days with high wind when hunting in suitable sections I saw them in numbers, though ordinarily the slight sounds that I made in passing through the monte were sufficient to cause the alert birds to hide. Frequently flocks descended to feed on or near the ground, but when alarmed rose at once into the treetops. Once I startled one badly in a forest path so that it rose with roaring wings like a tinamou or pheasant but usually the flight was silent. When alarmed, if low down they towered with rapidly

[2] Syn. Accipitres, ed. 2, pt. 1, Sept. 28, 1921, p. 58.

beating wings into the trees, or if found on high perches flew with a few quick strokes of the wings that terminated in a short sail in a direct line with long tail slightly spread and neck outstretched. They alighted as easily as jays, ran quickly along the large limbs and were lost at once to view. In the tops of trees they remained motionless to escape observation, but were often betrayed by the outline of the long tail or by a moving shadow caused by a head concealed behind leaves. When flocks chanced to alight overhead without having seen me approach the birds examined me curiously as they uttered low whining calls. When at ease they sank on the breast like pheasants and turned the head quietly from side to side. When excited the long neck was extended to full' length. Flocks came to feed in flowering lapacho trees standing at the border of the forest, apparently in search of the blossoms.

The mating season, heralded by the harsh calls of the males, began in September. Near at hand I found that the call began with a low resonant note followed by a harsh cackle that changed in tone and continued in rapid repetition for nearly a minute, thus, *bink, ka chee chaw raw taw, chaw raw taw, chaw raw taw*, a call that carried easily for half a mile. At once this was answered by another bird, another and another until perhaps half a dozen were calling from near at hand. At this season they were heard many times at night. On warm spring mornings the notes were heard on all sides—an odd chorus, barbarous and uncouth, but attractive in spite of its harshness. When engaged in calling males sought a commanding perch often in the top of some tall tree and were so engrossed in their challenges that it was possible to stalk them without great care. At times they were accompanied by females; copulation took place in the tree tops.

One morning I secured a bird just as a heavy rain came on. On skinning it later I found the cavity of the large oil gland entirely empty and judged that it had used the supply of oil to prepare for the coming storm.

The ancient Guarani name of Yacú-Caraguatá given to this bird in Azara's time has been abbreviated to Yacú though the species is called more frequently by the term Charatá. The Tobas in Formosa knew it as *Gua che na* (a cognomen of evident onomatopoeic origin) and the Anguetés in Paraguay as *Kin a tee*.

The trachea in females of this species is normal. In males it is elongated to form a loop over the breast muscles on the right-hand side that reaches to the keel of the sternum and then turns to pass back and enter the thoracic cavity. From the lower end of the loop a slender band of muscle passes back to insert on the connective tissue overlying the pectoralis major above the end of the keel on the sternum. The sternotrachealis muscles are inserted as usual on the costal processes of the sternum.

The flesh of these birds is excellent eating but the sport in their hunting lies entirely in the care and skill necessary to stalk them successfully. It is seldom that they give an open wing shot, but, on the contrary, offer snapshots as they run away along limbs.

In the series of seven fresh skins at hand individual variation seems to cover the phase described by Cherrie and Reichenberger from Suncho Corral, Santiago del Estero as *Ortalis canicollis grisea*,[3] though the skins in question come from Formosa and the Paraguayan Chaco near Puerto Pinasco. With the somewhat limited material at hand subspecies may not be recognized. Wagler based his description of *canicollis* on Azara's account of the Yacú-caraguatá so that his type locality must be located in southern Paraguay or the adjacent provinces of Argentina. (Azara remarks that the bird was not known south of 27° S. latitude.)

Order GRUIFORMES

Family RALLIDAE

FULICA ARMILLATA Vieillot

Fulica armillata VIEILLOT, Nouv. Dict. Hist. Nat., vol. 12, 1817, p. 47. (Paraguay.)

This large coot was common in the cañadones near Lavalle, Buenos Aires, where, near the Estancia Los Yngleses, an adult female was shot on October 29, 1920. The birds were found on open pools or among rushes, and swam about with nodding heads and other mannerisms typical of the genus. They were nesting and I believed that nests containing large handsomely marked eggs belonged to this species, but after considerable effort I was unable to identify the owners definitely, and did not take them.

At General Roca, Rio Negro, from November 30 to December 3 a band of nearly 100 of the present species, with a few *F. leucoptera* frequented an open space in a quiet channel near the river. I killed two males, and a male white-winged coot, at a single shot on November 30. (Pl. 16.) The birds fed in scattered company, but when alarmed gathered in a close flock. There was some mating activity among those of the present species, and males gave an occasional mating display in which they arched the neck, raised the tips of the wings, half closed the eyes, and opened the mouth. Though they turned toward their mates in these maneuvers they took care to guard against a savage bill thrust by remaining at a safe distance. Among themselves males, when not alarmed, were

---

[3] Amer. Mus. Nov., No. 27, Dec. 28, 1921, p. 2.

aggressive and fought savagely, often pursuing some vanquished competitor for some distance. A young bird two weeks old, seen on December 3, may have belonged to this species.

On January 31, 1921, on the Laguna Castillos, below San Vicente, Uruguay, *F. armillata* was found in bands that rested on the low shores and swam out into the lake as I came near. A number had been affected by alkali in the water so that I picked up a male in condition to skin and secured skulls from two more that lay dead on the shore.

At Concon, Chile, on April 24, I examined one that had been killled by a hunter.

An adult female shot on October 29 had the soft parts colored as follows: Bill in general slightly brighter than olive yellow; space on culmen from above nostril to center of frontal shield, and spot on base of bill immediately behind nostril burnt lake, shading on outer margin to Brazil red; iris Vandyke red, slightly clouded with duller markings; tarsus and toes in general olive lake; marginal webs and borders of scutes washed with deep neutral gray; spot on rear of crus dull English red; claws dull black.

The feet in the present species are enormous, much larger than in other species with which it is associated. and have proportionately broader lobes. The bird, in addition, is heavy in body and appears large when seen at a distance. In life adults of the three coots of this region may be distinguished without difficulty by the color of the frontal shield and bill. In *armillata* a dark-red mark on the lower margin of the shield and base of the bill crosses the otherwise light color of these parts. In *leucoptera* bill and shield are entirely light, while in *rufifrons* the shield is dark red. According to Doctor Dabbene,[4] *F. armillata* in swimming does not hold the tail erect over the back, as is customary in *rufifrons*, but drops it in the manner of a tinamou.

Though the three are separated without trouble in the field, or when freshly killed, there is often difficulty in naming dried skins. The following notes may be of assistance in separating them:

a.[1] Frontal plate produced posteriorly in a narrow acute point; tail longer (58 to 62 mm.) ; frontal plate dark red; outer web of tenth primary usually plain (occasionally with a faint white margin)__Fulica rufifrons.

o.[2] Frontal shield rounded or if pointed posteriorly not greatly elongated; tail shorter (47 to 56.5 mm.) ; frontal plate orange or yellow; outer web of first primary bordered with white.

b.[1] Feet relatively larger; crus more or less reddish; secondaries plain or very slightly margined with white at extreme tip_____Fulica armillata.

b.[2] Feet relatively smaller; crus greenish or yellowish; secondaries tipped prominently with white_____Fulica leucoptera.

---

[4] An. Mus. Nac. Hist. Nat. Buenos Aires, vol. 28, July 19, 1916, p. 190.

In *F. armillata* the tarsus is usually equal to one-third of the wing, the median under tail coverts are neutral gray, and the secondaries may be plain or may have a very small white tip. The crus is marked more or less with red. Measurements of the three adults taken follow: Two males, wing, 197–205; tail, 50–52; tarsus, 65–71; one female, wing, 179; tail, 52; tarsus, 60 mm.

FULICA RUFIFRONS Philippi and Landbeck

Tulica rufifrons PHILIPPI and LANDBECK, Anal. Univ. Chile, vol. 19, no. 4, October, 1861, p. 507. (Chile.)

The red-fronted coot was common near Lavalle, Buenos Aires, and two were taken on October 29, 1920, in a large cañadon on the Estancia Los Yngleses. The birds were found in the deeper parts of the marshes among the rushes or, less frequently, in open water. They have loud clucking notes and swim with rapidly nodding heads like others of the genus. They were so shy that when found in the open they swam back to cover to avoid any possible danger. Several swam up to examine a dead bird that I had killed as it lay in the water. An adult female had the bill lemon chrome, with a slight wash of light cadmium, changing at tip to jewel green; sides of bill at base, and frontal shield as far forward as anterior end of nostril diamine brown, becoming madder brown at outer margin; iris chocolate; tarsus citron green, toward margin of scutes verging to mignonette green; crus and toes mignonette green; margins of scutes, joints, and lobes on toes neutral gray; claws blackish. This specimen has the following measurements: Wing, 170; tail, 62.3; tarsus, 55.3 mm.

In *Fulica rufifrons* the crus is greenish and the base of the acutely elongated frontal plate dark red. The median under tail coverts are black or blackish slate, there is no white on the tips of the secondaries, and the outer web of the tenth primary is plain or very faintly bordered with white. The tail measures 58.2–62.3 mm. and the birds usually have more white on the abdomen than either *leucoptera* or *armillata*.

FULICA LEUCOPTERA Vieillot

Fulica leucoptera VIEILLOT, Nouv. Dict. Hist. Nat., vol. 12, 1817, p. 48. (Paraguay.)

Near Lavalle, Buenos Aires, the white-winged coot was recorded as common from October 31, 1920, when two were taken, until the middle of November, when I left this region. The birds frequented open pools in the marshes where they swam about with nodding heads, but at the slightest alarm disappeared behind the protecting screen of the rushes. From this secure retreat their clucking notes were always audible, but it was often difficult to see the birds. At

General 'Roca, Rio Negro, from November 23 to December 3 the birds were found in small sloughs, or, with bands of the larger bodied *F. armillata*, on open channels. From their actions they seemed to be breeding, but no nests were found and Doctor Dabbene states that they do not nest until the end of January or the first part of February.[5]

Near Carhue, Buenos Aires, white-winged coots were found from December 15 to 18 in a little fresh-water marsh that bordered an arroyo draining into Lake Epiquen. Truculent males grasped one another by the feet and then struck savage blows with their pointed bills. From March 3 to 8, 1921, bands of adults and young were found along the open shores of the Laguna del Monte at Guamini, where there were no growths of rushes of any kind. One was noted on the Rio Aconcagua near Concon, Chile, on April 28.

An adult female shot October 31 had the bill, eye, and legs colored as follows: Tip of bill Biscay green, shading inward to dull green-yellow; basal half of bill pale vinaceous fawn, becoming whitish at extreme base; base of mandible tinged with green; frontal shield slightly paler than strontian yellow; iris mars orange; tarsus and toes Paris green, with posterior face of tarsus and outer margin of lobes on toes dawn gray, shading on the outer margins of the lobes to castor gray.

Skins of the white-winged coot are marked by the greenish crus and the orange or yellow shade of the frontal shield. The median under tail coverts vary from black to dark neutral gray, the outer margin of the tenth primary is margined with white, and the secondaries are tipped more or less extensively with white. The frontal plate is rounded posteriorly and the tail measures from 48–56.5 mm. There is little or no white on the abdomen.

Measurements of a pair are as follows: Male, wing, 191.0; tail, 56.5; tarsus, 58.0; female, wing, 173; tail, 50.6; tarsus, 52.5 mm.

GALLINULA CHLOROPUS GALEATA (Lichtenstein)

Crex galeata LICHTENSTEIN, Verz. Ausgest. Säug. Vög., 1818, p. 36. (Brazil.)

The validity of this form, as distinct from *G. c. cachinnans* Bangs from North America, is sustained by a series of five males and two females secured August 9 and 16, 1920, at the Riacho Pilaga, Formosa. Individuals in adult stage are readily separated from similar specimens of *cachinnans* by their more olivaceous, less brownish backs. In addition the white of the abdomen, when birds are viewed in series, is less extensive in southern than in northern birds, and may be practically absent in adults of *galeata*. The white on this area, however, varies so with age as to be of little use in studying

---

[5] An. Mus. Nac. Hist. Nat. Buenos Aires, vol. 28, July 19, 1916, p. 184.

single birds. An immature individual, a male, is very brown above, and on first glance seems to be identical in color with the adult of *cachinnans*. In the northern bird immature specimens in addition to being more extensively white below are browner above than adults, a distinction that seems to hold in *galeata* as well. Though this immature is similar to the adult of *cachinnans* it is darker and more olivaceous than the immature of that form, in addition to being less extensively white below. A second specimen is somewhat intermediate in stage of plumage.

Measurements of these specimens, in millimeters, are as follows: Five males, wing, 170–186 (176.5); tail, 66–74.2 (69.9); tarsus 52.5–62.5 (57.2). Two females, wing, 164–177.6 (170.8); tail, 62.8–67.5 (65.1); tarsus, 50–56.2 (53.1).

On one of the large lagoons at the Riacho Pilaga gallinules were common, and when first seen as they were swimming about in open water at a distance I mistook them for coots. An Indian to whom I appealed for a boat quickly fashioned a crude pointed raft with three or four armloads of tall, green cat-tails, bound together with a few of the tougher stems, and on this somewhat precarious craft I paddled out to explore the lagoon. The gallinules were shy but by working up behind concealing points of rushes I succeeded in shooting several, as well as a grebe, before all had flown or swam into shelter of the rushes. A few days later an Indian brought me more that he had killed at the same place.

At Rio Negro, Urúguay, on February 18, 1921, the birds were common in the vegetation concealing the water of a small lagoon. Near Tunuyan, Mendoza, two were recorded March 26.

PARDIRALLUS RYTIRHYNCHOS RYTIRHYNCHOS (Vieillot)

Rallus rytirhynchos VIEILLOT, Nouv. Dict. Hist. Nat., vol. 28, 1819, p. 549. (Paraguay.)

Eight specimens of this species were secured, six of which were preserved as skins, one as a skeleton, and one in alcohol. An adult male from Lazcano, Uruguay, shot February 7, 1921, and an adult male from Rio Negro, Uruguay, taken February 18, differ constantly from a series from Buenos Aires in darker, duller coloration. It is possible that these should be separated as typical *rytirhynchos* and that the Argentine birds represent another form. These birds vary individually to such an extent that a considerable series will be needed to establish or disprove this point. An adult male from General Roca, Rio Negro, in northern Patagonia, taken December 3, 1920, and two males and a female from Tunuyan, Mendoza, secured March 23, 25, and 28, 1921, agree in color and are similar to others from Argentina.

The .type and one other specimen described by Peale [6] as *Rallus luridus* from Orange Harbour, Tierra del Fuego, belong to the form *Pardirallus rytirhynchos sanguinolentus* (Swainson) [7] so that *luridus* should be cited in the synonymy of that subspecies. Should *Rallus setosus* named by King,[8] it is presumed from the Straits of Magellan, prove the same this name will antedate *sanguinolentus*. From the description, however, it appears that *setosus*, like *rytirhynchos*, is of the type with dark centers in the feathers of the dorsal surface.

Pardirallus r. rytirhynchos was recorded at the following points: Lavalle, Buenos Aires, October 29 to November 15, 1920; General Roca, Rio Negro, December 3; Carhue, Buenos Aires, December 17; Carrasco (near Montevideo), Uruguay, January 16, 1921; Lazcano, Uruguay, February 7 and 8; Rio Negro, Uruguay, February 18; Tunuyan, Mendoza, March 22 to 28. One recorded near Concon, Chile, on April 26, that was not secured, is supposed to have been *Pardirallus r. sanguinolentus*.

Pardirallus during the summer season is a frequenter of rush or grass-grown marshes in the pampas, or ranges through swamps or along channels grown with low dense shrubbery, particularly where such growth stands in water. Though it resembles ordinary rails in general habits it must swim well as it is found many times where the water stands nearly a meter deep. When in its haunts one may be startled by a solemn hollow-sounding repetition of notes, *too too too-oo-oo*, an odd, lugubrious call suggestive almost of the super-natural, coming from the rushes almost at hand, although no sign of the bird may be seen. If one retreats a short distance and waits quietly a rail may run out to the edge of cover, but more often the only sign of its presence is the hollow repetition of its calls. In fact, for some time I was inclined to attribute these notes to a grebe as they came from rushes that stood in fairly deep water. After the breeding season the hollow notes are given less frequently, but a low, grunting sound, suggestive of the protest of the *tuco tuco* (*Ctenomys*) in its underground chambers, may be heard, or a sudden gunshot may startle the rails into emitting wheezing shrieks that are answered and repeated from every side. All are strange sounds, not at all birdlike in their nature, and so different in quality and tone as to make it seem almost impossible that they come from the same bird.

The brilliant colors of the bill are easily seen when the birds venture into the open. They suggest Virginia rails as they work about with the tail cocked over the back and twitched at intervals. At times deliberate in movement, again they traverse runs in the

---

[6] U. S. Explor. Exped., vol. 8, 1848, p. 223.

[7] *Rallus sanguinolentus* Swainson, Anim. Menag., 1838, p. 335, said to inhabit " Brazil and Chili "; type locality restricted to Chile by Chubb, Ibis, 1919, p. 51.

[8] Zoöl. Journ., vol. 4, April, 1828, p. 94.

grass more rapidly, occasionally pausing at some opening to peer about. It was unusual to have them venture more than 2 meters from shelter. In small swamps in Uruguay, where dense shrubbery grew in less than a meter of water, a tangle so heavy as to be almost impenetrable, these rails clambered about among the branches like gallinules, as much as 2 meters above water. Again, one was flushed from some scant cover of rushes or grass along a ditch through an alfalfa field, or one ran down to the water's edge at some river channel with comparatively high, brush-bordered banks.

A male taken at General Roca on December 3 and a female shot February 18 at Rio Negro, Uruguay, were breeding.

Near Tunuyan, Mendoza, toward the end of March, these rails were common, and were evidently in migration from colder regions in Patagonia. Marshes and cienagas were filled with them, while others were encountered in heavy growths of weeds, at the borders of hemp fields (one taken had hemp seed in the throat), or along irrigation ditches. At this time it was common for them when startled to flush from exceptionally heavy cover, almost certain proof that they were migrants, as no resident rail familiar with the runs and passages would think of leaving such excellent hiding places. The flight was rather swift and at times the birds rose 3 or 4 meters in the air. On the wing they appear almost black.

An adult male shot December 3 had the base of the mandible and the side of the maxilla, below and behind the level of the nostril, madder brown; small frontal shield and base of mandible yale blue; center of bill mineral green, shading to dusky green toward tip; iris slightly darker than ferruginous; tarsus and toes between coral pink and light coral red; posterior face of tarsus clouded with fuscous. Females taken seemed as brilliant, and birds of both sexes shot in fall were equally bright.

CRECISCUS MELANOPHAIUS (Vieillot

Rallus melanophaius VIEILLOT, Nouv. Dict. Hist. Nat., vol. 28, 1819, p. 549. (Paraguay.)

At the Riacho Pilaga, Formosa, an adult female taken in the rushes bordering a lagoon was brought to me on August 8, 1920. The bird was known locally as *canastita* or in Guaraní as *batuitui*.

This specimen has the throat, breast, and abdomen pure white, with the barring on the posterior underparts restricted to the flanks and the white bars wider than the dark ones. Four specimens seen from Bahia and São Paulo, Brazil, have a reddish wash on the under surface with a much broader barred area on sides and flanks, that extends over on the abdomen, where the dark bars are wider than the white ones.

NEOCREX ERYTHROPS (Sclater)

Porzana erythrops, P. L. SCLATER, Proc. Zool. Soc. London, 1867, p. 343, pl. 21. (Near Lima, Peru.)

At Tapia, Tucuman, on April 13, 1921, a peon brought me an immature specimen of *Neocrex* that he stated had been killed by a small weasel-like animal. The bird, apparently two-thirds grown, has the body plumage developed but wings and tail are not completely feathered. It is much darker above than an adult of *erythrops* from Lima, Peru, so that the *Neocrex* from Argentina may represent a distinct form. According to Lillo [9] the bird is common near the city of Tucuman.

ARAMIDES CAJANEA CHIRICOTE (Vieillot)

Rallus chiricote VIEILLOT, Nouv. Dict. Hist. Nat., vol. 28, 1819, p. 551. (Paraguay.)

The present species is more of a true wood rail in habit than *A. ypecaha*, as it frequented wooded swamps or small channels running through forests where dense cover was close at hand and did not venture into the broad *pajonales*, or saw-grass swamps, inhabited by its larger relative. My first one was seen September 30 on the forested bank of the Rio Paraguay, opposite Puerto Pinasco, where it was killed as it walked slowly along among dead weeds above the river's margin at the border of a thicket. This specimen, an adult male, when fresh had the tip of the bill bice green, shading to olive ocher at base; bare skin of eyelid, gape, and spot on the bare interramal space pompeian red; iris pecan brown; front of tarsus hydrangea red, shading to Corinthian red on posterior face; nails fuscous.

At La Paloma, near Rocha, Uruguay, on January 23, 1921, as I rounded a sharp turn in a brush-grown arroyo cut between low, clay banks, I surprised one of these rails at rest on the odoriferous carcass of a horse that lay partly submerged in a pool of water. The bird stood with one leg drawn up against the body with no apparent discomfort from the horrible stench that rose around it, until, sighting me, it flew ashore and ran off through the brush. Near San Vincente on January 28 one ran with long strides, neck extended, and twitching tail, along trails made by cattle through heavy brush bordering a swamp, and was so alert that it eluded me in short order. Several were noted at the Paso Alamo on the Arroyo Sarandi on February 2, and on February 6 in forest bordering a pool on the Rio Cebollati below Lazcano I killed an immature female about two-thirds grown but not fully fledged, as rusty downs persist on the crown and the foreneck. Others were recorded at Rio Negro, Uruguay, February 15 and 17.

---

[9] An. Mus. Nac. Buenos Aires, vol. 8, Oct. 2, 1902, p. 215.

ARAMIDES YPECAHA (Viellot)

Rallus ypecaha VIEILLOT, Nouv. Dict. His. Nat., vol. 28, 1819, p. 568. (Paraguay.)

The ypecahá, as this large wood rail is known, was fairly common about the saw-grass swamps in the Chaco where its strange notes, that suggested the combination wheeze and clank of a rusty windmill pump, came morning and evening in one of the strangest bird concerts that I have ever heard. Occasionally during the day one ran out through the rank growth to pause with twitching tail to look from the crest of some low bank before it disappeared over the rise and was lost in heavy cover beyond. At the Riacho Pilaga one evening a dog that had accompanied me while I set some traps, plunged into a swamp and immediately two wood rails came flying swiftly out and passed rapidly to safer cover. One morning at daybreak, while crossing from Lazcano to the Rio Cebollati in southern Uruguay, I saw two walking about with heads erect and twitching tails in an open pasture far from any cover, but on no other occasion were they observed save as they crossed ahead of me from one grass covert to another.

The only specimen taken was a female shot at Lazcano, Uruguay, on February 7, 1921. The species was recorded at the following points: Las Palmas, Chaco, July 30, 1920 (heard daily during my stay here, but not recognized during the first few days); Riacho Pilaga, Formosa, August 13 to 20; Puerto Pinasco, Paraguay, September 1 and 3; Lavalle, Buenos Aires, November 2 and 9; Lazcano, Urugaúy, February 7 to 9.

The species was kept often in captivity and was among the native birds offered for sale in the bird stores in the cities.

Family ARAMIDAE

ARAMUS SCOLOPACEUS CARAU Vieillot

Aramus carau VIEILLOT, Nouv. Dict. Hist. Nat., vol. 8, 1817, p. 300. (Paraguay.)

In the northern portion of the Chaco the limpkin, known by the appelation of carau, was fairly common in localities remote from habitation.

In Formosa the species was observed from the train on August 5, 1920, and again on August 21, in passing between the town of Formosa and the station in the interior at Kilometer 182. At times 40 or 50 were congregated on suitable marshes. Limpkins were noted at the lagoon at Kilometer 110, west of Puerto Pinasco on September 23, and a male was taken. Others were seen at Kilometer 200 on the following day. On September 30 I found two in flooded forest on the eastern bank of the Rio Paraguay opposite Puerto Pinasco.

At Lavalle, Buenos Aires, limpkins were noted on November 2 and 9, and several were recorded on November 16 when crossing to Santo Domingo. On February 2, 1921, I killed an adult male (preserved as a skeleton) at the Paso Alamo on the Arroyo Sarandi, north of San Vincente, eastern Uruguay. In the marshes bordering the Rio Cebollati near Lazcano, Uruguay, the birds were fairly common from February 6 to 9, and at Rio Negro, Uruguay, I found them from February 15 to 18.

In the Chaco limpkins ranged in open, rush-grown marshes, probably because the season was winter and water holes and swamps had become dry in wooded sections. Similar regions were inhabited on the pampas as these formed the only tracts in this area suited to the habits of this species. Elsewhere limpkins frequented wooded swamps, areas that seemed better suited to their needs. In traversing the open country the birds flushed frequently from small openings among the rushes, rising with the peculiar flight that marks them as far as they can be seen. When the bird is not hurried the wings are extended at an angle of 45° above the back, and are stroked quickly at short intervals down to the level of the body, but little or no farther, and then raised again. At the highest point of elevation there is a distinct pause before the wing is brought down again, so that the bird sails for a few feet with stiffly held raised wings. The whole wing motion suggests that of some huge butterfly save that the line of flight is direct rather than erratic. When startled the birds flap away as any crane or stork might with neck extended and legs trailed behind.

In dense wooded swamps my attention was frequently attracted to limpkins by abrupt explosive or clattering notes that often resembled the syllables *kop kop* or *kaup*. The ordinary call of *car-r-r-rau car-r-r-rau*, that gives this and the northern limpkin their common name throughout their range in the West Indies and Latin America, may be heard for a great distance. Two or three individuals calling at once may cause a tremendous noise; in fact one might well believe that the chorus was produced by a considerable congregation of birds concealed in the bushes.

Limpkins often sought elevated perches in the tops of low trees or rested concealed among heavier growth, where they turned the head from side to side and at short intervals twitched the tail upward with a quick jerk that suggested a similar motion common among rails. Their food consisted mainly of large fresh-water snails (*Ampullaria insularum* d'Orbigny). Empty shells of these mollusks were found in abundance resting on the mud, with the opening upward and the thin, corneous operculum lying a few inches away, where it had dropped after it had been pulled away. At rest the birds appear remarkably ibislike.

In the Angueté language the species is known as *allat*.

Bangs and Penard[10] have indicated that the limpkin from southern South America differs from that of northern localities in larger size, a contention that is upheld by the skins available in the National Museum. The male that I secured west of Puerto Pinasco has the following measurements: Wing, 355; tail, 170; culmen, 127; tarsus, 144 mm. The wing measurements given by Bangs and Penard for males of the southern form range from 341 to 343, so that this bird is of maximum size.

For a recent revision of the forms of *Aramus* the reader is referred to a paper by Peters (Occ. Pap. Boston Soc. Nat. Hist.: vol. 5, Jan. 30, 1925, pp. 141–144).

Family CARIAMIDAE

CARIAMA CRISTATA (Linnaeus)

Palamadea cristata LINNAEUS, Syst. Nat., ed. 12, vol. 1, 1766, p. 232. (Brazil.)

On September 13, 1920, near the ranch at Kilometer 80, west of Puerto Pinasco, Paraguay, one of these strange birds ran out in front of my horse as I rode up over the steep bank of a small stream, but traveled on through the brush so rapidly that I did not get a shot at it.

CHUNGA BURMEISTERI (Hartlaub)

Dicholophus burmeisteri HARTLAUB, Proc. Zool. Soc. London, 1860, p. 335. (Tucumán and Catamarca, Argentina.)

Near Tapia, Tucuman, from April 6 to 13, 1921, these birds were fairly common, but were so wary that no specimens were secured. Their high-pitched yelping calls were heard daily from low hilltops where the forest was rather open, but as noiseless approach through the thorny scrub was impossible the birds invariably took alarm and ran away before I was within sight or range of them. Once or twice I had a glimpse of one down some long opening in the brush, but had no opportunity for closer approach. An immature *Chuña* was examined that had been killed by a hunter who refused to part with it.

The generic name *Chunga*, ordinarily attributed to Reichenbach,[11] has been assigned correctly by Waterhouse[12] to Hartlaub.[13] There is some question as to whether the date of publication of Reichenbach's paper was 1860 or 1861, but as Reichenbach on page 160 refers

---

[10] Notes on a collection of Surinam Birds, Bull. Mus. Comp. Zool., vol. 62, April, 1918, p. 42.
[11] Vollst. Naturg. Tauben, 1861 (?), p. 159.
[12] Ind. Gen. Avium, 1889, p. 45.
[13] Proc. Zool. Soc. London, August, 1860, p. 335.

to the original description of the *Chuña* in the Proceedings of the Zoological Society of London his name must have appeared later than Hartlaub's notice of it. Hartlaub remarks that the subgenus *Chunga* is proposed by Burmeister, but no reference to such action is at present known.

Order CHARADRIIFORMES

Family STERCORARIIDAE

STERCORARIUS PARASITICUS (Linnaeus)

Larus parasiticus LINNAEUS, Syst. Nat., ed. 10, vol. 1, 1758, p. 136. (Coast of Sweden.)

Three specimens were taken 15 miles south of Cape San Antonio on the eastern coast of the Province of Buenos Aires, a female on November 4, 1920, and two males on November 7. The female is white below and on the hind neck, with slightly indicated grayish-brown streaks on the neck, both in front and behind. Broken bars of a similar color cross the breast and to a less degree the abdomen, and the under tail coverts are barred with fuscous and white. The two males represent the dark phase and are dull in color throughout. These specimens apparently are in their second year, as traces of the first-year plumage of the juvenile are evident. The primaries and rectrices are worn and have not yet been molted.

While adult jaegers may be named with ease, it is trite to remark that identification of specimens representing the immature stages is attended with more or less difficulty. With regard to the two smaller forms, *parasiticus* and *longicaudus*, it may be necessary to have recourse to more than one character in order to be definitely sure that identification is correct. The following brief summary (for which I make no claim of originality, as it is a combination of matter given by Ridgway, Coues, Saunders, and Mathews) may be of assistance in naming birds of this group:

a^1. Bill higher than wide at base; wing usually more than 350 mm.; in adult the middle pair of rectrices broad throughout, twisted. (Subgenus *Coprotheres*) _____ pomarinus.
a^2. Bill not higher than wide at base, wing less than 345 mm.; in adult the middle pair of rectrices straight.
　b^1. Length of horny cere (supranasal saddle) decidedly greater than length of dertrum; tarsi and feet black; three or more outermost primaries with shafts ivory yellow. (Subgenus *Stercorarius*.) _____parasiticus.
　b^2. Length of horny cere (supranasal saddle) not greater than length of dertrum; tarsi wholly or in part light in color, feet black; only two outermost primaries with shafts ivory yellow (the shaft of the third sometimes light, but with a brownish tinge for at least part of its length). (Subgenus *Atalolestris*.) _____longicaudus.

The first species given above, the pomarine jaeger, has in recent usage been recognized as a monotypic genus *Coprotheres*, a distinction proposed first by Reichenbach in 1850. The second species *parasiticus* is the type of the genus *Stercorarius* of Schaeffer 1789, while Mathews [14] has proposed *Atalolestris* as a subgenus for *longicaudus*. After consideration of the differences indicated it has seemed that the three small jaegers may be placed in one genus *Stercorarius*. In case any further degree of superspecific difference is desired it may be indicated by the use of subgeneric terms.

On November 4, south of Cabo San Antonio, two parasitic jaegers came beating down the sea beach from the northward, pausing at intervals for an agile pursuit of some tern, or to investigate some other source of food. I secured one and later saw another. On November 7 several more were seen and two were taken. Though less abundant than the long-tailed species the parasitic jaeger was well represented in the migration passing southward at this time, so that I estimated that they were present in a ratio of 1 to 15 among the bands of *longicaudus*. The notes made on the habits of the long-tailed jaeger apply equally well to the present species.

The female secured in light phase had the tip of the bill and the gape dull black, the remainder deep mouse gray; iris natal brown; tarsus and toes black.

STERCORARIUS LONGICAUDUS Vieillot

Stercorarius longicaudus VIEILLOT, Nouv. Dict. Hist. Nat., vol. 32, 1819, p. 157. (Northern Regions.)

Four females and two males were taken on the beach 25 kilometers south of Cape San Antonio, eastern Buenos Aires, on November 4, 6, and 7, 1920. Of these birds one is a nearly adult male, as it has only scattering feathers of the immature plumage on the white upper breast and on the flanks. The remaining five are immature but seem to be more than 1 year old. The feathers of the upper surface are more or less margined with white. All are light underneath, but vary from deep mouse gray on the throat to white streaked with deep mouse gray. The wing quills show more or less wear. A female had the tip of the bill and the gape dull black, the remainder of the bill deep mouse gray; crus, toes, and webs black; tarsal joint pale green-blue gray; tarsus pale olive gray, with the dark and light areas on tarsus and toes sharply delimited. The blotching of light and dark was conspicuous, so that it attracted attention in handling the birds. The extent of the light patches varied considerably, as in some individuals it covered the entire tarsus and the hind toe with its nail, while in others part of the

---

[14] Birds of Australia, vol. 2, pt. 5, Jan. 31, 1913, p. 500.

tarsus itself was black. One specimen taken had lost the distal portion of one foot in some ancient injury.

From November 4 to 7 long-tailed jaegers, acompanied by a few parasitic jaegers, were in migration southward along the broad sand beach extending southward from Cape San Antonio. On the afternoon of November 4 three came drifting slowly down to leave two of their number lying on the sand, while the third, more wary, kept out of range and continued southward.

During the two days following a tremendous gale of wind and rain made field work useless, so that I was confined to short excursions about camp. Occasional jaegers passed, keeping low down behind the shelter of the dunes, sweeping by at high speed, driven by the high velocity of a quartering wind. With fairer weather on November 7 the birds increased and were in sight constantly, all in silent passage toward some winter range in the south. The number that I actually saw during the period of my observations must have ranged between 1,200 and 1,500, while the total number of individuals that passed was far in excess of this.

The birds traveled alone or in little groups of three or four, some wary and others very tame. They drifted along, frequently scaling for long distances or occasionally flapping their wings, never more than 50 feet in the air, often only a few feet above the sand. At intervals one dropped lightly to the beach near the water mark to pick up a few beetles that had drifted ashore after the storm and then remained to rest for a few minutes. Others more energetic harried the Trudeau's terns with agile wing strokes until they disgorged their prey of fish on the sand, when the jaeger stooped easily to pick it up and then continued its flight. Their steady southward movement without pause to circle about or return was most impressive.

Family LARIDAE

LARUS DOMINICANUS Lichtenstein

Larus dominicanus LICHTENSTEIN, Verz. Doubl. Zool. Mus. Berlin, 1823, p. 82. (Coast of Brazil.)

In a recent paper Fleming[15] has named a form of this gull from the South Shetland Islands on basis of lighter color. Mathews and Iredale[16] list the black-backed gull of New Zealand as *Larus dominicanus antipodus* (Bruch) without comment as to the differences considered as a basis for this subspecific designation. After study of an insufficient series of *dominicanus* that includes specimens from both Atlantic and Pacific coasts of South America, New Zealand,

---

[15] *Larus dominicanus austrinus* Fleming, Proc. Biol. Soc. Washington, vol. 37, Dec. 29, 1924, p. 139. (Deception Island, South Shetland Islands.)
[16] Ibis, 1913, p. 248.

and Kerguelen Island, I do not detect any constant characters that may seem to serve to distinguish birds from these localities. The gonydeal angle in two specimens from Kerguelen Island is very prominent but is approached in this respect by birds from South America. Two specimens from New Zealand seem to have the wings and mantle somewhat blacker than others, but here again the difference may break down when recourse is had to a fair series from elsewhere. The bird from South Shetland I have not examined. On the whole, specimens from the scattered localities at hand seem remarkably constant in their conformity to one type of coloration.

The similarity in color and structure between *L. dominicanus* and *L. marinus* from North Atlantic and Arctic regions is striking, and in final analysis birds from the two regions seem separable by difference in size alone so that one may well question the degree of relationship between the two. In habit and distribution the two are complementary one to the other, and it seems logical to conclude that they have arisen from one parent stock. Differentiation in the two regions has apparently progressed to a point where we may recognize the two as full species though the propriety of calling them subspecies of one form may be considered.

Three specimens were taken near Lavalle, Buenos Aires, a male on October 25 and a male and a female on November 3. All are in partial immature dress. The male first mentioned above has molted in part into adult plumage though worn brown feathers are scattered over the dorsal surface, the primaries are still old, and only part of the tail has been renewed. The neck and lower surface are still more or less mottled. The two secured on November 3 are less advanced in stage of plumage though dark feathers are appearing on the mantle. The female is small, so that when I killed it I was under the impression that it was an individual of some other species.

In June the Dominican gull, known as *gaviota cocinera*, was common in the harbors of Rio de Janeiro (June 16), Montevideo, and Buenos Aires. At Berazategui, Buenos Aires, a number were recorded along the Rio de la Plata in company with smaller gulls on June 29. In the vicinity of the coast near Lavalle, Buenos Aires, the species was common from October 25 to November 13. The birds ranged along the tidal mouth of the Rio Ajo and in the Bay of Samborombon, or occasionally came a few leagues inland in search of refuse about the killing pens at the estancias. On the beach south of Cape San Antonio they were fairly common. The majority seen at this season were in immature dress and those taken were not in breeding condition.

On December 13 I saw a number in the bay at Bahia Blanca, where they gathered with harsh calls to feed on refuse cast overboard from ships at anchor in the harbor. The species was recorded near Montevideo and Carrasco, Uruguay, on January 9 and 16, 1921. At nightfall I observed them passing along the coast to some resting place to the eastward. A few were observed at La Paloma, Rocha, on January 23. Near Guamini, Buenos Aires, on March 5 I was rather surprised to observe three in company with flocks of *Larus maculipennis* about the large lake near town. *Larus dominicanus* was more usual in occurrence near the coast, yet came here about 120 miles inland.

The species is similar in habits and notes to large gulls of the Northern Hemisphere.

LARUS MACULIPENNIS Lichtenstein

Larus maculipennis LICHTENSTEIN, Verz. Doubl. Zool. Mus. Berlin, 1823, p. 83. (Montevideo.)

The brown-hooded gull is the most common of the lariform birds found on the open pampas. An adult male taken at Lavalle, Buenos Aires, on October 29, 1920, was in full plumage and about to breed. In this bird when freshly killed the bill was Vandyke red; margin of eye-lids dull Brazil red; iris Vandyke brown; tarsus and toes Vandyke red; and nails dull black. The colors of bill and legs have become somewhat duller in the dried skin. One of two females secured on the coast 15 miles below Cape San Antonio on November 3 is in winter plumage with brown feathers of the juvenal plumage present on the lesser wing-coverts. In the other the dark hood is indistinctly outlined on the crown and sides of the head, with scattering dark feathers on the throat. In both specimens the ends of the inner secondaries are grayish brown and the tail is tipped with dull black. Both have the feathers of wings and tail considerably worn. An adult female secured at the Laguna Castillos, near San Vicente, Department of Rocha, Uruguay, on January 31, 1921, is in full winter plumage except that the outer primaries are being renewed. The plumage of the breast has a faint rosy tint. Two adult males in full winter plumage (one prepared as a skeleton) were secured at Guamini, Buenos Aires, on March 5, and a skull was taken from a dead individual on the same date. The skin preserved shows a few pin feathers on the ventral surface. From other specimens at hand it appears that the post-breeding molt is completed about the end of March.

The genus *Chroicocephalus* that has been used for the hooded gulls (including *L. cirrocephalus* and *L. maculipennis*) is seemingly based entirely on differences in color and color pattern. Several years ago I examined the skeleton of *Larus franklini, philadelphia,* and *atricilla,* and several species considered to represent typical *Larus,* but after careful study was unable to make out structural differences other than those that serve to separate species. As there are no structural characters known, either internal or external that may be used to diagnose *Chroicocephalus* I prefer to include the two species of hooded gulls treated in the present paper in the genus *Larus*.

The call notes of the brown-hooded gull suggest those of *Larus franklini* and are entirely different from the cawing calls of *L. cirocephalus* that was associated with it in small numbers. In the Province of Buenos Aires *maculipennis* was common in late June, 1920, along the Rio de la Plata. Near Berazategui on June 29 these birds were abundant in flocks that rested on the muddy beaches or flew over the fields inland. Hunters decoyed them within range by waving some white object and killed them in numbers for food.

From October 22 to November 17 the brown-hooded gull was common in eastern Buenos Aires in the vicinity of Lavalle. At this season adults in full plumage were found in pairs that stood about in the pampa near little pools of water or that came circling overhead curiously with a scolding *Kek Kek Kek* to examine any intruder. At the same time I observed flocks of birds still in winter plumage both on the open plains and along the sea beach below Cape San Antonio. Apparently part at least of the young may require two years to reach sexual maturity and full plumage. Occasional adult birds were observed with these flocks. The species has a slow, flapping flight and with its short square tail, notes, and general appearance is strongly suggestive of Franklin's gull.

I observed a dozen gulls near General Roca, Rio Negro, on November 30, that may have been the present species. At Ingeniero White, near Bahia Blanca, on December 13, brown-hooded gulls were fairly common over the bay, and near Carhue, Buenos Aires, they were found on the shores of Lake Epiquen. Near Montevideo, Uruguay, in January the species was common.

Brown-hooded gulls breed in abundance on rocky islets along the coast of the Department of Rocha, eastern Uruguay, and formerly it was the practice to raid these colonies to secure eggs in large quantities. In recent years the commission charged with oversight of agricultural affairs (the Defensa Agricola) has afforded the gulls absolute protection, a step that has been well merited. Near La Paloma, the seaport town for the city of Rocha, I saw bands of these birds containing as many as 200 individuals feeding in the pastures on the abundant grasshoppers. The gulls were gathered in close flocks that flew slowly, barely above the earth, and as grasshoppers were discovered dropped to earth to secure them. Those from the rear rose continually to fly over their companions to the head of the column so that the band drifted slowly along, as though blown by the wind, in close though continually shifting formation.

Such flocks were a usual sight and to their activities may be attributed the comparative freedom of this region from the plagues of locusts that are frequently so destructive elsewhere. The useful habits of these gulls are recognized by many farmers, who object to hunters who kill the birds, an attitude that may be followed with great advantage in other regions.

Near San Vicente on January 31 I found brown-headed gulls common near the Laguna Castillos where I saw several individuals sick or dead from alkali poisoning. This gull was common near the large lakes at Guamini, Buenos Aires, from March 3 to 8, and was observed in abundance in the marshy region below Cañuelas. At this season adult birds were ragged and disreputable as they were in molt. In the majority the body was clothed in winter plumage, but wing and tail feathers, more notably the outer primaries, were still in process of renewal. Adult and immature individuals, gathered along the lake shore in flocks that contained as high as 100 individuals, were wary as they were shot by hunters at every opportunity. A thousand birds or more were seen daily. Flocks frequently passed out to feed in the pastures and were observed following plowmen at work in the fields to feed in the freshly turned furrows.

No gulls were observed during my work in the Chaco.

This species is known in Argentina simply as gaviota.

LARUS CIRROCEPHALUS Vieillot

Larus cirrocephalus VIEILLOT, Nouv. Dict. Hist. Nat., vol. 21, 1818, p. 502. (Brazil.)

The gray-hooded gull was observed only at the Estancia Los Yngleses, near Lavalle, Buenos Aires, from November 1 to 10, 1920, and at the mouth of the Rio Ajo in the same vicinity on November 15. An adult female was taken November 1, and an adult male, preserved as a skeleton, on November 10. After minute comparison of four of these gulls from Buenos Aires, with a series of 27 from Africa, all adult, I may only substantiate the observations of others and state that, anomalous as it may seem, there is no apparent difference between birds from the two localities. Those from South America seem very slightly larger but in the series at hand the distinction is too slight to be reliable. One may well ponder on the conditions that have brought about such a remarkable distribution in this species and on the length of time that the two groups of individuals have been separated.

The adult bird is easily distinguished from the brown-hooded gull that inhabits the same region by the much lighter head, a difference that may be detected in a favorable light at a considerable distance. The note of *cirrocephalus* is a strange *caw caw*, similar to that of a

crow, and entirely different from that of other gulls that I know in
life, a call so characteristic that it distinguishes it at once. I found
the birds in pairs, apparently mated though I saw no nests, that
frequented the vicinity of the killing pens at Los Yngleses where
they searched for waste scraps of meat. Others beat back and
forth across the open pampa or came to hover over a fallen companion. The flight is steady and direct. The adult female taken
had the soft parts colored, as follows: Bill madder brown, becoming
diamine brown at base; iris naphthalene yellow; bare eyelids dragon's-blood red; tarsus and toes dragon's-blood red; nails black. The
light iris in this species is peculiar.

GELOCHELIDON NILOTICA (Gmelin).

Sterna nilotica GMELIN, Syst. Nat., vol. 1, pt. 2, 1789, p. 606.   (Egypt.)

The gull-billed tern was found in small numbers. At the port
of Ingeniero White, a few kilometers from the city of Bahia
Blanca, Buenos Aires, I watched about 50 on December 13, 1920,
as they fed over shallow bays or rested on muddy points. They
circled about with chattering calls, frequently diving for small fishes
in the tidal channels. One flew over my head with one of the
abundant crabs (*Chasmagnathus granulata*) in its bill and after
alighting in shallow water pulled off the animal's claws and then
swallowed it. Six or eight gull-billed terns were observed at Lake
Epiquen near Carhue, Buenos Aires, on December 15 and 18, and
several were noted in company with Royal Terns below Carrasco,
near Montevideo, Uruguay, on January 9, 1921.

Mathews [17] has separated gull-billed terns from South America
as *Gelochelidon n. grönvoldi* stating that they differ from North
American birds *Gelochelidon n. aranea* (Wilson) in longer bill and
wing. It is unfortunate that I secured no specimens of the South
American bird as there are none available in the National Museum.
Gull-billed terns are said to nest on Mexiana Island near the mouth
of the Amazon, and along the coast of Brazil, while the winter home
of the North American bird is not certainly known. Southern
records may therefore not be allocated under subspecies without
study of specimens.

This tern was first properly designated by Linnaeus in Hassel-
quist's Reise Palästinum, German translation (1762, p. 325). According to opinion 57 of the International Commission on Zoological
Nomenclature names in this work are untenable as the first edition
appeared in 1757. It hardly seems that this attitude is proper,
however, since binomial nomenclature is taken as beginning on
January 1, 1758, and this German translation appeared four years

---

[17] Birds of Australia, vol. 2, pt. 3, Sept. 20, 1912, p. 331.

subsequent to that date. As Dr. L. Stejneger has pointed out in a dissenting opinion accompanying this action of the International Committee the names given in the edition for 1762 fulfill all of the conditions imposed by the code to make them eligible.

STERNA TRUDEAUI Audubon

Sterna trudeaui AUDUBON, J. J., Birds of America (folio), vol. 4, 1838, no. 82, pl. 409, fig. 2. (Great Egg Harbor, N. J.)

Trudeau's tern was locally common in eastern and southern Buenos Aires and near the coast in eastern Uruguay, but was not recorded elsewhere. It was first noted on October 21, 1920, near Dolores in eastern Buenos Aires, where I recorded a dozen or more in pairs that flew back and forth along a drainage canal cut through a marsh. As two passed near at hand I killed the female with a shot from my collecting pistol. Near Lavalle the species was fairly common from October 25 to November 15, and a second female was secured on November 4. One shot October 31 was preserved as a skeleton. Near Carrasco, east of Montevideo, Uruguay, one was seen January 9, 1921, and on January 16 the birds were common. The skull of a dead bird that had washed ashore was secured on the latter date. One was observed on the coast at La Paloma below Rocha, Uruguay, on January 23, and about 20 were recorded January 31 at the Laguna Castillos below San Vicente. One was secured there in a helpless condition from alkali poisoning. Near Guamini, Buenos Aires, from March 3 to 8 the species was fairly common on the borders of the large lagoons. It was not unusual to find 100 or more gathered in company with gulls. This point was the farthest inland at which I noted the species, as elsewhere it was found only along the coast or in level marshy areas near tide water.

Trudeau's tern in flight and general actions is similar to other smooth-headed terns. As the birds beat back and forth with zigzag flight along shallow channels they darted down at intervals to secure small fish that appeared in the water within striking distance. When they were not feeding they gathered in close flocks to rest on some sandy beach or point near water. Below Cape San Antonio parasitic and long-tailed jaegers harried them and made them disgorge. Once or twice I noted hooded gulls in similar attempts, but in each case the tern with seeming ease eluded its less agile pursuer.

The call notes of Trudeau's tern are sharp and explosive and suggest in many ways the sounds emitted by Forster's tern, a species that the present one suggests strongly in life. A usual call was a sharp *tik tik tik*, changed when birds became angry or excited to a drawn out *keh-h-h*. As birds in full plumage approach across the level expanses of the broad marshes or along some open sea beach they appear plain gray, with light head and a prominent dark mark through the eye. In winter the undersurface of the body is entirely white. At times they were wary but were enticed within range by a white bird or a handkerchief waved in the air. Near Lavalle through inquiry I located a small breeding colony of the *gaviotina*, as these terns were known locally, where the birds were associated with gulls, but before I was able to visit it the ternery was raided by boys who sold the eggs to the local baker for use in preparing cakes.

The two females secured for skins in October and November were in full breeding plumage. In one the back of the crown is washed with gray of the same shade as the back in the form of a transverse bar. In this specimen, an adult, the soft parts were colored as follows: Tip of bill cinnamon buff, base between zinc orange and tawny, band across distal third black; iris Vandyke brown; tarsus and toes zinc orange, the scutes clouded with fuscous; nails blackish. In winter plumage the bill is black tipped with yellowish, a condition that suggests Cabot's tern, from which the present species may be distinguished readily in the field by its lack of a nuchal crest. A female, apparently adult, secured on January 31, is in the winter plumage as the bill is black at the base and the undersurface of the body is entirely white. The primaries in part had been renewed recently but the outer ones were much worn and broken. Measurements of the two adult females in full plumage are as follows: Wings, 255–266; tail, 139–135; exposed culmen, 41.8–42.5; tarsus, 24–24.5 mm.

STERNA HIRUNDINACEA Lesson

Sterna hirundinacea Lesson, R. P., Traité d'Ornith., 1831, p. 621. (Coast of Brazil.)

On November 4, 1920, I found 20 or more on the beach below Cape San Antonio, eastern Buenos Aires, mixed among flocks of Trudeau's tern. The birds were wary and difficult to approach as they rested in close flocks in the sand. At rest or on the wing they suggested Forster's or common terns, but appeared larger. A female secured had the forehead and part of the lores white with slight mottlings of white throughout the otherwise black crown. The wing feathers and tail were considerably worn. The soft parts in this specimen were colored as follows: Bill slightly darker than jasper red, space behind nostril dusky neutral gray; iris natal brown; tarsus and toes jasper red, webs scarlet, nails black. A male secured at the same time is in worn immature plumage, with the nape and upper hind neck clouded with blackish, and the lesser wing coverts dusky. The wing feathers were considerably worn, while the elongate tips of the outer tail feathers had been entirely lost.

At Guamini, Buenos Aires, on March 7, 1921, I saw a flock of 10 slender, black-capped terns that seemed to be the present species, but I did not approach near enough to them to secure specimens.

STERNA SUPERCILIARIS Vieillot

Sterna superciliaris VIEILLOT, Nouv. Dict. Hist. Nat., vol. 32, 1819, p. 176. (Paraguay.)

This handsome little tern was seen first at Puerto Pinasco, Paraguay, on September 3, 1920, when one was observed feeding along an estero near the Paraguay River. On September 30 in the same vicinity I encountered two that rested on dead stubs in the water or circled about in the air. A carancho (*Polyborus p. brasiliensis*) passed overhead and one of the terns pursued it for some time with sharp metallic cries. A male in full adult plumage was tolled within range by waving a hankerchief. On January 9, 1921, on the beach near Carrasco, east of Montevideo, Uruguay, I encountered two pairs of these small terns, evidently on their nesting grounds as they darted constantly at my head with complaining cries until I had passed beyond their bounds. On January 31 near San Vicente, in the Department of Rocha, I found a dozen, all immature, beating back and forth over the Laguna Castillos in company with Trudeau's terns. Though young these birds were expert as fishermen, and were so wary that I had difficulty in shooting one. The bill in this specimen, a female, is not yet fully formed as it is only three-fourths as long as that of adults. The bird is still in mottled juvenal plumage.

In habits and form this species is suggestive of the least tern and frequents similar localities along large fresh-water streams or sea beaches.

The adult male secured had the soft parts colored as follows: Bill wax yellow; tarsus and toes, olive ocher; crus, yellowish olive; claws, black; iris, Rood's brown. Measurements of this specimen are: Wing, 189; tail, 82; exposed culmen, 36.6; tarsus, 16.8 mm.

In a small series birds from Colombia and British Guiana seem a trifle smaller than those from Paraguay.

THALASSEUS MAXIMUS MAXIMUS (Boddaert)

Sterna maxima BODDAERT, Tabl. Planch. Enl. d'Hist. Nat., 1783, p. 58. (Cayenne.)

On November 4, 1920, I shot a male royal tern on the beach 24 kilometers below Cape San Antonio, on the coast of the Province of Buenos Aires. The bird was found in company with smaller terns. November 15 several were noted in the mouth of Rio Ajo and in the small indentation of the Bahia de Samborombon into which it opens, below Lavalle, Buenos Aires. On January 9, 1921, 40 or more were recorded at Carrasco, a bathing resort a few kilometers east of Montevideo, Uruguay, and on January 16 others were seen in the same locality. The male secured had the soft parts colored as follows: Bill, between flesh-ocher and rufous; iris, natal brown; tarsus, black with a few of the scutes marked with vinaceous russet; underside of toes, ochraceous orange. The bird was in full plumage.

PHAETUSA SIMPLEX CHLOROPODA (Vieillot)

Sterna chloropoda VIEILLOT, Nouv. Dict. Hist. Nat., vol. 32, 1819, p. 171. (Paraguay.)

The large-billed tern was fairly common near the Rio Paraguay. At Las Palmas, Chaco, it was seen in small numbers about lagoons from July 23 to 31. On July 26 two were observed resting on a flat clump of grass and one, an adult female, was taken. At Puerto Pinasco several were observed over the Paraguay River on September 3, and from September 6 to 21 occasional birds were recorded west of that point at lagoons near the ranch at Kilometer 80. They were seen frequently in steamer travel along the river, the last being noted at Villa Concepcion, Paraguay, on October 3.

These birds on the wing appear more robust than other terns, an appearance heightened by the strong, heavy bill. They frequently swing up to an intruder and examine him curiously or scold vigorously with harsh raucous calls. At such times the light-colored bill is prominent.

The soft parts in the adult female secured were colored as follows: Bill lemon chrome, becoming light cadmium at base of culmen; tongue and inside of bill lemon chrome, becoming cress green toward the fauces; iris fuscous; tarsus and toes primuline yellow; claws dull black at tip, changing at base to gray number 7.

Large-billed terns from the northern portion of South America have the upper surface varying from darker than neutral gray to dark neutral gray, while in specimens from Paraguay and northern Argentina (Chaco) the hind neck, back, scapulars, lesser wing coverts, and tail are between light neutral gray and neutral gray. The evident differences separate birds from the two regions as well-marked subspecies.

Sterna simplex of Gmelin [18] has been referred doubtfully to the present species by several writers. It is based on the simple tern of Latham [19] from Cayenne. Turning to the original citation in Latham it is found that the bird described is evidently in immature plumage. The points given agree perfectly with those of the young

---

[18] Syst. Nat., vol. 1, pt. 2, 1789, p. 606.
[19] Gen. Syn. Birds, vol. 3, pt. 2, 1785, p. 355.

of the present species save that the crown is described as nearly white and the legs as red. The crown in the immature large-billed tern is pale gray, while the feet are yellow. The size, color of the wing coverts, large bill, and forking of the tail are those of *Phaetusa*, and the head, while not true white, is lighter than the back and might be characterized as "nearly white." It is my opinion that the present species may be recognized from this description, which can not refer to any other tern of this region. Gmelin in translating Latham's English into Latin wrote "vertice * * * alba," but this has no consequence, as Gmelin had no specimens but simply took what Latham had said regarding the bird. The "variety a" of Latham's simple tern which follows refers to some other species.

The name of the large-billed tern, therefore, becomes *Phaetusa simplex* Gmelin with the type locality Cayenne. The southern form will stand as *Phaetusa s. chloropoda* Vieillot. *Sterna brevirostris* Vieillot[20] based on the *hati pico corto* of Azara, a name that has been assigned doubtfully to the present bird, is based on the immature of some other species. Otherwise it would have priority over *chloropoda*.

Specimens of the northern form have been seen (in the collections of the United States National Museum, the Field Museum of Natural History, the Academy of Natural Sciences, and the Museum of Comparative Zoology) from Colombia (Barranquilla), Venezuela (Punta Caiman, Manimo River, Rio Uracoa, Aruba Island, Lake Valencia), British Guiana (Georgetown), Dutch Guiana (Braamspunt, Tygerbank, Diana Creek), and Brazil (Serra Grande and Conceicao, Amazonas, Santarem and Pernambuco). The southern form is represented in the Museum of Comparative Zoology by skins from Concepcion del Uruguay (collected by Barrows) in addition to the localities that have been noted. It seems probable that *P. s. chloropoda* is found in the Paraguay-Parana drainage and that *P. s. simplex* covers the river basins of northern South America.

Family JACANIDAE

JACANA JACANA (Linnaeus)

Parra jacana LINNAEUS, Syst. Nat., ed. 12, vol. 1, 1766, p. 259. (Surinam.[21])

The jacana, common in the Chaco and in parts of Uruguay, was recorded at the following points: Resistencia, Chaco, July 9 and 10, 1920; Las Palmas, Chaco, July 17 to 31; Riacho Pilaga, Formosa, August 9 to 17; Formosa, Formosa, August 23; Puerto Pinasco, Paraguay, September 1 to 30 (found from the Río Paraguay, west

---

[20] Nouv. Dict Hist. Nat., vol. 32, 1819, p. 166.
[21] See Berlepsch and Hartert, Nov. Zool., vol. 9, April, 1902, p. 129.

for 200 kilometers); Lazcano, Uruguay, February 7, 1921; Rio Negro, Uruguay, February 18; Rio Lules, near Tucuman, Tucuman. April 1. At Las Palmas, Chaco, a male and two females (one preserved as a skeleton) were shot July 23, a male was taken at the Riacho Pilaga, Formosa, August 9, a female at Puerto Pinasco, Paraguay, September 8, and a male at Lazcano, Uruguay, February 7.

Specimens from Demerara and north Brazil appear darker than those from the south, but the series available shows some variation in this respect.

The *aguapiasó*, a name for the jacana that has persisted since the days of Azara and his friend Noseda, was found at times where masses of floating vegetation choked narrow winding esteros through the forests but more frequently was seen at the borders of open lagoons. The birds frequented areas where aquatic plants covered the surface of the water so thickly as to present the deceptive appearance of solid ground, where with their long widely spread toes the jacanas sank only to a slight distance. In such situations, though they walked about with long strides, apparently preoccupied with momentous affairs, the sight of a hawk in the distance was sufficient to send all scurrying to cover in the rushes. At times 30 or 40 jacanas were found scattered in little groups over an extensive area. Though social and gregarious, they resented too close approach on the part of their fellows with cackling calls and threatening, upraised wings quivering above their bodies. Where washerwomen came daily to wash piles of clothing that they balanced expertly on their heads while working, in lieu of other dry places to pile them, jacanas became tame and paid little attention to men. In the central Chaco as the marshes dried during the winter season the birds were restricted in range, and it was not unusual to see them about ponds and mudholes surrounded by the thatch huts of peons where the birds mingled with domestic ducks and chickens of the door yards.

As jacanas walk about over the water hyacinth and other growths one may get a suggestion of red or brown in the plumage, but on the whole they are as inconspicuous as most other waders, so that the flash of greenish yellow as they spread their wings in flight or extend them, perhaps as a signal, above their backs is always a pleasant surprise. In flight the rapidly moving wings form a strongly contrasted color patch on either side of the body. In general appearance they suggest long-legged gallinules as they stalk about, a resemblance that remains as they fly with neck and legs extended. While feeding the birds often pull over bits of vegetation and then peer at the exposed leaves.

A female shot September 8 at Kilometer 80, west of Puerto Pinasco, Paraguay, had the ovaries enlarged and was near the period of oviposition. At Rio Negro, Uruguay, on February 18 a fully grown youngster was still under the anxious care of its parents. Although the adults, as demonstrative as avocets under similar circumstances, fearlessly ran or flew after me with clatter-ing scolding calls, or when I looked in their direction lay prostrate with feebly fluttering wings to draw my attention, the young, con-spicuous in its light plumage, watched me suspiciously and kept well out of range.

Jacanas are silent unless on the alert, when they utter a variety of grunting calls or a whistled alarm, or when bickering among themselves give vent to displeasure in scolding, clattering notes.

An adult male, when shot on July 23, had the base of the maxilla, rictal lappets, and the frontal leaflet mineral red; rest of bill cinna-mon, becoming more yellow below, and tinged with slate at tip; iris very dark brown; tarsus deep neutral gray; posterior face of crus tinged with vetiver green; toes dusky brown.

Family RECURVIROSTRIDAE

HIMANTOPUS MELANURUS Vieillot

Himantopus melanurus VIEILLOT, Nouv. Dict. Hist. Nat., vol. 10, 1817, p. 42. (Paraguay.)

Near Santa Fe, in the Province of Santa Fe, on July 4, 1920, the stilt was observed in flocks on marshy ground in the lowlands that border the Rio Parana and on the following day from the train was noted at intervals in suitable localities between Vera, Santa Fe, and Charadai, Chaco. On July 26 two came to a lagoon near Las Palmas, Chaco, but were too wild to allow near approach. Near Kilometer 80, west of Puerto Pinasco, Paraguay, stilts visited a lagoon near the ranch house on September 6, 13, and 20, but did not frequent it or others near by regularly. As this happened also at Las Palmas it would appear that they may wander during winter to some extent. Two were shot at Kilometer 80 on September 6, and two more on September 20. The Angueté Indians called them *keh tsay a nah*, while in Guaraní they were designated as *ta too*. In the Chaco stilts were found only about the more open lagoons and did not occur about those with borders heavily grown with rushes.

In the vicinity of Lavalle, eastern Buenos Aires, stilts were seen from October 22 to November 15, but were not very common, though the open ponds and marshes of that region were well suited for their needs. They were seen on the coastal mud flats at the mouth of the Rio Ajo. It is probable that increase in grazing and cultivation has caused a decrease in their numbers, and, as the birds are large and

conspicuous, that they have suffered extensively at the hands of gunners. On December 15 nearly 100 were noted in a close flock in an
indentation on the shore of Lake Epiquen near Carhue, Buenos
Aires. On the following day a few pairs seen on partly inundated
land on the lake shore seemed to be on their breeding grounds and
may have had eggs or young as they ran about with rapidly waving
wings and scolding calls. A few were recorded here on December
17 and 18. Near Guamini, in this same region, stilts were common
from March 3 to 8 in close flocks or scattered bands that fed in
shallow pools or bays. These flocks consisted of young and old
that apparently had banded together in preparation for migration.
Adults were still somewhat anxious about their young, though the
latter were fully grown, and scolded sharply with barking calls that
were answered by the whistled notes of their offspring. A pair of
adults taken were molting the primaries. The birds are similar in
appearance and carriage to the black-necked stilt (*Himantopus mexicanus*). They walk about slowly in mud or shallow water with
heads bent in search for food, seldom wading where the water is
deep in spite of their extraordinary length of leg. Though ordinarily inoffensive, they sometimes drive the young about after the
latter are fully grown, or may fly at them and force them to lie
prostrate to avoid being struck.

A male in first year plumage with gray crown and brownish gray
back, taken September 6, had the bill black; iris orange chrome;
tarsus and toes flesh pink, washed with pale quaker drab at joints.

In a small series there is no difference apparent in birds from
Chile, Peru, Paraguay, and Argentina.

Family HAEMATOPODIDAE

HAEMATOPUS PALLIATUS Temminck

Haematopus palliatus TEMMINCK, Man. Orn., ed. 2, vol. 2, 1820, p. 532.
(South America.)

Although on the morning of June 16, 1920, as the steamer came
in toward the wharf in the harbor of Rio de Janeiro, seven oyster
catchers circled past barely above the low waves, I did not have
opportunity to observe and watch these birds further until I reached
the coast of the Province of Buenos Aires in late October. Two
were seen on the mud banks at the mouth of the Rio Ajo below Lavalle on October 25, and from November 3 to 7 they were fairly
common on the broad sand beach that extends southward for many
miles below Cape San Antonio. Here oyster catchers in pairs fed
in the shallow sweep of the surf, often where waves of more momentum than usual came nearly to their bodies. The birds walked
slowly, with necks drawn in and heads inclined forward, seldom

extending the neck to full length unless on the wing. Their flight was swift and direct, usually only a few feet above the sand, but not infrequently, to avoid me, in a semicircle that carried them over the dunes or out over the sea.

They were difficult to kill at any great distance because of their dense plumage and heavily muscled bodies. A female and two males were shot on November 3, and a second female on the day that followed. The nesting season was about at hand, and it is probable that some had eggs at that season, as females shot were nearly ready to lay. One male in mating ardor pursued a female in swift flight that carried them turning and dodging over the dunes along the beach until the birds were lost to sight. On November 15, at the mouth of the Rio Ajo again, where several oyster catchers were seen, one pair had a nest somewhere on a small strip of sandy beach. I hid behind a clump of grass and watched from a distance, but though the birds returned in a short time, I failed to locate either eggs or young.

At Ingeniero White, on December 13, an oyster catcher was eating small crabs that it pursued quickly across the mud or secured by pulling them out of holes sunk in the clay. Four oyster catchers were recorded on the coast near Montevideo, Uruguay, on January 16, 1921, and several noted on the sandy beach at La Paloma, Rocha, on January 23, may have had young, as they circled past me with shrill whistles. The species is known locally as *teru de la costa*.

An adult female shot November 3 had the center of the bill between scarlet red and jasper red, shading at base to a color between bittersweet orange and flame scarlet, and at extreme tip to antimony yellow; bare eyelids slightly darker than orange chrome; iris cadmium yellow; tarsus and toes cartridge buff; nails buff.

The specimens taken, which have been placed in Doctor Murphy's hands for study, are assumed to be the subspecies *durnfordi* of Sharpe, but definite allocation is delayed pending his forthcoming revision.

Family PHALAROPODIDAE

STEGANOPUS TRICOLOR Vieillot

Steganopus tricolor VIEILLOT, Nouv. Dict. Hist. Nat., vol. 32, 1819, p. 136. (Paraguay.)

Wilson's phalarope was recorded first at the mouth of the Rio Ajo below Lavalle, Buenos Aires, on November 15, 1920, when four were seen feeding on a mud bank. Later a flock of a dozen circled past with soft honking calls and a female in full winter plumage was taken. At Carhue, Buenos Aires, 40 were recorded December 15 in company with lesser yellowlegs on mud bars in a brackish water marsh behind a fringe of rushes that bordered Lake Epiquen, where they had some protection from the gales that swept the pampa. A dozen were noted here on December 16 and more on December 18.

Family SCOLOPACIDAE

PHAEOPUS HUDSONICUS (Latham)

Numenius hudsonicus LATHAM, Index Orn., vol. 2, 1790, p. 712. (Hudson Bay.)

One was recorded on the beach at Concon, Chile, April 25, 1921.

BARTRAMIA LONGICAUDA (Bechstein)

Tringa longicauda BECHSTEIN, in Latham, Allg. Ueb. Vögel, vol. 4, pt. 2, 1812, p. 453. (North America.)

Formerly abundant, the upland plover is now rare in the region where it spends the period of northern winter. Its winter range on the open pampa is a region so vast that it is difficult to form a proper estimate of the actual number of individuals of the species that remain. . Among epicures the species has inherited in part the name and reputation of the Eskimo curlew and is sought constantly by gunners to supply that demand. The few that survive frequent remote regions on some of the large estancias where they are secure until they leave their seclusion and begin their return flight northward. The majority of those that I noted were identified by their liquid calls, heard, as is the case in Washington, as they passed at night.

They were noted first at Puerto Pinasco, Paraguay, on September 29, 1920, when several passed in the evening driving southward over the Paraguay River. Others were recorded at Villa Concepcion, Paraguay, on October 3, also in passage down river. During October, November, and December none were recorded, a significant indication of the present-day rarity of the species, as during this period I traversed hundreds of miles of pampa where the birds had formerly been abundant. Not until the spring migration northward began did I again note the upland plover.

The first was seen at La Paloma, below Rocha, Urugay, on January 23, 1921. On February 7, near Lazcano, Uruguay, two passed at daybreak driving directly northward, and another in similar flight was heard about 10 in the evening on February 22 at Concordia, Entre Rios.

During the latter part of February and the first half of March, *batitú*, as the bird is known locally, was a regular item on the bill of fare in the better class hotels and restaurants in the city of Buenos Aires. I was told that now they were difficult to secure as few were offered for sale. The game market was closed by law at this season so that the birds were not offered openly, but reached the hotels in a surreptitious manner. As game birds were served with head and tarsus intact it was a simple matter to determine that the birds offered were actually upland plover and not other shore birds, as I proved by ordering them on various occasions. Once my waiter brought two to show me that had been plucked and cleaned, but were still uncooked. The price charged for a portion was a peso and thirty or forty centavos, about 65 cents in our currency. I was told that the birds were so scarce that they were secured only by those gunners familiar with places where the upland plover alighted when in migration.

Two were seen near Ezeiza, Buenos Aires, on March 2; at Guamini, Buenos Aires, two were recorded in northward flight, high in the air on March 3, and two more on March 4. At Tucuman, Tucuman, five were heard early in the evening of April 1 as they passed over the city traveling due north during a slow rain accompanied by heavy mist. On the night of April 5 under similar conditions an extensive flight of shore birds began at a quarter of 10 and continued until half past 11. During this period. J. L. Peters, with whom I was traveling at the time, and I identified the call of the upland plover from 38 individuals. The birds were in company with yellowlegs, solitary sandpipers, and a few golden plover. How many passed unheard in the darkness there was no way to know. The calling of these birds when in northward migration was a phenomenon of common knowledge in Tucuman during that season in the year, but all commented upon the fact that the birds seemed to have decreased greatly in abundance in recent years.

In conclusion I may say that while the upland plover was recorded on various occasions this took place when the birds were in flight and that though special search was made I was not fortunate enough to discover an area where the birds were in residence. As, like the Eskimo curlew, a species that I did not meet, the Bartramian sandpiper inhabits the drier uplands it is probable that difference in ecological conditions due to intensive cultivation and grazing have wrought such great changes in the more primitive conditions found on the pampa in its original state that the birds are unable to adjust themselves to them and have been slowly crowded out, where other destruction has not overtaken them.

ACTITIS MACULARIA (Linnaeus)

Tringa macularia LINNAEUS, Syst. Nat., ed. 12, vol. 1, 1766, p. 249. (Pennsylvania.)

On October 25, 1920, near the mouth of the Rio Ajo below Lavalle, Buenos Aires, I saw the familiar form of a spotted sandpiper teetering on a projection at the base of a cut bank of clay. The bird proved to be an immature female. The species had been taken once previously in Argentina near Concepcion, Tucuman, on March 4, 1918,[22] but the present record is the farthest south at which the spotted sandpiper has been known. (The statement in El Hornero that my specimen was secured at Cape San Antonio was due to a misunderstanding on the part of Doctor Dabbene.)

TRINGA SOLITARIA CINNAMOMEA (Brewster)

Totanus solitarius cinnamomeus BREWSTER, Auk, vol. 7, 1890, p. 377. (San Jose del Cabo, Lower California.)

As has been said in the account under *Totanus melanoleucus*, the solitary sandpiper belongs with the wood sandpiper in *Tringa*, a genus of tringine sandpipers characterized by a two-notched metasternum, with the nasal groove extended for two-thirds or less of the maxilla.

The solitary sandpiper in its southward migration reached Formosa, Formosa, on the Rio Paraguay, on August 23, 1920, when three were found on overflowed ground along a slough tributary to the Paraguay. The birds were silent and walked so quietly along the borders of the pools, often where overhung by brush or grass, that they might easily have been overlooked. An adult female that I shot was thin in flesh, and from other indications I was certain that these birds had just arrived. At Kilometer 80, west of Puerto Pinasco, Paraguay, solitary sandpipers passed southward, stopping occasionally at the lagoons, from September 6 to 21, and on September 24 and 25 a number were seen at Laguna Wall at a point 200 kilometers west of the river. The species was not recorded during spring and summer on the pampas, and was not seen again until December 3, when a male was killed on the Rio Negro, near General Roca, Rio Negro, where it was found amid scattered willows on a muddy shore from which water had recently receded. Apparently this is the farthest south from which the species has been recorded. At Lazcano, Rocha, from February 2 to 8, solitary sandpipers were in migration in small numbers and were traveling northeastward along the Rio Cebollati toward the coast. A female was taken February 7. One was recorded at Rio Negro, Uruguay, on February 17, and one was seen at a roadside pool near General Campos, in Entre Rios, Argentina, on February 23. Another was noted at 25 de Mayo, Buenos Aires, March 2. During the night of April 5 at Tucuman, Tucuman, the call of this species was heard frequently among the notes from the great flight of waders that passed northward over the city. At this southern end of their range the solitary sandpiper frequents the margins of shallow pools as in the north, often in localities unsought by other waders. I found it far from common.

The specimens taken at Formosa and General Roca belong certainly to the western form, on the basis of size (male, wing, 134.3; female, wing, 136.7 mm.), dorsal coloration, and the presence of mottling on the inner web of the outer primary. A female from Lazcano, Uruguay, has molted the outer primaries, but on the basis of other measurements and on the presence of some dark buff mottling on the back seems within the limit of variation of *cinnamomea* and is identified as the same as the other two. Though the typical subspecies *solitaria* is recorded definitely from Colombia by Chapman,[23] these findings seem to cast a doubt on its presence as far south as Argentina.

A specimen taken August 23, newly arrived from the north, shows no indication of molt. One on December 3 has begun the renewal of feathers on the side of the breast and the wing coverts, but has not yet shed the flight feathers. On February 7 one has the wing feathers, including the coverts, renewed with the body plumage save on head, neck, and back mainly new. The species seems to have a complete molt during the period of northern winter.

TOTANUS FLAVIPES (Gmelin)

Scolopax flavipes GMELIN, Syst. Nat., vol. 1, pt. 2, 1789, p. 659. (New York.)

On July 31, 1920, at Las Palmas, Chaco, three lesser yellowlegs, the earliest of the northern migrants, appeared at a lagoon during a heavy wind. It may have been imagination, but to me it appeared that they flew slowly as though tired, suggesting that they had just arrived from the north. From September 5 to 21 the species was in steady migration southward at Kilometer 80, west of Puerto Pinasco, Paraguay, and on September 24 and 25 it was common with other shore birds at Laguna Wall, 120 kilometers farther west. Elsewhere the species was recorded as follows: Dolores to Lavalle, Buenos Aires, October 23, many; Lavalle, Buenos Aires, October 29 to November 15; General Roca, Rio Negro, November 23; Carhue, Buenos Aires, December 15 to 18; San Vicente, Uruguay, January 31, 1921; Lazcano, Rocha, February 5 to 9; 25 de Mayo, Buenos Aires, March 2; Guamini, Buenos Aires, March 3 to 8; Tunuyan Mendoza, March 23 to 28; Simoca, Tucuman, April 1; Tucuman, Tucuman, April 5; Tapia, Tucuman, April 13; and Tafi Viejo, Tucuman, April 15.

The lesser yellowlegs was wide spread in distribution after October and was more abundant on the whole than *Totanus melanoleucus*. The birds frequented the shores of open lagoons, shallow pools, or coastal mud flats, and though found distributed singly or two or three together it was not unusual to encounter them in larger bands

---

[23] Bull. Amer. Mus. Nat. Hist., vol. 36, 1917, p. 223.

that might contain 100 individuals.   On their wintering grounds they were rather silent, but with the opening of northward migration resumed their habit of uttering musical though noisy calls when disturbed in any manner.   On the pampas they congregated during drier seasons about lagoons and flocks often sought refuge from the violent winds that swept the open plains behind scant screens of rushes.   After any general rain these flocks dispersed to pools of rain water in the pastures, where insect food was easily available. The winter population was thus not stationary, but shifted constantly with changes in the weather.   By the first of March the lesser yellowlegs had begun their northward movement and numbers were found near Guamini, where they paused to rest after a northward flight from Patagonia.   In their case, as in that of other migrant species from North America, it was instructive to note that the migration southward came in September and October when the birds traveled southward with the unfolding of the southern spring and that the return northward was initiated by the approach of rigorous weather in faraway Patagonia.   Migrant flocks, many of whose members offered sad evidence of inhospitable treatment at the hands of Argentine gunners in the shape of broken or missing legs, were noted on the plains of Mendoza, near the base of the Andes, in March.   And during early April the migration became a veritable rush so that on the night of April 5, at Tucuman, the air was filled with the cries of these and other waders in steady flight northward above the city.

The lesser yellowlegs seems to undergo a complete winter molt while in the south.   Two females secured September 6 and 21 at Kilometer 80, Puerto Pinasco, Paraguay, have both body and flight feathers worn.   A female from Lazcano, Uruguay, taken February 7, is renewing the two outermost primaries in either wing.   The other primaries, the secondaries, and the tertials, as well as the wing coverts, are new feathers, and the body plumage is in process of renewal.   Another female killed at Guamini March 8 has the wing feathers entirely replaced and new plumage appearing on the body.

Like the greater yellowlegs, *Totanus flavipes* has been reported in Argentina from May to August, but it must be assumed, on the basis of crippled individuals.

The generic relationships of this species have been discussed in the account of its larger brother *T. melanoleucus.*

TOTANUS MELANOLEUCUS (Gmelin)

Scolopax melanoleuca GMELIN, Syst. Nat., vol. 1, pt. 2, 1789, p. 659. (Chateau Bay, Labrador.)

The generic relationship of the two yellowlegs to one another and to related shore birds has been subject to considerable difference of opinion. Mathews [24] used the genus *Iliornis* of Kaup for the little greenshank, and remarks (p. 199) that "the species *T. flavipes* seems to fall easily into the genus *Iliornis*." Further [25] he unites the greater yellowlegs with *Glottis nebularius* in the genus *Glottis*. Mr. Ridgway [26] has segregated the two yellowlegs in the genus *Neoglottis*. Doctor Hartert [27] has united them in the genus *Tringa* with eleven other species, namely, *incana, totanus, guttifer, erythropus, nebularia, ochropus, solitaria, hypoleucos, macularia, stagnatilis*, and *glareola*.

Comparison of the greater and lesser yellowlegs fails to reveal characters of generic value that may serve to separate them. Save that the bill may be a trifle shorter in relation to the length of tarsus the lesser yellowlegs is practically a miniature of the greater. Mathew's suggestion that the two are not congeneric may be dismissed as untenable.

With due respect to Mr. Ridgway's opinion I do not believe that the two yellowlegs may be separated successfully in a generic sense from *Totanus totanus*. Examination of *melanoleucus, flavipes, totanus, nebularius, erythropus*, and *stagnatilis* reveals much of interest. *Glottis* has been considered as a distinct genus for *nebularius* on the basis of the recurved bill in that species. In this character it is approached by *melanoleucus* and furthermore varies in amount of curvature so that in series *melanoleucus* and *nebularius* may not be separated on this basis. Some specimens of *Totanus totanus* have a distinct web between middle and inner toes, while in others this web is reduced in extent. Development of this web gradés down in unbroken series from *totanus* through *erythropus* where it is distinct but small, to *melanoleucus, flavipes*, and *stagnatilis*, in which it is faintly indicated. The bill is shorter than the tarsus in *melanoleucus, flavipes*, and *stagnatilis*, from somewhat shorter to as long as the tarsus in *totanus* and longer than the tarsus in *erythropus*. Here, again, there is no line of demarcation. The little greenshank, *stagnatilis*, has the bill somewhat more slender than the others but in insufficient amount to validate its separation as a distinct genus. There is no reason apparent for not including in *Totanus* the following species, *totanus, erythropus, nebularius, melanoleucus, flavipes*, and *stagnatilis*. In addition it seems doubtful if *Pseudototanus* may be successfully maintained for *guttifer*, a matter that is here left in abeyance since I have seen only one skin of this species.

The course followed by Doctor Hartert in lumping 13 species under the genus *Tringa* seems ill advised. Four of the included species, *hypoleucos, macularia, solitaria*, and *ocrophus* have but two notches

---

[24] Jirds Australia, vol. 3, pt. 2, May 2, 1913, p. 197.
[25] Jol. 3, pt. 3, Aug. 18, 1913, p. 224.
[26] Birds North and Middle America, vol. 8, 1919, p. 129.
[27] Vög. Paläark. Fauna, vol. 2, Heft. 13, Feb., 1921, pp. 1607–1608.

in the posterior border of the metasternum, while in the nine remaining there are four such indentations. In *hypoleucos* and *macularia* the maxilla is grooved nearly to the tip, while in *ocrophus* and *solitaria* this groove extends less than two-thirds of the length of the maxilla. *Actitis* may be used for the first two leaving *ocrophus*, the type of *Tringa*, and *solitaria* united in the genus *Tringa*. Of the two species that remain of the 13 mentioned by Hartert *Heteroscelus* may be used for *incanus* (and also for *brevipes*) because of its difference in tarsal scutellation, while *glareola*, closely allied to the species here placed in *Totanus*, differs in that the tarsus is decidedly less than one and one-half times the middle toe without the claw and so may be maintained in *Rhyacophilus*.

During my work in South America the greater yellowlegs was recorded as follows: Kilometer 80, west of Puerto Pinasco, Paraguay, September 8 to 21, 1920; Kilometer 200, in the same region, September 24; Dolores, Buenos Aires, October 22; Lavalle, Buenos Aires, October 23 to November 15; Carhue, Buenos Aires, December 15 to 17; San Vicente, Uruguay, January 31 and February 2; Lazcano, Rocha, February 5 to 8; 25 de Mayo, Buenos Aires, March 2; Guamini, Buenos Aires, March 3 to 8; Tunuyan, Mendoza, March 25 to 28; Tucuman, Tucuman, April 5; Concon, Chile, April 24 and 26.

After their arrival in September greater yellowlegs were distributed throughout the open pampa wherever shallow ponds offered suitable feeding places. Occasionally 10 or 20 gathered in a flock, especially when northward migration was under way in March and April, but when on their wintering grounds it was usual to find two or three in company, seldom more. They are rather silent during the winter season but when the northward journey begins are as noisy as is their custom in the north. The species is large so that it is attractive to pot hunters and many are killed. I saw a number of crippled birds during the last two months of my stay in Argentina and consider that it is these injured individuals, unable to perform the necessary flight, or without desire to do so from their injuries, that are recorded on the pampas from May to August when all should be in the Northern Hemisphere. Reports of their breeding in Argentina, based on the presence of these laggards in migration are wholly unauthenticated.

An adult female shot February 2 at San Vicente, Uruguay, and another taken March 6 at Guamini, Buenos Aires, were molting the feathers of the forepart of the body and the neck. The primaries were fresh and unworn and appear to have been newly grown.

CROCETHIA ALBA (Pallas)

Trynga alba PALLAS, in Vroeg, Cat. Rais., 1764, Adumbr., p. 7. (Coast of North Sea.)

An adult female of the sanderling was taken November 6, 1920, on the exposed outer beach 24 kilometers south of Cape San Antonio, Buenos Aires, one of the few birds that cared to brave the severe gale then in progress. On the following day 20 in three flocks passed in southward migration, flying about a meter above the sand near the line marked by the wash of the waves. The species was not recorded again until April 29, 1921, when 25 were seen near Concon, Chile, in flight northward along the coast.

PISOBIA MELANOTOS (Vieillot)

Tringa melanotos VIEILLOT, Nouv. Dict. Hist. Nat., vol. 34, 1819, p. 462. (Paraguay.)

Azara, with his usual meticulous care, described his *chorlito lomo pardo*—the basis of Vieillot's *Tringa melanotos*—so minutely that there is no mistaking it for one of the two larger *Pisobia*, while his note to the effect that the tarsus was greenish points unmistakably to the pectoral sandpiper, since, as is well known to observant field naturalists, Baird's sandpiper, the only other species that may here be confused, has the tarsus black. The dimensions given by Azara are also those of the pectoral sandpiper. Sadly enough, *Tringa melanotos* on page 462 of Vieillot's work has priority over *Tringa maculata* on page 465 and so must supplant it.

The pectoral sandpiper was recorded as fairly common. At Kilometer 80, west of Puerto Pinasco, Paraguay, the species arrived ˙ on September 9, 1920, and passed in small numbers until the close of the month. On September 24 and 25, thirty or more were seen on muddy areas at the Laguna Wall at Kilometer 200. In crossing from Conessa to Lavalle, Buenos Aires, October 22, two flocks, containing in the aggregate 20 individuals, were seen near pools on the grass-grown pampa. Four were noted at the mouth of the Rio Ajo October 25. At Carhue, Buenos Aires, one was found in company with lesser yellowlegs December 15. On January 15 I observed four or five captive in the zoological gardens in Montevideo, Uruguay, and was informed that they had been captured that season. Near Lazcano, Uruguay, two were seen at a small lagoon, and on the following morning I found a flock of 16 resting on mud lumps in a pool in a road. These latter were evidently tired, as all rested quietly in the sun, several crouched on their breasts. When flushed the flock flew on to the westward instead of following down the Rio Cebollati, as was the custom of other migrating shore birds noted here. At Guamini, Buenos Aires, a pectoral sandpiper was noted on March 4 and another March 5. Near Tunuyan, Mendoza, six were recorded March 26 in company with the two species of yellowlegs. Two pectoral sandpipers that I shot here were extremely fat—in fact, one could not be preserved as a skin for this reason—but flew easily in spite of their heavy bodies. None were recorded later than this date.

Two adult females shot at Puerto Pinasco, Paraguay, September 9, 1920, were in worn breeding plumage, with no indication of molt. A female shot at Lazcano, Uruguay, February 8, had renewed the entire plumage save that new feathers in small amount were still in sheaths on breast and back. A male taken at the same time had the outer primary in either wing barely appearing and the ninth, the adjacent one, not quite fully grown. Nevertheless the bird was apparently in northward migration. A male taken at Tunuyan, Mendoza, March 26, was in full plumage.

PISOBIA BAIRDII (Coues)

Actodromas bairdii Coues, Proc. Acad. Nat. Sci. Philadelphia, 1861, p. 194. (Fort Resolution, Great Slave Lake, Canada.)

Three Baird's sandpipers were observed March 5, 1921, near Guamini, Buenos Aires, in company with white-rumped sandpipers. The species was at this time in northward flight from a wintering ground in Patagonia.

PISOBIA FUSCICOLLIS (Vieillot)

Tringa fuscicollis Vieillot, Nouv. Dict. Hist. Nat., vol. 34, 1819, p. 461. (Paraguay.)

In addition to the characters of the white or dark upper tail coverts and other color differences usually assigned to distinguish the white-rumped and Baird's sandpipers, the two may be easily separated by the form of the bill. In *P. bairdii* the bill tip is little expanded, the maxilla is elongately pointed, and the dorsal surface of the tip is hard and smooth. In *P. fuscicollis*, on the contrary, the tip of the bill is sensibly widened and has the surface distinctly pitted. These distinctions, perceptible under a low magnification lens, when once seen are recognized easily with the unaided eye and form a valuable identification character when, for example, one has specimens of the white-rumped sandpiper in which the dark centers of the white feathers are somewhat more extensive than usual, or in which some of the light tail coverts have been lost and not yet renewed in molt. In fact, in an extensive series it is not difficult to find specimens that may not easily be separated from *P. bairdii* by color alone but that are readily identified by the bill.

The same differences that have been pointed out between the bills of the white-rumped and Baird's sandpipers serve to distinguish

Baird's sandpiper from the pectoral, since *P. melanotos* has the tip of the bill heavily pitted. The appearance of the bill will thus separate these two readily where size or the color of the rump and upper tail coverts are not sufficiently distinct.

The white-rumped sandpiper was the most abundant of the migrant shore birds in the regions visited in southern South America. The species was not recorded until September 6, 1920, when it appeared in abundance in southward migration on the lagoons at Kilometer 80, west of Puerto Pinasco, Paraguay. The first flocks from which specimens were taken were adult females, and two taken on the date when they were first recorded had laid eggs a few weeks previous as was shown by the appearence of the ovaries. The southward migration came with a rush as the birds passed through the night as witnessed by their calls. The flight continued until September 21, when a dozen, the last seen here, were recorded. The birds circled about lagoons in small compact flocks or walked along on muddy shores, where they fed with head down, probing rapidly in the soft mud; anything edible encountered was seized and swallowed and the bird continued without delay in its search for more.

Farther south this species was encountered in abundance in its winter range on the pampa. Ten were recorded at Dolores, Buenos Aires, October 21, and from October 22 to November 15 the species was found in numbers on the coastal mud flats on the Bay of Samborombom. A few were seen at pools of water in the sand dunes below Cape San Antonio. Along the Rio Ajo white-rumped sandpipers were encountered in flocks of hundreds that came up stream to search the mud flats at low tide or were concentrated on bars at the mouth when the water was high. In early morning there was a steady flight of them passing to suitable feeding grounds. The birds flew swiftly with soft notes from 3 to 15 feet from the earth. In feeding they scattered out in little groups that covered the bare mud systematically. It was not unusual to record as many as 2,000 in a day.

About 200 were observed in the bay at Ingeniero White, the port of Bahia Blanca, on December 13, and at Carhue, Buenos Aires, from December 16 to 18, white-rumped sandpipers were noted in fair numbers on inundated ground back of the shore of Lake Epiquen or about fresh-water ponds on the pampa inland. None were found in Uruguay during February.

At Guamini, Buenos Aires, from March 3 to 8, white-rumped sandpipers were encountered in northward migration from a winter range in Patagonia. The species was fairly common on March 3 and increased greatly in abundance on the two days that followed. The northward journey was apparently as concerted as the movement that carried the birds southward, as on March 6 there was a noticeable decrease in their numbers, and by March 8, though the birds were still common, the bulk of individuals had passed. They arrived in flocks from the southward, often of several hundred individuals, that whirled in and circled back and forth along the lake shore to decoy to birds feeding on the strand or to rise again and continue swiftly northward. Those that paused kept up a busy search for food along the muddy beaches in or near shallow water, or in company with little parties of buff-breasted sandpipers on the drier alkaline flats back of the shore line. In early morning they were especially active and were in continual movement. Occasionally they worked out into comparatively deep water where in feeding it is necessary to immerse the head over the eyes nearly to the ear openings. When disturbed flocks rose with soft notes that resembled *tseet tseet* or *tseup* to circle to new feeding grounds on the lake shore.

Occasional parties of males, animated by the approaching breeding season, broke into soft songs and called and twittered, often for several minutes, in a musical chorus in low tones that had so little carrying power that they merged in the strong wind, and it was some time before I succeeded in picking out the sweet individual songs *tsep a tsep a tsep a* or *twee twee tee tee ty tee* given as the head was bobbed rapidly up and down. Occasionally when the fall sunlight came warmly I sat in the mud and let little bands of whiterumps work up around me until they were feeding and calling within a meter or so, eyeing me sharply for any cause of alarm. At such times their twittering choruses came sweetly and pleasantly, clearly audible above the lap of waves and the rush of the inevitable winds of the pampas. Between songs the search for food continued without cessation. At short intervals, activated by the warmth of the sun, they suddenly indulged in dozens of combats with their fellows, bloodless affrays, of bluff and retreat, where they lowered their heads and with open mouths ran at one another pugnaciously. The one attacked sidled quickly away or fluttered off for a short distance, save where two of equal temperament chanced to clash when first one and then the other threatened with raised wings in alternate advance and retreat until the fray was concluded to their mutual satisfaction. At such times the movements of these otherwise plain little birds were sprightly and vivacious to a degree. Their loquacity at this season was marked as it contrasted strikingly with their silence and quiet during the resting period of southern summer. Flocks frequently rose to perform intricate evolutions and then returned with a rush to sweep along the shore and join less ambitious comrades. As they passed the white rump flashed plainly, certain advertisement of the species. At times the chattering of these active flocks reminded me of the twittering of swallows.

One adult female, taken September 6 is still in worn breeding plumage, another has replaced a part of the plumage of breast and back. Both have worn rectrices. An immature female in full plumage was taken September 21, while one shot November 7 has lighter tips on the feathers of the dorsal surface partly worn away. Two more females, shot March 4 and 5, are in full prenuptial molt, a change that has involved the upper tail coverts so that these are as much brown as white and, though in migration, have the outer primaries still not quite grown. A complete molt seems to take place in February and March.

CALIDRIS CANUTUS RUFUS (Wilson)

Tringa rufa WILSON, Amer. Ornith., vol. 7, 1813, p. 43, pl. 57, fig. 5. (Shores of Middle Atlantic States.)

On November 7, 1920, a sanderling in winter plumage was killed at a pool of fresh water in the dunes 24 kilometers south of Cape San Antonio on the coast of Buenos Aires. The bird was feeding in company with white-rumped sandpipers.

TRYNGITES SUBRUFICOLLIS (Vieillot)

Tringa subruficollis VIEILLOT, Nouv. Dict. Hist. Nat., vol. 34, 1819, p. 465. (Paraguay.)

In addition to the generic characters usually cited, *Tryngites* may be recognized by the conformation of the nostril, which is elongate and has a median lobe projecting from the upper margin so as to lend the appearance of a median division. In addition, a single line of tiny plumes extends forward from the frontal antiae along the lower side of the nasal slit to a point anterior to the dividing lobe. These feathers may be worn away in some specimens but are present in the majority. *Aechmorhynchus* has an approach to this condition in a slightly swollen flap on the upper margin of the nostril, but this extends for the full length of the slit, and there are (in the four specimens seen) no feathers below the nostril. *Prosobonia* I have not seen. The nasal lobe in *Aechmorhynchus* is suggestive of the condition found in the plovers.

The buff-breasted sandpiper was first recorded in fall when an adult male was found on September 21, 1920, standing a little apart from other sandpipers on the open shore of a lagoon at Kilometer 80, west of Puerto Pinasco, Paraguay. On November 13 another was seen with other sandpipers on the tidal flats below Lavalle, Buenos Aires. No others were noted until, from March 3 to 8, 1921, a few were encountered in northward migration near Guamini,

Buenos Aires, where they arrived from regions farther south, tarried for a time, and then continued their northward flight.

The birds ranged in small flocks that occasionally fed with other sandpipers in the shallows or on muddy shores, but more frequently worked a short distance farther back on alkaline barrens where the surface was damp from the salt in the soil, but there was no standing water, and where vegetation was reduced to stumps of herbaceous growth that had been killed by concentration of alkali. They walked nervously, picking at the ground, and were active and quick in all their movements, constantly in motion, occasionally running a few feet to join others that had passed on ahead. When in the air or on the ground they are distinctly buff in color, with a glimpse of the marbled underwing surface as they rise or pass, and a flash of the gray tail with its darker markings as they alight. On the ground in profile, they show a long neck and long legs, while the short bill is suggestive of that of a pigeon. The neck is drawn in during flight. As they rise they may give a low call that resembles *chwup*, somewhat robinlike in tone; a second call note is a low trilled *pr-r-r-reet*. The species is to be confused in the field with no other shore bird.

An adult male shot September 21 had the bill black, shading to deep olive gray at the base of the maxilla; iris cameo brown; tarsus olive ocher, changing to dark olive buff on the toes, with a shading of the same color on the crus and the tarsal joint; nails black. Another male taken March 3 had the tarsus yellow ocher, shading to honey yellow on the toes.

A male shot September 21 was in worn breeding plumage. Others, secured March 3 and 5 that were completing the molt, had the outer primaries not quite grown and new contour feathers still developing. Specimens secured in March were extremely fat and difficult to prepare.

MICROPALAMA HIMANTOPUS (Bonaparte)

Tringa himantopus BONAPARTE, Ann. Lyc. Nat. Hist. New York, vol. 2, 1826, p. 157. (Long Branch, New Jersey.)

The stilt sandpiper was encountered only in the Chaco, west of Puerto Pinasco, Paraguay, though it has been said that it is common in some parts of the Province of Buenos Aires in winter. At Kilometer 80, on September 20, 1920, the first arrivals, a flock of a dozen, were recorded at the border of a lagoon; as I watched they rose suddenly to whirl rapidly away to the southward. On the following day about 20 were seen and an adult female was taken. At Kilometer 170 on September 24 a small flock passed down the nearly dry channel of an alkaline stream known as the Riacho Salado, while at Laguna Wall (Kilometer 200) about 30 were seen September 24. and 40 on the day following. The birds were found in little flocks, often mingled with other waders that walked or waded through shallow water on muddy shores where they probed with their bills for food.

The specimen taken had molted and renewed the wing feathers and was in winter plumage save for a few old feathers on the back.

LIMOSA HAEMASTICA (Linnaeus)

Scolopax haemastica LINNAEUS, Syst. Nat., ed. 10, vol. 1, 1758, p. 147. (Hudson Bay.)

A reconsideration of my previous statement[23] as to the validity of Mathews' proposed genus *Vetola* for *Limosa haemastica, fedoa*, and *lapponica* confirms my belief that the structural characters in which the four species of godwits differ *inter se* are too tenuous to warrant division of the genus *Limosa*.

Save for a record to be mentioned later the Hudsonian godwit was first recorded on November 13, 1920, when four, in winter plumage, were found with small sandpipers on the tidal flats near the mouth of the Rio Ajo, below Lavalle, Buenos Aires. Two more were seen here on November 15. The species was not noted again until March 3, 1921, when two were seen along the Laguna del Monte in the outskirts of Guamini, Buenos Aires. Four more were found on March 4, one in brown dress and the others still in winter plumage. On March 5 eight were recorded, one only showing distinct signs of breeding plumage. On the day following three passed swiftly northward over the lake without pausing to alight, while on March 7 eight were seen together and a single bird later, and by a lucky shot I secured one, a male. March 8 twelve that fed in a small bay were so slow in rising that I secured three. At dusk 12 more came to roost on a mud bar in company with golden plover. Though reported 50 years ago as found in great bands and among the most abundant of shore birds in this region, the small number that I have recorded here are all that were observed in continued field work throughout the winter range of the species. I was fortunate in seeing these, as by chance I found a spot where they tarried in northward migration from some point to the south.

In plain gray winter plumage this godwit is as inconspicuous and nondescript in appearance as a willet. In general size it suggests a greater yellowlegs but can be distinguished at any distance by its quiet carriage, for it does not practice the constant tilting that is the habit of the yellowlegs. These godwits sought company with scattered flocks of stilts or smaller shore birds, and in feeding walked rapidly, at times in water nearly to their bodies or again in the shallows. As they moved they probed rapidly and constantly in the mud with a nervous thrusting motion, often with the beak immersed clear to their eyes. Morsels of food that were encountered were passed rapidly up the length of the bill and swallowed. When their movements carried them too near the stilts the latter hustled them about, and made them run rapidly to escape their bills, but in spite of this discouragement the godwits remained in as close proximity as permitted to their belligerent neighbors perhaps because of similarity in feeding habit. Some Hudsonian godwit gave a low chattering call when flushed, a low *qua qua* that resembled one of the notes of *L. fedoa*. As they extend the wings to fly the dark axillars show as a patch of black and in flight the white tail, with black band across the tip, is prominent. The birds are hunted to such an extent that they were exceedingly wary. When opportunity offered I took only a few for specimens.

A male shot March 7 is in full winter plumage with worn primaries but newly grown tail feathers and lesser wing coverts. Two females shot March 8 have renewed the flight feathers and tail and have the breeding plumage growing rapidly on the body.

Reports that the Hudsonian godwit nest in the Southern Hemisphere are without foundation and the presence of large flocks in eastern Buenos Aires as early in the season as July 2, as recorded by Gibson,[29] may be explained only by considering them possible early migrants or by supposing that many did not breed each year, as from my own experience I know to be the case with some other shore birds, and that flocks of these nonbreeders may have failed to migrate northward. It may be added that Gibson's records of the birds in large flocks, though no year is given, must refer to his early observations, since the species has been rare for many years.

On my first day afield in Argentina, on June 29, 1920, a holiday when dozens of gunners were along the Rio de la Plata, near Berazategui, Buenos Aires, I am satisfied that I saw a Hudsonian godwit in the hands of a gunner but circumstances were such that I could not secure or handle the bird. It was a specimen in full winter plumage.

The passing of this fine bird must be a cause for regret among sportsmen and nature lovers alike, to be attributed to the greed of gunners and to the fact that its large size and gregarious habit made it desirable to secure and when opportunity offered easy to kill in large numbers. There is little hope even under the most rigorous protection that the species can regain its former numbers. It would appear that the small number that remain winter mainly in Patagonia, as the species was encountered in any number only when in migration from that region.

---

[29] Ibis, 1920, p. 70.

CAPELLA PARAGUAIAE (Vieillot)

Scolopax paraguaiae VIEILLOT, Nouv. Dict. Hist. Nat., vol. 3, 1816, p. 356.
(Paraguay.)

Mathews and Iredale [30] call attention to an overlooked generic
name in *Capella* of Frenzel,[31] which as it was published in 1801 has
precedence over *Gallinago* Koch 1816, and must be used for the true
snipes. The only copy of Frenzel's work known seems to be the one
in the library of the late A. Newton at Cambridge.

The two true snipe found in the level country of eastern and
southern South America are similar in general appearance and are
difficult to distinguish on casual inspection. After examination of
a small series it appears that they may be separated by the following
characters:

a^1. Markings of foreneck and upper breast broader, indistinct, especially on
    lower foreneck; more buffy on breast and above (especially in fresh fall
    plumage) ; outer rectrix tapering at tip; longer tertials more or less
    acuminate at distal end_____Capella paraguaiae.
a^2 Markings of foreneck and upper breast finer, blacker, more sharply defined;
    breast and dorsal surface blacker, less buffy; outer rectrix rounded,
    almost truncate at tip; longer tertials more or less rounded at distal
    end_____Capella braziliensis.

The relative length of outer secondaries and primary coverts
seems to be a variable character upon which one should not place
too strong reliance. On the whole, *C. braziliensis* is darker and
C. paraguaiae paler in general tone.

Capella andina Taczanowski has been considered a subspecies of
braziliensis, but, on the basis of two specimens, seems best considered an offshoot of the same stock that has produced *paraguaiae*, as it agrees with that species in pale tone of coloration, in
pointed outer rectrix, and in acuminate tertials. It is thus the
andean representative of a species that in the South Temperate
Zone ranges at lower altitudes.

On June 29, 1920, near Berazategui, Buenos Aires, several Paraguayan snipe were flushed in marshy spots along the Rio de la
Plata, and one that had been killed by a hunter was examined. One
was recorded at Dolores on October 21 and another seen near Conessa, between Dolores and Lavalle, on the day following. The
species was far from common here at this season, as none were recorded in nearly three weeks' work around Lavalle. At Zapala,
Neuquen, on December 8, one flushed from a boggy seep at a tiny
spring where the spot of surrounding marshy vegetation at its
greatest dimensions was not more than 10 by 30 feet and arid slopes

---

[30] Austr. Av. Rec., vol. 4, Dec. 16, 1920, p. 131.
[31] *Capella* Frenzel, Beschr. Vög. und Eyer Geg. Wittenberg, 1801, p. 58. (Type,
by monotypy, *Scolopax coelestis* Frenzel.)

stretched for miles. At an altitude of 1,900 meters above Potrerillos, Mendoza, several were flushed at swampy spring holes on March 19, 1921, and at Tunuyan, Mendoza, on March 25, 26, and 28 the birds were common around the muddy borders of lagoons and cienagas. They gathered in favored spots, and along certain muddy channels it was not unusual to flush a dozen together. The birds fed behind what cover offered, but were not averse to walking in the open, even on bright days, but if startled crouched flat on the mud. Little flocks frequently circled high in passing from one part of the marsh to another. In habit and action they resemble Wilson's snipe and have the same swift, erratic flight. The note with which they rise is harsh, but is flat and not so abrupt or startling as that given by the snipe of North America. A male was shot on March 25.

Near Holt, in Entre Rios, on October 9, while waiting for a train ferry to cross the Rio Parana, snipe, apparently of this species, were seen in a mating display in which they flew swiftly 12 or 15 meters above the ground and suddenly extended the wings stiffly in a V-shaped angle above the back and fell laterally through the air for a considerable distance.

CAPELLA BRAZILIENSIS (Swainson)

Scolopax braziliensis SWAINSON, Fauna Bor.-Am., vol. 2, 1832, p. 400. (Equinoctial Brazil.)

Near the ranch at Kilometer 80, Puerto Pinasco, Paraguay, on September 12, 1920, one of these snipe flushed in front of my horse as I rode along the border of an estero. I marked it down and dismounted and when it rose again secured it. As I waded in to retrieve it another flushed and was taken. They were found on floating vegetation in a scant growth of rushes where the water was knee-deep. Both were adult females. In Guarani they were known as *jacaberé*.

From February 7 to 9, 1921, this species was common in the bañados between Lazcano and the Rio Cebollati, in eastern Uruguay, where short, marshy vegetation covered a black, mucky soil. Two females were taken on February 7. The birds flush with a low harsh note, flatter in tone and less startling than the explosive call of the Wilson's snipe, dart off across the marsh in swift zigzags for a space, and then start straight away. It appeared that they were slower and heavier than the jacksnipe of North America and were easier to kill. They seem dark in color on the wing. Adults were in molt at this season and young were fully grown. On a few occasions I saw them standing erect or walking about in the grass, but at the slightest alarm they crouched with the head extended on the ground. Other snipe that appeared the same were seen at Rio Negro, Uruguay, on February 16, 17, and 18.

Totanus semicollaris VIEILLOT, Nouv. Dict. Hist. Nat., vol. 6, 1816, p. 402. (Paraguay.)

The painted snipe of South America differs so in structural characters from the Old World species as to warrant its generic separation, and indeed to such an extent as to cast some doubt on belief in the near affinity of these two groups. The bird under discussion here has the bill more curved at the tip, the tip expanded on both upper and lower mandibles, the distal end distinctly pitted, a median groove to distal end of gonys, a slight web between outer and middle toes, the tail strongly wedge-shaped, the median feathers tapered, and soft in structure at the tip with the median upper and lower coverts longer than the lateral rectrices. In *Rostratula*, as here restricted, the bill is less curved, with no distal expansion or pitting, no median groove on the gonys, no web at the base of the outer and median toes, tail only slightly rounded, of stiff blunt feathers with all of the rectrices longer than the tail coverts. The distinctions are easily evident on examination of specimens. For the South American bird Wetmore and Peters have erected the genus *Nycticryphes*.[32]

The South American painted snipe was fairly common near Lavalle, Buenos Aires, from October 28 to November 9, 1920, in boggy, fresh-water marshes where partly submerged areas grown with rushes afforded shelter and clumps of grass standing in from 75 to 150 mm. of water gave them footing. At that season the birds were found in pairs or alone though several might be startled near one another. In general habits, flight, and appearance they suggested jacksnipe but seemed more averse to bright sunlight, as, though they flushed readily in cloudy weather, it was often difficult to start them when the sky was clear and the light intense. They rose always near at hand with a sudden spring accompanied by a low rattle of wing quills that was stilled at once as they darted rapidly away. After a flight over the rushes they hesitated for an instant as though uncertain of the ground below and then dropped suddenly to cover. It frequently required considerable tramping to start them a second time. As they rose the light lines on either side of the back and the curved bill showed plainly, but as they traveled away they appeared wholly light and dark in color. Though in form and action their flight was not unlike that of *Gallinago*, they pursued a less erratic course and were silent. However, it required quick work to shoot them, so that their local cognomen of *corre correro* was well warranted. As they were killed over dense cover it was often difficult to locate those that had fallen.

---

[32] Proc. Biol. Soc. Washington, vol. 36, May 1, 1923, p. 143.

A female shot October 28 showed well-developed ovaries. On November 2 I flushed one from a nest and killed it, but unfortunately lost it, so that I did not learn the sex. The nest was placed under a tuft of dead grass on a dry, open island a hundred meters in extent, surrounded by a broad expanse of marsh. The site selected was on dry, open ground 15 meters from water. A few grass stems had been broken down to form a little protected cavity, entirely covered save in front, in which the two eggs lay with no nest lining. In its lack of definite structure the whole reminded me of the nest of a Wilson's phalarope. Incubation had begun. The eggs, suggestive in a way of those of the black tern, have a ground color slightly brighter than pale olive buff, spotted with heavy irregular spots of black, and less extensively with buffy brown and Saccardo's umber. The markings are much bolder and heavier in one than in the other. The two eggs measured 34.7 by 24.6 mm. and 34.2 by 24 mm.

Near the Laguna Castillos below San Vicente, Rocha, Uruguay, an adult male was taken among rushes on January 31, 1921.

The species was next encountered in the cienagas near Tunuyan, Mendoza, where it was common on March 25 and 26. Here painted snipe frequented scattered clumps of grass or *Scirpus* at the border of dense stands of cat-tails or, less often, marsh vegetation dead or living that bordered channels of almost bottomless black mud. At times a dozen or fifteen birds flushed together from some sheltered opening among the cat-tails, but more often they were encountered alone. Once or twice one darted in to alight near me and instantly assumed a motionless attitude, standing with legs erect but with the head and body inclined forward with the bill almost touching the earth. Occasionally one broke this tense attitude by a jerky bow and then became motionless once more. Two adult males were taken on March 26 and one on March 28. Though it was fall, testes in these birds were 8 mm. long, as large as white navy beans.

An adult female shot October 28 had the tip of the bill cinnamon buff; base strontian yellow; the intermediate space on maxilla water green, and on mandible celandine green; iris Rood's brown; tarsus and toes vetiver green, shading to deep grape green on toes and inside of tarsus.

Family CHARADRIIDAE

CHARADRIUS COLLARIS Vieillot

Charadrius collaris VIEILLOT, Nouv. Dict. Hist. Nat., vol. 27, 1818, p. 136. (Paraguay.)

The widely ranging collared plover was recorded in small numbers at several localities. Two were recorded at Santa Fe, Santa Fe. on July 4, 1920, and one July 8 at Resistencia, Chaco. On the whole the species seemed rare in the Argentine Chaco, perhaps because of a lack of suitable range for it, as it was not found again until I reached Kilometer 80, west of Puerto Pinasco, Paraguay, where it was recorded in small numbers about an open lagoon from September 6 to 21. A pair was taken September 6, and I noted that the breeding season was near. Two were seen on the shore of a small pond near Lavalle, Buenos Aires, November 13. The species was common along the sandy beaches on the coast of southern Uruguay, and in January was nesting. A number were recorded between Montevideo and Carrasco, January 9, 1921, and east of Carrasco, January 16. Others were seen at La Paloma in the Department of Rocha, January 23. At this season all seemed to have well-grown young but still showed much anxiety as I passed, and forced the young to hide. The parents circled around me with low calls, their light bodies often difficult to distinguish against the sky in the brilliantly reflected light of the sun. The birds were found on the outer beaches or through the bare dunes a short distance inland, where they ran about in scattered companies. The alarm note was a sharp, metallic *tsee* and occasionally they uttered a slightly rolling *tur-r-r*. In winter they were more silent and only uttered a low whistled *chap* or *cherp* as they rose and darted rapidly away. Near Rio Negro, Uruguay, on February 18, while crossing an area of high prairie where the soil was water-soaked from recent rains, I found about 20 of these plover, both adult and young, and judged that they had forsaken their coastal breeding grounds to wander inland as the young were fully grown. All were very wild. On March 3 I saw two near Guamini on the open shore of the Laguna del Monte. Two were recorded March 22 along the Rio Tunuyan at Tunuyan, Mendoza, and on March 25, 26, and 28 I found several in company with other shore birds along a small, muddy arroyo near some extensive cienagas. The bed of this channel was sunk about 4 meters below the surrounding level and was barely 30 meters wide, a greatly restricted area for these birds when the open areas that they frequent ordinarily are considered. A female was taken here March 25. At Concon, Chile, April 25, about 25 were found on a sandy beach and when flushed flew off in close flock formation.

In general habits this species suggests the snowy plover, but seldom runs for such long distances as is the habit of that species.

CHARADRIUS FALKLANDICUS Latham

Charadrius falklandicus LATHAM, Index Orn., vol. 2, 1790, p. 747. (Falkland Islands.)

The Falkland plover is easily distinguished in the field from companion species by the two distinct bands on the breast. One was seen on the shore of the Rio de la Plata near Berazategui, Buenos

Aires, on June 29, 1920. At Zapala, Neuquen, on December 8, I shot a juvenile specimen just able to fly from a family that was running about in the closely cropped grass of a pasture near a tiny stream. Two broods were recorded at Ingeniero White, Buenos Aires, on December 13, while on December 15 and 16 several were seen on the shore of Lake Epiquen, near Carhue. An adult male was taken on the 15th. Between Montevideo and Carrasco, Uruguay, a few were recorded on January 9 and 16, 1921, on sandy beaches where they fed at the water line by thrusting the bill quickly in the sand. One was noted at La Paloma, Uruguay, January 23. Near Guamini, Buenos Aires, they were fairly common on the muddy shores of the Laguna del Monte from March 3 to 7, in company with sandpipers. Two immature females were shot March 4. Half a dozen were seen at Concon, Chile, April 25.

The Falkland plover inhabits sandy beaches on the seashore or the borders of open lagoons inland. In habits and appearance it is similar to related species and like them frequently squats and hides to avoid detection. The ordinary call is a sharp *pit pit*.

PLUVIALIS DOMINICUS DOMINICUS (Müller)

Charadrius dominicus MÜLLER, Natursyst., Suppl., 1776, p. 116. (Santo Domingo, West Indies.)

Golden plover arrived at Kilometer 80, west of Puerto Pinasco, Paraguay, on September 6, 1920, and continued in southward passage until September 25. The birds came to the open shores of lagoons with other sandpipers, but were more often seen in flocks of 30 or 40 scattered over open savannas where the grass was not too long. At this season they were rather silent and were very wild. On September 16 cold weather in the south drove many back on their route, and birds passed north during the entire afternoon, not pausing to alight though the weather at the point of observation was not unfavorable. The return southward began two days later. On September 24 and 25 flocks were seen at Laguna Wall, 200 kilometers west of the Paraguay River. The Angueté Indians called this species *pill wit*.

On October 23, near Conessa, Buenos Aires, small flocks were scattered over the open pampa and the number seen was estimated at 260. On November 6, 7, and 8, golden plover were scattered over the open camp back of Cape San Antonio and a number arrived from the south. On November 13 and 15, I found a considerable number near the mouth of the Rio Ajo and on November 16 about 30 were recorded in crossing from Lavalle to Santo Domingo.

December 13 a golden plover was seen on the mud flats near Ingeniero White, Buenos Aires; December 14 thirty were seen near Saavedra, and from December 15 to 18 a few were noted near Carhue.

The species was not found west of these points and seems to be restricted during the resting period to the better watered grass-grown eastern pampa.

The northward migration began with a flock of nine seen January 23, 1921, at a little fresh-water pool on the beach near La Paloma, Uruguay; when flushed these passed on to the west. Single individuals were seen near San Vicente, Uruguay, in flight toward the northwest on January 24 and 30. At Lazcano, Uruguay, birds in passage north were seen in early morning on February 7 and 8, and one was recorded February 18 at Rio Negro, Uruguay. On March 8 at Guamini, Buenos Aires, 15 came in at dusk to roost on a little mud bar in company with Hudsonian godwits. The migration seemed almost at an end then, as later I saw only four at Tunuyan, Mendoza, on March 23; and on April 5 only a few were heard calling with other shore birds in flight northward over Tucuman, Tucuman.

An adult female shot September 6 had renewed part of the body plumage but had the flight feathers still worn. Another, shot November 15, was nearly in winter plumage and had begun the molt of the inner primaries.

OREOPHOLUS RUFICOLLIS RUFICOLLIS (Wagler)

Charadrius ruficollis WAGLER, Isis, 1829, p. 653. (Canelones, Uruguay.)

The spelling of the generic name for this plover as proposed by Jardine and Selby [33] is *Oreopholus*, as given above.[34] Brabourne and Chubb [35] state that no type locality has been given for this species and suggest Patagonia. In Wagler's original description, however, is the statement " Habitat in *America.* (*Canelonnes.*) (*Mus. Berol.*)," which indicates Canelones, Uruguay, as the source of his type. Uruguay has not been included usually in the range of this species, but *Oreopholus ruficollis* was seen by Aplin [36] at Santa Ana, and is recorded by Tremoleras [37] in Montevideo and Canelones.

On December 8, 1920, near Zapala, Neuquen, while traversing a sandy area with a thin cover of low bushes my attention was attracted by a low plaintive whistle, *whees tur tur.* As I watched to determine the source of the sound one of these plover ran forward a few steps and then stopped to watch me quietly. In this bird, an adult male, the tarsus was pinkish vinaceous, and the toes black.

The subspecies *O. r. simonsi* Chubb from southwestern Peru, Bolivia, and Tarapaca is a strongly marked race that differs from

[33] Ill. Orn., vol. 3, December, 1835, pl. 151.
[34] See Ridgway, Birds North and Middle America, vol. 8, 1919, p. 63.
[35] Birds South America, 1912, p. 38.
[36] Ibis, 1894, p. 207.
[37] El Hornero, vol. 2, no. 1, July, 1920, p. 13.

the typical form in larger size and in more buffy coloration. The light margins on the feathers of the back and wings are more rufescent and are broader than the dark central streaks, the head, neck, and rump are more buffy, less grayish, and the undersurface has a rufescent wash. It is probable that birds from central Chile may be differentiated from eastern specimens with a more extensive series than is available to me, as two specimens from near Santiago have the gray of the breast more restricted than some from Argentina.

BELONOPTERUS CHILENSIS CHILENSIS (Molina)

Parra Chilensis MOLINA, Sagg. Stor. Nat. Chili, 1782, p. 258. (Chile.)

The use of the name bestowed by Molina on the Chilian lapwing has been disputed but after due consideration it seems that it may be recognized. In the first edition of his work on the Natural History of Chile (cited above), Molina gave a brief diagnosis in Latin as a footnote for each species of bird treated. On referring to his account of the present bird we find it given as number 23, Il Theghel, *Parra Chilensis*, with the diagnosis " Parra unguibus modicis, pedibus fuscis, occipite subcristato." The description that follows with a considerable account of the habits refers to *Belonopterus* save that he states " la sua fronte è guernita di una carnosità rossa divisa in due lobi," a condition found in the Jacana and not in the lapwing. Evidently he was endeavoring to describe the plover as he gives an excellent account of its habits but had confused with it the lobed forehead of the Jacana—probably because both birds possess a spur on the wing. In the second edition of Molina's work, printed in 1810 (p. 205), is a duplication of the account of *Parra Chilensis* save that it is numbered 7; has the Latin diagnosis omitted, and has included a reference to a Chilian vocabulary. On page 206 is added the following statement: " Questa proprietà, che gli è comune col Vanello, e la maggior parte de' caratteri sopra-espostí, me avevano da prima determinato a porlo nel medesimo genere, denominandolo *Tringa Chilensis*, ma la piccola carnosità della sua fronte m' ha obbligate a lasciarlo nel genere *Parra*, dal quale però si scosta per la modicità delle sue dita."

Here, again, there is confusion regarding a supposed fleshy lobe on the head, but again attention is called to the short toes of the bird in mind. Molina's description therefor is composite, but when it is carefully considered will be found to apply in the main to *Belonopterus;* there can be no question but that *Belonopterus* is intended. It appears that the use of *chilensis* as the subspecific name of the western form of the South American lapwing is warranted.

Should any decide that this name is not properly identified, then they must fall back on *Vanellus occidentalis* Harting,[38] a new name

---

[38] Proc. Zool. Soc. London, 1874, p. 450. (Chile.)

proposed for the Chilian bird, with *Párra chilensis*, concerning whose identity Harting was apparently uncertain, cited in the synonymy. Dr. P. R. Lowe when he considered Molina's name untenable [39] and proposed to call the western form *Belonopterus cayannensis molina*, evidently overlooked this action of Harting's though the type of *occidentalis* [40] is supposed to be in the British Museum. As I have noted beyond the name *Vanellus grisescens* Pražák,[41] based on a specimen from North Chile is also a synonym under this form.

The Chilian lapwing is even handsomer, in its larger size and more extensive black markings contrasted boldly with the gray head, than the teru teru of the eastern and northern pampas, to which it is similar in habits and appearance. Like that fine bird, though it exasperates one with the ceaseless iteration of its calls, it may be forgiven much for confiding interesting traits that it exhibits at times and for its showy coloration. No lover of birds can recall days afield in the pampas without seeing in recollection the contrasted markings of these fine plovers.

The call notes of the Chilian lapwing, though similar to those of the eastern bird, are harsher and are pitched in a higher tone so that there is no difficulty in recognizing it when one enters its haunts. I encountered it first at Zapala in western Neuquen on December 8, 1920, but secured no specimens. At Tunuyan, Mendoza, males, taken on March 25 and 26, 1921, are similar in color and size to Chilian specimens from the western side of the Andes. They were observed here in small numbers from March 23 to 28. At Concon, Chile, they were noted from April 25 to 27. East of the Andes this form occurs on the somewhat broken plains at the base of the mountains where it ranges eastward for an unknown distance. It is possible that there is a gap between the areas inhabited by the eastern and western forms.

The specimens from Tunuyan (both males) have the following measurements, in millimeters: Wing, 256–258; tail, 114.5–129; exposed culmen, 30–30.5; tarsus, 74.5–73.5.

BELONOPTERUS CHILENSIS LAMPRONOTUS (Wagler)

Charadrius Lampronotus WAGLER, Syst. Av., pt. 1, 1827, p. 74. (Paraguay.)

Study of the South American lapwings available has shown that they may be divided into three well-marked races, as follows:

1. BELONOPTERUS CHILENSIS CHILENSIS (Molina).

Range: Chile, Patagonia, western Argentina along the eastern base of the Andes, and Peru (no Peruvian specimens seen). Char-

---

[39] Bull. Brit. Orn. Club, vol. 41, Apr. 13, 1921, p. 111.
[40] Cat. Birds Brit. Mus., vol. 24, 1896, p. 735.
[41] Ornith. Monatsber., vol. 4, 1896, p. 23.

acters: Larger size, gray head and neck, more extensive black markings on head and neck, and further posterior extension of black on breast. (Wing 240 to 258 mm.)

2. BELONOPTERUS CHILENSIS CAYANNENSIS (Gmelin).

Range: Colombia, Venezuela, the Guianas (type locality Cayenne), and northern Brazil (south at least to Diamantina, near Santarem). Characters: Brownish head and sides of neck, forming an unbroken collar on foreneck, and greater length of tarsus (77.5 to 82.5 mm.).

3. BELONOPTERUS CHILENSIS LAMPRONOTUS (Wagler).

Range: Southern and eastern Brazil,[42] Paraguay, Uruguay, and Argentina west to the plains at the eastern base of the Andes south into northern Patagonia. Characters: Brownish gray head and sides of neck, with a black line passing from black of throat to breast, and shorter tarsus (69.5 to 75.5 mm.).

The third form has been recognized as *Belonopterus grisescens* Pražàk by Brabourne and Chubb.[43] As *Vanellus grisescens* Pražák [44] was described from a single specimen secured by Richard Materna in " North Chile," it can not refer to the bird of the eastern pampas and must be considered a synonym of *chilensis*. As a matter of fact, the name *lampronotus* of Wagler cited above is available for the southern form, and does not refer to the northern typical subspecies, as in his description Wagler states that *lampronotus* has a black line leading down the middle of the foreneck to the breast, sides of the head, hind neck, and sides of neck ashy and tarsus 3 inches long, characters that indicate a bird from the south. He cites the range as Paraguay, Brazil, and Cayenne. The type locality is hereby restricted to Paraguay.

The large, conspicuously colored teru teru is one of the most prominent birds found on the Argentine pampas, a species that the traveler meets almost at once on reaching open country. Where the birds are common one is never free from their insistent espionage, and though the birds are pleasing in color, their clamor soon becomes tiresome even when they do not alarm desirable game. Where herdsmen pass continually through the fields on horseback the plovers are tame; elsewhere, where hunting is prevalent, they may be more shy, but it is seldom difficult to call them within gun range by sitting down in the open or by waving a white handkerchief. During the breeding season the birds fly out to meet all comers and with clamorous calls conduct intruders across their chosen domain. On moonlit nights their barking stiltlike calls may be heard continuously, while they call at any time when disturbed, even though

---

[42] A skin in the Field Museum, from Cidade da Barra, Rio San Francisco, Bahia, is representative of this form.
[43] Birds of South America, December, 1912, p. 38.
[44] Ornith. Monatsber., vol. 4, 1896, p. 23.

it may be pitch dark. In olden times they were prized for their watchfulness at night, which gave warning of the approach of any possible enemies, and it is to this that the sentiment in which they are held at present is due. Though peons rob their nests continually, comparatively few of the terus are killed, which may account for their abundance in settled regions.

The species was found in practically all of the regions visited. In the Chaco where the country is broken by frequent tracts of monte they were less common than farther south in more open country. During winter they were frequently observed in pairs that often seemed to have a restricted range where they were observed daily. They traveled to some extent, however, and were frequently seen in strong, direct flight, passing high overhead. In the Chaco back of Puerto Pinasco they were seen at a distance of 200 kilometers west of the Paraguay River. A female that I secured on September 6, 1920, at Kilometer 80, was about ready to breed. It is possible that another form may be described from this region as this specimen, in common with one or two seen from southern Brazil, has the black line down the foreneck considerably restricted.

The well-watered eastern pampas form the true metropolis of the teru teru, as though the species frequents open, grassy plains, it seeks always the vicinity of water. In eastern Buenos Aires the birds were especially abundant and at the end of October apparently were breeding. At this season they became especially pugnacious in pursuit of passing hawks and storks, while one even dashed repeatedly at an inoffensive European hare that loped along ahead of me. It was frequent to see two pairs of terus high in air in a display flight in which the fully extended wings, marked prominently with black and white, were waved slowly. The birds were observed west to Carhue, Buenos Aires, but none were noted at Victorica, Pampa, though I was told that they occurred there in wet seasons. A few seen at General Roca, Rio Negro, were supposed to be the present bird but may have been *chilensis*.

In Uruguay terus were common. On January 22 between San Carlos and Rocha I observed several bands of 20 or 30 gathered on open spaces on the banks of little arroyos running with clear water. These bands, observed commonly until the end of February, were composed of old and young, all less noisy than earlier in the season. The birds rested or fed in loose flocks that ran aside to permit passage of vehicles or men on horseback. At this season terus were in molt and their resting places were strewn with cast feathers. Grasshoppers, present in great abundance, were a favorite food, and the birds on the whole must have a considerable economic importance. A young bird only a few days old was captured near Lazcano on February 7. An adult taken on the following day is representative of the pampas race. Flocks were noted near Guamini, Buenos Aires, as late as March 3.

In Paraguay the lapwing was called *taow taow* by the Angueté Indians, an interesting similarity to the *teru teru* of the Guaranís by which the bird is known almost universally in the southern republics. Both names are bestowed in imitation of the bird's notes.

The downy young bird [45] secured is avellaneous on the dorsal surface with irregular spottings and markings of black, a band of black across the nape, and broken streaks of heavy black in the center of the back and on the flanks; undersurface white, with a broad black band across foreneck and upper breast that is variegated with white in a median longitudinal line, forming a faint stripe; thighs vinaceous buff; tail mixed avellaneous and black. This bird has no wing spur but, like adults, possesses a prominent claw on the pollex.

Attention is called to the fact that the plate used as the frontispiece for volume 2 of Hudson's Birds of La Plata (London, 1920) represents the northern typical form of teru teru with undivided grayish-brown breastband, and not the subspecies that inhabits the pampas.

Family THINOCORIDAE

THINOCORUS RUMICIVORUS RUMICIVORUS Eschscholtz

Thinocorus rumicivorus ESCHSCHOLTZ, Zool. Atlas, pt. 1, 1829, p. 2, pl. 2. (Concepcion Bay, Chile.)

Several subspecies of the small seed snipe have been proposed from various parts of its extensive range; so far as I may perceive from material now at hand those from Chile, Argentina, and Uruguay belong to one form. Rothschild [46] has described *Thinocorus rumicivorus venturii* from Barracas al Sud, Buenos Aires (based on winter migrants from the south, since in eastern South America the species does not breed north of Patagonia). Series including birds from Chile, from near Buenos Aires (Conchitas), Uruguay, and Patagonia (Zapala, Neuquen, and Coy Inlet, Santa Cruz) show considerable individual variation but no differences that may be correlated with range. Should it prove on the basis of more extensive material that an eastern subspecies may be recognized *Tinochorus swainsonii* Lesson [47] named from Buenos Aires [48] must be used for it.

Lowe's *Thincorus peruvianus*,[49] from Islay, Peru, must be considered a synonym of Peale's *Glareola cuneicauda*,[50] named from San

---

[45] See also description of this plumage by Henninger, Auk, 1923, p. 122.
[46] Bull. Brit. Orn. Club, vol. 41, Apr. 13, 1921, p. 111.
[47] Ferussac's Bull. Sci. Nat. Geol., vol. 25, June, 1831, p. 344.
[48] See Lesson, Ill. Zool., Livr. 6, pl. 16, dated June 1, 1831 (according to Mathews. Nov. Zool., vol. 18, 1911, p. 12, published in February, 1833).
[49] Bull. Brit. Orn. Club, vol. 41, Apr. 13, 1921, p. 111.
[50] U. S. Expl. Exp., vol. 8, 1848, p. 244.

Lorenzo near Callao. Peale's type is an immature male, now considerably worn and stained by time. In small size, however, it agrees with Lowe's diagnosis, so that the form found on the Peruvian coastal region must be known as *Thinocorus r. cuneicaudus* (Peale). The type specimen (in the United States National Museum) has the following measurements: Wing, 97.5; tail, 52.3; culmen, 8.5; tarsus, 16 mm.

At Zapala, Neuquen, on December 8 and 9, I encountered the small seed snipe on its breeding grounds on the closely grazed slopes of an open valley in which there was a tiny stream and occasional little seeps or spring holes. As I came suddenly over the top of a high bank above the little rill that drained the valley a half-grown chick, that I recognized instantly as a seed snipe, ran out with wings spread and low piping calls, and after some difficulty I captured it. The mother flushed only a few feet away. Farther on in the valley adults were fairly common and, though they were wild, on the two days mentioned four males were taken.

The area had an alkaline soil that supported scant herbage through which were scattered hillocks a few inches high. Male seed snipe rested quietly on the tops of these, at a distance resembling some curious lark or sparrow. As I approached they ran quickly away or crouched and hid. When flushed suddenly they rose swiftly and darted away in swift zigzags, uttering a low harsh call. The markings of their wings and their appearance at these times bore a striking resemblance to those of a small snipe or sandpiper. Males when at rest occasionally uttered a plaintively whistled *whew* with slightly expanded pulsating throat. To escape pursuit they ran rapidly, with head slightly forward like little plover, and when out of my path crouched with head and neck extended on the ground. When not alarmed they walked slowly, with short steps, frequently with nodding head like a dove. Occasionally males darted off to mount high in air and circle over the valley. On their return they set their wings and came down rapidly, checking their descent every few feet so that they descended in a series of "steps." The performance was accompanied by a curious chuckling double note.

The bill in these birds was usually stained by adherent bits of vegetation on which they had been feeding.

April 25, 1921, near Concon, Chile, about 25 seed snipe were found at the mouth of the Rio Aconcagua, on a sandy area where vegetation was scant and there was much gravel mixed with the soil. A part had scattered over a wide tract, but a dozen or so ranged together and flew in unison. On alighting they spread somewhat in search of food. At intervals males towered and called as in the breeding season.

A young bird half-grown, collected December 8 at Zapala, has most of the original covering of down replaced by the immature plumage, though much down still persists on the head. Adult males in somewhat worn plumage were taken December 8 and 9. One of these had the gonys, culmen, and tip of the bill dull black; remainder of maxilla deep olive buff; remainder of mandible vetiver green; nasal operculum deep neutral gray; iris Rood's brown; tarsus and toes chamois. A male and a female preserved as skins from Concon are in full fall plumage. Specimens were preserved in alcohol and as skeletons.

The curious opercular flap that covers the nostril in the seed snipe is an undoubted protection against wind and sand to a bird that much of the time inhabits regions where protective cover is scant and strong winds carry clouds of dust. It may be remarked that the Pteroptochids *Rhinocrypta* and *Teledromas* that inhabit the same areas have developed a very similar structure.

Order COLUMBIFORMES

Family COLUMBIDAE

LEPTOTILA OCHROPTERA OCHROPTERA (Pelzeln)

Leptoptila ochroptera Pelzeln, Ornith. Brasiliens, pt. 3, January, 1870, p. 278. (Sapitiba, Brazil.)

After comparison of a fair series of pigeons of this species from southern Brazil, Paraguay, and Uruguay, I am able to distinguish the typical subspecies from *chlorauchenia* only by size, as in the series at hand color differences seem variable. Northern birds usually are duller in color on the abdomen than those from the south, a difference that perhaps is constant in fresh specimens, but is subject to variation with age, as younger individuals are duller than those that are older. Specimens also discolor with time, so that those taken 30 or 40 years ago are not comparable with fresh material. Males of *L. o. ochroptera* have the wing 146 to 148.5 mm. (specimens from Kilometer 182, Formosa, and upper Paraguay); females 135.5 to 142.5 mm. (Jaboticabel, São Paulo; Puerto Pinasco, Paraguay; Rio Bermejo, above its mouth, Argentina).

In *Leptotila o. chlorauchenia* the wing in males measures from 153 to 158 mm. (specimens from Rio Grande do Sul, Brazil; Lazcano, Uruguay; Las Palmas, Chaco; Conchitas, Buenos Aires); in females from 150.5 to 153.3 (San Vicente, Uruguay; Las Palmas, Chaco.) Specimens from Las Palmas, Chaco, near the Rio Paraguay, are the southern form. Two from Kilometer 182, in the interior of the Formosan Chaco, have the measurements of the northern form but are intermediate in color, since one is light and the other dark on the ventral surface. Intergradation apparently takes place through this region. The skins listed from above the mouth of the Rio Bermejo, from the old Page collection, are somewhat open to suspicion as to locality.

An immature female from an altitude of 2,000 meters on the slopes of the Sierra San Xavier, above Tafi Viejo, Tucuman, is placed tentatively with *L. o. ochroptera* on the basis of small size. Concerning the validity and relationships of *L. callauchen* and *L. saturata* named by Salvadori[51] from San Lorenzo, in eastern Jujuy, I am uncertain.

These pigeons inhabit rather heavy growths of low woods and, though their habits are somewhat like those of quail doves, are not difficult to secure, as they come out frequently in openings or are seen walking in trails or in sections free of undergrowth. When not alarmed they walk steadily about with nodding heads, or if frightened may remain motionless. When alarmed they flush rapidly, with darting flight, often with a rattle of wings. On infrequent occasions their initial flight is accompanied by a shrill whistling like that made by a woodcock, produced probably by the attenuated tip of the outermost primary. They may dart away to heavy cover, or after a flight of a few yards may perch on some low limb where they are partly screened from view.

Occasionally one bowed low with elevated tail and suddenly flashed the white tips of the rectrices, an action observed more fre-frequently during the breeding season, and one that suggested a similar habit in *Melopelia asiatica*. The call was a low, resonant *who whoo-oo*, a sound similar to that produced by blowing across the opening of a wide-mouthed bottle, and one that suggested the note of *Oreopeleia montana*. I was surprised to find that the female as well as the male gave this curious note. More rarely a bird gave a low *coo-oo*, barely audible at the distance of 10 meters. The white tail tip and reddish-brown undersurface of the wings are prominent in flight.

At the Riacho Pilaga, Formosa, from August 9 to 21, these birds were fairly common; males were taken on August 9 and 18. In very early morning they were found in open roads or trails but later in the day sought the seclusion of the forest. Individuals recorded at Formosa, Formosa, on August 23, may have been of the subspecies *chlorauchenia*.

Near Puerto Pinasco the species was fairly common and was encountered at Kilometer 80, and beyond to Laguna Wall on the westward, from September 1 to 25. An adult male was taken September 3 near the port, and females on September 16 and 23 near Kilome-

[51] Boll. Mus. Zool. Anat. Comp. Univ. Torino, vol. 12, no. 292, May 12, 1897, p. 33.

ter 80. On September 30 I found the species common on the heavily forested hill known as the Cerro Lorito, on the eastern bank of the Paraguay River.

Notes on *Leptotila* made in Tucuman are placed here with reservation. At Taipa from April 7 to 13, 1921, the birds ranged in fair numbers through the dry forests. At this season they were silent and their presence was unsuspected save when they chanced to flush with a rattle of wings. Above Tafi Viejo on April 17, occasional wood pigeons flushed near the winding trail that traversed the lower slopes of the Cumbre San Xavier. Between 2,000 and 2,500 meters 15 or 20 were scattered through a small, rather open grove of tree alders, where they flushed from the ground with a rattle of wings, and flew up to concealed perches among the yellowed leaves that still clung to the branches, or when driven from the shelter of the grove darted swiftly down the steep slopes to more secure cover in the denser forest below. Many of them, like a female that I killed, were in dark brown immature plumage. In the immature bird secured, though the outer primary is narrow, the incision at the tip is much less pronounced than in adults, as the extremity of the feather measures 5 mm. in width. In adult birds it is barely more than 2 mm. at this point.

LEPTOTILA OCHROPTERA CHLORAUCHENIA Giglioli and Salvadori

Leptoptila chlorauchenia GIGLIOLI and SALVADORI, Atti. Roy. Acad. Scienz. Torino, vol. 5, pt. 2, 1870, p. 274. (Estancia Trinidad, Montevideo, Uruguay.)

Names for the southern wood pigeon are in confusion and the usage followed, while that of custom, is considered tentative. *Leptoptila ochroptera* was published in the third part of Pelzeln's Ornithologie Brasiliens, dated (on the original cover) 1870. As this part of Pelzeln's work is mentioned in the abstract of the meeting of the Deutsche Ornithologische Gesellschaft of Berlin for February 1, 1870,[52] we may assume that it appeared in January of that year. *Leptoptila chalcauchenia* was proposed by Sclater and Salvin before a meeting of the Zoological Society of London for December 9, 1869, and appeared in the last part of the Proceedings of that organization for 1869, which, if the usual custom was followed, was printed in March or April, 1870. *Leptoptila chlorauchenia* Giglioli and Salvadori read before a meeting in Turin on January 2, 1870, was published in the issue of Atti Royale Accademie Scienze for January, 1870, a number that includes a Summary of meetings from January 2 to January 30, so that it probably appeared in February or later. It was also published in the Ibis for April, 1870 (p. 186). From

---

[52] See Journ. für Ornith., 1870, p. 153.

available information it is not apparent whether *chalcauchenia* or *chlorauchenia* has priority, and it is not impossible that one of them may antedate *ochroptera*. In a recent number of El Hornero (vol. 3, 1923, p. 200), Peters has stated that *chalcauchenia* is valid which may well be true but requires further substantiation. When sufficient material is available for a review of these pigeons it seems probable that what is here considered as the species *ochroptera* may be merged as a part of the wide ranging *Leptotila verreauxi*.

The southern wood pigeon was fairly common in the southern portion of the Chaco and, wherever thickets offered shelter, among the hills or near the streams of southern Uruguay. On the pampas it was found in small numbers in groves, and though driven out in some areas where the tala forests have been destroyed has spread in other sections where groves have been planted on estancias to furnish shelter to stock from the severity of storms. The habits of this form are similar to those detailed under *L. o. ochroptera*.

At Las Palmas, Chaco, from July 16 to 30, 1920, the species was common and specimens were taken July 16 and 17. This was the beginning of the breeding season and birds taken were sexually active. They called constantly, and were especially noisy during dull, rainy weather when they sat huddled on low perches calling in resonant tones at frequent intervals. An adult female had the bill black; bare skin about eye gray number 7; an irregular semilunar mark before and behind eye acajou red; tarsus and toes acajou red; claws black.

Two pairs inhabited the tala woods at the Estancia Los Yngleses, near Lavalle, Buenos Aires, as was recorded on November 8. Near San Vicente, Uruguay, from January 25 to 31, 1921, the birds were fairly common on the brush-covered slopes of the hill known as the Cerro Navarro. They called frequently and an adult female taken January 25 was about to lay. At Lazcano, Uruguay, from February 5 to 8, the species was found in the heavy woods along the Rio Cebollati where a male was taken February 6. At Rio Negro, Uruguay, the birds were recorded from February 14 to 19.

Leptotila ochroptera has a diastataxic wing, thus agreeing with . what Miller[53] has recorded in *L. verreauxi*.

METRIOPELIA MELANOPTERA MELANOPTERA (Molina)

Columba Melanoptera MOLINA, Sagg. Stor. Nat. Chili, 1782, p. 236. (Chile.)

Specimens secured in Mendoza and Neuquen agree with two from Santiago, Chile, in the National Museum, which are taken as representative of the typical form. J. L. Peters has called my attention to a note in the Journal für Ornithologie, 1913 (p. 401), where

---

[53] Bull. Amer. Mus. Nat. Hist., vol. 34, 1915, p. 130.

Reichenow states that *Columba melanoptera* of Molina should be placed in the genus *Zenaida*. Since *Zenaida zenaida* has 14 rectrices and a diastataxic wing,[54] while in *Metriopelia melanoptera* I find 12 rectrices and a eutaxic wing (verified in two specimens), Reichenow's action has no basis other than that of superficial resemblance. At Zapala, Neuquen, on December 7, 1920, I flushed a male in open brush on the slope of a hill, and killed it as it darted away. Near Potrerillos, Mendoza, a male was taken March 15, 1921, two females on March 16, and a fourth specimen on March 17.

The birds were found, rather rarely, on arid hill slopes, grown with open brush, or on the gravelly flood plains of small streams. It is seldom that they are seen on the ground, as there they are concealed by the rocks and brush, among which they walk with nodding heads; they become motionless at any alarm. They flush swiftly with a peculiar, almost metallic, rattle of the wings, that resembles exactly the winnowing whistle of a blackbird's flight when part of the primaries are missing in molt. They climb for a few feet in rising and then dart swiftly away. The black under wing surface is prominent in flight. On March 19 several were seen at El Salto at an altitude of 1,800 meters, while on March 24 one flushed in a dry wash on the flats near the base of the foothills 25 kilometers west of Tunuyan. The bird is known as *paloma de la sierra*.

An adult male taken December 7 had the bill black; bare skin before eye salmon color; iris yellowish glaucous; tarsus and toes black. Another shot March 15 had the bill dull black; cere deep neutral gray; iris light dull glaucous blue; eyelids light Payne's gray; margin of lower lid, anterior canthus, and space before eye, extending as a crescent below the lower lid slightly brighter than salmon color; tarsus and toes dark quaker drab; nails black. Females shot March 16 did not differ from the one last described.

COLUMBINA PICUI (Temminck)

Columba Picui TEMMINCK, Hist. Nat. Gen. Pig. Gall., vol. 1, 1913, pp. 435, 498. (Paraguay.)

Material of this pigeon at hand includes specimens from southern Brazil, Paraguay, central Argentina, Mendoza, Chile, and intermediate localities, in which I can find no differences that warrant subdivision of the species. The status of a race in northeastern Brazil seems somewhat uncertain, so that I have not attempted to use a trinomial for my specimens.

This small pigeon was widely distributed throughout the region that I visited and was recorded at many points. The species is social, and where food is abundant decidedly gregarious. Two or three to

---

[54] Miller, Bull. Amer. Mus. Nat. Hist., vol. 34, 1915, p. 130.

half a dozen were nearly always found in company, while bands of 25 or 30 were not unusual save when the birds were breeding. They frequented the borders of thickets, weed patches, open prairies, or old fields seldom far from trees that might give them shelter. Where high grass covered the open savannas the birds gathered in any little open space at the border of forest, while recent burns or plowed fields were always attractive to them. At the Riacho Pilaga, Formosa, a little band came daily to feed on sorghum heads stored under a porch at the ranch. The species is one that comes frequently about houses and that may be found in the plazas of the larger cities. I found it even in the Plaza San Martin, in the heart of Buenos Aires, so that it was almost a surprise to encounter the birds in comparative abundance in the wilder sections of the Chaco far from human habitation. In March, near Tunuyan, Mendoza, extensive fields of hemp called the birds in abundance. As the hemp was grown solely for the fiber that it produced, the seed was allowed to ripen thoroughly and shelled out to lie on the ground. Doves gathered literally in hundreds, especially where lines of willows bordered the fields to offer resting places. The birds flush quickly with a darting flight and fly rapidly, but as they are small in size are not hunted, save by the pothunters, who kill all small birds. In the air the black and white in the wing flash alternately, while there is an additional line of white visible in the tail. In habits and general appearance the birds are suggestive of ground doves.

Males began their monotonous cooing calls in October. On December 17 a female, shot near Carhue, Buenos Aires, contained an egg with the shell partly formed. Near Montevideo, Uruguay, on January 16, 1921, two young, not more than three-fourths grown but strong in flight, were seen. A nest recorded January 29, near San Vicente, Uruguay, was placed at the border of a little thicket, more than 2 meters from the ground, where two small limbs crossed and offered firm support. The nest, a slight platform of grass and weed stems, contained two eggs nearly ready to hatch. A second nest found February 19 near Rio Negro, Uruguay, at the border of a thicket was placed in a shrub among thickly laced branches more than 2 meters from the ground. The nest, a slight structure of grass and fine twigs with a few feathers from the bodies of the owners, contained two slightly incubated eggs that are white in color and rather dull in texture. These measure, respectively, 23.6 by 17.7 and 23.1 by 18.2 mm. To my surprise, when I collected the bird that was incubating, it proved to be the male. Males were calling and were still in breeding condition at Tunuyan, Mendoza, on March 24. At Tapia, Tucuman, in the second week of April young birds, fully grown, were common and the breeding season appeared to be at an end. With allowance for differences in climatic conditions it appears from these notes that two broods are reared each season, while in some localities three may be produced.

These pigeons were recorded at the following points: Buenos Aires city, June, October, and January, in the Plaza San Martin; Resistencia and Las Palmas, Chaco, July, 1920; Riacho Pilaga, Formosa, August 7 to 21 (male taken August 7); Formosa, Formosa, August 23 and 24; Puerto Pinasco, Paraguay, from the Rio Paraguay west for 200 kilometers, September 3 to 25 (male taken at Kilometer 80, September 17); Lavalle, Buenos Aires, November 2, 9, and 13; Bahia Blanca, Buenos Aires, December 14 (one in the main plaza); Carhue, Buenos Aires, December 15 to 18 (female shot December 17); Victorica, Pampa, December 23 to 29; Carrasco, near Montevideo, Uruguay, January 9 and 16, 1921; La Paloma, Uruguay, January 23; San Vicente, Uruguay, January 25 to February 2; Lazcano, Uruguay, February 3 to 8; Rio Negro, Uruguay, February 14 to 19 (male and set of eggs taken February 19); Guamini, Buenos Aires, March 6 and 8; Mendoza, Mendoza, March 13; Potrerillos, Mendoza, March 17 and 19; Tunuyan, Mendoza, March 22 to 28 (three males taken March 24); Tapia, Tucuman, April 7 to 13.

A male taken September 17 had the maxilla and tip of mandible dusky neutral gray; base of mandible olive buff; iris Payne's gray, with paler margin; bare skin about eye storm gray; tarsus and toes dull Indian purple; claws blackish.

ZENAIDA AURICULATA AURICULATA (Des Murs)

Peristera auriculata DES MURS in Gay, Hist. Fis. Pol., Chile, Zool., vol. 1, 1847, p. 381. (Central provinces of Chile.)

Study of a considerable series of *Zenaida auriculata* shows the validity of the well-marked form described by Bangs and Noble[55] from Huancabamba, Peru, as *Z. a. pallens* (marked by smaller size and darker posterior underparts), but does not reveal other forms that may be recognized at present. There is a tendency for birds from Buenos Aires and Mendoza southward to be larger than specimens from Uruguay and Paraguay northward to Colombia, while specimens from Neuquen and Mendoza, at the base of the Andes, are slightly grayer as well as larger. The species is one in which there is considerable change in color as specimens age, while measurements at times seem contradictory to the statements made above. Chilian examples at hand are insufficient to establish the characters of the typical form, so that I do not care to attempt any further subdivision. It may be remarked that specimens of *pallens* from Pisac, Chospiyoc, the Tomba Valley, and Lima, Peru, are browner above than true *auriculata*, rather than paler and grayer as stated in the original description. Otherwise the diagnosis is correct in

---

[55] Auk, 1918, p. 446.

pointing out the brighter, pinker coloration of the breast and the paler, buffier tint of the posterior underparts.

The Zenaida dove has a wide distribution in open or semi-open country in southern South America and, like the mourning dove of the United States, is one of the species of birds seen constantly during travel, whether by train or other conveyance. It was present in the Chaco in winter, but was less abundant than in drier, more open country farther south. At the Riacho Pilaga, Formosa, on August 19, 1920, I noted these doves in close flocks, containing from a dozen to twenty-five individuals, that fed in open fields with other pigeons. Two were seen near the town of Formosa August 23. At Puerto Pinasco, Paraguay, one was observed on September 1 at Kilometer 25 in company with a flock of blackbirds, *Gnorimopsar chopi*. Elsewhere in this region the species was observed on September 24 and 25 at Laguna Wall (Kilometer 200). The Zenaida dove was abundant on the pampa in the vicinity of Lavalle, Buenos Aires, from October 22 to November 13, and was found in equal numbers at General Roca, Rio Negro, from November 23 to December 3, and near Zapala, Neuquen, from December 7 to 9. Adult females were taken at Lavalle October 25, at General Roca December 2 (preserved as a skeleton), and at Zapala December 7. At Carhue, Buenos Aires, several were observed December 17, and an adult male was taken. The species was common at Victorica, Pampa, from December 23 to 29, often in yards in town. It was noted near Carrasco, Uruguay, January 9 and 16, and was abundant at La Paloma January 23. It was noted in numbers at San Vicente January 25 to February 2 (a female taken January 28), Lazcano, February 3 to 8, and Rio Negro, February 14 to 19. One was seen at Guamini, Buenos Aires, March 4, and occasional birds were reported at Potrerillos, Mendoza, March 17, 19, and 21. On the plains near Tunuyan the birds were gathered in considerable flocks from March 22 to 29. Near Tapia, Tucuman, they were observed in small numbers from April 7 to 13, and others were recorded at Concon, Chile, April 26 to 28 (a female taken April 26).

Like the Zenaida dove of the West Indies, *Z. auriculata* suggests in habits and even in appearance (save for shorter tail) the equally abundant mourning dove (*Zenaidura macroura*) of our northern continent. The birds feed on the ground in pairs or flocks, often in plowed fields where seeds are abundant, but more frequently in pastures where there is cover of grass or weeds. In the more arid sections they walk about in shelter of open scrubby bushes with the quick, short steps and nodding heads usual in pigeons. When approached they suddenly stop motionless and then flush with loudly clapping wings that become silent as the birds dart away in swift, direct flight. On cool mornings single birds or little bands gathered in the top of some open tree, frequently an ombú, to rest in the sun. They were common in groves of tala or ombú trees in the eastern pampas, and flushed constantly with clapping wings as I passed near them. Others fed along the walks or in the extensive open grounds surrounding estancia houses.

From October to February birds were found in pairs, though flocks gathered to feed in suitable localities. Males called at this season, giving a low, sad-toned *whoo whoo whoo whoo-oo* in a guttural tone with little carrying power or volume of sound. When about to coo the upper part of the throat was expanded with air. Males often sailed with set wings in short circles above the trees, with the throat distended with air, so that with their short tails they presented an odd, rounded appearance—display flights that ended by a descent to some perch. As flocks of birds flew past me individuals suddenly darted sideways to produce a rattling sound with their wings. A nest found February 2, 1921, near San Vicente, Uruguay, was composed of a few weed stems and twigs placed on a foundation of an old nest of some oscinine, in the fork of a small tree more than 2 meters from the ground. It contained two eggs too hard set to save. A second nest, built like the first on fragments of the old nest of some perching bird, was a slight mat of fine twigs in the limbs of a small tree 2 meters from the earth. A third, recorded February 14 at Rio Negro, Uruguay, placed more than 2 meters from the ground amid dense limbs of a shrub, was a slightly cupped platform of twigs so loose in construction that the eggs were visible through it from below. All nests examined contained eggs so hard set that they could not be blown. February 6 to 8, while working near Lazcano, I found young three-quarters grown, able to fly, in thickets growing in sandy soil near the Rio Cebollati.

After the breeding season Zenaida doves congregated wherever food was abundant and frequently forsook areas where they had been common earlier in the year. Grain or hemp fields were especially attractive, as were tracts where certain large-seeded weeds were common. The species is hunted for game, but no more so than other small birds. The flesh is similar to that of other wild pigeons.

PICAZUROS PICAZURO PICAZURO (Temminck)

Columba Picazura TEMMINCK, Hist. Nat. Pig. Gall., vol. 1, 1813, pp. 111, 449. (Paraguay.)

At the Riacho Pilaga, Formosa, this large pigeon was common during my stay in August, 1920, and two adult males were killed on August 16. The birds were gregarious while feeding or resting, but in flight across country were found alone or in bands of small size. They frequented the open savannas where recent burns attracted them, and also congregated on occasional newly plowed fields. When one or two drop in to some feeding ground, others that chance to pass decoy to them until in a comparatively short time from 50 to 100 may· be found in company. On the ground they walk quickly with characteristic jerking, nervous, pigeon gait, pecking at any food that may offer. When satisfied they leave, a few at a time, to fly to water to drink, and then come stringing back singly or in little groups. One or two may alight in low trees and all that pass that way are sure to join them. Some of the birds rest on open branches and others among leaves, where in spite of their large size they are well concealed. They usually gather in close proximity where they may catch the warmth of the sun.

They flush with loudly clapping wings, but the ensuing flight is noiseless. In the air their wings appear broad and heavy, and only slightly pointed, so that with their short tails they have a heavy, fore-shortened appearance. The flight was direct but only moderately swift. The birds roosted somewhere to the south of the ranch where I was stopping and during the day were observed frequently, in early morning as they came from their roost, and later as they passed to feeding grounds or watering places. They are heavy in body and furnish an abundance of excellent meat. At intervals I heard males give an odd call *kōh kŭh kwaoh*, given rapidly four or five times, and then, after a brief rest, repeated once more.

Others were seen at Formosa, Formosa, on August 23 and 24 (perhaps of the following form), and at Kilometer 110, west of Puerto Pinasco, I saw a number on September 23.

In Guarani they were known as *picazuró*.

A male killed August 16 had the bill dawn gray with a wash of fuscous at the tip; cere neutral gray obscured by a whitish, flaky encrustation; bare skin about eye acajou red, save for the lids which are Hathi gray, with the margin about the eye acajou red; iris flesh ocher; scutes on tarsus and toes acajou red; nails fuscous.

Since no one appears to have examined specimens of this pigeon from Paraguay, the type-locality, as Temminck's description is based on that of Azara, there is a reasonable doubt as to whether birds from that region belong to the pale northern form, as was assumed by Hartert when he described the dark southern bird as a distinct subspecies. Should it prove true that it belongs to the southern form it will be necessary to describe the northern bird. In the present case I find that an adult male from the interior of Formosa is paler than others from central Argentina and Uruguay. Hence I identify it as the typical form.

PICAZUROS PICAZURO REICHENBACHI (Bonaparte)

Crossophthalmus reichenbachi BONAPARTE, Consp. Gen. Av., vol. 2, 1857, p. 55. (Patagonia.)

When Doctor Hartert[56] pointed out that the picazuro pigeon from northern Argentina differed from that from more northern regions and named it *Columba picazuro venturiana*, he overlooked the fact that another name was available for it, probably because the name here used, that of Bonaparte, had been placed by Salvadori[57] in the synonymy of *Notioenas maculosa*. *Crossophthalmus reichenbachi* was described by Bonaparte from an adult from Patagonia and a juvenile (at least so characterized) from Paraguay. On scanning the original reference it will be seen that the adult refers to *Picazuros picazuro* and the supposed young bird to *Notioenas maculosa*, since the former is said only to have the wing coverts margined with white, while the latter has the feathers of the back and the superior wing coverts terminally spotted with white. As the adult of this composite is *picazuro* the name should be restricted to that bird. It will apply to the form that has been described as *venturiana* and must supplant that name. The type-locality given as Patagonia may be erroneous, but may perhaps be determined by examination of Orbigny's original specimen if still extant.

An adult male of the dark southern form of the picazuro pigeon was taken at San Vicente, Uruguay, on January 26, 1921. Since a specimen in the collection of the United States National Museum from Corrientes is also representative of the dark southern bird, the question of the identity of the bird from Paraguay, the type-locality of the typical form, naturally arises, as Corrientes is not far from the Paraguayan border. The fact that a bird from the Formosan Chaco is pale does not necessarily indicate that one from east of the Paraguay River in the same latitude would be the same, since the specimen in question came from the interior of the Chaco in a region where pale forms, similar to those from the Paraguayan Chaco, occur. In other words, it is not improbable that a pale bird may range in the Paraguayan Chaco and that a dark one may occupy eastern Paraguay.

Several of these pigeons were recorded at the Estancia Los Yngleses, near Lavalle, Buenos Aires, October 23, 1920, and one was seen on November 9. In southern Uruguay I found them common. One was seen at La Paloma, January 23, while near San Vicente, from January 25 to February 2, the birds were common and were breeding. They were most common in the extensive *palmares*, where forests of palms covered broad marshy areas in the lowlands. The pigeons rested in the tops of the palms and flew out with loudly

[56] Nov. Zool., vol. 16, December, 1909, p. 260.
[57] Cat. Birds Brit. Mus., vol. 21, 1893, p. 273.

clapping wings as I passed, or if at a reasonable distance watched me alertly with jerking heads. The note of a wounded individual was a curious, low, growling call. The display flight of the males at this season was interesting. The birds sailed out with the wings broadly extended, elevated at an angle above the back, and thrown slightly forward so that there was a space between the tips of the tertials and the sides. In this manner the pigeon described a grace-ful curve to another perch or returned to the one that it had left. The action was similar to that of the domestic pigeon under similar cir-cumstances. A nest, found January 29, was placed at the border of a little thicket on a small horizontal limb a little more than 2 meters above the ground. This was a slight platform of grass, weed stems, and little twigs, irregular in outline, and from 60 to 70 mm. in diameter. An adult pigeon flushed from the nest, and in it rested a young bird about half grown, with contour feathers partly cover-ing the body. As usual in pigeons of this group, the incoming con-tour feathers are deeper in color, and more reddish than in the adult. Hairlike filaments of down that still adhere to head and breast are chamois color.

At Lazcano, farther north in the Department of Rocha, from Feb-ruary 5 to 8, these pigeons were found in flocks of half a dozen that fed in weed patches or rested in the shade of coronillo trees in open pastures, or in dense thickets near the Rio Cebollati. A few were recorded at Rio Negro, Uruguay, on February 15 and 19.

NOTIOENAS MACULOSA MACULOSA (Temminck)

Columba maculosa TEMMINCK, Hist. Nat. Pig. Gall., vol. 1, 1813, pp. 113, 450. (Paraguay.)

At Lazcano, Uruguay, the spotted-winged pigeon was found from February 7 to 9, 1921, feeding in little flocks in weed patches, often in company with *Picazuros p. picazuro*. The mottled shoulder of the present species shows plainly in favorable light, and, with slightly smaller size, is sufficient to distinguish the bird from its rela-tive, which it resembles closely in form and habit. Two adult males were taken February 7. One of these has the following measure-ments: Wing, 203.5; tail, 97.5; culmen from cere, 11; tarsus, 27.8 mm.

NOTIOENAS MACULOSA FALLAX (Schlegel)

Chloroenas fallax SCHLEGEL, Mus. Hist. Nat. Pays-Bas, vol. 4, 1873, p. 80. (Rio Negro, Patagonia.)

The spotted-winged pigeon from central and western Argentina is darker on head, neck, upper back, and ventral surface, and is larger than what is assumed to be typical *maculosa* from the Río Bermejo, Argentina, and Lazcano, Uruguay. *Chloroenas fallax* as originally described by Schlegel was based on a specimen said to have come from Mexico, and a female shot in June, 1871, in Rio Negro. Since the description is that of *Notioenas maculosa* the type should be restricted to the second specimen as is done here, and the specimen said to have come from Mexico if (as seems from what Schlegel has written) this species, considered as bearing an erroneous locality.

The present subspecies was recorded December 28, 1920, at Victorica, Pampa, in fair numbers, and an adult male was taken. The birds at this season were in pairs that ranged through the open monte. During the heat of the day they rested on the broad limbs of caldén or algarroba trees where they found comfortable perches well shaded from the intense rays of the sun. At evening they came to drink from a water hole near town. When approached they flushed with loudly clapping wings and darted swiftly away. As I passed a tree containing a large hollow, I heard a strange growling call in a low tone that I attributed to the young of some monte cat. On investigation I found nothing in the hollow, though I noted that a pair of spotted-winged pigeons flushed from the tree, a circumstance to which I paid no attention until later when a wounded pigeon in my hand uttered the same queer call.

The male secured, when first killed, had the bill dull black; iris slightly darker than pearl gray; tarsus and toes neutral red; claws black.

Order CUCULIFORMES

Family CUCULIDAE

CROTOPHAGA ANI Linnaeus

Crotophaga ani LINNAEUS, Syst. Nat., ed. 10, vol. 1, 1758, p. 105. (Eastern Brazil.[58])

At Las Palmas, Chaco, the ani was fairly common and a male was taken on July 19, 1920. Others were recorded July 22 and 24. Several were seen at Formosa, Formosa, August 24, and on September 30 a male and a female were shot from a flock of a dozen on the Paraguay River at Puerto Pinasco, Paraguay. In Guarani the species is known as *ano-i* or little *ano*, so that the name ani is perhaps a contraction of this term.

The three taken seem to offer no tangible differences from specimens from the northern range of the species.

GUIRA GUIRA (Gmelin)

Cuculus guira GMELIN, Syst. Nat., vol. 1, pt. 1, 1788, p. 414. (Brazil.)

In a small series of specimens from Paraguay, Uruguay, and the following Provinces and Territories of Argentina, namely, Buenos

---

[58] See Berlepsch and Hartert, Nov. Zool., vol. 9, April, 1902, p. 98, and Hellmayr Nov. Zool., vol. 12, September, 1905, p. 299.

Aires, Mendoza, Santa Fe, and Chaco, no definite variation is apparent save that specimens from Mendoza appear somewhat paler and whiter below. Birds of this species easily become soiled and most of the skins from eastern localities are considerably discolored by dirt so that on careful examination the lighter color of skins from semiarid Mendoza seems more apparent than real. Cory has stated that specimens from Ceara average smaller than those from the south, but a single skin from Marajo is as large as those from Buenos Aires.

Though a bird of open country, the guira, like the ani, prefers regions of open savannas diversified with thickets and groves so that, though fairly common on the pampas, it was most abundant in the partly wooded areas farther north. In southern Uruguay and central Argentina the species is called *urraca*, signifying properly a magpie, but in the north where jays occur the term *urraca* is applied to them, and the guira is known as *pirincho*. It is interesting to note that *guira*, in the Guaraní tongue signifies bird as a group designation.

Guiras are social and range in pairs or flocks that frequently number 20 individuals. The birds feed on the ground, usually with one member of the band perched as guard where it may survey the country. Open pastures or savannas are frequented and the birds are attracted by recent burns in grasslands. As they alight they throw the tail up and the head down, and then walk or run rapidly with long legs fully extended. At any alarm they utter a curious rattling call that may be represented as *kee-ee-ee-ee*, from which they derive their Guaraní name of *piriri* (meaning a crackling noise, as the crackle of a brush fire, or the noise produced in walking through dry weeds), and then fly off in a loose, stringing flock to alight in company on a tree, post, or fence. The flight is slow and weak, accomplished by a brief beating of the wings, followed by a short sail in which the wings are held stiffly extended. On the wing, head and tail are held at a slightly higher level than the back. The long tail is held at various angles when the birds are at rest, while the wings may be drooped and the crest lowered or raised. Curiosity, interest, or indifference are expressed in constantly changing attitudes, many of which are bizarre and unusual so that the long, slenderly formed guiras are always of interest. On cool mornings flocks rest in the rays of the sun with hanging tail, drooping wings, and fluffed out feathers, a sight so frequent that *in el sol como un pirincho* is a common saying for any one who basks in the sun's rays on a cold morning.

The song of the guira, if such it may be called, is a series of discordant notes of great carrying power, exactly like the noise produced by blowing on a grass blade held taut between the thumbs;

in fact, for some time when I heard this sound I mistook it for some child in noise-making play and wondered mildly that the method of producing the sound, common in the United States, should be known to the youth of the south. The food of this species is mainly insects, and the species is a valuable aid to agriculture in its destruction of injurious grasshoppers. On one occasion I saw one with a cicada in its bill. The birds are considered excellent for domestication since they are said to rid houses of all of the creeping and running insects that pester man, while it was rumored, probably without basis in fact, that they might learn to imitate a few words of human speech. They have a parrotlike habit of searching with the bill through the plumage of companions, perhaps for parasites. Their feathers are long and not very abundant, while the skin is thick and strong. The body exhales a strong, pungent odor, similar to that of the ani, and cuckoos of the genus *Coccyzus*, to me a disagreeable smell that if endured for any length of time produces headache.

Guira guira was definitely recorded as follows: Santa Fe, Santa Fe, July 4, 1920; Resistencia, Chaco, July 9; Las Palmas, Chaco, July 13 to August 1 (a male taken July 15); Riacho Pilaga, Formosa, August 8, 15, and 20; Formosa, Formosa, August 23 and 24; Puerto Pinasco, Paraguay (from the river west to Kilometer 80, September 1 to 30 (a female taken at Kilometer 80, September 16); Dolores, Buenos Aires, October 21; Lavalle, Buenos Aires, October 27 to November 13; Carhue, Buenos Aires, December 17 (an adult female shot); Victorica, Pampa, December 23 and 27; Carrasco, Uruguay, January 9 and 16, 1921; La Paloma, Uruguay, January 23; San Vicente, Uruguay, January 25 to February 2 (a male shot January 27); Lazcano, Uruguay, February 3 to 9 (one taken February 6); Guamini, Buenos Aires, March 3 and 4; Tunuyan, Mendoza, March 24 and 29; Tapia, Tucuman, April 6 to 13.

In the museum of the University of Kansas are two specimens taken near Bahia Blanca, Buenos Aires, by H. T. Martin and S. A. Adams, one in November and one on December 10, 1903. This is about the southern limit of the species from information at present available.

A male taken July 15 had the tip of the bill varying from apricot orange on the culmen to salmon orange on the mandible; base of bill and bare skin on side of head reed yellow; iris cadmium orange; tarsus and toes dark olive gray, becoming olive gray at margins of scutes.

TAPERA NAEVIA CHOCHI (Vieillot)

Coccyzus chochi VIEILLOT, Nouv. Dict. Hist. Nat., vol. 8, 1817, p. 272. (Paraguay.)

On April 8, 1921, near Tapia, Tucuman, one of these birds flushed from the ground at the border of a thicket and alighted on a low branch among leaves, where it peered out with extended neck, raised crest and slowly vibrating tail. It proved to be a juvenile female barely grown. The culmen and base of the maxilla were blackish mouse gray; remainder of maxilla buffy brown; mandible tea green; iris smoke gray; tarsus and toes vetiver green, shaded on side of tarsus with castor gray.

It is recorded by Hartert and Venturi [59] that this strange cuckoo foists its domestic cares on certain smaller birds, notably on species of *Synallaxis*. Dr. H. von Ihering [60] reports a young bird secured from the rounded stick nest of *Synallaxis spixi*, while in another nest of this same species he secured an incubated egg, larger and duller in color than others in the set, that contained an unmistakable embryo of *Tapera*. Fonseca [61] also records it as parasitic on *Synallaxis spixi*, and says that as the cuckoo is too large to enter the globular inclosed nest of its host it tears a hole in one side to give access to the nest cavity where it deposits its egg.

The specimen from Tapia is in juvenal plumage, with spotted crown and barred throat. The usage of Bangs and Penard [62] has been followed in recognizing a large southern form of *Tapera*, though I find specimens from Venezuela as large as those from the south. Individual variation in color in this species, with condition of plumage is extensive.

MICROCOCCYX CINEREUS (Vieillot)

Coccyzus cinereus VIEILLOT, Nouv. Dict. Hist. Nat., vol. 8, 1817, p. 272. (Paraguay.)

The first of these cuckoos was recorded at Victorica, Pampa, on December 23, 1920, when an adult female was shot as it rested in the sun in the top of a tree. The note of this individual was a sonorous *cow-w cow-w cow cow*, in tone like the call of the yellow-billed cuckoo but without the rattling, clucking termination usual in the song of that bird. At Tapia, Tucuman, on April 7 and 8, several were seen and three taken, including an adult pair and a juvenile female not fully grown. They were found in rather dense, dry scrub in a region of barrancas. The birds were alert but silent

---

[59] Nov. Zool., vol. 16, 1909, p. 230.
[60] Rev. Mus. Paulista, vol. 9, 1914, pp. 391–395.
[61] Rev. Mus. Paulista, vol. 13, 1923, pp. 785–787.
[62] Bull. Mus. Comp. Zoöl., vol. 62, April, 1918, p. 50.

and worked rapidly through the branches of the trees. Their flight was direct, and seldom high above the earth. The small wings and long tail give them the appearance in the air usual in allied species. The young bird taken, barely from the nest, had the inside of the mouth ornamented with a series of tubercles dead white in color that outlined the throat cavity in a startling manner when the mouth was opened. Four lay at the angles of the pharynx, one ornamented the tongue, and five somewhat less prominent were found on the palate. Against a duller background their pure white color stood out prominently. Such ornaments, visible only when the mouth is fully opened must serve as directive markings to assist the adults in placing food properly, when feeding the young.

The young bird, not yet fully fledged, has the throat and breast grayish white and the wings, tail, and dorsal surface in general with a faint rufescent wash that is absent in the adults. The adult male, taken April 8, had renewed all of the flight feathers save the tenth primary. The adult female secured the day previous has renewed all save the third and sixth primaries and some of the secondaries.

The female, taken December 23, had the bill black; bare skin around eye hydrangea red; iris pomegranate purple; tarsus and toes deep neutral gray; claws black.

COCCYZUS MELACORYPHUS (Vieillot)

Coccyzus melacoryphus VIEILLOT, Nouv. Dict. Hist. Nat., vol. 8, 1817, p. 271. (Paraguay.)

The present cuckoo is a thicket-haunting species found in low scrub, so shy that it is difficult to observe. Near Rio Negro, Uruguay, from February 14 to 19, 1920, I found these birds on their breeding grounds in heavy brush that bordered streams. Here they were observed searching for food among leafy branches from 3 to 6 meters from the ground, or were heard calling, a low, guttural *cuh-h-h*, audible only a short distance away. They were shy and nervous and sought safer cover whenever they found that they were seen. On February 15 one was seen carrying food to young, while on February 18 two in juvenal plumage, fully grown, were taken. Adult males taken February 15 and 19 and females shot on February 14 and 18 are in full breeding plumage. The two young (male and female), secured February 18, are slightly duller brown above than adults and have the light markings on the tail diffuse and not sharply delimited. The young apparently molt the retrices before the next breeding season.

At Tapia, Tucuman, two males were taken in the thickets frequented by *Micrococcyx cinereus* on April 7 and 12. These, both adult, are in process of molt on head, neck, and breast and are renewing the quill feathers. In one molt of the primaries has begun with renewal of the fifth, ninth, and tenth in one wing and the sixth and eighth in the other. The second specimen has molted the second primary in either wing.

The species does not seem to have been recorded previously from Uruguay.

COCCYZUS AMERICANUS AMERICANUS (Linnaeus)

Cuculus americanus LINNAEUS, Syst. Nat., ed. 10, vol. 1, 1758, p. 111. (Carolina.)

On February 15, 1921, an adult female was taken near Rio Negro, Uruguay, as it worked slowly and silently through willows in an open thicket along a small stream, in the same type of country as that frequented by the black-billed *C. melacoryphus*. The specimen is completing the molt and has the last of the new contour feathers appearing on head and breast. The fifth, eighth, and tenth primaries are not yet fully developed, though all of the old feathers in the wing (and also in the tail) have been cast. As this bird is in molt, measurements of wing and tail are not reliable. The culmen equals 26.2 mm.; the tarsus, 26.5 mm. The coloration of the dorsal surface is the same as that in early spring migrants of *C. a. americanus* when they arrive in the United States, and the inner webs of the primaries are distinctly rufescent, not buffy as in *C. a julieni* (Lawrence). The yellow-billed cuckoo has not been recorded previously from Uruguay, but the present record is not surprising, as the bird has been noted from Buenos Aires.

Order PSITTACIFORMES

Family PSITTACIDAE

AMAZONA AESTIVA (Linnaeus)

Psittacus aestivus LINNAEUS, Syst. Nat., ed. 10, vol. 1, 1758, p. 101. (Southern Brazil.)[63]

An adult male taken at Kilometer 80, Puerto Pinasco, Paraguay, has the shoulder partly red and partly yellow. As comparative material is not available the subspecific identity of this specimen is uncertain though on geographic grounds it may be supposed to represent *A. a. xanthopteryx* of Berlepsch.[64]

At Kilometer 80, west of Puerto Pinasco, Paraguay, these amazons were common from September 7 to 21, and a male was taken September 7. On September 23 a few were observed farther west at Kilometer 110. These parrots, called *pilh' pul'* by the Angueté Indians, were found in pairs, usually in groves of palms where they

---

[63] See Hellmayr, Abh. Kön. Bayer. Akad. Wiss., Klass. II, vol. 22, 1906, p. 593.
[64] *Chrysotis aestiva xanthopteryx* Berlepsch, Ornith. Monatsb., vol. 4, 1896, p. 173. (Bueyes. Bolivia.)

fed on the palm seeds. It was common to see them in flight across country, their passage announced by-noisy calls that at a distance have a ludicrous resemblance to calls for *help help*. In early morning when the sun chanced to strike them at the proper angle their beautifully variegated colors of red, yellow, and green showed plainly, but at midday they appeared as dark silhouettes or, if near at hand, plain green.

The male taken, when fresh, had the bill dusky neutral gray; cere dusky green gray; iris orange chrome, at inner margin shading to light orange yellow; bare skin surrounding eye pale olive buff; tarsus and toes deep mouse gray, the scales outlined with grayish white.

AMAZONA TUCUMANA (Cabanis)

Chrysotis tucumana CABANIS, Journ. für Ornith., 1885, p. 221. (Tucuman.)

Between 1,800 and 2,000 meters elevation on the Sierra de San Xavier, above Tafi Viejo, Tucuman, on April 17, 1921, these parrots were common in bands that passed screeching over the forested slopes or worked about in dense forest growth, well concealed by heavy limbs and abundant foliage. The flocks were wild and seldom permitted near approach, and only once did a shot offer, when a female, apparently an immature bird, was taken. The bird secured has no trace of the red on the tibial region described by Salvadori [65] in two specimens taken at Lesser, Salta.

PIONUS MAXIMILIANI SIY (Souancé)

Pionus siy SOUANCÉ, Rev. et Mag. Zool., 1856, p. 155. (Paraguay and Bolivia.)

From the description of *Pionus bridgesi* of Boucard [66] it is apparent that it is a bird in immature plumage of the species known as *maximiliani*. It appears from available material that specimens of *maximiliani* from central Brazil are smaller than those from southern Brazil, Paraguay, the Argentine Chaco, and eastern Bolivia. The distinction between these has been pointed out by Souancé, who restricted the name *maximiliani* to the bird of Brazil and called the larger specimens from Paraguay and Bolivia *Pionus siy* from the vernacular name given by Azara. The name of Souancé thus antedates the designation of Boucard, and *bridgesi* becomes a synonym of *siy*.

What I assume to be *Pionus m. maximiliani* is represented in material seen by skins from Macaco Secco (near Andarahy) and Santa Rita (State of Bahia), Jacareinha and Rio de Janeiro, Brazil; it seems to be distinguished from *P. m. siy* only by size.

---

[65] Boll. Mus. Zool. Anat. Comp. Torino, vol. 12, no. 292, May 12, 1897, p. 27.
[66] Hummingbird, Apr. 1, 1891, p. 27. (Bolivia, and Corrientes, Argentina.)

A male from near Andarahy has a wing measurement of 167.2 mm., while in three others, with sex not indicated, from the other localities listed, the wing measures 170, 176, and 179.2 mm. Hellmayr [67] records the wing of a bird from Bahia as 170 mm. A fair series of *P. m. siy* shows a variation in wing measurement from 183.5 to 197 mm. The specimens seen are marked as males or do not have the sex indicated. One bird from Santa Catherina, Brazil, has the wing 192 mm.; two males from Fazenda Cayoa, E. Parana, measure 183.5 and 186.8 mm.; and two from Puerto Pinasco, Paraguay (both males), 190 and 191 mm. respectively. A bird (male) from Las Palmas, Chaco, measures 196 mm., and one from Corrientes 197 mm. Four males from Puerto Suarez, Bolivia, range from 187.4 to 193 mm. Another from Santa Cruz, Bolivia, measures 185.8 mm.

P. m. lacerus Heine from Tucuman is said to be larger and darker than *siy*. A male seen in the Field Museum (No. 48990) from Metan, Salta, is probably best referred to this form since it has a wing measurement of 202 mm., though it is no darker than *P. m. siy*.

Pionus m. siy was found near Las Palmas, Chaco, from July 14 (when a male was taken) to July 31, 1920. At Puerto Pinasco, Paraguay, on September 1 it was common on the hill at Kilometer 25 (specimen), and at Kilometer 80 was recorded from September 13 to 20 (a male shot September 15). Several were seen at Kilometer 200 on September 25. The birds ranged through the forest in little bands that were seen frequently in swift flight to water, or from one tract of monte to another. Their passage was heralded in most cases by strident shouts of *chulp chulp* that were redoubled when one of their number chanced to receive an injury. On the wing they appear very dark.

The Angueté Indians called them *yeht a pilh' pul'*.

AMOROPSITTACA AYMARA (d'Orbigny)

Arara aymara d'ORBIGNY, Voy. Amér. Mér., vol. 2, 1839, p. 376. (Palca, Province Cochabamba, Bolivia.)

These little mountain parrakeets were recorded first on March 13, 1921, on the slopes above the city of Mendoza, when by following back on the line of flight of a small flock we found a water hole, perhaps the only one in an otherwise wholly arid tract. On March 19 near El Salto, at an altitude of nearly 2,000 meters above Potrerillos, Mendoza, these parrakeets were common. Here they ranged over the hills in small bands that fed in berry bearing bushes, or descended to search for fallen fruit in the grass below. The birds were highly social and were found always in parties. Their flight

[67] Abhandl. Kon. Bayer. Akad. Wiss., Klass. II, vol. 22, 1906, p. 590.

was swift and direct. Their chattering notes were high pitched and at times suggested the excited calls of barn swallows. Three were taken. On March 21 a flock was recorded at an altitude of 1,800 meters near the hotel at Potrerillos.

MYIOPSITTA MONACHUS MONACHUS (Boddaert)

Psittacus monachus BODDAERT, Tabl. Planch. Enl., 1783, p. 48. (Montevideo.)

No locality is cited by Boddaert in connection with his *Psittacus monachus*, though Brabourne and Chubb [68] give Montevideo as the type locality, apparently taking this from Latham.[69] Latham under his gray-breasted parrakeet includes reference to Buffon, Daubenton, and Pernetty, the latter of whom says that the species was found at Montevideo. As it is proposed to divide the species into three geographic races, Montevideo, Uruguay, is here accepted as the type-locality. The three subspecies recognized in available material will stand as follows:

MYIOPSITTA MONACHUS MONACHUS (Boddaert).

Bill large and heavy; abdomen more yellowish; dorsal surface bright green.

Culmen from cere, 18.6–22; wing, 140.5–157; tarsus, 17–19.4 mm. Buenos Aires, Santa Fe, Entre Rios, and Uruguay (at least in southern part).

MYIOPSITTA MONACHUS COTORRA (Vieillot).

Bill small; abdomen less yellowish; dorsal surface bright green. Culmen from cere, 16–17.3; wing, 127–140.6; tarsus, 15.2–16.9 mm. Formosa (Kilometer 182) and Paraguay (Puerto Pinasco).

MYIOPSITTA MONACHUS CALITA (Jardine and Selby [70]).

Bill small, wing short, dorsal surface distinctly duller green. Culmen from cere, 16.8–17.7; wing, 131–137.5; tarsus, 15–16.4 mm. Mendoza, and San Luis (Nueva Galia). According to Hartert,[71] also at Rio Colorado, Tucuman, and La Banda, Santiago del Estero.

The large-billed form of the monk parrakeet was recorded near Lavalle, Buenos Aires, from October 23 to November 15, 1920, and a male and two females were preserved as skins on October 30. A small colony inhabited a clump of eucalyptus trees in town, while at the estancias in the surrounding country the birds were common wherever there was tree growth to furnish them shelter. At Los Yngleses large stick nests of this parrakeet were placed in the higher

[68] Birds South America, 1912, p. 85.
[69] Syn. Birds, vol. 1, pt. 1, 1781, p. 247.
[70] *Psittaca calita* Jardine and Selby, Ill. Orn., vol. 2, pt. 6, August, 1830, pl. 82 (Province of Mendoza). The name *calita* is an evident *lapsus calami* for *catita*, the common name of this parrakeet in Argentina and Uruguay. Las Catitas in the Province of Mendoza, given by Jardine and Selby as *calitas*, is today an important center in the grape district.
[71] Nov. Zool., vol. 16, December, 1909, p. 234.

BORDER OF A BAÑADO, OR LOWLAND MARSH
Near San Vicente Uruguay January 27, 1921

A VIEW IN ONE OF THE LOWLAND PALM FORESTS OF EASTERN URUGUAY
Near San Vicente, Uruguay, January 27, 1921

A CROSSING ON THE ARROYO SARANDI (IN EASTERN URUGUAY) KNOWN AS THE PASO ALAMO

Taken February 2, 1921

VIEW OF THE DENSE, LOW FOREST BORDERING THE RIO CEBOLLATI

Near Lazcano, Uruguay, February 8, 1921

eucalypts where companies of screeching birds clambered about over them, or, as this was spring, added new material to the mass already accumulated. The nests were rough and unfinished externally, and as the larger ones were 2 meters in diameter often contained material sufficient to fill a wagon. The majority were placed 14 or 18 meters from the ground. After a tremendous storm in early November, I found that several of the nests had been dashed to the ground. The birds frequented trees, save when once or twice a day they flew out to drink at some channel in the marshes. On November 3 and 6 small bands were noted in a clump of isolated trees near the coast below Cape San Antonio, where they had apparently settled recently, and on November 16 between Lavalle and Santa Domingo a flock was seen in the iron work of a high wagon bridge.

In southern Uruguay the monk parrakeet was common wherever trees offered shelter. A few were seen near La Paloma January 23, and at San Vicente from January 25 to February 2 the species was common in extensive palm groves in the lowlands. In the low forested tract that bordered the Rio Cebollati below Lazcano from February 5 to 9 monk parrakeets frequented open pastures studded with trees, and were also found in the dense forest.

Where scattered palm trees grew in small openings nearly every palm held one of the large stick nests of this bird, usually with a pair of parrakeets clambering over it. At this season the parrots fed on thistle heads in rank growths of these weeds that here have ruined thousands of acres of pasture. The thistle heads were nipped off from the stem and held in one foot while the bird extracted the seeds with the aid of bill and tongue. It was reported that monk parrakeets damaged maize extensively when the grain was ripening. Two were shot at Lazcano on February 8.

Small flocks of monk parrakeets that were noted at Victorica, Pampa, on December 24, 27, and 29, were probably *M. m. calita*, as a specimen in the National Museum taken at the Estancia El Bosque near Nueva Galia, San Luis, a short distance farther north, belongs to that form. Since none were shot at Victorica the identity of the bird from that region is not wholly certain.

MYIOPSITTA MONACHUS COTORRA (Vieillot)

Psittacus cotorra VIEILLOT, Nouv. Dict. Hist. Nat., vol. 25, 1817, p. 362.
. (Paraguay.)

Psittacus cotorra of Vieillot based on Azara is a composite of observations by Azara on the monk parrakeet in Buenos Aires and in Paraguay. Since the bulk of the notes refer to Paraguay, and the measurement of the bill, given as 8 lines, indicates the small northern bird, the type locality is here fixed as Paraguay. The name is thus available for the small northern subspecies.

At the Riacho Pilaga, Formosa, monk parrakeets of the northern form were abundant from August 10 to 21, and a number were preserved as specimens. The birds roosted somewhere in the extensive monte, and in early morning flew across in pairs or small flocks to spend the day in a large field where recent plowing had brought to the surface discarded sweet potatoes left from last year's harvest. The parrots alighted in close company, a hundred or more together, in search of food, often in company with cowbirds, chopi blackbirds, and large pigeons. As they fed they maintained a constant conversational chatter, while at the slightest alarm the flock rose shrieking and screaming to circle about in the air. As I worked at specimens or notes through the long afternoons their rather disagreeable uproar came constantly to my ears, but in compensation for this discord I found a tree filled with resting birds, that nestled against one another in pairs or clambered singly through the branches, a beautiful and pleasing sight when the light fell properly to bring out the contrast between the bright green of their plumage against the dull gray green of the foliage. Indians, armed with old single-barreled shotguns, at intervals potted these flocks either to secure food, or in expectation that they might sell the birds to me for specimens, and a carrancho (*Polyborus p. plancus*) swooped down a dozen times a day and flushed the flocks in screaming chatter, in hope that he might encounter a cripple, less agile than its companions, that might not escape his talons. The passing of an Aplomado falcon brought consternation to the parrots so that they rose and circled a hundred yards in air, and as hawks were common the shadow of a turkey vulture was frequently sufficient to throw them into screaming confusion. During a heavy storm far to the southward that continued for an entire day with barely audible thunder, the parrakeets were very nervous, and at each low rumble rose with screams to circle in the air. They seemed inordinately afraid of the flashing of lightning, perhaps from the destruction wrought by storms to their nests.

The Toba Indians called this species either *il-lit* or *Ki likh*.

At Kilometer 80, west of Puerto Pinasco, Paraguay, from September 6 to 18, the species was fairly common and was observed to feed on the seeds of an algarroba. Three were taken on September 18. At Kilometer 110 on September 23 a dozen were observed eating the ripening seeds of an algarroba. As two parrakeets passed overhead an Indian cast a throw stick at them and the birds barely escaped it.

A male shot September 18 had the sides of the bill fawn color, shading to avellaneous on gonys and culmen; iris benzo brown; tarsus and toes deep mouse gray, the scales outlined with whitish.

PYRRHURA FRONTALIS CHIRIPEPE (Vieillot)

Psittacus chiripepe VIEILLOT, Nouv. Dict. Hist. Nat., vol. 25, 1817, p. 361. (Paraguay.)

As indicated by Salvadori,[72] since *Psittacus vittatus* Shaw, 1811, is preoccupied by *vittatus* Boddaert 1783, *Pyrrhura vittata* (Shaw) must be replaced by *Pyrrhura frontalis* (Vieillot).[73] *Psittacus frontalis* was stated by Vieillot to have come from Cayenne; but since it was based on the *perruche-ara a bandeau rouge* of Levaillant,[74] the type locality must, according to Salvadori,[72] be cited as Brazil.

Females were shot at Las Palmas, Chaco, on July 15 and 21, 1920, and 25 kilometers west of Puerto Pinasco, Paraguay, on September 1. These differ from skins from eastern Brazil in small size, and are taken as representing the subspecies *chiripepe*. None of the three has the reddish spot on the back ascribed to this species, so that this character is variable.[75] These specimens have the following measurements:

No.	Sex	Locality	Date	Wing	Tail	Culmen from cere	Tarsus
283752	Female	Las Palmas, Chaco	July 15, 1920	128.1	133.3	15.0	14.0
283751	...do	...do	July 21, 1920	131.5	138.2	16.2	13.0
283750	...do	Puerto Pinasco, Paraguay	Sept. 1, 1920	131.8	139.0	17.0	13.0

At Las Palmas, Chaco, these parrakeets were common from July 14 to 31, and were observed frequently in passage across the sky. They were wild and difficult to approach in most cases, but like many other birds of such habit were surprisingly tame when I came upon them suddenly near at hand. They were known locally as *loro naranjero* and were said to do considerable damage in orange groves, an allegation that I verified by personal observation. Their flight was swift and darting, and on the wing they often suggested pigeons. Their screaming calls may be represented as *kree-ah kree kree kree ah*.

On September 1 these birds were common on a low wooded hill 25 kilometers west of Puerto Pinasco, but were not found farther inland here, nor were they recorded in the interior of Formosa. To the Angueté Indians they were known as *yem a seet i gwi*.

NANDAYUS NENDAY (Vieillot)

Psittacus nenday VIEILLOT, Tabl. Enc. Méth., vol. 3, 1823, p.1400. (Paraguay.)

A male was killed near the Rio Paraguay at Puerto Pinasco, Paraguay, on September 3, 1920, and another near Kilometer 80

---

[72] Ibis, 1900, p. 669.
[73] *Psittacus frontalis* Vieillot, Nouv. Dict. Hist. Nat., vol. 25, 1817, p. 361.
[74] Hist. Perr., 1801, pl. 17.
[75] See Hellmayr, Abh. Kön. Bayerischen Akad. Wiss., II. Kl., vol. 22, 1906, p. 585.

on September 13. Through this region these parrakeets, known in Angueté as *chi to gwi*, and in Guaraní as *ñenday*, were fairly common, especially among open palm forests where palm nuts offered food. Occasionally they fed on the ground under trees that had dropped their seeds. Like other parrots they fly regularly to water and alight in bushes where these stand in pools and sidle down until they can reach the fluid. They travel in flocks of ten to a dozen individuals, that feed, or move in company, with fairly swift, direct flight. Their approach is heralded by loud squalling calls and should one of their number be killed, those remaining redouble their outcry. The common call may be represented as *kree-ah kree-ah*. The species was recorded at Kilometer 200 on September 25. A few were noted during August at the Riacho Pilaga, in central Formosa, but were so wary that none were secured.

A male taken September 13 had the bill sooty black; iris chamois; tarsus pale vinaceous fawn.

THECTOCERCUS ACUTICAUDATUS (Vieillot)

Psittacus acuticaudatus VIEILLOT, Nouv. Dict. Hist. Nat., vol 25, 1817, p. 369. (Lat. 24° S., Paraguay.)

At the Riacho Pilaga (Kilometer 182), Formosa, on August 11, 1920, a male parrakeet of this species flew out from the border of a forest to rest in the sun in a dead tree, where I shot it after a long stalk. On August 18 another was taken from a flock of 20 that passed swiftly overhead. Two others were seen on August 21. The Toba Indians called this bird *ta tas*.

On December 28 I found these birds fairly common in the open forest near Victorica, Pampa, and collected a male. At this season they seemed to be nesting for frequently as I passed through the timber single birds darted swiftly around with shrieking calls. They were feeding here on the piquillín and other berry-bearing bushes.

At Tapia, Tucuman, from April 6 to 13 the species was common in the forests, where they ranged in considerable bands wherever seeds, berries, or the fruit of large tree cacti offered food. Morning and evening bands flew down to the river to drink, often flying high in the air, and then returned to the cover of the forest. It was said that they destroyed much corn at certain seasons. The long tail readily distinguished this parrot from others and gave them somewhat the appearance of *Cyanoliseus*. As they passed overhead the light colored maxilla and feet were often visible. Three that were killed on April 8 were preserved as specimens. The birds are not bad eating where meat is scarce.

Specimens from Pampa and Tucuman seem slightly duller green than those from Formosa, but the series at hand is too small to establish the difference definitely. The acute tip of the maxilla in this species in contrast with the broadened mandible is remarkable; in most specimens the distal end of the bill shows considerable wear, indicating that it is used extensively in work that requires heavy cutting.

CYANOLISEUS PATAGONUS PATAGONUS (Vieillot)

Psittacus patagonus VIEILLOT, Nouv. Dict. Hist. Nat., vol. 25, 1817, p. 367. (Buenos Aires.)

Since Azara,[76] from whom Vieillot took his description of the Patagonian parrot, says that he had four specimens of this bird from Buenos Aires, and continued with the statement that he was informed that the bird ranged from latitude 32° S. to the Patagonian coast, it seems logical to assume that the type locality should be in the Province of Buenos Aires, where the birds were common formerly, and not in Patagonia as is usually stated.

The barranca parrot or *loro barranquero* was found near General Roca, Rio Negro, from November 23 to December 3, 1920. In this arid region the parrots frequented the flood plain of the Rio Negro in the main, though occasionally a small flock ranged inland among the gravel hills that bordered the valley. In early morning barranca parrots were astir an hour or two after daybreak, when the air had been warmed by the sun, and remained abroad until dark. In early morning flocks were encountered near the river, where they came for water, and later worked inland wherever berries or seeds offered them food. At such times they traveled rather low, ranging from 2 to 10 meters in the air. As customary with parrots, they fly steadily, in direct line, with the usual accompaniment of screeching calls. Their food consisted of berries that chanced to be ripe at that season, among which may be noted *Lycium salsum* and *Discaria*, species. The birds resemble macaws in appearance, a suggestion that is furthered by the flashes of color that appear in their plumage during flight.

An adult female when killed had the bill deep neutral gray; bare skin around eye pale olive buff; iris light buff; tarsus and toes cartridge buff; claws black.

Order CORACIIFORMES

Family TYTONIDAE

TYTO ALBA TUIDARA (J. E. Gray)

Strix tuidara J. E. GRAY, in E. Griffith, ed. Cuvier's Anim. Kingd., vol. 6, 1829, p. 75. (Brazil.)

At Puerto Pinasco, Paraguay, on September 1, 1920, a pair of barn owls had a nest in the roof of a store building, where the screech

---

[76] Apunt. Hist. Nat. Paxaros Paraguay, vol. 2, 1805, p. 420.

of the young was heard constantly at night. A barn owl was noted in the town of Tunuyan, Mendoza, about 3 in the morning on March 26, 1921.

Strix perlata Lichtenstein, 1819 [77] is antedated by *Strix perlata* Vieillot,[78] 1817 for another species so that the name for the South American barn owl becomes *tuidara* as indicated by Mathews.[79]

Family STRIGIDAE

GLAUCIDIUM BRASILIANUM BRASILIANUM (Gmelin)

Strix brasiliana GMELIN, Syst. Nat., vol. 1, pt. 1, 1788, p. 289.   (Brazil.)

An adult female was taken February 5, 1921, near Lazcano, Uruguay, in a heavy thicket of low trees and shrubs that bordered the Rio Cebollati. The bird was frightened out as I forced my way through the dense cover, and flew to another perch a few meters away to turn and peer at me. The tail had been entirely molted and partly renewed and new feathers were appearing on the body. Molt of the wing quills had not yet begun.

GLAUCIDIUM NANUM VAFRUM Wetmore

Glaucidium nanum vafrum WETMORE, Journ. Washington Acad. Sci., vol. 12, August 19, 1922, p. 323.   (Concon, Intendencia of Valparaiso, Chile.)

A female was taken April 27, 1921, on a brush-grown hillside near Concon, Chile, as it sunned itself on an open limb in the cool air of early morning. The bird crouched with wings slightly extended and feathers fluffed out so that it appeared twice natural size. The eyelike spots in the back of the head were very prominent so that their appearance was curious to an extreme.[80] The tip of the bill in this bird was deep olive buff; base puritan gray, shading to deep olive buff, the gray clear below, indistinct above; iris pale greenish yellow.

The form of this owl from central Chile differs from typical *nanum* from the Straits of Magellan in broader, heavier dark bars on the tail. The southern subspecies, typical *nanum*, appears to range north through the forested region to near Temuco, though there it shows strong evidence of intergradation toward *vafrum*. *Glaucidium nanum* is closely allied to *G. brasilianum* so that examination of the subspecies composing the two groups, as they are now understood, reveals that they are separated by a difference in depth of color alone. The two subspecies composing *nanum* differ from those attributed to

---

[77] *Strix perlata* Lichtenstein, Abh. Kön. Akad. Wiss. Berlin, 1816–17, 1819, p. 166. (Brazil.)

[78] *Strix perlata* Vieillot, Nouv. Dict. Hist. Nat., vol. 7, 1817, p. 26.

[79] Birds Austr., vol. 5, pt. 4, Aug. 30, 1916, p. 371.

[80] For a striking representation and description of this peculiarity see J. Koslowsky, El Hornero, vol. 1, 1919, pp. 229–235, pl. 3.

brasilianum in darker, more suffused coloration on the dorsal surface and in heavier markings on the underparts. The two groups are complementary in range and it seems highly probable that in time they will be found to intergrade.

ASIO FLAMMEUS FLAMMEUS (Pontoppidan)

Strix flammea PONTOPPIDAN, Danske Atlas, vol. 1, 1763, p. 617, pl. 25. (Sweden.)

The short-eared owl was fairly common in marshy areas on the pampas, and elsewhere was found in tracts of low greasewoods or other small bushes. It was recorded as follows: Formosa, Formosa, August 24, 1920; Dolores, Buenos Aires, October 21; Lavalle, Buenos Aires, October 29 to November 15; General Roca, Rio Negro, November 25 and December 3; Zapala, Neuquen, December 9; Ingeniero White (Bahia Blanca), Buenos Aires, December 13; Carhue, Buenos Aires, December 15 to 18; Guamini, Buenos Aires, March 3.

At Carhue I heard the high-pitched hooting call of this owl and one circled about my head, giving a curious barking note.

A female was taken at Lavalle, Buenos Aires, October 29, 1920. According to Bangs's outline of this group,[81] this bird should stand as *Asio flammeus breviauris* (Schlegel), but in the series available to me from South America there is nothing apparent to distinguish them from North American birds either in color or size of bill.

SPEOTYTO CUNICULARIA CUNICULARIA (Molina)

Strix Cunicularia MOLINA, Sagg. Stor. Nat. Chili, 1782, p. 263. (Chile.)

At Las Palmas, Chaco, a burrowing owl was seen in a little open prairie near the fonda where I was lodged during July, 1920, but no others were observed in the north. Near Dolores, Buenos Aires, one was noted October 21, and in the vicinity of Lavalle the species was common from October 27 to November 15. An occupied nest hole, recorded November 4, was dug in the side of a sand dune, while another, seen November 13, was in a level open pasture. In both the entrance had been decorated with broken bits of dried cow dung. Near General Roca, Rio Negro, burrowing owls were fairly common from November 23 to December 3, mainly on alkaline plains near the Rio Negro. Two were seen at Zapala, Neuquen, on December 9. They were recorded in fair numbers at Carhue, Buenos Aires, from December 15 to 18, and near Victorica, Pampa, from December 23 to 27. Near Carrasco, Uruguay, young fully grown were seen January 9, 1921, and others were recorded January 16.

---

[81] Notes on South American Short-eared Owls, Proc. New England Zool. Club, vol. 6, Feb. 18, 1919, p. 96.

At La Paloma, Uruguay, the species was noted January 23. Near San Vicente, Uruguay, from January 25 to February 2, and in the vicinity of Lazcano, from February 3 to 9, the birds were common in open country, and were among the prominent forms of the region. Single birds or little groups were noted constantly, and it was amusing to see them drop prudently down a hole as I approached instead of taking to wing. The species must be counted among valuable enemies of the locust. In certain country districts in Uruguay the flesh of the burrowing owl is served as a delicacy to those convalescing from illness in the belief that it produces appetite for other food. At Guamini, Buenos Aires, these owls were noted from March 3 to 8, and at Tunuyan, Mendoza, they were recorded from March 24 to 29. Near Tapia, Tucuman, they were heard calling occasionally at night from April 6 to 13, and one was seen at Concon, Chile, April 28.

Males were collected at General Roca, Rio Negro, on November 30, 1920, and at Tunuyan, Mendoza, March 27, 1921. This species exhibits even more variation in color from light to dark on the southern continent than in the United States. Specimens that were very light, in fact almost white, were observed frequently, at times using the same holes as dark individuals. In revision of subspecies this must be borne in mind as otherwise confusion will result. The two specimens secured agree fairly well in size and color with birds from Chile, though there is a tendency for Argentine birds to average larger. Those from northwestern Argentina are particularly large and with abundant material may prove to belong to a distinct race.

Family NYCTIBIIDAE

NYCTIBIUS GRISEUS GRISEUS (Gmelin)

Caprimulgus griseus GMELIN, Syst. Nat., vol. 1, pt. 2, 1789, p. 1029. (Cayenne.)

A female *Nyctibius* was taken September 30, 1920, on the heavily forested slopes of the Cerro Lorito on the eastern side of the Paraguay River opposite Puerto Pinasco, Paraguay. The form of the bird as it rested on a gall projecting from the trunk of a tree 150 mm. in diameter caught my eye by chance. The claws grasped the perch firmly while the body stood erect parallel to the tree, and separated from it only by the space of 25 mm. The tail hung straight down, the eyes were closed, and the feathers in front of and above the eye on either side were erected to form projecting horns. The bird resembled a bit of stick that had fallen to become lodged on the tree trunk.

The specimen taken is grayish in tone and has the following measurements: Wing, 267.0; tail, 183.5; exposed culmen, 14.6; tarsus, 15.2 mm. Specimens of *griseus* from the type locality are not at hand, but a bird from the Para River is assumed to represent the typical form. The bird from Paraguay is very similar to it in size, and in color pattern differs only in having slightly heavier black streaks on the under tail coverts. For the present, therefore, *Caprimulgus cornutus* Vieillot must be considered as a synonym of *Nyctibius g. griseus*. Examination of a small series of potoos seem to show that, like many of the Caprimulgidae, they have two types of coloration, one dark and more or less rufescent, and the other pale and gray, a fact that makes the proper designation of geographic races difficult. Skins from Peru and Ecuador referred doubtfully by Mr. Ridgway to *cornutus* [82] are larger and darker than the specimens mentioned from Brazil and Paraguay. Whether they represent an unnamed form or whether they should be referred to the bird from Panama, which they resemble closely, it is not possible at present to decide.

The female shot September 30 in Paraguay had the bill black; margin of mandible vinaceous buff; iris deep chrome; tarsus and toes drab.

Family CAPRIMULGIDAE

THERMOCHALCIS LONGIROSTRIS (Bonaparte)

Caprimulgus longirostris BONAPARTE, Journ. Acad. Nat. Sci. Philadelphia, vol. 4, pt. 2, no. 12, 1825, p. 384. (Brazil.[83])

South of Tunuyán, Mendoza, on March 27, 1921, a male of this species flushed with a chattering, whistling call among low bushes on a sandy hill slope, and darted swiftly and erratically away to drop to fresh cover. As it rose a second time it was secured. The large white wing patches give a resemblance to *Chordeiles* when the bird is on the wing. Another was seen April 8, near Tapia, Tucuman, in dry forest on a steep rocky slope.

On an evening in mid-October one of these birds flew from tree to tree along the Avenida de Mayo, in the heart of the business district of Buenos Aires, an individual that had become bewildered during migration. At Lavalle on November 12, 1920, I was taken to view a curious bird, described as "possessing a moustache like a Christian," that had been captured in a garden about three weeks ago, to find that it was the present species. According to popular belief a feather of this bird was a potent love charm, and the fortunate owner of the bird had been charging 10 centavos for a view of the bird to those of the populace whose curiosity regarding the anomalous creature was uncontrollable, while feathers retailed at a peso each,

---

[82] Birds North and Middle America, vol. 6, 1914, p. 587.
[83] Bonaparte described his bird as from South America without known locality, but Brabourne and Chubb, Birds of South America, 1912, p. 101, cite "Brazil" without comment.

apparently a thriving business, since the captive had been denuded of nearly a third of its plumage.

SETOPAGIS PARVULUS (Gould)

Caprimulgus parvulus GOULD, Proc. Zool. Soc. London, 1837, p. 22.   (Parana River near Santa Fe, Argentina.)

Near the ranch at Kilometer 80, behind Puerto Pinasco, Paraguay, these goatsuckers began to call suddenly on the evening of September 17, 1920, and it was supposed that they were migratory as none had been recorded previously.   On the following night, by means of an electric headlight, two of the nocturnal songsters were taken, so that their identity was established beyond question.   Their song resembles *you cheery chu chu chu chu chu chu*, the first two notes uttered in a clear tone and the rest forming a bubbling, rattling trill.   They called from leafy trees in open pastures, or from the forest, or came out to rest on the limbs of fallen trees along the borders of the monte, or in paths cut among the trees.   Here the headlight caught their eyes with a reflected glow of deep burning red, like a coal of fire but more intense in color, a beautiful object against the dark background.   At intervals this light disappeared, apparently as the bird turned its head, and then came into view again.   Never more than one point was seen at a time so that vision seemed to be entirely monocular.   The species continued its notes until my departure. It was recorded September 23 at Kilometer 110, and September 24 and 25 at Laguna Wall (Kilometer 200).   None were found at Puerto, Pinasco, itself.

In Spanish this species was known as *cuatro cuero*, in Guarani as *uro-ooh*, while the Angueté Indians called it *ka jee vay ta ta nee nai*.

For various reasons one of the specimens secured was preserved entire in alcohol.   The other, an adult male, seems large as it measures: Wing, 141.2; tail, 103; exposed culmen, 9.7; tarsus, 15.6 mm. Comparative material is not at present available to decide the status of this bird.

The rictal bristles in this species are remarkably short.

HYDROPSALIS TORQUATA FURCIFERA (Vieillot)

Caprimulgus furcifer VIEILLOT, Nouv. Dict. Hist. Nat., vol. 10, 1817, p. 242. (Paraguay.)

Specimens of this fork-tailed goatsucker from Argentina, Uruguay, Paraguay, and southern Brazil are distinguished from those of northern and eastern Brazil by slightly larger size, and darker coloration, with more buff on the abdomen so that the southern birds should rank as a subspecies of *H. torquata*.

At Victorica, Pampa, from December 23 to 29, 1920, this goatsucker was fairly common in low open forests of caldén and algarroba, where the undergrowth of shrubs grew in thicket formation, leaving small openings dotted with tufts of grass. The birds rested in shade, on little spots of bare earth, singly, or occasionally as many as three together. When startled they flushed with a rattle of wings produced in part by striking the wings above the back, and after a graceful, somewhat erratic darting flight of a few meters, dropped suddenly to earth. After alighting they frequently bobbed up to the full length of the legs and then dropped suddenly down again, or opened and closed their great mouths in silent protest at my intrusion. Adult males in flight swung erratically from side to side with a flashing of the long, deeply forked tail as they turned. It was not uncommon for them to elude me completely in dense brush without offering a shot. One that fell disabled emitted a low growling call that at times terminated sharply in a croak intended to startle an enemy. Of six collected here, four were preserved as skins. Three of these are adult males in partial molt, and one an immature female. The elongated lateral rectrices had been dropped in two of the males, and in the other were very loosely attached, while molt of the body plumage was beginning. Two specimens show an interesting molt of the rictal bristles in which the separate bristles are being shed and renewed irregularly. The immature female had developed the posterior bristles first, and then six of the anterior ones had grown in simultaneously, and have the bases still inclosed in sheaths.

An adult male had the bill dull black; iris Hay's brown; tarsus and toes deep brownish drab; claws dull black.

At Lazcano, Uruguay, a female, apparently adult, was taken February 5, 1921, among small open thickets near the Rio Cebollati. Three were seen and an adult female taken near Rio Negro, Uruguay, on February 14.

One flushed in a dry wash on March 13, above the city of Mendoza, in the Province of Mendoza, was probably a migrant as it was in a drier, more arid region than usual.

Specimens from western Argentina have the nuchal collar slightly paler, more buffy, less rufescent than those from Brazil, Paraguay, and Uruguay, but this character is slightly variable and may be due to age or condition of plumage.

PODAGER NACUNDA (Vieillot)

Caprimulgus nacunda VIEILLOT, Nouv. Dict. Hist. Nat., vol. 10, 1817, p. 240. (Paraguay.)

The ñacunda, a summer visitant in southern South America, is now apparently rarer than in the days of Hudson, a condition caused perhaps by increased cultivation and extensive grazing in its haunts on the open pampa. Early writers speak of flocks containing from 40 to 200 individuals, but the bird in my experience was so rare that to see an individual was a treat, while a flock of a dozen gathered by some condition of favorable food left a thrill that is still remembered. At Formosa, Formosa, on August 23, 1920, one passed down the Paraguay River at dusk, evidently the first of the spring migrants. Others passed Kilometer 80, west of Puerto Pinasco, Paraguay, on September 6, 7, 8, 9, and 10, all bound southward in evening, or once in early morning after a storm. The migration ceased abruptly before there was opportunity to secure specimens, and in work in the pampas I failed to find the birds, until on February 7 at Lazcano, Uruguay, one passed the patio of the hotel on the evening of February 7. At Rio Negro, Uruguay, from February 13 to 17, the species was present in fair numbers and a male was collected on February 16. Another was taken on the following day. In early evening ñacundas hawked about high in air, but at dusk circled low over patches of weeds at the edge of town, attracted by myriads of beetles that filled the air. On the 16th I watched for them at the border of a small lagoon, and, after one or two alarms from teru terus coming in silently to roost, was rewarded by the sight of a goatsucker hawking low over the grass. In a few minutes I had the bird in my hand where I could admire the beautiful contrast of color in the plumage and the large, lustrous eyes.

The wings in this species are short and the body heavy, so that at times it presents an owllike appearance. The birds quarter back and forth when feeding like nighthawks, and though strong fliers are not as graceful on the wing as birds of that group. They are strong in body and often difficult to kill. When they alight they may stand erect for a moment to look about and then sink to the crouching position ordinary to goatsuckers. The leg is long and fairly strong, and in one taken the feet were very muddy, an indication that the birds walk about more or less, as might be supposed from the structure of their legs. As they flush from the ground they may give a low rattling call and a wing-tipped bird opened its mouth threateningly and hissed like a nighthawk in similar condition.

Near Rivas, Buenos Aires, eight were noted while passing in a train on March 11.

Family ALCEDINIDAE

MEGACERYLE TORQUATA TORQUATA (Linnaeus)

Alcedo torquata LINNAEUS, Syst. Nat., ed. 12, vol. 1, 1766, p. 180. (Mexico.)[84]

Two specimens of the ringed kingfisher, both males, taken, respectively, on July 17, 1920 at Las Palmas, Chaco, and August 16 at

---

[84] See Berlepsch and Hartert, Nov. Zool., vol. 9, April, 1902, p. 104.

Kilometer 182 (Riacho Pilaga), Formosa, appear similar to skins from northern South America and Mexico. In a fair series I am not able to make out the distinctions assigned by Bangs and Penard [85] when they recognized the form from Paraguay as *M. t. cyanea* (Vieillot).

The species was encountered only in the northern portion of the region that I visited. At Resistencia, Chaco, from July 8 to 10, the ringed kingfisher was recorded in fair numbers, and it was noted that the breeding season was near, as the birds frequently rose in pairs or threes to circle about from 60 to 100 meters from the earth, with slow wing beat and constant calls. At Las Palmas, Chaco, they were fairly common from July 13 to July 31. Mating was completed during this period and by the close of the month their pairing evolutions in the air became infrequent. On July 21 an occupied nest was found in a cut bank above a small stream, the Riacho Quia, placed beneath a mass of roots. The entrance hole was about 140 mm. in diameter and showed along the bottom a double furrowed track with a ridge 16 mm. high, separating the two channels made by the feet of the owners. The nest was nearly 2 meters deep and may have contained eggs as both parents circled near in excitement. The location was such that time did not permit an excavation.

On August 2 in traveling by steamer up the Rio Paraguay above Puerto Las Palmas, the large kingfisher was common in pairs. Frequent cut banks along the streams offered nesting sites. At the Riacho Pilaga the species was recorded from August 9 to 17, and several were seen at Formosa, Formosa, August 23 and 24. It was common along the Paraguay river from September 1 to 3, and on September 30, near Puerto Pinasco, Paraguay and was recorded occasionally at Kilometer 80 in the interior from September 7 to 21. A few were noted on February 6 and 8, 1921, near Lazcano, Uruguay; along the Rio Cebollati, and one was seen February 19 near Rio Negro, Uruguay.

The usual call note of the ringed kingfisher is a harsh *chuck* or *check* varied at times by a rattle, both calls so similar to notes of some of the woodpeckers as to suggest those birds whenever they are heard. The flight is strong and vigorous and in actions and appearance these birds are suggestive of the belted kingfisher of the north.

CHLOROCERYLE AMAZONA (Latham)

Alcedo amazona LATHAM, Index Orn., vol. 1, 1790, p. 257. (Cayenne.)

Near Las Palmas, Chaco, this species was recorded at intervals from July 17 to 30, and an adult male was taken July 17. The

---

species was fairly common on the Paraguay River above Puerto Las Palmas on August 2. At Puerto Pinasco, Paraguay, the birds were found in pairs along the Rio Paraguay on September 3 and 30, and were recorded at lagoons near Kilometer 80 on September 8, 10, and 18. On February 2, 1921, two were taken at the Paso Alamo, on the Arroyo Sarandi north of San Vicente, Uruguay. These birds call in a high-pitched steely rattle or a low *chuck*. They are found over water where they have the habits common to their larger relative, the ringed kingfisher, but differ from that bird in that they are frequently found about little ponds in the savannas, where the water is shallow and small in extent.

A wounded bird exhibited a peculiarity, that I have not observed previously, in that a small air sac, capable of distension at will, lay underneath the skin of the lower eyelid. When this sac was filled with air the distended lid half covered the eye, while the free margin of the lid was pressed out until it was 2½ mm. from the eyeball. The sac was rudely oval, was pointed at either end, and extended across from inner to outer canthus. Its distension was greater below than at the margin of the lid. The sac was distended and deflated several times during my examination of the bird, and the chamber was readily seen on dissection. The function of this curious structure is evidently to protect the eye during impact with water when the bird dives.

The wing in this species is eutaxic as has been recorded by W. D. Miller.[86]

CHLOROCERYLE AMERICANA VIRIDIS (Vieillot)

Alcedo viridis VIEILLOT, Nouv. Dict. Hist. Nat., vol. 19, 1818, p. 413. (Paraguay.)

This form of the green kingfisher differs from *C. a. americana* in larger size, and lighter green above, while in the female the dark breastband is narrower and is more obscured by pale tips on the feathers. In the areas visited the bird was found in the same localities as its larger relatives—in fact, I collected my first specimens of all three of the species found in northern Argentina within the space of fifteen minutes—but is less abundant. Like *C. amazona* it often frequents small pools in the savannas. At Las Palmas, Chaco, a female was taken July 17, 1920, just after it had flown gracefully out to capture a passing insect on the wing. July 27 another fished in a small estero, hovering over the water, and then darting down to strike the water with great dash and speed. At the Paso Alamo on the Arroyo Sarandi, north of San Vicente, Uruguay, a female was taken on February 2, 1921. Another was shot near

---

[86] Auk, 1920, p. 427.

Lazcano, Uruguay, on February 7, and one was seen near Rio Negro, Uruguay, on February 19. The ordinary call of this species is a sharp *click click* given as the tail is twitched and the anterior portion of the body is raised. This species has an air sac in the lower eyelid similar to that described in *C. amazona.*

Family BUCCONIDAE

NYSTALUS MACULATUS STRIATIPECTUS (Sclater)

Bucco striatipectus SCLATER, Ann. Mag. Nat. Hist., vol. 13, ser. 2, May, 1854, p. 364. (Bolivia.)

Though Dr. P. L. Sclater read a description of *Bucco striatipectus* before a meeting of the Zoological Society of London on December 13, 1853, this was not published in the Proceedings until November 14, 1854, so that the reference to the first published description must be cited from the Annals and Magazine of Natural History for May, 1854, where it is included in a synopsis of the Bucconidae.

This species was observed only near Tapia, Tucuman, from April 7 to 11, 1921, save for several seen near Rio Colorado, Tucuman, on April 1, from a train. Buccos were found at rest on low perches in the trees, often in rather brushy.localities, in regions cut by steep-walled barrancas. During clear weather that followed rains they occasionally came out to more open perches or even rested in the sun on telegraph and telephone wires. They perched with head drawn in, bill held level or pointed slightly upward and tail elevated, the latter from its slender form appearing as if stuck into the body. The birds were stolid and allowed close approach, though at times they turned toward an intruder with a half-threatening air that was ludicrous. When startled they flew for short distances with a loud rattling flutter made by their small, rounded wings. They were wholly silent. Locally the species was known as *durmi-durmi* or *durmili-durmili.*

Five specimens collected on April 7, 8, 9, and 11 include adults and young of both sexes. A male when first killed had the bill, in general, dull black; sides of maxilla, and mandible at cutting edge, etruscan red; iris straw yellow, becoming neutral gray at outer margin; tarsus and toes grayish olive; claws dull black.

The small size of the brain with slight development of the cerebellum in this species is particularly noticeable, while in the skeleton the huge skull in contrast to the tiny plate of the low-keeled sternum is even more striking.

There are two specimens in the United States National Museum collections marked as secured by Capt. T. J. Page on the Bermejo River in February, 1860, a region from which the bird does not appear to have been reported previously.

Family RAMPHASTIDAE

RAMPHASTOS TOCO Müller

Ramphastos Toco MÜLLER, Natursyst., Suppl., 1776, p. 82.   (Cayenne.)

A few of these toucans were seen at Resistencia, Chaco, during early July, while at Las Palmas, in the same Territory, they were common from July 13 to 31.   An adult male was taken July 20. The birds frequented the border of open forest or were found in trees scattered through small openings.   They ranged in little companies of from three to six individuals and were rather wary. In early morning it was common for them to perch in the top of some tree to enjoy the heat of the sun, when if chance brought them between the observer and the bright light their bills appeared translucent.   Trees of various sorts that bore berries were frequented, and in spite of the apparently clumsy bill, drupes were seized and swallowed adroitly.   On one occasion one descended to a perch on a tree root fully 15 inches above the inky water of a lowland stream in order to drink.   It bent over gingerly, hestitating several times before dipping the tip of the bill in the water, a caution directed by the presence of savage fish and *jacarés* (alligators).   When a few drops of water had been secured the head was thrown back and the fluid swallowed.   The call note of this species is a harsh rattling grunt.   Flight is accomplished by a succession of beats of the wings followed by a short sail.

The specimen taken had the line of culmen, mandible, and side of maxilla scarlet red; upper part of mandible wax yellow; tip of mandible and line around entire base of bill black; bare skin around eye orange chrome; eyelids smalt blue surrounded by a line of light cadmium; iris hazel; front of tarsus parrot green; posterior face of tarsus and toes Alice blue; claws black.

Family PICIDAE

PICUMNUS CIRRATUS PILCOMAYENSIS Hargitt

Picumnus pilcomayensis HARGITT, Ibis, 1891, p. 606.   (Rio Pilcomayo.)

Eight skins of the piculet that I have listed under this name represent two distinct forms, but at the present moment these can not be separated successfully because of confusion existing with regard to Hargitt's designation of type for his *Picumnus pilcomayensis*.   In the series mentioned three skins (and one alcoholic specimen) from Resistencia and Las Palmas, Chaco (two males and two females), differ constantly from five from the Riacho Pilaga (Kilometer 182), Formosa, and Puerto Pinasco, Paraguay (four males and one female), in heavier barring of the undersurface that covers the entire foreneck and lower throat, leaving the chin alone white.   In the series from Formosa and Paraguay the barring on the foreneck is restricted or absent, a distinction seen without difficulty in comparing specimens. In describing *pilcomayensis*, Hargitt chose a specimen without a label, and assumed that it came from the " Rio Pilcomayo." He mentions a second male secured near Empedrado (a short distance below Corrientes), and therefore not far from Resistencia and Las Palmas, both this bird and the type being of the form with barred neck as nearly as may be determined from his description. He also had a third specimen, a female from Fortin Page on the Pilcomayo. Kerr [87] lists only two specimens, No. 6, which by reference to Hargitt we learn comes from an island opposite Empedrado, and No. 125, a female shot at Fortin Page on the north fork of the Pilcomayo. Kerr mentions that he found this bird on the Pilcomayo, at Puerto Bermejo, and on an island in the Parana opposite Empedrado. Obviously Hargitt's unlabeled type may have come from any of these regions. Since on geographical grounds the bird from Fortin Page should resemble the light-breasted form from the interior of Formosa and from Puerto Pinasco, by inference it is possible that the type of *pilcomayensis*, if of the barred-throated form, may have come from near either Empedrado or Puerto Bermejo. Until this specimen is examined with the female from Fortin Page, the two forms, one of them undescribed, here included can not be successfully differentiated.

Measurements of the skins secured are appended.

No.	Sex	Locality	Date	Wing	Tail	Exposed culmen
284629	Male	Resistencia, Chaco	July 9, 1920	51. 2	33. 0	10. 8
284627	...do	Las Palmas, Chaco	July 21, 1920	49. 2	32. 5	10. 2
284624	...do	Riacho Pilaga, Formosa	Aug. 8, 1920	50. 4	32. 5	11. 0
284626	...do	....do	Aug. 18, 1920	47. 6	30. 0	9. 5
284628	...do	Puerto Pinasco, Paraguay	Sept. 1, 1920	50. 4	32. 8	10. 0
284622	...do	....do	Sept. 30, 1920	48. 6	31. 0	10. 2
284625	Female	Las Palmas, Chaco	July 14, 1920	48. 5	29. 2	10. 3
284623	...do	Riacho Pilaga, Formosa	Aug. 11, 1920	49. 5	29. 5	11. 0

This piculet was found in tangles of low brush and vines at the borders of tracts of forest, often in company with the little bands of flycatchers, tanagers, wood hewers, and formicariids that traveled in congenial companies in search for food. The piculets were found frequently in pairs, though the season was winter. They ranged in such tangled cover, and clambered, climbed, and hopped about so actively, that one seldom caught more than a glimpse of them, as their tiny size made it difficult to follow their course. Attention was usually drawn by their high-pitched calls *tse-re tse-re*, or by their pecking at some bit of bark. In general appearance and

[87] Ibis, 1892, p. 138.

actions they suggested titmice or nuthatches, but lacked the habit of traveling head down found in the latter birds, though the piculets climbed back down along the underside of limbs without difficulty. The tail was not used as a brace and did not touch the limbs on which the bird traveled save by accident or for an instant when the clinging attitude common to titmice was assumed. The birds ordinarily ranged from 1 to 3 meters from the ground. The flight is undulating, and on the wing the bird appears thick set and heavy. The feathers exhale the strong, rank odor characteristic of hole-roosting woodpeckers.

In Paraguay the Angueté Indians called the piculet *kehlanke moh*. These birds were recorded as follows: Resistencia, Chaco, July 9, 1920; Las Palmas, Chaco, July 14 to 31; Riacho Pilaga, Formosa, August 8 to 18; Puerto Pinasco, Paraguay, Kilometer 25 West, September 1, Kilometer 80, September 11, and the Cerro Lorito, September 30.

A male shot July 9 had the maxilla and the tip of the mandible blackish slate; base of mandible gray number 6; tarsus and toes slate; iris dark brown.

In the male the red-tipped feathers on the forehead have broader, stiffer shafts than in the plainer feathers of the remainder of the crown, or in the less decorative head feathers of the female.

DYCTIOPICUS MIXTUS BERLEPSCHI (Hellmayr)

Dryobates mixtus berlepschi HELLMAYR, Verh. Orn. Ges. Bayern, vol. 12, July 25, 1915, p. 212. (Mangrullo, Neuquen, Argentina.)

Near the Rio Negro, below General Roca, Rio Negro, two of these woodpeckers were encountered in a grove of large willows on December 3, 1920, and an adult female was taken. In the open forest in the vicinity of Victorica, Pampa, this form was fairly common, so that two immature birds were shot on December 27 (one a male, the other with sex not determined), an adult male was secured on December 28, and an immature female on December 29. In form and habits these birds suggest *Dryobates nuttalli*. They search persistently for food on the rough bark of low trees, and though rather shy, as they often rest motionless on the side of a limb to avoid detection, are after all easy of approach. The flight is undulating and the call note a low rattle. At the end of December immature birds, recently from the nest, were observed in company with adult males.

Adults of this form may be distinguished at a glance from typical *mixtus* (as represented by a series of four from near Buenos Aires, taken in March, September, and October) by the much longer bill, while on closer comparison the darker auricular spot of *berlepschi* is readily apparent. The adult female secured at General Roca, with a culmen measurement of 23 mm., comes from near the type-locality, as Mangrullo, Neuquen, lies between 80 and 100 kilometers northwest. The culmen in the adult male from Victorica measures 24.5 mm. and in coloration the bird is in agreement with the specimen from Roca and with Hellmayr's description. Since Hellmayr described this subspecies from specimens from Mangrullo, Arroytos, and "Rio Limay" in the Territory of Neuquen, the record from Victorica, Pampa, represents a considerable extension of range. It is probable that *berlepschi* is found throughout the belt of open forest that extends from southern San Luis south through Pampa. Conditions in the tract mentioned are more favorable to its existence than in the scanty willows that border the Rio Negro and its tributaries, the Neuquen and Limay.

DYCTIOPICUS MIXTUS MALLEATOR Wetmore

Dyctiopicus mixtus malleator WETMORE, Journ. Washington Acad. Sci., vol. 12, Aug. 19, 1922, p. 326. (Las Palmas, Chaco, Argentina.)

In the Chaco this woodpecker was only fairly common. On July 23, 1920, an adult male (type of the subspecies) was taken near Las Palmas, Chaco, as it worked busily in a dead fall in dense, swampy forest. A female was shot July 27 as it hopped about in low, scattered trees on an open prairie. Another female was shot at the border of forest, near Kilometer 80, Puerto Pinasco, Paraguay, on September 11, 1920. Near Tapia, Tucuman, a female was taken April 8, 1921, and another was observed April 9. The birds when feeding hammer busily at the bark and trunks of trees and in manner suggest small *Dryobates*. Their call is a low rattle.

A single specimen from Tapia is slightly more heavily streaked on the breast than others. The present subspecies is easily distinguished from typical *mixtus* by the heavier black markings on the lower surface and the restriction of the white on the back.

DYCTIOPICUS LIGNARIUS (Molina)

Picus Lignarius MOLINA, Sagg. Stor. Nat. Chili, 1782, p. 236. (Chile.)

Near Concon, Chile, this woodpecker was observed on April 25, 26. and 27, 1921, and a female was collected on April 26. The birds were found in low growths of dense brush and in habits resembled *D. mixtus*.

VENILIORNIS OLIVINUS (Malherbe)

Picus olivinus MALHERBE, Mem. Soc. Roy. Sci. Liege, vol. 2, pt. 1, 1845, p. 67. (Brazil.)

Woodpeckers of this species, like those of the other two forms of the same genus here discussed, were quiet inhabitants of dense forests where they worked industriously, often in concealed situations in which they were discovered with difficulty. In mannerisms

and in the persistence with which they searched for food they suggested the smaller members of the genus *Dryobates*, but frequented
denser cover than is usual in birds of that group. Usually they were
found working steadily along, tapping the limbs over which they
passed as if to test them. When food was discovered they worked
quietly and rapidly, hammering steadily without apparent attention
to their surroundings. When insects were abundant considerable
areas on dead trunks were denuded of bark before the bird had ex-
hansted possibilities and had moved to other feeding grounds. The
only note heard from them was a low *chuh chuh*. On September 20
one of these quiet little birds under the influence of spring began to
drum. The rattle produced was rather short and was made somewhat slowly, with a decrease in speed toward the end that produced
a drawling sound of little carrying power.

The species was recorded as follows: Resistencia, Chaco, July 8
and 10, 1920; Las Palmas, Chaco, July 14 to 31; Riacho Pilaga,
Formosa, August 18; Formosa, Formosa, August 23; Puerto Pinasco,
Paraguay, September 1, 11, and 20. A female was taken at Resistencia July 8; a male at Las Palmas, July 14, and a male at Kilometer
80, west of Puerto Pinasco, September 11.

VENILIORNIS FRONTALIS (Cabanis)

Cloronerpes (Campias) frontalis CABANIS, Journ. für Ornith., 1883, p.
110. (Tucuman.)

April 17, 1921, between 1,600 and 1,800 meters on the slopes of the
Sierra San Xavier above Tafi Viejo, Tucuman, several of these
woodpeckers were seen, and a female was taken. They worked
busily along limbs or small trunks, often in localities near the
ground where they were entirely concealed by dense vegetation, and
their presence was indicated merely by their low calls or their hammering in search for food.

VENILIORNIS SPILOGASTER (Wagler)

Picus spilogaster WAGLER, Syst. Av., 1827, p. 33. (Brazil and Paraguay.)

An adult female was taken near San Vicente, Uruguay, on January 30, 1921, in heavy tree growth in a gulch on the side of the
Cerro Navarro. The bird was in such heavy cover that I should
not have found it save for its steady hammering on a dead limb.
Another was shot at Rio Negro, Uruguay, on February 15, as it
worked quietly through a lowland thicket near a small stream.

Both specimens are in molt on the body. The first one taken
had the maxilla and tip of the mandible dull black; base of mandible
storm gray; iris bone brown; tarsus and toes dark-grayish olive;
claws dusky neutral gray.

SCAPANEUS LEUCOPOGON (Valenciennes)

Picus leucopogon VALENCIENNES, Dict. Sci. Nat., vol. 40, 1826, p. 178.
·· (Brazil.)

The present species was first recorded at Las Palmas, Chaco, when a male was collected on July 14, 1920. One was seen on July 24, and a female was shot on July 27. At the Riacho Pilaga, Formosa, the species was fairly common from August 7 to 16, and four were taken on August 7, 11, and 18. The birds inhabited tracts of forest, where the trees were tall, and fed over the trunks and larger limbs as do pileated woodpeckers. On August 11 males began to drum, indicating that the breeding season was at hand, and a female shot on this day had a fully formed egg in the oviduct. It measures 35.6 by 24.6 mm., and, like other woodpeckers' eggs, is white in color. The drumming of the male was a curious performance, entirely different from what one would expect from so strong and robust a bird. The dead limb chosen for a resonator was struck twice with great rapidity, *ta tat*, the two sounds almost blending into one so quickly did they come. After a rest of a few seconds the drum was given again, and so on, frequently for considerable periods, especially during the hours of early morning. The sound produced, though short, was audible for a considerable distance, so that I heard it frequently when working about lagoons far from forests. A male taken on August 18 must have shared in · the duties of incubation, as the entire abdomen was bare of feathers, while the skin of the denuded area was wrinkled and thickened.

Near Puerto Pinasco, Paraguay, in the vicinity of Kilometer 80, this woodpecker was heard drumming on. September 9, and individuals were seen on September 15 and 20. In this region they were far from common. Near Tapia, Tucuman, a pair was taken on April 12, 1921, in open, dry forest on the slope of a hill. At intervals these two gave subdued chattering calls. In flight this species progresses in strong bounds. The skin is thick and tough and adheres so closely to the body that the preparation of specimens is difficult. On the head it is necessary to separate the skin from the skull with a knife, so firmly is it attached to the bone.

The bill in an adult male, shot July 14, was olive buff, browner toward the tip; iris pinard yellow; tarsus chaetura drab.

The Toba Indians called this species *ne on rah.*

CEOPHLOEUS LINEATUS LINEATUS (Linnaeus)

Picus lineatus LINNAEUS, Syst. Nat., ed. 12, vol. 1, 1766, p. 174.
(Cayenne.)

A fine adult male was taken on July 16, 1920, near Las Palmas, Chaco, in a grove of tall trees on an open prairie. Attention was attracted to it by its steady hammering as it knocked flakes of bark from the dead limbs. The bill in this bird was pale smoke gray, with the culmen lined with hair brown, the base of the mandible and the sides of the maxilla gray number 8; tarsus and toes slate gray number 5; iris dull white.

This specimen differs from *C. l. improcerus* Bangs and Penard in larger size, as the wing measures 193 mm., but is similar to a male from Diamantina (near Santarem), Brazil, and to a series from Surinam (in the Museum of Comparative Zoölogy) the nearest approach to skins from the type-locality available.

CELEUS KERRI Hargitt
•

Celeus kerri HARGITT, Ibis, 1891, p. 605. (Fortin Donovan,[88] Rio Pilcomayo.)

Males of this fine species were secured at Las Palmas, Chaco, on July 31, 1920, and near Kilometer 80, west of Puerto Pinasco, Paraguay, on September 20. The species has been recorded previously from the Rio Pilcomayo, Sapucay, and Curuzu Chica, Paraguay, and Pan de Azucar, Brazil, so that the first of these constitutes a southward extension of range from information in published records. The bird from Puerto Pinasco is distinctly browner in tone than the one from Chaco. This species frequents the heaviest growth of the swampy forests in the Chaco, where, save for its persistent hammering as it chisels its food from decaying wood, it might readily pass unnoticed. On careful approach through the tangle of vines and thorny scrub the birds were found within a short distance of the ground, often under such somber conditions that even their light-colored, crested heads were hardly to be distinguished. In spite of Kerr's derogatory remarks regarding the soiled appearance of his specimens, I found this a strikingly marked and beautiful bird, the more so since its finely contrasted colors, viewed amid its somber surroundings, came as a distinct surprise. A fully adult male had the maxilla light mouse gray, becoming dark quaker drab along the line of the culmen; mandible yellowish glaucous, shaded at base with neutral gray; iris morocco red; tarsus storm gray; toes gray number 6.

PICULUS CHRYSOCHLOROS CHRYSOCHLOROS (Vieillot)

Picus chrysochloros VIELLOT, Nouv. Dict. Hist. Nat., vol. 26, 1818, p. 98. (Paraguay.)

Oberholser[89] has indicated that *Chloronerpes* Swainson, 1837, is to be replaced by *Piculus* Spix, 1824.

The present species was encountered in the Chaco on only a few occasions in heavy woods in the vicinity of streams. Two males

---

[88] See Ibis, 1892, p. 136.
[89] Proc. Biol. Soc. Washington, vol. 36. Dec. 19, 1923, p. 201.

were taken at Las Palmas, Chaço, on July 17, 1920, and another (preserved in alcohol) on July 27. A female was secured August 11 near Kilometer 182, Formosa, and another was seen on August 18. A male was shot at Kilometer 80, west of Puerto Pinasco, Paraguay, on September 11. These birds work quietly about the larger limbs and trunks of trees, often descending on dead stubs to within a few feet of the ground. They spend much time in digging for food in deadwood, hammering as steadily as a *Dryobates*. Though silent they showed some curiosity and decoyed readily within range.

The Toba Indians called them *kwi rah*.

This form shows considerable variation in length of wing, as two males from Las Palmas measure 115 and 127 mm., respectively, while in one from Puerto Pinasco the wing equaled 122.5 mm.

PICULUS RUBIGINOSUS TUCUMANUS (Cabanis)

Chloroncrpes tucumanus CABANIS, Journ. für Ornith., 1883, p. 103. (Tucuman.)

On April 17, 1921, two were seen at an elevation of 1,700 meters in the heavy forest covering the Sierra San Xavier above Tafi Viejo, Tucuman, and a female was taken. The birds fed among the higher tree limbs. Their usual call was a thin *pick*, a curious call resembling that of a rose-breasted grosbeak, and in addition they gave a rattling chatter.

The specimen taken has a wing measurement of 120 mm.

TRICHOPICUS CACTORUM (d'Orbigny)

Picus cactorum d'ORBIGNY, Voy. Amér. Mérid., Ois.. 1835–1844, p. 378, pl. 62, fig. 2. (Mizque, Bolivia.)

This genus belongs among the melanerpine woodpeckers and appears to be allied to *Tripsurus*. In its characters it is intermediate between *Tripsurus* and *Centurus*, particularly in respect to the restriction of feathering about the eye, as in *Trichopicus* the ocular apterion while large is less extensive than in *Tripsurus*, while there is a distinct line of feathers on the lower lid. Until more is known of the structure of these birds the present species must be considered of doubtful generic distinctness.

At Las Palmas, Chaco, the yellow-throated woodpecker was encountered and specimens taken on July 19, 27, and 30. They ranged in open groves or in trees growing scattered through the savannas, and were found in bands of four to half a dozen that roved about and were not settled in any particular region. In actions they suggested *Balanosphyra* strongly as they worked about on the upper limbs of the trees. Their common call was a harsh *yak-ah yak-ah* and their alarm a scolding rattling *chuh-h-h chuh-h-h check-ah*. At times they were rather vociferous, particularly when excited. When alarmed they hid motionless among the limbs and remained thus for some time. One of the specimens taken had the third toe on one foot only partly developed with no claw, while the fourth toe on the opposite side had a deformed claw. Yet this bird climbed without difficulty.

At Tapia, Tucuman, these woodpeckers were fairly common from April 9 (when one was taken) to 13, 1921. They ranged through the scrubby forest in little bands that contained from three to six individuals, usually in the vicinity of the giant cactus that grew abundantly in this region. Frequently they clambered about or perched on the cacti, and holes of some small woodpecker in the cactus trunks were attributed to this species.

An adult male, taken July 19, had the bill dull black; iris natal brown; tarsus and toes deep-grayish olive.

An immature male from Tapia, not wholly in adult plumage, has the light dorsal line smoky gray instead of white as in specimens from Las Palmas.

LEUCONERPES CANDIDUS (Otto)

Picus candidus OTTO, Nat. Vög. Buffon, vol. 23, 1796, p. 191. (Cayenne.)

This handsome woodpecker was recorded first at Las Palmas, Chaco, on July 28, 1920, when half a dozen passed with low chattering notes. At the Riacho Pilaga, August 12, three were seen and collected, and others were recorded on August 21, near Fontana, on the railroad. Two were noted at Formosa, August 23. Near Puerto Pinasco, Paraguay, one was observed September 3, while near Kilometer 80, west of Puerto Pinasco, they were fairly common on September 7, 17, and 18. Three were taken on the date last mentioned. These birds are gregarious, at least in winter, and seem to wander, as they were seen in flight in long looping bounds over the open country, when their contrasted colors of black and white were sure to attract attention. When at rest they were encountered in regions of scattered trees in the savannas, or in the open, straggling growth of palmars. Their gregariousness was marked, and when in flight if one alighted the others came down at once to join it, while they hovered over dead companions, or rested near by with scolding notes. Their usual call was a drawn out *kee-ee-ee* or *kee-ee-ah*, uttered with a mournful cadence, given with greater vehemence when the birds were excited. When approached they had the usual woodpecker habit of working around to the opposite side of a tree trunk, but in some cases paid little attention to me. In mid-September, with approach of the mating season, males extended their wings above the back and then fairly danced up the tree trunks with raised crest, mouth opened, and excited calls. In many of their ordinary mannerisms they suggested ant-eating woodpeckers.

In external characters this genus seems to be only slightly differentiated from *Tripsurus*, with which it agrees in the broadly naked area about the eye.

CHRYSOPTILUS MELANOLAIMUS MELANOLAIMUS (Malherbe)[00]

Chrysopicus melanolaimus MALHERBE, Mon. Pic., vol. 2, 1862, p. 185, pl. 89, figs. 7 and 8. (Bolivia and Chile.)

A male in partial molt secured at Tunuyan, Mendoza, on March 27, 1921, is taken as representing the typical race of this woodpecker, though specimens from Bolivia, the type-locality, have not been available for comparison. The black area posterior to the malar stripe is broad and extensive, so that it passes down on the side of the neck well below the level of the ear. The underparts are marked with heavy black spots and bars, and small spots cover the entire abdomen. The rump is very light. Measurements of this specimen are as follows: Wing, 148; culmen, 37.2; tail, 95.6; tarsus, 32 mm. Scattered new feathers are still in the sheaths on the head, neck, breast, back, and wing coverts. The outer primaries have been renewed, but the inner ones and some of the secondaries are still of old growth.

An adult female from Victorica, Pampa, taken December 27, 1920, is somewhat intermediate toward *C. a. perplexus*, but in the small series of specimens at hand seems to be nearer to those that I have called *melanolaimus*. The bird in question is in worn breeding plumage. The light markings of the upper surface are bleached until they are nearly white, and the black post malar mark is extensive. The markings of the lower surface are less heavy than in birds from Mendoza, but the abdomen is distinctly spotted. This bird measures as follows: Wing, 161.2; culmen, 37.1; tail, 104; tarsus, 31 mm.

A male in the United States National Museum, taken at Santiago del Estero, July 29, 1922, by D. S. Bullock, is also representative of this form, as it has the bold markings and large size characteristic of *melanolaimus*. It measures as follows: Wing, 156.2; tail, 102.2; culmen from base, 38.2; tarsus, 30 mm.

At Victorica, Pampa, these woodpeckers were found from December 27 to 29, at times in parties of five or six, through the dry, open forest of caldén, algarroba, and similar trees prevalent in this section. The presence of two in the arid region near Tunuyan, Mendoza, on March 27, on a low hill covered with bushes, was a surprise to me, as large tree growth in this neighborhood was confined to poplars, cottonwoods, and willows growing along irrigation ditches.

---

[00] Though the bird is indicated as *melanolaimus* in the text the plate is marked *melanolaemus*.

CHRYSOPTILUS MELANOLAIMUS PERPLEXUS Cory

Chrysoptilus melanolaemus perplexus Cory, Cat. Birds Americas, pt. 2, no. 2, December 31, 1919, p. 442. (Conchitas, Buenos Aires.)

An adult male of this recently recognized form was secured at the Estancia Los Yngleses, near Lavalle, Buenos Aires, on November 9, 1920. Study of a small series of specimens from northern Buenos Aires shows that the subspecies rather doubtfully characterized as *perplexus* by the late Mr. Cory may be distinguished from birds from Mendoza, taken as representing true *melanolaimus*, by shorter bill, somewhat smaller black markings on the ventral surface with the abdomen more nearly immaculate, and more extensive light markings on the inner web of the second rectrix. The culmen in two males of *melanolaimus* measures 33.7–37.2 mm.; in two females, 37.1–37.5 mm. In three males of *perplexus* the culmen ranges from 30.7–31, in four females from 29.5–30 mm. The black area posterior to the malar stripe is slightly more restricted than in the Mendozan birds. The wing is slightly shorter in *perplexus*, but the distinction here seems rather slight. In the bird from Lavalle the wing measures 141.5 mm., but I note that the primaries are somewhat worn.

These birds were found in fair numbers in the grove at the Estancia Los Yngleses on November 9. One was observed on November 13.

When comparing woodpeckers of this species it must be borne in mind that the bird is much brighter colored and has the light markings much more yellow or orange, as the case may be, when in fall or winter plumage than later in the year. The bright colors, through wear and fading, become paler and less intense, so that summer and winter specimens are frequently very different in appearance.

CHRYSOPTILUS MELANOLAIMUS NIGROVIRIDIS C. H. B. Grant

Chrysoptilus nigroviridis C. H. B. Grant, Ibis, 1911, p. 321. (Fortin Nueve,[91] Rio Pilcomayo, Paraguay, lat. 24° 53′ S., long 58° 30′ W.)

The skin of a male secured at Las Palmas, Chaco, on July 16, 1920, is taken as representative of the present bird. A second male preserved as a skeleton was secured on July 31. Grant in his original description considered *Chrysoptilus nigroviridis* as a distinct species, somewhat intermediate between *C. melanolaimus* and *C. melanochlorus*. He described it as "rather larger" than *melanolaimus*, but with the black behind the malar stripe more restricted, the rump golden yellow, the ear coverts washed with golden buff and somewhat more greenish below. The single skin that I have from Las Palmas fits his description both in color

---

[91] From Kerr, Ibis, 1892, pp. 122, 135.

and measurements with sufficient accuracy, and is identified as the bird that Grant described. From my comparisons it appears that *nigroviridis* is a fairly well marked subspecies of *melanolaimus* that may be distinguished from specimens from Mendoza, taken as representative of typical *melanolaimus*, and from *perplexus*, by the restriction of the black marking behind the malar stripe, the smaller size of the spots on the undersurface, and slightly more greenish cast below. In measurements this form seems somewhat intermediate between true *melanolaimus* and *perplexus*. Measurements of two male birds from Las Palmas (including the one prepared as a skeleton) are as follows: Wing, 155–157.5 mm.; culmen, 31.6–34 mm. The type of *nigroviridis* collected by Prof. J. G. Kerr was taken in the month of April (1890), so that we may presume that it was in winter plumage, which would account for the golden wash on ear coverts and rump mentioned by Grant. On the basis of my present studies (on very insufficient material) it would seem that *nigroviridis* is the form of *melanolaimus* found in the Chaco. In restriction of the black behind the malar stripe and in the greenish tinge of the undersurface, the Las Palmas specimen, while undoubtedly closely related to *melanolaimus*, suggests *Chrysoptilus melanochlorus*, and it may be that further collecting will produce intermediate specimens that will link the forms of the two groups as geographic variants of one wide-ranging species.

These woodpeckers, handsomely marked and of good appearance, were common in the Chaco near the Rio Paraguay, but did not seem to penetrate far inland. In general habits they resemble flickers, but are more partial to wooded sections than the South American species of that group, and to not range far into open country unless trees are near at hand. They fed on the ground in little openings, and in settled districts were observed on plowed ground between rows of trees in orange groves. Burns in open savannas were attractive to them. They are gregarious to the extent that five or six may be found together, though it is not unusual to see single birds at rest quietly in the open top of a tree, or to have one fly up to hitch along the larger limbs of a tree under shelter of leaves. The flight is bounding like that of any flicker and in general appearance the species suggests the pampas flicker; in fact, I killed my first specimen on a dull, gloomy, rainy day under the impression that it was an ordinary flicker. The ordinary call note is a loud scolding *keah keah keah kah.*

From July 16 to 31, 1920, the species was fairly common at the borders of open savannas near the small stream known as the Riacho Quia (Guarani for "dirty creek"), at Las Palmas, Chaco.

54207—26——15

In a male taken July 16 I noted that the iris was garnet brown; in a second male secured July 31 the bill when fresh was dark neutral gray, the iris Vandyke brown, the tarsus and toes tea green.

No specimens were taken elsewhere, so that the following observations are allocated under the present subspecies on the basis of probability. These birds may be quite local in their distribution, as during my collecting at the Riacho Pilaga, Formosa, I did not meet with them, but on August 21 several were seen among standing dead trees in cut-over lands that I traversed in coming out to the railroad at Kilometer 182, to the station known locally as Fontana. Near the town of Formosa I found them fairly common, on August 23 and 24, in palm forests that grew in swampy localities. One that I observed drumming on the trunk of a palm varied the pitch of its music by a shift of position on the tree. The roll or drum produced began slowly but increased in rapidity to its close. It continued for a period of three seconds and then stopped abruptly. It suggested the sound produced by *Colaptes auratus*, but was delivered more slowly. At Puerto Pinasco the birds were seen near the river, on September 3, in fair numbers. At this time they seemed to be mating, and when calling from a perch had the habit of opening and closing the wings suddenly, to flash the vivid yellow concealed beneath. None were observed inland to the west of Puerto Pinasco.

CHRYSOPTILUS MELANOCHLORUS CRISTATUS (Vieillot)

Picus cristatus VIEILLOT, Nouv. Dict. Hist. Nat., vol. 24, 1818, p. 98.
(Paraguay.)

A male and a female fully grown but in immature plumage, shot near San Vicente in the Department of Rocha, Uruguay, on January 26 and 29, 1921, are taken to be representative of this form. These birds are pale greenish yellow below, heavily marked with sharply defined black spots that become bars on the sides and flanks. These specimens measure as follows: Male, wing, 149; tail, 98; culmen, 31; tarsus, 26.8 mm. Female, wing, 148; tail, 97; culmen, 28.4 tarsus, 26 mm.

These handsome flickerlike woodpeckers were observed only near San Vicente, Uruguay, from January 26 to 31, 1921. They ranged in little family parties of five or six through the extensive palm forests of the lowlands, attracting attention by their loud calls. On January 29 the muffled chatter of young attracted attention to a nest at a height of 7 feet from the ground in a living hard-trunked tree (not a palm) that grew in a small grove in the bottom of a shaded gulch on a hill slope. The woodpeckers had drilled through living wood into a spot that had decayed, and then had dug out a large irregular cavity a foot deep. The four young in nakedness, long necks, and general appearance suggested flickers of the same age.

They gave the chattering call usual to young woodpeckers in begging for food, and in addition emitted a low wheezing note. The smallest was entirely devoid of down. All had prominent heelpads. These birds were preserved in alcohol.

COLAPTES CAMPESTRIS CAMPESTRIS (Vieillot)

Picus campestris VIEILLOT, Nouv. Dict. Hist. Nat., vol. 24, 1818, p. 101. (Paraguay.)

A male collected at Kilometer 80, west of Puerto Pinasco, Paraguay, on September 13, 1920, was the only one secured. Two were recorded near the Rio Paraguay at Puerto Pinasco on September 3. At Villa Concepcion, Paraguay, on October 3, an individual at rest in a tree continually flashed one wing to display the yellow markings below.

The Augueté Indians called this species *upaikú*.

The male taken had the following measurements: Wing, 159; tail, 106.4; exposed culmen, 34.5; tarsus, 32.1 mm.

COLAPTES CAMPESTROIDES (Malherbe)

Geopicos (Colaptes Swainson) *campestroides* MALHERBE, Rev. Mag. Zool., 1849, p. 541. (South Brazil.)

In habits and general appearance the pampas flicker differs little from the familiar *Colaptes auratus* of the eastern United States. The *carpintero del suelo*, as *C. campestroides* is usually known, is found frequently in little bands that feed on the ground in open country, dotted with trees to which the birds may fly for shelter. Through the Chaco they were common in the open savannas, and were also abundant through the undulating pampas and the lowland palm forests of eastern Uruguay. Recent burns were always attractive to them. Though formerly reported as common on the level plains of Buenos Aires and near-by Provinces, the species may now be almost extinct there as I saw none in extended travels through that region. (Pl. 3.)

One of the call notes of the pampas flicker is a loud call strongly suggestive of the whistle of a greater yellowlegs, especially when heard at a distance across a marshy savanna where conditions of situation favor such a deception. When several gather on the trunk of a tree or a fence post, they go through many gesticulations with nodding heads, the whole accompanied by loud ejaculations of *whick whick whick*. Often one or both wings are extended and retracted quickly with a sudden flash of yellow, as the undersurface of the flight feathers is displayed. Another call is a harsh *kiu*, a signal that carries far across the open country. In the palm groves of Uruguay, when young flickers had recently left their nests, this species joined the oven birds in railing at my intrusions.

Though this flicker fed constantly on the ground, and in the trees often perched on a limb like other birds, it climbed with ease, and often emulated other woodpeckers in clambering over the trunks or limbs. The flight was bounding and was marked by the display of the white rump and the flashing of the undersurface of the wings. In Uruguay, as in Brazil, they were known as *pico pao* or less often as *pico pico*.

The species was recorded at the following points: Las Palmas, Chaco, July 13 to 31, 1920; San Vicente, Uruguay, January 26 to February 2, 1921; Lazcano, Uruguay, February 5 to 9; Rio Negro, Uruguay, February 15 to 19. The lack of records from the southern part of the range of the species is notable. An adult male from San Vicente, taken January 26, has the undersurface of the tail washed strongly with yellow. A male and a female from Las Palmas, taken July 13, lack this marking entirely.

COLAPTES PITIUS PITIUS (Molina)

Picus pitius MOLINA, Sagg. Stor. Nat. Chili, 1782, p. 236. (Chile.)

The typical form of the Chilian flicker is distinguished from *C. p. cachinnans* Wetmore and Peters [92] of southern Patagonia by longer, broader bill, and less heavily barred underparts, in particular on the sides.

After careful consideration of the smaller generic groups into which the flickers recently have been separated, I do not consider that the genera *Pituipicus* and *Soroplex* are well founded. *Pituipicus* has been distinguished from *Soroplex* and *Colaptes* on the grounds that it possesses a bill longer than the head, gonys longer than mandibular rami, and length of tail equal to less than two-thirds of wing. The last-named character has no weight, as in a series of nine specimens I find that the tail is universally equivalent to two-thirds or more of the length of wing. As regards the other characters it is found that they do not hold true in the short-billed southern subspecies of *pitius*, in which the gonys is no longer than the mandibular rami and the bill equal to or shorter than the head. *Pituipicus*, therefore, may not be maintained. To continue, after careful study it is found that the South American flickers included in the genus *Soroplex* differ from true *Colaptes* only in heavier bill, in black rather than in highly colored undersurface of tail, and in lack of a black breast crescent. Alleged characters based on the form and character of the gonys are unstable. On consideration of this matter I do not hold *Soroplex* as a valid group to be distinguished from true *Colaptes*.[93]

---

[92] Proc. Biol. Soc. Washington, vol. 35, Mar. 20, 1922, p. 43. (Bariloche, Gobernacion de Rio Negro, Argentina.)

[93] *Nesoceleus* may be readily separated from *Colaptes* by the open, exposed nostrils, with no covering of forward projecting plumes.

Colaptes p. pitius was encountered near Concon, Chile, in small numbers. Three of these birds frequented a small valley with slopes grown with low bushes studded with occasional trees along a watercourse, where they fed on the ground among the bushes in small openings grown with grass. In spite of this cover they were so wary that it was difficult to approach them, as at the slightest alarm they rose and traveled away with strongly bounding flight that displayed alternately the white rump and the yellow undersurface of the wings. Their call was a high-pitched double note, flickerlike in nature, but of different character than the species encountered in Argentina. Like other flickers, I found that they decoyed readily to a " squeak " when they had no cause to suspect danger, a trait that brought two within range and added them to my collection.

A specimen secured on April 28, 1921 (probably a male), had the bill dark neutral gray, iris chalcedony yellow, and tarsus and toes deep olive gray.

Family TROGONIDAE

TROGONURUS VARIEGATUS BEHNI (Gould)

Trogon behni GOULD, Mon. Trog., ed. 2, 1875, pl. 20, with text. (Bolivia.)

Two skins preserved from the vicinity of Puerto Pinasco, Paraguay, exhibit the characters assigned to the subspecies *behni*. A male differs from specimens from eastern Brazil in larger size, restriction of white on the tips of the lateral rectrices, and more greenish cast to head and breast. The female is larger in size. The white breastband appears wider in both sexes than in the typical form. The male from Puerto Pinasco measures: Wing, 128; tail, 124; exposed culmen, 16.5; tarsus, 14 mm.; the female, wing, 120; tail, 120; exposed culmen, 15.4; tarsus, 13.6 mm.

On September 1, 1920, a female was secured in heavy timber on a low hill at Kilometer 25, west of Puerto Pinasco, Paraguay. The bird flew from some unseen perch to alight as easily as a jay on an open limb, 3 meters from the ground. When it observed me the crest feathers were slowly raised. On September 21 two were taken in the border of somewhat open forest, near Kilometer 80. The call note of this species is a fairly loud regular *coo coo coo* varied to *coh coh coh coh coh*. The flight is direct and darting, and on the wing the birds suggest cuckoos in manner and actions. The body in this species exhales the same strong odor found in cuckoos, particularly in the Crotophaginae.

An adult male, when fresh, had the bill court gray; lower mandible washed with light celandine green; margin of eyelids yellow ocher; iris carob brown; toes storm gray, with the scales outlined in whitish. In an adult female the bill was light celandine green;

culmen lined faintly with fuscous; margins of eyelids around eye isabella color; iris carob brown; toes as in male.

In Guarani this species was known as *su ru cu áh*, in Anguetê as *tsa lakh.*

TROGONURUS SURRUCURA (Vieillot)

Trogon surrucura VIEILLOT, Nouv. Dict. Hist. Nat., vol. 8, 1817, p. 321. (Paraguay.)

An adult female taken at Las Palmas, Chaco, on July 21, 1920, was found at rest on an open limb in rather open forest near a stream. It perched with tail hanging straight down and at intervals uttered a low *plick.*

The sides of the bill were court gray; base of maxilla dark neutral gray; gonys light celandine green; iris light seal brown; lower end of tarsus and toes dark neutral gray, with scales outlined in grayish white; underside of toes yellowish.

This specimen has the following measurements: Wing, 134; tail, 149.5; exposed culmen, 15.2; tarsus, 12.6 mm.

Family TROCHILIDAE

LEUCIPPUS CHIONOGASTER (Tschudi)

Trochilus chionogaster TSCHUDI, Fauna Peruana, Orn., 1845–1846, p. 247, pl. 22, fig. 2. (Peru.)

Two immature males secured at Tapia, Tucuman, on April 12 and 13, 1921, agree with descriptions of the present species save that in one there are a few tiny, light-vinaceous cinnamon feathers between the mandibular rami and that in the other a few flecks of the same color occur on the throat and sides of the throat. Both of these specimens are obviously only recently grown and are in molt on the foreneck. It is suggested that the brownish color described is a juvenal plumage that is entirely lost in the adult. These specimens should perhaps bear the subspecific name *longirostris* of Schlüter,[94] but in the absence of comparable material this is not certain. Schlüter states that his form has the following measurements: Wing, 60; culmen, 25–27 mm. In my specimens the wing measures 54.8 and 55 mm. and the culmen 23 and 23.6 mm. (It must be borne in mind that these are evidently immature birds.) Simon[95] places the more southern birds under the subspecific name *hypoleucus* of . Gould.[96] Gould gives the length of bill in his type as 28 mm., while Simon records from 23 to 25 mm. only for specimens that he considers *hypoleucus.* As a further complication, Mr. Ridgway[97] considers the generic name *Leucippus* Bonaparte, type *Trochilus*

[94] *Leucippus leucogaster longirostris* Schlüter, Falco, 1913, p. 42. (Province of Salta.)
[95] Hist. Nat. Troch., 1921, p. 103.
[96] *Trochilus hypoleucus* Gould, Proc. Zool. Soc. London, 1846, p. 90. (Bolivia.)
[97] Birds North Middle America, vol. 5, 1911, p. 305.

fallax Bourcier, as not applicable to *T. chionogaster* Tschudi, which he says should probably be placed in *Talaphorus* Mulsant.

The two or three individuals of this species seen were observed at the flowers of a common red-flowered epiphyte (*Psittacanthus cunei-folius*) that was attractive to other hummers in the region. They hovered with humming wings in order to probe the long tubed blossoms, and at short intervals paused to rest on some convenient perch. Their call note was a low *chit chit* suggestive of that of *Sappho.*

One of the birds when taken had the base of the mandible deep corinthian red, and the remainder of the bill and the feet black. In the other the base of the mandible was vinaceous fawn; maxilla and tip of mandible black; tarsus and toes aniline black.

The plate in Tschudi's work on which the species is depicted was made previous to the writing of the text as it is lettered *Trochilus leucogaster,* a name proposed for another hummer, so that in the text change was made to *chionogaster.* No definite type locality is given by Tschudi so that Peru is assumed.

HYLOCHARIS CHRYSURA (Shaw)

Trochilus chrysurus SHAW, Gen. Zool., vol. 8, pt. 1, 1812, p. 335. (Paraguay.)

Four specimens of this interesting hummer were taken as follows: A female at the Riacho Pilaga on August 14, 1920, a female at Kilometer 25, a male at Kilometer 80, and a second male on the Rio Paraguay near Puerto Pinasco, Paraguay, on September 1, 15, and 30. These four specimens present a puzzling combination of differences that with available material may not be treated satisfactorily. The two females from Formosa and Paraguay, with a culmen measurement of 20 and 21.5 mm., respectively, and a wing 52.5 mm. long, agree fairly well with birds from Buenos Aires and offer nothing worthy of comment. A male taken on the Cerro Lorito on the eastern bank of the Paraguay River opposite Puerto Pinasco has the culmen 20.2 mm. and the wing 50 mm. It differs from any others in the small series examined in having the entire maxilla dull black, as well as the distal half of the mandible. In skinning this bird I marked it as sexually fully adult, but it may be that the black bill is still an indication of an immature condition. The last specimen, a fully adult male, from the Chaco at Kilometer 80, with the culmen 18.6 mm. and the wing 51.4 mm., in its small bill suggests the form named *maxwelli* by Hartert [98] from the plains near Reyes on the Rio Beni in northern Bolivia. Hartert records that specimens from Matto Grosso seem intermediate between his form and the typical

---

[98] *Hylocharis ruficollis maxwelli* Hartert, Nov. Zool., vol. 5, December, 1898, p. 519.

one, and it is not improbable that my specimens from the Chaco and from northern Paraguay are in the same category.

The present species seemed to be partially migratory. In the Chaco it was recorded near the Riacho Pilaga, Formosa, on August 13 and 14, 1920, while on the wooded hill at Kilometer 25, Puerto Pinasco, I found it common on September 1. The species was found only on September 15 near Kilometer 80, but was recorded again on the Cerro Lorito opposite Puerto Pinasco on September 30. The birds were usually found about flowers in heavy forests, though occasionally they searched for insects over the bark of trees or came to the blossoms of lapacho trees (*Tecoma obtusata*) that grew at the border of the monte. They were nervous and excitable, and on several occasions darted swiftly at my head when I was squeaking to call up other birds. In feeding they worked actively at flower clusters for several minutes and then rested on perches protected by overhanging leaves. In flight their wings produced a loud rattle, and in addition the hummers made a metallic sound, composed of a series of rapid notes that were plainly vocal since the throat was in movement as they were uttered. Often when the birds scolded from a perch the wings were extended wide for a few seconds and then drawn in again to the body. A male shot September 15 had the testes enlarged about one-half.

The bird was known in the Guarani tongue as *mainumbú*.

CHLOROSTILBON AUREO-VENTRIS (d'Orbigny and Lafresnaye)

Ornismya aureo-ventris d'ORBIGNY and LAFRESNAYE, Mag. Zool., vol. 8, 1838, cl. 2, p. 28. (Moxos, Cochamba, Bolivia.)

An adult male of this hummer taken at Rio Negro, Uruguay, February 14, 1921, lacks the distinct coppery reflections of the undersurface found in specimens from Argentina, and is smaller so that it agrees with what is currently known as *egregius*.[99] This bird has the following measurements: Wing, 50.4; tail, 30.6; exposed culmen, 18.8 mm. An adult male from Lazcano, Uruguay, shot February 6, and an adult female taken January 26, near San Vicente, Uruguay, while not wholly similar to skins from Argentina, agree with them closely. Of a pair from the Riacho Pilaga, Formosa, secured August 11 and 14, 1920, the male is smaller than those from Buenos Aires, as its wing measures 51 mm. and the exposed culmen, 17.7 mm. Three immature birds from Tapia, Tucuman (April 7, 8, and 13, 1921), have the bill distinctly duller in color than adults. One male has an extensive area of the green adult

[99] *Chlorostilbon egregius* Heine, Journ. für Ornith., 1863, p. 197. (Sao Joao del Rey, Minas Geraes, Brazil.) The type locality of this bird has been cited as Taquara, though Heine states distinctly that he described it from two skins in the Berlin Museum secured by Sellow at Sao Joao del Rey.

plumage on the throat, with more feathers of the same color in process of growth around it. The male until it dons adult plumage has a white spot behind the eye like that found in females. Immature birds of both sexes are duller above than adults.

Present understanding of the geographic forms of *Chlorostilbon aureo-ventris* is highly unsatisfactory. Simon has described a southern form,[1] including a range from southeastern Brazil to Buenos Aires, but has complicated matters by calling *egregius* a form of *C. prasina*. The subspecific variations of *aureo-ventris* must for the present remain clouded in doubt.

Chlorostilbon aureo-ventris was recorded at the following points: Riacho Pilaga, Formosa, August 11, 14, and 18, 1920; Lavalle, Buenos Aires, October 29 to November 15; San Vicente, Uruguay, January 26 to February 2, 1921; Lazcano, Uruguay, February 3 to 9; Rio Negro, Uruguay, February 14 to 19; Tapia, Tucuman, April 7, 8, and 13. Individuals were seen during July, 1920, near Las Palmas, Chaco.

During winter these hummers were encountered occasionally at the border of forests and thickets, where they received the warmth of the sun and were protected from cold winds. Small shrubs that were in flower were frequented, and I saw them gleaning insects from the limbs of trees. In the pampas during summer hummers came in small numbers to flowers about the estancias or occasionally dropped down into the inclosed patios of small country hotels. On January 26, near San Vicente, Uruguay, a female flashed by me in a forest of palms, and on glancing up I caught sight of her nest placed on a swinging bit of fern root, 12 feet from the ground, below the crown of leaves that formed the top of the tree. To my great disappointment the nest was empty. It was of the usual hummer type, a soft, cup-shaped structure made of plant downs and fine bark, covered with fragments of brown bark fastened in place with spider webbing.

Near Tapia, Tucuman, this hummer, with others, frequented the red flowers of an abundant epiphyte (*Psittacanthus cuneifolius*). A female, taken August 14, had the tip of the bill black, and the basal half orange-cinnamon, the two colors blending at the point of junction; iris Rood's brown; tarsus and toes fuscous; nails black.

OREOTROCHILUS LEUCOPLEURUS Gould

Oreotrochilus leucopleurus GOULD, Proc. Zool. Soc. London, 1847, p. 10. (Chilian Cordillera.)

Near Potrerillos, Mendoza, two females were secured on March 16 and 19, 1921. This species is more sluggish in its movements than

---

[1] *Chlorostilbon aureiventris tucumanus* Simon, Hist. Nat. Troch., 1921, p. 65. (Tucuman.)

most hummers, and though it fed at the same plants (the red-flowered *Psittacanthus cuneifolius*), it flew directly to the flower clusters and clung to them with its strong feet while it probed the blossoms, instead of hovering in the air before them. The flight was comparatively slow and the wing motion far from rapid. At El Salto (altitude 1,800 meters) these birds sought the warm shelter of hill slopes, where they rested in low bushes among fragments of rock, and at intervals darted down into the valley below to feed. During the flight the white of the tail is prominent.

SEPHANOIDES GALERITUS (Molina)

Trochilus galeritus MOLINA, Sagg. Stor. Nat. Chili, 1782, p. 247. (Chile.)

Since G. R. Gray in 1840[2] listed *Sephanoides* (which he attributes to Lesson) as a genus, with *Trochilus kingii* of Vigors as type, this name must replace *Eustephanus*, erected by Reichenbach in 1850. The specific name of this hummer may also be open to question since Molina in his Latin diagnosis says "*Trochilus curvirostris*" and his entire description, as usual, is somewhat vague.

This species was common near Concon, Chile, from April 25 to 28, where a male was taken April 25 and females on April 25 and 27. The male, in fresh fall plumage, is dark green above, with little of the coppery reflection found in most skins, so that it suggests the condition in the bird described by Boucard[3] as *Eustephanus burtoni* on the basis of one specimen from Chile. It is possible that *burtoni* represents the fresh plumage of *galeritus*, since its measurements agree with those of the common bird.

Near Concon, *S. galeritus* was common among open, brushy growths that covered ranges of low, sandy hills. The birds fed at the blossoms of flowering shrubs, searched old yucca heads for insects, or snapped at gnats dancing in the air. Their flight was rather slow, accompanied by a subdued, barely audible humming; at intervals they closed the wings for a second, allowing the body to sink for a foot or so, and then with renewed motion continued on their course. The legs, for a bird of this group, were long and the feet and claws strong. They often clung to flowers with their claws, while probing them for food, in the manner noted in *Oreotrochilus leucopleurus*. The male had a twittering song with a metallic rattle that suggested some finch. The usual call was a high-pitched *tsee-ee* that changed to a steady rattle, as, with a flash of reflected light from the brilliant crown, the bird darted away in pursuit of some intruder.

An adult male had the bill black; iris Vandyke brown; tarsus and toes fuscous black.

An adult male (preserved as a mummy), and a male and two females in alcohol, of the rare Juan Fernandez hummer generously

[2] List Gen. Birds, 1840, p. 14.
[3] Hummingbird, vol. 1, Mar. 1, 1891, p. 18.

presented by Dr. Edwyn Reed, of Valparaiso, demonstrate that this species is generically separable from *Sephanoides* under the name *Thaumaste* of Reichenbach, on the basis of broader, shorter bill, stronger tarsi and feet, broader rectrices, and relatively longer tail. These distinctions, though most apparent in males, are readily seen when females are compared. The four specimens donated by Doctor Reed form a valuable accession to the collections in the United States National Museum, since the Juan Fernandez hummer had been represented there previously by the skin of a male alone.

SAPPHO SAPHO (Lesson)

Ornismya sapho LESSON, Hist. Nat. Ois.-Mouch., 1829, p. 105. ("Interior of Brazil.")

Trochilus sparganurus of Shaw,[4] the name in common use for the present species, must be transferred to the bird described by Gould[5] as *Cometes phaon*, since Shaw's plate, though crude, shows distinctly the long bill and the light line extending beneath the eye of this bird, while he describes the gold crimson bar on the otherwise black tail that also is characteristic of it. The locality assigned by Shaw, Peru, is also the one inhabited by this species. The bill in the hummer, that has been called *sparganurus* in the past, is shorter, a white line, where present, extends only to the anterior margin of the eye, not below it, and the tail is coppery red instead of crimson; in addition, the bird is of more southern range. *Cometes phaon* Gould, therefore, takes the name *Sappho sparganura* (Shaw) and *Ornismya sapho* must be used as the name for the other species. The locality assigned by Lesson, "interior of Brazil," is doubtless incorrect. Cory[6] recently has used *Lesbia* of Lesson[7] for the species under discussion. Though Lesson did not designate a type for this genus, Gray[8] subsequently selected *Ornismya kingii* of Lesson,[9] a species not now considered congeneric with *Ornismya sapho*, so that *Sappho* of Reichenbach[10] must be used for the present species.

Near El Salto, at an elevation of between 1,500 and 1,800 meters above Potrerillos, Mendoza, this beautifully marked hummer was fairly common on March 19, 1921. A red-flowered epiphyte (*Psittacanthus cuneifolius*), parasitic on creosote bush, was common, and *Sappho* came with other hummers to feed at it. The plant grew in clumps from 1 to 2 meters from the ground, with the massed color of the flowers against the gray green of the surrounding vegetation prominent at a considerable distance. On the rock-strewn

---

[4] Gen. Zool., vol. 8, pt. 1, 1812, p. 291.
[5] Proc. Zool. Soc. London, 1847, p. 31. (Peru.)
[6] Cat. Birds Amer., pt. 2, no. 1, Mar., 1918, p. 281.
[7] Ind. Gen. Syn. Troch., 1832, p. xvii.
[8] List Gen. Birds, 1840, p. 14.
[9] Troch., 1832, p. 107, pl. 38.
[10] Av. Syst. Nat., 1850, pl. 40.

hill slopes above were little nooks protected from wind and warmed by radiation from the sun-heated stones, to which the hummers resorted to rest, and from which they darted down at intervals to feed on the flower clumps below.

Near Tapia, Tucuman, from April 7 to 13, this species, found as before about the brilliant flower clusters of *Psittacanthus*, was common. The birds were especially active on days of bright sunshine. Males and females alike poised with vibrating wings in feeding, and at frequent intervals paused to rest on open twigs, usually in the sun as the weather was cool. When three or four gathered at one flower clump there was much fighting among the long-tailed males, while any intruder was greeted with a low chattering call, *chit-it*, that often came from birds prudently concealed behind the dense shelter of thorny branches. When on the wing both sexes frequently expanded the deeply forked tail, a display, of course, most prominent in the long-tailed males, and males at rest often jerked the tail up and down in gnat-catcher fashion. The flight was rapid and direct, though the bird had the usual hummer habit of swinging up or down with an irregular bounding motion. Because of the long tail it appeared large and was easier to follow with the eye than most hummers. A few were seen near 2,100 meters on the Sierra San Xavier above Tafi Viejo, Tucuman, on April 17.

A male, taken March 19 above Potrerillos, was molting on the throat and had the tail almost grown anew. A male (April 9) and four females (April 8, 9, and 13) from Tapia were in partial molt on head, tail, and body. Immature females differed from the single adult taken in larger green spots on the throat and a wash of cinnamon on this area. In an adult male when freshly taken the bill and tarsus were black; iris liver brown.

HELIOMASTER FURCIFER (Shaw)

Trochilus Furcifer SHAW, Gen. Zool., vol. 8, pt. 1, p. 280. (Paraguay.)

A female taken near Las Palmas, Chaco, on July 31, 1920, like other hummers seen here (in winter) was found near the border of forest, in a region protected from cold winds. The bird, attracted by squeaking, alighted with a subdued humming of its wings on a limb near at hand.

The bill, tarsus, and toes, in life, were black.

Family MICROPODIDAE

MICROPUS ANDECOLUS DINELLII (Hartert)

Apus andecolus dinellii HARTERT, Bull. Brit. Orn. Club, vol. 23, December 31, 1908, p. 43. (Angosta Perchela, altitude 2,550 meters, Jujuy, Argentina.)

Dinelli's swift was recorded first in the valley of the Rio Negro south of General Roca, Rio Negro, on November 27, 1920, when 10 or more were noted and 5 were collected. These birds present a strange appearance in the air, as their long, thin, narrow wings seem as broad at the tip as near the body, while in color they appear wholly light-brownish gray or white, veritable ghosts of birds with wings barely thicker than paper. During the strong wind that prevailed they sailed constantly with set wings cutting the air rapidly; when one did choose to fly, it passed with lightninglike speed. Males and females, the latter paler in color on the back, were taken, and I supposed that the birds had drifted across from breeding stations on the high rock escarpment on the southern side of the valley. Frequently they were seen in trios. Others of these swifts were recorded about a rocky point in the valley of the Rio Blanco at Potrerillos, Mendoza, on March 18, 19, and 20, 1921. From about 30 that were seen, 4 were taken on March 18 and 1 on the day following. The call of this species is a high-pitched laughing chatter that does not carry far in the wind.

Immature birds, as represented in the fall series, have the forehead darker than adults secured in spring and are somewhat more buffy below. Young females are somewhat darker on the back than those taken in spring, but are still noticeably paler than males.

This swift does not appear to have been recorded previously south of the Province of Mendoza.

STREPTOPROCNE ZONARIS (Shaw)

Hirundo zonaris SHAW, Cim. Phys., 1796, p. 100, pl. 55. (Chapada, Matto Grosso,[11] Brazil.)

The collared swift was recorded above Mendoza, Mendoza, on March 13, 1921, and on the slopes of the Sierra San Xavier above Tafi Viejo, Tucuman, on April 17. No specimens were secured.

CHAETURA ANDREI MERIDIONALIS Hellmayr

Chaetura andrei meridionalis, HELLMAYR, Bull. Brit. Orn. Club, vol. 19, Mar. 30, 1907, p. 63. (Isca Yacu,[12] Santiago del Estero, Argentina.)

The present species was found only in the vicinity of Puerto Pinasco, Paraguay, where an adult female was taken September 20, near Kilometer 80, and a male September 23, near Kilometer 110 (the latter preserved as a skeleton). Swifts were found over the forest in certain localities, where they seemed to have selected breeding stations in hollow trees. Though seen on September 1 near the low hill at Kilometer 25, and on September 30 over the Cerro Lorito on the Rio Paraguay, I found them also in certain areas in the level country.

[11] Doctor Chapman's action (Bull. Amer. Mus. Nat. Hist., vol. 33, Nov. 21, 1914, p. 605) in selecting Chapada, a point far distant from the coast, as the type locality of Shaw's *Hirundo zonaris*, may perhaps be questioned, since it is doubtful if interior specimens had been seen or described as early as 1796.
[12] See Dabbene. El Hornero, vol. 1, 1917, p. 7.

Ordinarily they circled about high overhead, where it was impossible to reach them, and when they did descend to lower altitudes it was necessary usually to take a snapshot at them overhead as they dashed past little openings in the tree tops. Under these circumstances considerable time and ammunition were expended before one was finally secured. The birds spent much time in circling high in air, but descended at intervals to flash past dead stubs in which I supposed they would nest later. Often they flew about in trios and frequently were seen in pairs. Their wing motion was extremely rapid, and when sufficient momentum had been gained they scaled rapidly along with set wings. During cold, rainy weather a few appeared about the lagoon at Kilometer 110, perhaps driven in here by unfavorable feeding grounds in other regions. Their usual call note was a low *chu chu chu chu*, followed by a rattling chipper. On September 30, on the Cerro Lorito, I found them circling about a little clearing, in which there were one or two dead trees. The Angueté Indians called this species *mee tset tse he*.

The skin preserved, an adult female, measures as follows: Wing, 127.5; tail, 37; exposed culmen, 4; tarsus, 11.5 mm. Swifts of this genus have been recorded seldom from Paraguay, while the names under which they are given are so involved that it is difficult to place them. My specimens have been identified in accordance with Hellmayr's treatment of the South American forms of *Chaetura*.[13]

Order PASSERIFORMES

Family DENDROCOLAPTIDAE

DENDROCOLAPTES PICUMNUS Lichtenstein

Dendrocolaptes Picumnus LICHTENSTEIN, Abh. Kön. Akad. Wiss. Berlin for 1818–19 (pub. 1820), p. 202. (Brazil.)

An adult female shot in heavy forest on the Cerro Lorito opposite Puerto Pinasco, Paraguay, was the only one seen. This specimen has fine shaft streaks of whitish on the feathers of the back, a character that, according to Hellmayr, is one that differentiates *D. picumnus* from the closely allied *D. intermedius* Berlepsch. The latter species is said also to have the head browner and back more reddish brown. Hellmayr[14] has recorded *D. picumnus* from Bernalcue, Paraguay, and there are other less definite records for Paraguay.

My specimen was secured as it clung to the base of a tree above a large colony of ants, on which it fed eagerly, hopping and sidling

[13] Verb. Ornith. Ges. Bayern, vol. 8, 1908, p. 145.
[14] Abh. Kön. Bayerischen Akad. Wiss., II Kl., vol. 22, 1906, p. 632.

rapidly about, perhaps to prevent the insects from climbing into its feathers.

XIPHOCOLAPTES MAJOR MAJOR (Vieillot)

Dendrocopus major VIEILLOT, Nouv. Dict. Hist. Nat., vol. 26, 1818, p. 118. (Paraguay.)

Individuals of this species were seen twice near Las Palmas, Chaco, during the first part of July, 1920, but none were collected until I arrived at the Riacho Pilaga, Formosa. Two secured there, on August 18, 1920, were the only ones seen. At Kilometer 80, west of Puerto Pinasco, Paraguay, two were taken on September 15 and another on September 17. The skins preserved, while somewhat variable in color, have the abdomen distinctly barred so that they show no approach to *X. m. castaneus*, described by Ridgway from Bolivia, in which the abdomen is said to be plain. A skin from the interior of Formosa is decidedly darker above than three from Paraguay. One of the latter, however, is much deeper in color below than any of the others, so that variation in shade of brown seems to be an individual character.

In accordance with Azara's observations I did not find the present species common, though it was recorded on several occasions. The birds were found in or near the heavier timber in the Chaco, and from my limited records seem to feed to a considerable extent on the ground, where the groves were fairly open and the vegetation below not too dense. Often they were restricted in such haunts to the borders of trails or cattle paths. When flushed they flew up to cling to a tree trunk sometimes near the ground and again among the higher branches where their attitude and actions were similar in a way to those of a woodpecker. When clinging in this fashion the feet usually were placed wide apart, and the bird progressed in a series of long hitches, with the head and neck erect. They seemed to range in pairs, though none of those taken were breeding. One that I wounded called harshly, while its mate appeared a few feet away to call *kway kway* in an inquiring tone. The birds are heavily muscled and in form are robust. Many of the tendons in the muscles of the lower leg are more or less ossified, so that in preparing skins it is noticeably difficult to cut them.

An adult male, taken September 17, had the tip of the bill dark neutral gray, shading on median portion of maxilla to light-grayish olive; rest of maxilla and mandible light neutral gray; iris vinaceous rufous; tarsus and toes between deep and dark olive gray.

Near Puerto Pinasco those who spoke the Guaraní tongue called this species *uravo-vahí*.

LEPIDOCOLAPTES ANGUSTIROSTRIS ANGUSTIROSTRIS (Vieillot)

Dendrocopus angustirostris VIEILLOT, Nouv. Dict. Hist. Nat., vol. 26, 1818, p. 116. (Paraguay.)

An adult male shot at Resistencia, Chaco, on July 10, 1920, and an immature female taken at Las Palmas, Chaco, July 13, are supposed to represent the typical form of the present species. These two have the undersurface definitely but not heavily streaked and give the following measurements for wing and culmen: Male, wing, 98.2; culmen from base, 32.7; female, wing, 92.3; culmen from base, 31 mm. Two males secured at Tapia, Tucuman, on April 9 and 11, 1921, have the streaks on the ventral surface blacker and much heavier, and are referred to this form with reservation as it is probable that they represent a distinct subspecies. The specimens in question (in molt) have the following measurements: Wing, 94.4 and 99.6 mm.; culmen from base, 34 and 35.7 mm. In coloration of the dorsal surface they are rather close to the birds from Chaco.

At Resistencia, Chaco, this bird was recorded only on July 10, 1920, when a male was taken. Near Las Palmas, where forests were more extensive, they were found in fair numbers from July 13 to 31, while near the Riacho Pilaga they were recorded on August 11 and 18. No specimens were taken here, and it is possible that the birds noted were *L. a. certhiolus* found in the Paraguayan Chaco. At Tapia, Tucuman, they were fairly common from April 7 to 13.

This wood hewer in the Chaco frequented the heavier growths of timber that grew in swampy localities, where it ranged in pairs that frequently joined company with little traveling bands of other brush and forest hunting birds, and accompanied them on their rounds in search for food. The flight of the bird under discussion is undulating, and is seldom continued for any long distance. They alight on a tree trunk, to which they cling with sharp claws and firmly braced tail, and begin immediately to hitch upward, often assisting their progress by a rapid flit of the wing. They continue up the trunk and over the larger branches and then fly to another tree or drop to the base of the trunk they have just examined and cover the ground once more. Their long bills were frequently thrust into the recesses of small air plants, or under moss and loose bark, which was pried away with a quick twist of the head to expose any animal life concealed beneath. The call notes of the present species in all its forms are loud and musical.

The tongue is small and undeveloped in proportion to the size of the bill, in contrast to what is found in such long-billed groups as humming birds, honey creepers, and honey eaters.

LEPIDOCOLAPTES ANGUSTIROSTRIS PRAEDATUS (Cherrie)

Picolaptes angustirostris praedatus CHERRIE, Bull. Amer. Mus. Nat. Hist., vol. 25, May 20, 1916, p. 187. (Concepcion del Uruguay, Entre Rios, Argentina.)

The present subspecies, seemingly differentiated from the typical form by greater size, longer bill, and the extension of the stripes on head and hind neck down to the upper back, is the southernmost representative of the species, as it ranges through the scanty wooded areas of the northern pampas, in western Uruguay, Buenos Aires, and in the band of forest that crosses the Territory of Pampa in the vicinity of Victorica.

There is in the United States National Museum an old specimen taken near Buenos Aires by the Page expedition, while I secured two females near Victorica, Pampa, on December 24 and 27, 1920, and an immature male at Rio Negro, Uruguay, on February 15, 1921. If the skin from Buenos Aires be considered as typical of the form, a logical procedure, as the type came from Concepcion, in similar country 300 kilometers to the northward, it is found that the skins from Victorica have a shorter wing, are grayer, and that one has a much longer bill. (The mandibles in the second skin from Victorica are imperfect.) The last are placed with *praedatus* on the basis of size of bill, and because they agree with *praedatus* in the extension of the head and neck streaks over the upper back. The specimen from Rio Negro, Uruguay, is only recently from the nest, and its bill is short and undeveloped. It is darker above, more blackish on the head, and more heavily streaked below than the others. Pertinent measurements, in millimeters, of these four skins are as follows:

Catalogue No.	Locality	Wing	Culmen from base
12358	Buenos Aires	102.0	39.7
283878	Victorica, Pampa	93.5	42.5
283879	----do----	91.0	
284572	Rio Negro, Uruguay	103.0	

Near Victorica, Pampa, several wood hewers of the present form were recorded on December 24 and 27, 1920, in the low, open forest of caldén, algarroba, and similar trees characteristic of the region. The dry atmosphere had a tendency to make feathers hard and brittle so that the tails in all the *Lepidocolaptes* seen were worn short and blunt from abrasion on the rough-barked trees. Near Rio Negro, Uruguay, on February 15, 1921, I found a brood of fully-grown young in low woods, clambering actively about with low, rather rapid, whistled notes, with all of the mannerisms of adults.

LEPIDOCOLAPTES ANGUSTIROSTRIS CERTHIOLUS (Todd)

Picolaptes bivittatus certhiolus TODD, Proc. Biol. Soc. Washington, vol. 26, Aug. 8, 1913, p. 173. (Curiche Rio Grande, eastern Bolivia.)

Four specimens from west of Puerto Pinasco, Paraguay, a female shot at Kilometer 25 on September 1, and three from Kilometer 80 (two males secured September 9 and 13, and a female taken September 17), seem best referred to the present form, described from eastern Bolivia. *L. a. certhiolus* differs distinctly from *L. a. angustirostris* in the fainter streaking on the ventral surface that in some specimens becomes nearly obsolete. The four skins from near Puerto Pinasco agree closely with the type of *certhiolus* (that through the kindness of W. E. Clyde Todd, has been compared directly with them) in color and amount of streaking on the breast and abdomen. The Paraguayan specimens are slightly variable in this respect, some representing a closer approach to typical *angustirostris* than others. They are slightly duller above than the type of *certhiolus*, and are also a trifle smaller (in males, wing, 91.6–95.2 mm.). In Mr. Todd's type-specimen the wing measures 100.6 mm.

Examination of a small series of these wood hewers shows conclusively that the bird with unstreaked breast, known as *bivittatus*, intergrades through *certhiolus* with the heavily marked group that has been maintained as *angustirostris*. In fact, intergradation seems so complete as to render division into forms difficult, since, when the vast range occupied is considered, comparatively few localities have been represented in the series seen. Since *angustirostris* of Vieillot is the older name, the forms of the species as at present known will stand as follows:

LEPIDOCOLAPTES A. ANGUSTIROSTRIS (Vieillot).

Dendrocopus angustirostris VIEILLOT, Nouv. Dict. Hist. Nat., vol. 26, 1818, p. 116. (Paraguay.)

Lower surface heavily streaked; streaks of head terminated on nape.

Eastern (?) and southern Paraguay, northern Argentina (Chaco, Tucuman), probably south to the limits of the Chaco in the Province of Santa Fe.

LEPIDOCOLAPTES A. PRAEDATUS (Cherrie).

Picolaptes angustirostris praedatus CHERRIE, Bull. Amer. Mus. Nat. Hist., vol. 25, May 20, 1916, p. 187. (Concepcion del Uruguay, Entre Rios, Argentina.)

Larger than *angustirostris* with longer bill, and with the streaks on the head and nape extended onto the lower back.

Central Argentina (Entre Rios, northern Buenos Aires, and central Pampa) and Uruguay (Rio Negro).

LEPIDOCOLAPTES A. CERTHIOLUS (Todd).

Picolaptes bivittatus certhiolus TODD, Proc. Biol. Soc. Washington, vol. 26, August 8, 1913, p. 173. (Curiche Rio Grande, eastern Bolivia.)

Less heavily streaked below than *angustirostris*, in some with streaks nearly obsolete, more rufescent.

Eastern Bolivia and the Chaco of Paraguay for an indeterminate distance southward.

LEPIDOCOLAPTES A. BIVITTATUS (Lichtenstein).

Dendrocolaptes bivittatus LICHTENSTEIN, Abh. Berlin Akad., 1820–21 (1822), p. 15S, pl. 2, fig. 2. (São Paulo, Brazil.)

Lower surface uniform or nearly so, lighter, more rufescent above. Southern Brazil (São Paulo to Matto Grosso?).

Four skins in the Carnegie Museum from the Province of Lara, Bolivia, have the streaks on the undersurface nearly obsolete and are somewhat brighter above than the average of *L. a. certhiolus* as here taken. These may represent intermediates toward *bivittatus* or may be that form, but with available material I am at a loss where to place them. It is possible that *bivittatus* ranges across Matto Grosso into the lower regions of north central Bolivia, while *certhiolus* may be restricted to the Chaco region of eastern Bolivia, and Paraguay west of the Rio Paraguay.

LEPIDOCOLAPTES A. CORONTUS (Lesson).

Picolaptes coronatus LESSON, Traité Orn., 1831, p. 314. ("Brésil.")

Similar to *L. a. bivittatus*, but undersurface, save throat slightly duller than a shade between cinnamon buff and clay color.

Eastern Brazil (Bahia, Piauhy, Ceara.)

Six skins of this form that I have examined in the Field Museum come from Jua and Quixada in Ceara. Doctor Hellmayr informs me that *coronatus* of Lesson, based on Plate 90 in the first volume of Spix's Avium Brasiliam (vol. 1), is the same as the form that he separated under the subspecific name of *bahiae*.[15]

L. a. certhiolus, like other forms of this species, inhabited forested regions, often where the growth was dense and almost impenetrable, but was not averse to working out through the more open palm groves, or *palmares*, that covered swamp regions, or open groves of hardwoods where brush fires had prevented undergrowth. Like others of its family, it climbed steadily in long hitches, woodpecker fashion, always traveling up the tree trunks or limbs on which it rested. One observed posed before its mate braced firmly with its

---

[15] *Picolaptes bivittatus bahiae* Hellmayr, Verh. Zool.-bot. Ges. Wien, 1903, p. 219. (Bahia.)

tail against a tree trunk, with wings spread, mouth open, and crest raised.

It was recorded west to Kilometer 110, west of Puerto Pinasco.

CAMPYLORHAMPHUS RUFODORSALIS (Chapman)

Xiphorhynchus rufodorsalis CHAPMAN, Bull. Amer. Mus. Nat. Hist., vol. 2, July 5, 1889, p. 160. (Corumba, Matto Grosso, Brazil.)

An adult male taken at Las Palmas, Chaco, July 19, 1920, and a female from the Riacho Pilaga, Formosa, secured August 11, are referred to the present species. Menegaux and Hellmayr [16] consider *rufodorsalis* indistinguishable from *lafresnayanus* (d'Orbigny) from Bolivia on the basis of examination of a series from Mattogrosso, d'Orbigny's original specimens marked Chiquitos, and a skin from Rio de la Plata. Through the kindness of W. E. Clyde Todd, I have seen five skins from the collections of the Carnegie Museum from Bolivia, three from Guanacos, Province of Cordillera, one from Palmarito, Rio San Julian, Chiquitos, and one from Curiche Rio Grande, eastern Bolivia. These differ constantly from the Argentine specimens in shorter, more slender bill (culmen from base, 67–73.5 mm.), and in duller, less rufescent coloration on the ventral surface, so that on the basis of this material I must hold *rufodorsalis* valid. The two skins from Chaco and Formosa have the culmen from base 90 and 87.8 mm., respectively. A third specimen from the Rio Bermejo (Page expedition), with the ends of both mandibles shot away, has the base of the bill much heavier than is true in the Bolivian birds. The three Argentine skins agree in being more rufescent below than the others. They are listed here under a specific name, though it is probable that *rufodorsalis* will prove to be a geographic race of *lafresnayanus*.

This curious form was encountered in low, swampy lands in heavy growths of timber, where it frequented dense cover. The birds clambered alertly up sloping tree trunks, with the tail braced to aid their ascent, and at any alarm disappeared in the jungle. From their actions I judged that the grotesque curved bill was employed to search for insects among the stiffened leaves of bromeliaceous epiphytes that grew abundantly on trees and shrubs; the clasping stems of the leaves of these plants formed cups that harbored considerable animal life, and often contained considerable water in which unfortunate insects were drowned, a feeding ground inaccessible save to the curious beak of this wood hewer.

A male, taken July 19, had the bill orange cinnamon, shaded with fuscous at tip and base of culmen; iris deep-brownish drab; tarsus and toes deep olive; underside of toes shaded with yellowish.

---

[16] Mem. Soc. Hist. Nat. Autun, vol. 19, 1906, pp. 78–79.

DRYMORNIS BRIDGESII (Eyton)

Nasica bridgesii EYTON, Jardine's contrib. Orn., 1849 (pub. 1850), p. 130, p. 38. (Bolivia.)

Near Victorica, Pampa, Bridge's wood hewer was fairly common from December 23 to 29, 1920, so that two males (one prepared as a skeleton) were taken December 23 and a female on December 24. They were found in little flocks of four or five in open forest where they fed on the ground where it was more or less free from undergrowth, often in company with *Pseudoseisura lophotes*. When startled they flew away with undulating flight to alight on some tree trunk, up which they climbed until they came to rest on a sloping limb. Their ordinary call was a loud *whee whee whee*, to which they added a chattering note when excited. Though they climbed readily, they seemed to prefer to rest on a sloping limb of good size and in such situations frequently ran along the branches instead of hitching about with the aid of the tail. A bird with a broken wing ran along on the ground as rapidly and easily as a plover or a lark, so that I had considerable difficulty in capturing it.

A few were recorded near Tapia, Tucuman, from April 7 to 13, 1921, but none were taken.

A male, secured December 23, had the base of the mandible ecru drab; rest of bill black; iris natal brown; tarsus and toes dull black.

SITTASOMUS SYLVIELLUS CHAPADENSIS (Ridgway)

Sittasomus chapadensis RIDGWAY, Proc. U. S. Nat. Mus., vol. 14, 1891, p. 509. (Chapada, Matto Grosso, Brazil.)

Five specimens of *Sittasomus sylviellus* from three rather widely separated localities offer some variation in color, but, until better series of skins are available, may be referred to the subspecies *chapadensis*. A male and a female shot west of Puerto Pinasco, the male at Kilometer 25, September 1, and the female at Kilometer 80, September 15, are slightly grayer, less yellowish both above and below than two females taken at Las Palmas, Chaco, on July 15. An adult male secured April 17, 1921, at an altitude of nearly 1,700 meters on the Sierra San Xavier, above Tafi Viejo, Tucuman, is much more yellowish than those last mentioned, especially on the lower breast and abdomen, while the bill appears larger and heavier. The three seem to represent phases that may eventually be recognized as subspecies. Hellmayr[17] refers skins from Tucuman to *chapadensis* and considers this to range in Goyaz and Matto Grosso, Brazil, eastern Bolivia, Paraguay (Colonia Risso), and northern Argentina (Jujuy, Salta, and Tucuman).

---

[17] Nov. Zool., vol. 15, June, 1908, p. 64.

I recorded this *Sittasomus* at Las Palmas, Chaco, from July 15 to 30, 1920; Riacho Pilaga, Formosa, August 11 and 18; Kilometer 25, Puerto Pinasco, Paraguay, September 1; Kilometer 80, in the same vicinity, September 8, 15, and 20; and on the Sierra San Xavier, above Tafí Viejo, Tucuman, on April 17, 1921.

The *taquarita*, as the species is known, inhabits heavy forest and is not found in open areas. In search for food it creeps and climbs over tree trunks and limbs, bracing with the tail to assist it in progress, and moving so actively that it is difficult to follow. In its method of climbing and the nervous activity that keeps it continually moving, it is suggestive of *Certhia*. The birds were found often in company with little groups of other forest birds that travel in social flocks. According to my brief observations, they were entirely silent.

Family FURNARIIDAE

GEOSITTA CUNICULARIA CUNICULARIA (Vieillot)

Alauda cunicularia VIEILLOT, Nouv. Dict. Hist. Nat., vol. 1, 1816, p. 369. (Near the Rio de la Plata and Buenos Aires.)

The common miner was recorded at the following points: Lavalle, Buenos Aires, November 7 and 8, 1920 (adult and immature males taken November 7); Carhue, Buenos Aires, December 15 to 18; Carrasco, Uruguay, January 9 and 16, 1921; La Paloma, Uruguay, January 23; San Vicente, Uruguay, January 31 to February 2 (adult shot at Paso Alamo, February 2); Lazcano, Uruguay, February 3 (adult female taken) to February 9; and Guamini, Buenos Aires, March 3 and 4. In the series at hand skins from the Province of Buenos Aires and Uruguay average slightly smaller than those from Zapala, other localities in northern Patagonia, and Chile. Northern skins are browner, and those from Chile grayer. Birds from Zapala are intermediate in this respect and are referred to *G. c. hellmayri*. Two males from Uruguay have a wing measurement of 88 and 88.2 mm., respectively (a third from Quinta, Rio Grande do Sul, that I have seen is similar), while males from Buenos Aires range from 89.2–93.5 mm. *Geositta c. frobeni*, as at present understood, distinguished by larger size (wing, 101.5–103.5 (and paler outer rectrix, is represented in the United States National Museum by three specimens from the Province of Mendoza.

Adults have much longer bills than immature individuals even when the latter appear fully grown. In immature plumage, birds are distinguished by the paler margins on the feathers of the dorsal surface, and by the narrow, buffy tips of the longer primaries. An adult male secured February 2 has begun to molt on the breast. Others are in full plumage.

The miner was local in its distribution and was found only where areas of open sandy soil offered it a suitable habitat. Near Lavalle, Buenos Aires, it was encountered among the dunes near the coast in a broad area that intervened between the bare wind-swept, constantly shifting sand hills just back of the beach, and the inner landward tract that had been covered entirely by vegetation. Others were recorded at sandy blow-outs farther inland where open sand was exposed for more limited areas; elsewhere they were found in one or the other of two types of country, among dunes or along sandy hill slopes. Frequently the birds were shy and from their inconspicuous coloration were overlooked. Their undulating flight carries them in sweeping bounds a meter or so above the earth, while when they alight they walk or run about among the scant growths of vegetation that maintain a precarious foothold in the soil. With head and body well erect they walk with nervous hesitant strides like *Furnarius*, or run for several steps and then pause. The tail in old and young is constantly vibrated when the birds are otherwise at rest. Among the dunes they disappear constantly over distant ridges, so that it is difficult to follow them. The only note that I heard them give was a curious song, attributed to the males, a high pitched *he he he he he he he*, uttered in a laughing tone as the birds rose from 4 to 10 meters in the air, and then descended with tremulously vibrating wings, or as they circled and swung about in erratic dips and curves over the undulating surface below. In the high winds that usually prevailed in their haunts their calls seemed ventriloquial, and may be confused with some of the notes of the more common burrowing owl. This song seems to be given without regard to the season of the year, as I heard it from the first week in November to the end of April whenever I chanced to encounter the species. From its habit of wagging the tail this bird is often called *menio-cola*.

An adult male shot November 7 showed developed testes, and had the abdomen bare, indicating that it had been incubating. An immature bird, fully grown, was taken at the same time.

GEOSITTA CUNICULARIA HELLMAYRI Peters

Geositta cunicularia hellmayri PETERS, Occ. Pap. Boston Soc. Nat. Hist., vol. 5, Jan. 30, 1925, p. 145. (Huanuluan, Rio Negro, Argentina.)

Two males from Zapala, Neuquen, an adult taken December 8 and an immature specimen shot on the following day, are referred to the present form, described recently by Peters. These two have wing measurements of 94.2 and 96 mm., respectively, and, in addition to larger size, are slightly grayer than skins from Buenos Aires and Uruguay.

Near Zapala the miner frequented sandy areas along the slopes of little valleys. On December 9 I noted a male standing in the entrance of a nesting burrow, excavated in the face of a low-cut bank, a tunnel without apparent end, as some mammal had attempted to dig it out without success, and after considerable labor I abandoned the task myself without having reached the nest cavity.

GEOSITTA CUNICULARIA FISSIROSTRIS (Kittlitz)

Alauda fissirostris Kittlitz, Mem. Acad. Imp. Sci. St. Petersbourg, Div. Sav., vol. 2, 1835, p. 468, pl. 3. (Valparaiso.)

An adult female, taken at Concon, Chile, April 25, 1921, is allotted here on geographic grounds. In the poor series at hand I am unable to assign definite characters differentiating this alleged race from typical *cunicularia*, except that it appears faintly grayer.

GEOSITTA RUFIPENNIS (Burmeister)

Geobamon rufipennis Burmeister, Journ. für. Ornith., 1860, p. 249. ("Parana." Dabbene [18] has substituted Cordillera de Mendoza.)

A female of this species, shot at an elevation of about 1,500 meters above Potrerillos, Mendoza, on March 18, 1921, has been difficult to place subspecifically in the light of available material. Burmeister in his original description assigned Parana as the locality for his specimens, although, as Doctor Dabbene has shown, the bird is not known to visit the central pampas. For this reason Dabbene substitutes the Cordillera of Mendoza as the type-locality. Furthermore, Dabbene distinguishes a form from Tucuman with cream-colored underparts as *G. r. burmeisteri*. The bird described by Burmeister was said to be "rothlichgrau" on the undersurface. Dabbene states that the Museo Nacional in Buenos Aires has a specimen collected in Mendoza during the time of Burmeister, and on this apparently bases his assignment of the type-locality. Menegaux and Hellmayr [19] write that Burmeister's types preserved at Halle are "blanc grisatre." Dabbene considers these as representative of his *G. r. burmeisteri*. A series of skins is needed to straighten out the forms involved successfully. My specimen from Potrerillos has a very slight wash of vinaceous buff on the otherwise grayish-white breast and abdomen and is distinctly paler than birds from Chubut. At the same time it is not cream colored below. It is possible that it is near the northern form.

At Potrerillos this bird was seen occasionally among low brush on rocky slopes, from which it flew out when alarmed with an undulating flight to seek other cover. The reddish brown of wings

---

[18] An. Mus. Nac. Hist. Nat. Buenos Aires, vol. 30, July 11, 1919, p. 133.
[19] Mém. Soc. Hist. Nat. Autun, vol. 19, 1906, p. 46.

and tail were prominent in flight. The bird taken was in active pursuit of some insect over the rocks. The specimen is in partial molt on the head.

FURNARIUS RUFUS RUFUS (Gmelin)

Merops rufus GMELIN, Syst. Nat., vol. 1, pt. 1, 1788, p. 465. (Buenos Aires.)

The typical form of the ovenbird, characterized by large size (wing 96.3–104 mm. in a series of 16 specimens), and by general grayish tone in color, ranges from the Province of Buenos Aires north into Uruguay and in Argentina as far as the Chaco. It was recorded and collected as the following points:

Berazategui, Buenos Aires, June 29, 1920 (adult female taken); Santa Fe, Santa Fe, July 4; Dolores, Buenos Aires, October 21; Lavalle, Buenos Aires, October 23 to November 15 (two adult males, shot October 30); Lavalle to Santo Domingo, Buenos Aires, November 16; Montevideo, Uruguay, January 9 to 16, 1921 (in parks and outskirts of the city); La Paloma, Uruguay, January 23; San Vicente, Uruguay, January 25 to 31 (two immature males, shot January 26, and one adult male, January 29); Lazcano, Uruguay, February 3 to 9 (immature female, taken February 6); Rio Negro, Uruguay, February 14 to 19 (immature male, shot February 15); and Guamini, Buenos Aires, March 6 and 7 (a male taken March 6).

Specimens from Uruguay are very slightly darker than those from Buenos Aires, but are the same size as birds from near the type-locality. During summer the plumage of this bird through wear, becomes considerably lighter, especially on the underparts, than when it is in fresh winter feather.

Through the pampas of Buenos Aires the ovenbird is restricted to the neighborhood of scattered groves of trees, and as these are usually located about houses the bird is one of the most domestic and beloved of Argentine birds. It is recorded south to Bahia Blanca, but in western Buenos Aires it is rare so that I was interested in obtaining it at Guamini. Though groves of tala, coronillo, and other trees were common near the Rio de la Plata in the early settlement of the Province of Buenos Aires, tremendous expanses of open plain were wholly without cover for birds that sought the shelter of trees. With the settlement of the country, eucalyptus were introduced and, with native species of trees, were planted in groves about the estancia houses, so that now one is seldom out of sight of them in crossing the pampa. As this has increased the shelter available, the effect must have been to increase the numbers of certain birds, among them the present species. It was interesting to note their occurrence in scattered tree clumps planted about wells and water holes miles from larger groves. In eastern Uruguay, where groves of palms cover kilometers of lowland, ovenbirds fairly swarm. There, though they came familiarly about houses, they are less domestic, as there is abundance of range for them, so that instead of being concentrated near houses they range through the countryside. At San Vicente any alarm was sufficient to bring ten or a dozen to shriek their disapproval of my intrusion.

The loud calls of the ovenbird never fail to announce its presence. One, presumably the male, with bill thrown up and wings drooped, gives vent to a series of shrieking, laughing calls, with distended, vibrating throat, and quivering wing tips. At about the middle of this strange song its mate chimes in with shrill calls of a different pitch, and the two continue in duet to terminate together. The loud notes may be audible at half a mile.

Ovenbirds remain paired throughout the year and mated birds may usually be found near one another. Though they seek shelter in trees they feed on the ground, frequently far distant from cover, where, with their plump bodies and short tails, they suggest a thrush in form. Search for a livelihood is a serious affair that absorbs every attention, so that they have a preoccupied air as they walk about in the herbage with nervous hesitant strides and slightly nodding heads. The calls of a neighbor or a mate are certain to bring response even from a distance.

The strange, domed, mud nest of this bird is certain to attract attention from the least observant, as it is built in the most conspicuous situations without the slightest attempt at concealment. Hundreds may be seen without effort during travel in the pampas. The usual structure averaged 300 mm. long by 200 mm. wide, though the dimensions varied according to circumstances, some being nearly globular and others more elongate. A dome-shaped roof with walls 25 to 35 mm. thick was elevated on a level mud platform until it was entirely arched over with an irregular hole in one side. An inner wall of curving outline was then constructed leading to the back of the inclosure at one side. This cut off the nest cavity from the entrance hall, and entry to the nest was through a small opening, with a raised threshold, below the roof. When the margins of the opening were rounded off the structure was complete. As the mud used for building material was mixed with vegetable fibers, grass, or hair, the whole made a structure of great firmness. When mud was scarce the birds sometimes utilized fresh cow dung as building material, making a structure that when dry, was as strong as a model of papiermâché. To examine the interior in any nest it was necessary to cut a hole in it with a heavy knife, as the hand could not be introduced into the nest cavity through the entrance.

DOMED NEST OF HORNERO, OR OVENBIRD (FURNARIUS R. RUFUS), BUILT
OF MUD. THE GATE ON WHICH THIS WAS PLACED WAS IN DAILY USE

Estancia Los Yngleses, LaValle, Buenos Aires, November 11, 1920

NEST OF HORNERO, OR OVENBIRD (FURNARIUS R. RUFUS). ON SUMMIT OF
TELEGRAPH POLE

Near Lazcano, Uruguay, February 9, 1921

TWO NESTS OF THE HORNERO, OR OVENBIRD (FURNARIUS R. RUFUS), ON
FACE OF CLAY BANK. NOTE THAT ONE IS ENTERED FROM THE RIGHT
AND THE OTHER FROM THE LEFT

La Paloma, Uruguay, January 23, 1921

HORNERO, OR OVENBIRD (FURNARIUS R. RUFUS), AND NEST
ON SIDE OF PALM

San Vicente, Uruguay, January 27, 1921

Though normally nests were placed on more or less horizontal limbs of trees, the birds were so abundant that almost any available site was utilized. Cross arms on telegraph poles were favored situations, while many of the mud structures were placed on the summits of poles. Through eastern Buenos Aires I recorded dozens of nests set as capstones on the tops of fence posts; any irregularity on the top of the post was filled in with mud and the nest, placed on the platform thus made, formed an ornamental ball that capped the pillar. Many nests were perched on the roofs of houses or on cornices. On one occasion I noted on a house with a gable roof two ovenbirds' nests placed one at either end at the very summit of the gables, where they resembled ornaments as symmetrically placed as though by the hands of the human occupants of the dwellings. On the coast of southern Uruguay, near La Paloma, ovenbirds sometimes placed their homes on projecting points on the abrupt faces of clay banks that bounded deep cut arroyos, where the rounded nests, perched like the structures of ancient cliff dwellers, were practically inaccessible. As they were built of the same clay as the banks on which they rested, they were almost indistinguishable save when shadows threw the openings into relief. Farther to the eastward, in Uruguay, it was usual to see nests stuck on the sides of palm trunks, where some slight roughness or projection offered support.

As the nests were plainly visible the birds made no point of stealth in visiting them; frequently if one stopped to look up at a nest the owner came down to rest upon the top of it.

Ernest Gibson many years ago commented upon the fact that in eastern Buenos Aires the nest of the *hornero*, as the ovenbird is known, had the opening invariably at the left side. I was interested in observing 'that birds in that region adhere to the same custom to-day, as in considerably more than 200 nests that I saw in the region of Dolores, Lavalle, and Santo Domingo all had the entrance at the left. In Uruguay and elsewhere right or left hand openings were made without evident choice. On one occasion in passing from Rocha to La Paloma I had opportunity to see about 100 of these ovens and found that they were more or less evenly divided as to position of the entrance.

As the nests are durable they last for more than one year, so that old ones are available for use of other birds. The band-breasted martin, *Phaeoprogne tapera*, appeared to choose these for nesting sites, and at times may have attempted to oust ovenbirds from domiciles still in use, as I recorded squabbles between the two species over the possession of ovens. (Pls. 14 and 15.)

The eggs of the *hornero* are white, without gloss, and with the shell somewhat roughened. In many cases the eggs are covered with mud. At Lavalle, Buenos Aires, on October 30, 1920, in one nest I found five eggs, all pierced by the bill of some bird. On November 16 I collected a set of four fresh eggs, with one of *Molothrus bonariensis* five leagues east of Santo Domingo, Buenos Aires, from a nest placed on top of a fence post. The side of one of the eggs had been broken when the nest was opened. Three of the eggs in this set are normal, while one is considerably dwarfed. They measure, in millimeters, as follows: 30.9 by 21.6; 30.7 by 21.7; 29.8 by 21.2; and 21.4 by 17.6.

FURNARIUS RUFUS PARAGUAYAE Cherrie and Reichenberger

Furnarius rufus paraguayae CHERRIE and REICHENBERGER, Amer. Mus. Nov., no. 27, Dec. 28, 1921, p. 5. (Puerto Pinasco, Paraguay.)

The Paraguayan ovenbird was recorded at the following localities: Resistencia, Chaco, July 5 to 10 (adult female, taken July 8); Las Palmas, Chaco, July 13 to 31 (adult male and female, shot July 31); Formosa, Formosa, August 5, 23, and 24; Riacho Pilaga, Formosa, August 7 to 21 (adult female, August 7, adult male, August 11); Puerto Pinasco, Paraguay, September 1, 3, and 30; Kilometer 80, west of Puerto Pinasco, Paraguay, September 6 to 21 (adult female, on September 17).

This recently described form is smaller and darker colored than *F. r. rufus*, and, though similar in measurements to *F. r. commersoni*, is much duller (less rufescent), especially on the back. In *F. r. rufus* (15 specimens) from the Province of Buenos Aires, Uruguay, and extreme southern Rio Grande do Sul (one specimen from Quinta, between Rio Grande do Sul and Pelotas) the wing measures from 97–104 mm. Seven skins of *F. r. paraguayae* from Chaco, Formosa and Paraguay (Sapucay and 80 kilometers west of Puerto Pinasco) range from 90–95.2 mm., while four *F. r. commersoni* from Urucum, Matto Grosso (loaned for examination by the American Museum of Natural History) measure from 90.4–94.4 mm. Specimens from Resistencia and Las Palmas are only slightly darker than birds from Buenos Aires, but are so small (wing, 92.4, 94.3, and 95 mm.) that they are best placed with *paraguayae*. Skins from the interior of Formosa are darker even than one from near Puerto Pinasco, the type locality. It would appear that *F. r. paraguayae* ranges in the Argentine and Paraguayan Chaco, and in an indeterminate area east of the Rio Paraguay in Paraguay, while *F. r. rufus* extends from Bahia Blanca and Guamini northward into central Uruguay (Rio Negro), and probably into southern Santa Fe.

Allocation of records for ovenbirds that I made at Tapia, Tucuman, from April 7 to 14, 1921, is uncertain, since Cherrie and Reichenberger consider birds from Perico, Jujuy, and Embarcacion, Salta, intermediate but nearer *F. r. commersoni*.

In the Chaco the ovenbird, while it came regularly about the older estancia houses and the small towns, was more of a bird of the country than in the pampas. The birds ranged through tracts of open groves near savannas, and did not penetrate far into the denser forests. Open borders of lagoons were favorite feeding places, and in such places *horneros* sometimes congregated until several were feeding in a relatively small area. When thus engaged they often suggested sandpipers, particularly when seen at a distance. In walking they frequently take several long steps, pause with one foot raised for an instant, and then continue. Their gait is easy and, when desired, as rapid as that of a blackbird. On September 3, at Puerto Pinasco, one was observed carrying food to young in the nest.

The Toba Indians called this species *kwo ti ih.*

UPUCERTHIA DUMETARIA Is. Geoff. Saint-Hilaire

Upucerthia Dumetaria Is. GEOFF. SAINT-HILAIRE, Nouv. Ann. Mus. Hist. Nat. (Paris), vol. 1, 1832, p. 394. (Patagonia.)

A male shot at General Roca, Rio Negro, November 23, 1920, and another taken April 28, 1921, at Concon, Chile, are difficult to place subspecifically with the comparative material at present available. The bird from Roca, in worn breeding plumage, is slightly more rufescent than specimens that may represent true *dumetaria* from farther south in Rio Negro. It thus shows approach to *Upucerthia dumetaria darwini* Scott,[20] according to the original description. The specimen has the following measurements: Wing, 105.2; tail. 76.2; culmen from base, 34; tarsus, 25.2 mm.

The bird from Concon, Chile, is much darker in color throughout and belongs to another form. It would appear that this is *Upucerthia dumetaria saturatior* Scott,[21] the type of which may have come from the vicinity of Valparaiso. *Upucerthia tamucoensis* Chubb[22] is doubtfully distinct from *saturatior*. The specimen from Concon measures: Wing, 96.8; tail, 70.6; culmen from base, 31.5; tarsus, 24 mm. It seems to possess the smaller measurements attributed by Chubb to *tamucoensis*.

Near General Roca this species was found among the heaviest growths of *Atriplex* and other shrubs in the lowland flood plain of the Rio Negro, where it was recorded on November 23, 24, and 27.

[20] *Upucerthia darwini* Scott, Bull. Brit. Orn. Club, vol. 10, Apr. 30, 1900, p. lxiii. (Mendoza.)
[21] *Upucerthia saturatior* Scott, Bull. Brit. Orn. Club, vol. 10, Apr. 30, 1900, p. lxiii. (Central Chile, " ex Berkeley James Coll.")
[22] (*Upucerthia tamucoensis* Chubb, Bull. Brit. Orn. Club, vol. 27, July 13, 1911, p. 101, (" Tamuco " southern Chile.")

The birds were shy and difficult of approach. Attention was attracted to them by their song, a rapid *chippy chippy chippy chip,* given as they sat on the top of a bush or post, with tail slightly raised. At the slightest alarm they made long flights low among the bushes, where it was difficult to follow their course. Near Zapala, Neuquen, a second specimen (preserved in alcohol) was shot on December 8, 1920. The bird was encountered there in heavy tracts of thorny brush in an arroyo leading toward the lowlands. At Concon, Chile, two were seen on April 28 in open brush on a sloping hillside.

The curved bill of this species gives it a thrasherlike appearance, a suggestion heightened by its habits and choice of haunts.

The male shot April 28 had the bill, in general, dull black, shading at base of cutting edge of maxilla and at base of gonys to hair brown; gape isabella color; iris bone brown; tarsus and toes clove brown.

UPUCERTHIA VALIDIROSTRIS (Burmeister)

Ochetorhynchus validirostris BURMEISTER, Reise La Plata-Staaten, vol. 2, 1861, p. 464. (Sierra de Mendoza.)

The present species was observed occasionally near Potrerillos, Mendoza, from March 15 to 21, 1921, and an immature female was secured March 15 at an altitude of 1,500 meters. The birds were found singly among bushes scattered over sloping hillsides, or on the gravel flood plains of small streams, where they walked about on the ground. They were secretive and were difficult to detect until they rose and flew with a strong undulating flight above the bushes. Occasionally one uttered a low *chwit*, but as a rule they were silent.

The specimen taken is fully grown, though immature. The throat is somewhat whiter than the breast, but the tail is distinctly rufescent, so that the specimen does not seem to agree with Scott's *Upucerthia fitzgeraldi*,[23] which is supposedly a subspecies of *validirostris*. The feathers of breast and throat have very faintly marked darker tips, an indication of the more prominent breast markings in *U. dumetaria.* The specimen measures as follows: Wing, 86.5; tail, 75.5; culmen from base, 33.7; tarsus, 26.7 mm.

The bill in this bird in life was black, shading to storm gray at base of mandible; iris deep quaker blue; tarsus and toes dull black.

UPUCERTHIA CERTHIOIDES (d'Orbigny and Lafresnaye)

Anabates certhioides d'ORBIGNY and LAFRESNAYE, Mag. Zool., 1838, cl. 2, p. 15. (Corrientes.)

The present species was recorded at Las Palmas, Chaco, from July 14 to 31, 1920 (an immature male, taken July 14, and an adult

---

[23] *Upucerthia fitzgeraldi* Scott, Bull. Brit. Orn. Club, vol. 10, Apr. 30, 1900, p. lxiii. (Puente del Inca, Mendoza.)

female, July 26); Riacho Pilaga, Formosa, August 11 and 18 (two males, taken on the latter date); and Tapia, Tucuman, April 7 to 13 (an immature female, shot April 7). There is no appreciable difference in appearance in specimens from the three localities represented by the five birds preserved as skins. The association of this and allied straight-billed forms in the genus *Upucerthia* with species of the *U. dumetaria* type is questionable.

These birds of wrenlike appearance and action inhabited heavy brush where they worked about on or near the ground, in such dense cover that it was difficult to observe them. At any alarm they gave vent to loud whistled calls, suggestive of those of a canyon wren, and at times were. called out by squeaking noises. Their notes are loud and might easily be attributed to a bird of greater bulk.

In an immature male the maxilla and tip of the mandible were dull black; base of mandible pallid brownish drab; tarsus and toes fuscous.

UPUCERTHIA LUSCINIA (Burmeister)

Ochetorhynchus Luscinia BURMEISTER, Journ. für Ornith., 1860, p. 249. (Mendoza.)[24]

The present species was encountered only on a dry flat above the city of Mendoza, Province of Mendoza, western Argentina, on March 13, 1921, when a female was taken. The few noted were found in low brush along a dry wash.

This bird has been treated as a geographic race of *U. certhioides*, a usage not borne out in my opinion by examination of specimens, since *luscinia*, in addition to larger size, much more robust form, and more grayish coloration, has a decidedly longer tail and broader rectrices. The difference between the two is so extensive that any intergradation, indicating subspecific relationship, must be considered extremely doubtful unless it may be definitely proved by specimens.

CINCLODES FUSCUS FUSCUS (Vieillot)

Anthus fuscus VIEILLOT, Nouv. Dict. Hist. Nat., vol. 26, 1818, p. 490. (Montevideo and Buenos Aires.)

At Berazategui, in the Province of Buenos Aires, several were seen and a male was taken on June 29, 1920, on low ground near the Rio de la Plata. An immature male was shot at El Salto, at an elevation of 1,600 meters above Potrerillos, Mendoza, on March 19, 1921. This second specimen has several white feathers in the crown, an albinistic tendency. It is darker brown than the one shot near Buenos Aires.

These birds walk on the ground with constantly wagging tails, and when flushed may fly, with a flash of the light band in the wings, to a

---

[24] According to Hartert (Nov. Zool., vol. 16, December, 1909, p. 208).

perch on a post or dead branch.   The species is reported as common on low, wet ground in the pampas in winter.

CINCLODES OUSTALETI OUSTALETI Scott

Cinclodes oustaleti SCOTT, Bull. Brit. Orn. Club, vol. 10, Apr. 30, 1900, p. lxii. (Central Chile.)

After the middle of March the present species was fairly common in the foothills of the Andes, in the Province of Mendoza, though it has not been recorded previously in any numbers in this region. The first one seen was found along an irrigation ditch in the out-skirts of Mendoza, on March 13, 1921.   At Portrerillos a male was shot and another seen on March 18, at an elevation of 1,500 meters along the Rio Blanco, while on the following day at El Salto, 300 meters higher, the birds were fairly common, and a male and a female were taken.   They were found feeding near the swift-running streams, where they clambered agilely over the steep rock surfaces, or along quieter channels and irrigation ditches.   At a small estancia several walked about on the beams supporting the roof of a shed.   When flushed they frequently flew up along dry hillsides to rest for a few minutes on huge bowlders.   Near Tunuyan, Mendoza, three or four were noted on March 25, and an immature male was taken along the muddy border of a small lagoon.   Others were seen here on March 26 and 28.

In life the present species, with its dark brown coloration, light superciliary stripe, and habit of constantly wagging the tail, is strikingly similar to the water thrushes of North America, a resemblance heightened by the haunts frequented by the *Cinclodes*, and its sharp emphatic call note.   The birds are continually in movement, as the tail wags constantly even when the body is quiet.   The flight is strongly undulating.

A male, taken at Portrerillos March 19, is in partial molt from juvenal to first fall plumage.   Others are in full fall plumage.   One individual has a few spots of white on the tips of the secondaries, indicative of albinism.   The rump feathers in this specimen are margined with whitish, a marking absent in others.   Measurements of the four birds secured are as follows: Males (three specimens), wing, 89.5, 91.1, and 93.4 mm.; tail, 64.5, 66, and 66 mm.; culmen from base, 18.1, 18.2, and 18.2 mm.; tarsus, 23.2, 24.2, and 25.1 mm.; female (one specimen), wing, 92 mm.; tail, 68 mm.; culmen from base, 17.2 mm.; tarsus, 25 mm.

ENICORNIS PHOENICURUS (Gould)

Eremobius phoenicurus GOULD, Zool. Voy. Beagle, Part 3, Birds, November, 1839, p. 69, pl. 21. (Santa Cruz, Patagonia.)

Near Zapala, Neuquen, an adult female of the present species (with another specimen that was preserved in alcohol) was taken

December 7, 1920. Others were noted December 8 and 9. If the genus *Eremobius*, of Gould (1839) be considered preoccupied by *Eremobia* Stephens,[25] applied to a group of insects, then the bird under discussion here must bear the name *Enicornis* Gray.[26]

Henicornis wallisi Scott [27] in all probability is a synonym of *phoenicurus* since the type locality, Arroyo Eke, is in central Santa Cruz, not far distant from the coast. Salvadori [28] states that, according to Hellmayr, Scott's form is of doubtful validity, as three specimens in the Tring Museum differ from *phoenicurus* only in having the middle rectrices "wholly brown or with but a small ferruginous patch at the base of the inner web." The types of *E. phoenicurus* in the British Museum are said to have the whole basal portion of the middle rectrices, on both webs, ferruginous. The skin from Zapala has the bases of the central tail feathers mottled faintly with ferruginous. Until more material is available this faint distinction is considered merely individual variation.

The type of *Enicornis striata* Allen (Bull. Amer. Mus. Nat. Hist., vol. 2, Mar. 22, 1889, p. 89), which I have seen in the American Museum of Natural History, described from a specimen brought back by Doctor Rusby, of unknown locality, ascribed questionably to Valparaiso, on examination proves to be *Upucerthia ruficauda*.

Near Zapala these birds were found amid patches of low thorny brush that grew on the slopes of rolling hills, where the soil was composed of sand and stones. Here they worked secretively under cover or ran along on the ground with the tail cocked at an angle over the back. Occasionally one flew with tilting flight to a secure retreat passing only a meter above the ground. In general appearance they suggested long-tailed wrens but were more terrestrial. Their call was a low clicking note. (Pl. 17.)

LOCHMIAS NEMATURA NEMATURA (Lichtenstein)

Myiothera nematura LICHTENSTEIN, Verz. Doubl. Zool. Mus., 1823, p. 43. (São Paulo, Brazil.)

Near San Vicente, Uruguay, on January 29, 1921, as I came down to a small stream in a rocky, heavily wooded gulch a small bird came out curiously to meet me, and then retreated to the somber shadows behind. With its sooty brown coloration it was difficult to distinguish in the cover, now dank and dripping from heavy rains, that it frequented, so that several times it had moved along while I was still trying to make out its dull-colored form on the perch recently occupied. Wrenlike in form and wrenlike in actions it

[25] Ill. Brit. Ent., Haust., vol. 3, 1829, p. 94.
[26] List Gen. Birds, 1840, p. 17.
[27] Bull. Brit. Orn. Club, vol. 10, Apr. 30, 1900, p. 63.
[28] Ibis, 1908, p. 453.

hopped silently among limbs and roots, its very quietness lending a sense of mystery heightened by the dark shadows of its haunts.

This individual, a female, is distinctly duller in color on the dorsal surface than two specimens in the United States National Museum marked Brazil without definite locality. It probably represents a southern form, at present not recognized. Measurements of this specimen are as follows: Wing, 63.7; tail, 45.5; exposed culmen, 17.4; tarsus, 23.1 mm.

PHLEOCRYPTES MELANOPS MELANOPS (Vieillot)

Sylvia melanops VIEILLOT, Nouv. Dict. Hist. Nat., vol. 11, 1817, p. 232. (Paraguay.)

The present species, an inhabitant of fresh-water marshes, was recorded as follows: Lavalle, Buenos Aires, October 28 to November 9, 1920 (three adult males, taken October 28, November 2 and 9, and an adult female, October 29); Carhue, Buenos Aires, December 15 to 18 (adult male, shot December 15, two adult females, on December 16 and 18); Lazcano, Uruguay, February 7 and 8, 1921 (immature female, taken February 7); Tunuyan, Mendoza, March 25 to 28 (two females, secured March 26). The series of nine specimens taken, compared with others from Chile and Patagonia, offers no constant differences and, save for seasonal variation, is quite uniform in color. Birds in fresh fall plumage are considerably browner than others. As the breeding season comes on they become paler through wear, so that birds secured in December are frequently almost white on the breast. There is considerable variation in length of bill. *Phleocryptes melanops schoenobaenus* Cabanis and Heine [29] is darker above and below and larger than true *melanops*. The generic name has been usually emended to *Phloeocryptes* but was written *Phleocryptes* originally.

The North American, viewing Neotropical birds for the first time, finds among them many striking similarities to birds from his own land, among which the tracheophone *Phleocryptes* is as striking as any, since in general appearance, notes, and haunt this frequenter of cat-tail and rush-grown marshes is similar to the oscinine long-billed marsh wren, a bird of an entirely different group. As one approaches the rushes of some cañadon in the eastern pampas, a small wrenlike bird may come near to hop about excitedly among the rushes to the accompaniment of clicking notes like those made by striking two pebbles together, and in a short time half a dozen of these *Phleocryptes* may be gathered about. Their alarm is soon over and it is not unusual to have them come almost within reach to look about confidingly. Where the aquatic

---

[29] *Phleocryptes schoenobaenus* Cabanis and Heine, Mus. Hein., pt. 2, 1859, p. 26. (Lake Titicaca, Peru.)

vegetation is composed of large-leaved floating forms the birds hop about on the stems and leaves, frequently with feet and tarsi immersed in the cold water, while they seize eagerly any insects or other life that appear on the plants or in the water. At other times they cling to the stalks of vertical reeds and reach out as far as possible to dig with their bills among the small floating plants, resembling duckweed, that cover the surface.

Their nests are curious globular structures 6 inches in diameter, suspended among dead rushes from about one-half to a little over a meter above the water. *Phleocryptes* frequently was seen transporting tremendous loads of wet stalks and leaves from dead marsh growth that were molded rapidly into their round nests. As it dried, this material was cemented firmly together by the hardening of the slime engendered by the dampness in which it had previously laid so that the walls of the nest were firm and strong. A small opening led into the interior near the top, and the structure was warmly lined with soft feathers, gathered where they had been dropped by other avian denizens of the marsh, and plant downs from the cat-tails and other marsh vegetation. Though the nest of *Phleocryptes* was like that of a marsh wren, there the similarity ceased, as the eggs were clear blue like those of a robin. Three eggs appeared to be the usual number, though I noted nests that contained only two young. Breeding begins early as hard set eggs were taken on November 2, and two nests containing young a week old were recorded at the same time. Others were nest building on this same date. The three eggs taken have a distinctly granular surface, and in color are slightly duller than lumiere blue. They measure as follows, in millimeters: 22.1 by 15.7, 22.4 by 15.3, and 22.6 by 15.5.

The young had prominent orange margins on the opened bill that showed plainly in the darkened interior of the nest. Their ordure was inclosed in a capsule as in oscinine Passeriformes. *Phleocryptes* is one of the few tracheophone species that decoys easily to the loud squeaking attractive to most of the Oscines, and came almost invariably to search for the source of the curious noise. From their general appearance one might consider them as sedentary, but Hudson records that they are migrant near Buenos Aires.

LEPTASTHENURA FULIGINICEPS PARANENSIS Sclater

Leptasthenura paranensis P. L. SCLATER, Proc. Zool. Soc. London, 1861, p. 377. (Argentina.)

The present bird was encountered only near Potrerillos, Mendoza, where it was found from 1,500 meters altitude to about 1,600 meters in the vicinity of the Estancia El Salto. The four specimens secured, including one male and three females, all immature (in fact, barely grown), were taken March 15, 19, and 21. In the identification of these I have not had true *fuliginiceps* for comparison. According to Hellmayr [30] the typical form differs from *paranensis* in larger size (wing, 64 to 65 mm.) and in brighter brown on the dorsal and ventral surfaces. The single male of *paranensis* secured has a wing measurement of 56.8 mm. The three females measure, respectively, 56, 56.5, and 57 mm.

These birds were observed in little family parties that ranged near the ground in the dense scrub that covered the hill slopes.

It has been stated that the original specimens of this bird, secured by Burmeister, came from Parana, a statement that must be incorrect, since *paranensis* is known only from the mountains of western Argentina. In the United States National Museum are two skins, secured from Burmeister, marked " Buenos Aires," while Burmeister himself, in his Reise durch die La Plata-Staaten (1861, p. 469), remarks under *fuliginiceps*, the name that he gave to this bird, " Bei Parana." It must be supposed, however, that he took his specimens during his travels in west Argentina.

LEPTASTHENURA PLATENSIS Reichenbach

Leptasthenura platensis REICHENBACH, Handb. Spec. Ornith., 1851, p. 160. (Rio de la Plata.)

This bird was encountered only at Victorica, Pampa, from December 24 to 29, 1920, and Rio Negro, Uruguay, on February 18, 1921. An immature male, taken at Victorica on December 29, and two females of the same age on December 24 and 28, were preserved as skins. A third immature female was secured at Rio Negro on the date mentioned. *L. platensis* seems to constitute a distinct species, as it differs constantly from *L. aegithaloïdes*, its nearest relative, in the rufous instead of grayish tips on the outer rectrices and in the more pointed crest. Immature examples of *platensis* are darker above than adults and have shorter, more bushy crests; though they resemble *L. a. pallida* in color of the dorsal surface they may be distinguished by the strongly rufescent tips on the outer tail feathers.

Leptasthenura platensis was noted in small flocks—family parties of old and young—in trees of the densest foliage, such as the coronillo and sombre todo (*Iodina rhombifolia*), where they clambered like titmice in a leisurely manner through the dense growths of limbs. Their notes, a faint *tsee-ee-ee*, were weak, and it was difficult to follow them to their source, so that with their retiring habit this species must often escape detection. When individuals were located it was often a matter of several minutes before they could be seen clearly.

---

[30] Nov. Zool., vol. 28, September, 1921, p. 260

LEPTASTHENURA AEGITHALOÏDES AEGITHALOÏDES (Kittlitz)

Synnalaxis Aegithaloïdes KITTLITZ, Mém. Acad. Imp. Sci. St.-Pétersbourg, Div. Sav., vol. 1, 1831, p. 187. (Valparaiso.)

Near Concon, in the Intendencia of Valparaiso, Chile, where this bird was fairly common from April 24 to 28, 1921, a male was secured on April 24 and three females on April 24, 26, and 27. As Kittlitz remarks that he secured his specimens " auf dem Höhen um Valparaiso," these may be considered as topotypical specimens. The present species is distinguished from *L. platensis* by somewhat more bushy crest, darker coloration, and grayish white on the inner webs of the rectrices. The genus *Leptasthenura*, of which *aegithaloïdes* of Kittlitz is the type, is distinguished among the Synallaxis group by the possession of a long, slender graduated tail of 12 rectrices and a more or less developed crest. The head feathers in certain other species are often full and long, but are kept closely appressed to the head. In *Leptasthenura* the development of the crest is observed at once when birds are handled in the flesh, though in the dried skin it is sometimes difficult to distinguish.

Near Concon, *L. aegithaloïdes* was encountered, often in company with other small brush-inhabiting birds, in open thickets of low growth that covered the slopes of rolling hills, or in growths of weeds and thorny shrubs near water. The birds clambered about among the limbs, occasionally uttering low complaining notes, in actions resembling titmice. They were gregarious and were not seen alone. The long, slender tail was a prominent character that served to identify them as they passed with tilting flight across small openings between clumps of trees.

A female, taken April 24, had the maxilla and tip of the mandible dull black; base of mandible dusky green gray; iris natal brown; tarsus and toes dull black.

LEPTASTHENURA AEGITALOÏDES PALLIDA Dabbene

Leptasthenura aegithaloides pallida DABBENE, El Hornero, vol. 2, no. 2, January, 1921, p. 135. (Puesto Burro, Maiten, Chubut, alt. 700 meters.)

The present form is similar to *L. a. aegithaloïdes* of Chile, but is easily distinguished by its general paler coloration. On December 3, 1920, near General Roca, Rio Negro, I found two of these birds resting in the sun in the tops of thick bushes in a region where the atriplex and other growth typical of alkaline flats was tall and dense. The birds rested quietly with long tails hanging straight down, at intervals uttering a low buzzing trill very similar to the songs of some *Synallaxis*. I was surprised to find that the bird taken, secured during the act of singing, was a female. When fresh the extreme base of the mandible in this specimen was olive buff; rest of bill dull black; iris natal brown; tarsus and toes black.

What I presume was this same form was recorded among low bushes covering the dry hills above the city of Mendoza, on March 13, 1921.

SCHOENIOPHYLAX PHRYGANOPHILA (Vieillot)

Sylvia phryganophila VIEILLOT, Nouv. Dict. Hist. Nat., vol. 11, 1817, p. 207. (Paraguay.)

The present species was recorded only in the Chaco, where it was noted near Las Palmas, Chaco, from July 16 to 31 (adult males taken on July 16 and 23), at the Riacho Pilaga, Formosa, from August 8 to 21 (an immature female, shot August 10), at Formosa, Formosa, August 23 (an adult female secured) and 24, at Puerto Pinasco, Paraguay, September 3, and at Kilometer 80, west of Puerto Pinasco, Paraguay, from September 8 to 20 (an adult male, shot September 15). The specimen from west of Puerto Pinasco is somewhat more heavily streaked above than others.

This species frequented saw-grass swamps, particularly where the clumps of grass were interspersed with low bushes or where low palms were scattered through the marshes, though occasionally it was encountered in open savannas. The birds flushed to fly with tilting flight and gracefully undulating tail to new cover from which it was usually difficult to dislodge them. The manner in which they were able to conceal themselves among the limbs of bushes, bare of leaves, was little short of miraculous, so that all in all they were difficult to secure, though fairly numerous. In heavy winds their tails were troublesome as they were blown about, even when at rest, unless in the densest of cover, while on the wing the long feathers were a serious handicap to flight.

During September they were quite noisy and evidently were preparing to breed, as males were seen posing, from perches below the females, with shaking wings and raised tail, while they gave their chuckling songs. The usual call is a low grating rattle.

The Toba Indians called them *to to likh.*

A male, taken July 16, had the maxilla dull black; mandible and sides of maxilla at base plumbeous; tongue whitish, with two indistinct blackish spots at base on either side; iris dragon's blood red; tarsus light cinnamon drab; toes pale mouse gray.

SYNALLAXIS CINNAMOMEA RUSSEOLA (Vieillot)

Sylvia russeola VIEILLOT, Nouv. Dict. Hist. Nat., vol. 11, 1817, p. 217. (Paraguay.)

This yellow-throated, marsh-inhabiting Synallaxis was found at Las Palmas, Chaco, on July 22 (adult male secured), 27 (specimen preserved in alcohol) and 30; at the Riacho Pilaga, Formosa, August 9 (adult male taken) to 17; at Formosa, Formosa, on August 23;

and near Kilometer 80, west of Puerto Pinasco, Paraguay, on September 17 (adult male taken). The three skins preserved are distinctly duller, grayer brown above than skins from Diamantina, Brazil, supposed to be near typical *cinnamomea*, or than two from Fundacion, Colombia, and "Bogota," that have the gray forehead distinguishing *S. c. fuscifrons Madarász*. A skin from Cachoeira, São Paulo, is darker above than skins from more northern localities but is much more refescent than *S. c. russeola*. It represents a distinct form that apparently should bear the name *S. c. ruficauda* Vieillot.[31] In spite of assertions to the contrary, measurements of the four forms here tentatively recognized agree so closely as to have no significance in separating the subspecies. Measurements of the three males that I secured are given below:

Locality	Sex	Wing	Tail	Culmen from base	Tarsus
Puerto Pinasco, Paraguay	Male ad	61.2	65.2	15.5	21.3
Riacho Pilaga, Formosa	do	62.0	66.4	15.1	20.0
Las Palmas, Chaco	do	62.7	64.5	15.5	21.5

The birds under discussion were found among rushes that bordered lagoons, and were doubtless more common than the few records given indicate, since it was difficult to observe them in their rather inaccessible haunts. At some slight alarm one might come through the reeds with rattling, scolding notes to peer for an instant between the stems of the rushes, and then drop down again to be lost to sight in the dense growth. Otherwise they passed wholly unnoted. Their notes were exactly like those of wrens of the genus *Telmatodytes*, so that their white breasts, reddish backs, and large size when they appeared always come as a surprise. Occasionally I had a glimpse of one feeding on the wet surface of vegetation floating among the reed stems, but here they kept so low that it was difficult to follow them for any distance.

The Toba Indians were acquainted with them under the name of *ve on reh*.

An adult male, killed July 22, had the maxilla and tip of the mandible black; base of mandible gray number 8; iris Rood's brown; tarsus gray number 6, with a slight greenish tinge in front at the lower end.

SYNALLAXIS SPIXI Sclater

Synallaxis Spixi SCLATER, Proc. Zool. Soc. London, Aug. 13, 1856, p. 98. (Brazil.)

The three more common forms of *Synallaxis* of central and northern Argentina and Uruguay, *S. spixi*, *S. a. frontalis*, and *S. a. albescens*, are often confused, but with a little care may be readily

---

[31] *Synallaxis ruficauda* Vieillot, Nouv. Dict. Hist. Nat., vol. 32, 1819, p. 310. ("Bresil.")

distinguished. *S. f. frontalis* has the tail reddish brown and the underparts distinctly gray. *S. spixi*, while resembling *frontalis* in color of the undersurface, has the tail plain brown, not decidedly rufescent, and, in addition, usually may be told by the lack of gray across the forehead, as the rufescent crown cap extends to the base of the bill. *S. a. albescens*, like *spixi*, has a brown tail, but is distinctly whiter below than either *frontalis* or *spixi*.

Doctor Oberholser has separated a form of *spixi* from Conchitas, Buenos Aires, under the subspecific name of *notius*, on the basis of grayer dorsal coloration than is found in birds from Brazil. Hartert [32] has considered this a synonym of *spixi*, but more lately Brabourne and Chubb [33] have cited it as a valid form. Examination of the type-specimen shows it to be in very worn plumage, while a second specimen from Conchitas is in similar condition. After careful comparison of these two with six other skins of supposedly typical *spixi* (Sapucay, Paraguay; Quinta, Rio Grande do Sul; Santa Catherina, and two without locality) I can distinguish no valid difference when due allowance is made for change in color due to wear.

The present bird was found only at Rio Negro, Uruguay, where on February 19, 1921, I secured a male in immature plumage in lowland thickets inhabited by *S. a. frontalis*. This bird has a mere trace of rufous in the crown and less on the wings than in adults. It is darker above and somewhat more brownish below than those in full plumage. It is distinguished from *S. a. frontalis* in the same plumage by the absence of rufous in the tail.

SYNALLAXIS FRONTALIS FRONTALIS Pelzeln

Synallaxis frontalis PELZELN, Sitz. Akad. Wiss. Wien, Math.-Naturw. Kl., vol. 34, 1859, p. 117. (City of Goyaz, Engenho do Cap Gama and Cuyaba.)

Synallaxis azarae, while it has a wing measurement about equal to *frontalis*, has a much longer tail; Zimmer [34] considers *frontalis* specifically distinct from *azarae*, an opinion in which I concur. Hellmayr [35] considers *Synallaxis frontalis* of Pelzeln as based on the female of *Parulus ruficeps* of Spix, [36] and uses Spix's locality, Rio São Francisco, as that of the type. Pelzeln, however, employs descriptive terms with his name, although *frontalis* is given as a manuscript name of Natterer, so that although he refers to Spix, the name is based on description and must take as type locality the

[32] Nov. Zool., vol. 16, 1909, p. 211.
[33] Birds of South America, 1912, p. 229.
[34] Field Mus. Nat. Hist., Zool. Ser., vol. 12, May 20, 1925, pp. 105–107.
[35] Nov. Zool., vol. 28, September, 1921, p. 264.
[36] Av. Bras., vol. 1, 1824, p. 85, pl. 86, fig. 2.

regions cited in the original reference. These are the city of Goyaz, Engenho do Cap Gama, and Cuyaba.

The limited series in the United States National Museum does not permit the description of new forms among the birds placed under this name, although one or more are probably represented. It may be noted that specimens that I secured at Rio Negro, Uruguay (in worn plumage), are slightly darker above than skins from the Chaco, while examples from the Chaco in turn seem slightly darker than those from Brazil.

This form was recorded and secured as follows: Resistencia, Chaco, July 9, 1920 (adult male secured); Las Palmas, Chaco, July 19 to 30 (female, July 19, and male, July 30); Riacho Pilaga, Formosa, August 7 to 21 (males, August 7 and 9); Kilometer 80, west of Puerto Pinasco, Paraguay, September 15 (adult male); Rio Negro, Uruguay, February 14 to 19 (male and female immature, February 14, adult male, February 17, and immature male, February 18); Tapia, Tucuman, April 9 (immature female) and 11 (adult male). This form is recorded by Hellmayr from Buenos Aires west to Cordoba, Tucuman, and Salta, and north probably to the plains of Bolivia. Tremoleras has noted it from Montevideo and Canelones in Uruguay.

Synallaxis f. frontalis has chosen as its haunt dense growths of weeds or thorny plants and heavy thickets, usually in low areas shaded by groves. It was especially common in the Chaco and in the dense thickets along the Rio Negro, in western Uruguay, and elsewhere was found in smaller numbers. The birds were found throughout the winter in bands containing from three to eight, apparently family parties still in company from the previous season, that often fed in growths of heavy grass at the borders of thickets. When startled they flew a short distance with tilting flight to some secure cover and at times paused in the open for a few seconds before disappearing among the branches. Frequently they fed among the leaves of bushes or dense herbage, usually near the ground, where they were constantly in motion. When not hopping about restlessly they peer quickly from side to side, at the twigs below or the leaves above, flitting the wings and twitching the long tail. All of these activities were carried on behind a screen of leaves so that only occasionally did a glimpse of them offer through some little opening.

They appear much richer in color in life than in the form of a skin, as the dark browns and grays of their plumage enlivened by touches of rufous on crown, wings, and tail form a pleasing combination. The feathers of the throat in life are puffed out so that their slaty black bases form a shield-shaped patch that appears almost black.

They have a variety of sputtering, scolding calls that they utter on occasion of alarm or interest, and in the spring and summer sing a pleasing little song. A male taken September 15 at Puerto Pinasco, Paraguay, was approaching breeding condition, while near Rio Negro, Uruguay, fully grown young were common during the middle of February. Young in juvenal plumage lack the rufous crown cap of adults but gain it at the post-juvenal molt. In first dress they are browner both above and below and have the wing coverts less bright than after the first molt. In an adult male the maxilla was dull black; mandible gray number 6; iris pecan brown; tarsus and toes grayish olive. Another differed in that the maxilla and tip of the mandible were blackish slate and the iris cinnamon rufous.

SYNALLAXIS ALBESCENS ALBESCENS Temminck

Synallaxis albescens TEMMINCK, Nouv. Rec. Planch. Col. Ois., vol. 3, 1838, pl. 227, fig. 2. (Province of São Paulo, Brazil.[87])

Birds of this species were recorded at the following points: Resistencia, Chaco, July 10, 1920 (immature female taken); Las Palmas, Chaco, July 16 to 31 (males, July 17 and 31, females on July 16, 17, and 28); Formosa, Formosa, August 24; Puerto Pinasco, Paraguay, September 3 (adult male); Victorica, Pampa, December 24 to 29 (males on December 24, 27, and 29, female, December 29).

The specimens secured are allocated to the typical form without direct comparison with skins from southern Brazil. The small series from Victorica, Pampa, in somewhat worn breeding plumage is slightly grayer than skins secured during winter in the Chaco, but seems otherwise similar. The remainder of the skins taken are quite uniform. Specimens in fresh winter plumage have the rufous color of the crown slightly obscured by brownish tips.

During the winter months, in the Chaco, this spine tail was abundant in saw grass and bunch grass at the borders of thickets, or in little openings among scattered trees and bushes on the savannas. On sharp frosty mornings comparatively few were encountered until about 11, when as the day became warmer these small birds appeared in numbers in brushy pastures where none had been visible two hours before. They flew out with quick, tilting flight to new cover or dodged in and out among the clumps of grass or low branches with quick scolding notes, but seldom paused to perch in the open. Their choice of haunt in fairly open savanna regions was in decided contrast to the habitat of *Synallaxis a. frontalis* that frequented the spiny growths of caraguatá amid denser, darker

---

[87] Designated by Berlepsch and Hartert, Nov. Zool., vol. 9, April, 1902, p. 59.

thickets. In a way the areas inhabited by the two are reflected in their tone of plumage, the bird of the grass clumps being paler than the one that frequents the darker forest border. *S. a. albescens* was gregarious, so that a number were found together, often in mixed flocks with little groups of finches. It was not unusual to find them in tall grass in wet localities rather distant from protecting shrubbery.

At Victorica, Pampa, *albescens* was of different habit, as here it frequented the bushes and low trees that formed a heavy ground cover in the open, scrubby forest. Toward the end of December the birds had completed breeding and were encountered in little parties that comprised adults and young. They were social and searched rather quietly through the limbs, often in close proximity to one another. Though quiet and deliberate in movement they clambered rather actively through the thorny twigs, always under cover. They ranged here from 1 to 10 meters from the earth, and were not found feeding on the ground. This difference in habit from what I had observed in the Chaco led me to suppose that the birds from Pampa were different, but such does not seem to be the case.

In a female, taken July 10, the maxilla was blackish brown number 1; mandible mineral gray; iris honey yellow around pupillar opening, becoming lighter toward outer margin; tarsus and toes deep grayish olive.

SYNALLAXIS ALBILORA Pelzeln

Synallaxis albilora PELZELN, Sitzungsb. Math.-Nat. Cl. Kais. Akad. Wiss. (Wien), vol. 20, 1856, p. 160. (Cuyaba.)

Near Kilometer 80, west of Puerto Pinasco, Paraguay, on September 16, 1920, I found one of these birds in heavy, low forest perched on a stick nest as large as a hat, while its mate hopped about in heavy brush near by. Attention was drawn to the pair by the curious song of the bird at the nest *tas pit who wé*, a peculiar succession of notes with rather slow nasal cadence. Both birds were slow and deliberate in their actions.

In the adult female the maxilla was black; mandible gray number 6; iris liver brown; tarsus neutral gray and toes storm gray. These birds may represent a new form, as the skin preserved is somewhat paler, less rufescent below than the average of a small series in the American Museum of Natural History from Matto Grosso, Brazil. The species is known from Brazil and Bolivia, but does not appear to have been recorded previously from Paraguay. Measurements of the bird secured are as follows: Wing, 58; tail, 74.5; culmen from base, 12.8; tarsus, 21.3 mm.

SIPTORNIS MALUROIDES (d'Orbigny and Lafresnaye)

Synallaxis maluroides d'ORBIGNY and LAFRESNAYE, Mag. Zool., 1837, p. 22. (Buenos Aires.)

S. maluroides is a bird of local distribution in the marshes of Argentina and Uruguay, and extends north along the coast into Rio Grande do Sul. It is often difficult to find and I met with it personally at only three localities. On October 25, 1920, I found it common in growths of rushes in the tidal marshes near the mouth of the Rio Ajo, and collected two males and two females. Superficially the birds resemble marsh wrens in appearance and in choice of habit, and, save for their long tails, might easily be confused with *Cistothorus*. Occasionally they were observed on the ground searching for food near clumps of grass but more often were encountered only in the heavier growths of *Juncus*. When startled they flew with tilting flight low over the grass, often blown about in the stiff breeze that swept the marshes. In alighting they sometimes perched for a short space with the head projecting above the grass cover so that they were able to look about though their bodies were entirely concealed. Like related species they were usually frightened by the squeaking noise attractive to most small oscinine birds and either flew to some safe retreat or hid in the densest cover available. Females taken were about to lay. On a second visit on November 15 I found these curious birds again common and secured another specimen that was preserved as a skeleton.

On February 7, 1921, an adult male *maluroides* was secured in a fresh-water swamp in the valley of the Rio Cebollati, near Lazcano, Uruguay. This bird was molting the feathers of wings and tail. An immature male secured in the rushes of a cienaga near Tunuyan, Mendoza, on March 26, is in full juvenal plumage.

Adults from Buenos Aires, when compared with the single specimen from Uruguay, show no appreciable differences, though it is possible that Uruguayan specimens may have larger bills. The bird in immature plumage from Tunuyan has the anterior portion of the crown slightly duller than dark olive buff, with only one or two incoming feathers to suggest the rufescent crown of the adult. The tail is also duller in color than in birds in full plumage. Adult females have the crown patch slightly paler than males though the distinction here is slight.

SIPTORNIS SULPHURIFERA (Burmeister)

Synallaxis sulphurifera BURMEISTER, Proc. Zool. Soc. London, 1868, p. 636. (Buenos Aires.)

In the vicinity of the Rio Cebollati near Lazcano, Uruguay, the present species was common from February 5 to 9, 1921, in marshes grown with saw grass. Though the birds were rather wrenlike in action they were less shy than some other marsh-inhabiting Furnariids, perhaps because this was their breeding season. One brood of young was already on the wing, as two birds in juvenal plumage were shot on February 5 and 7, but the adults were preparing for a second breeding period, as a female taken February 5 was laying, and males secured February 5 and 7 were in breeding condition. These spinetails rested often in the tops of the sawgrass clumps where their light-colored breasts were easily visible at some distance, or at any alarm came out with jerking tails on lower perches to chip at me anxiously. Males had a harsh little song that may be represented by the syllables *chree-a chree-a chree-a chree-chree-chree*, given indifferently on the wing or from a perch. Adults taken were in very worn plumage, probably from abrasion among the stiff grass stems among which they lived. Specimens in juvenal plumage lack the yellow throat patch and rufous wing coverts of adults, and have a more distinct buffy wash over the entire plumage, noticeable especially in the superciliary stripe and the breast.

SIPTORNIS PYRRHOPHIUS PYRRHOPHIUS (Vieillot)

Dendrocopus pyrrhophius VIEILLOT, Nouv. Dict. Hist. Nat., vol. 26, 1818, p. 118. (Paraguay.)

The present form is one of rather general distribution in suitable localities in central and northern Argentina and in Uruguay. It was encountered at the following points: Las Palmas, Chaco, July 15 to 31, 1920 (females taken July 15 and 26); Riacho Pilaga, Formosa, August 7 to 18 (adult male, August 7); Victorica, Pampa, December 23 to 29 (immature male and female, December 23 and 24); San Vicente, Uruguay, January 27 to 31, 1921 (immature male, January 28, adult and immature females, January 29, immature female, January 30); Lazcano, Uruguay, February 5 to 8 (adult female, shot February 5); Tapia, Tucuman, April 6 to 13 (immature male, April 9, female, April 8). No specimens of *S. p. striaticeps* from Bolivia have been seen, but according to Hellmayr in this form the streaking of the head extends down on the nape, the secondaries and tertials are bordered with cinnamon or russet, and the flanks are buffy brown. Typical *pyrrhophius* has the streaking on the head extended only over the occiput, the secondaries and tertials without conspicuous rufous margins and the flanks grayish brown. The coloration of the tail in the series of *pyrrhophius* at hand is variable. In many the inner web of the median rectrix is almost wholly brown, but the extent of this color is variable as in several it is restricted to a faint terminal spot. I can distinguish no constant differences among the birds from the different areas represented, though two skins from Tucuman appear very slightly darker than others. Birds in juvenal plumage have the tips of breast, neck, and upper back feathers more or less faintly vermiculated with fuscous; the dark markings on the breast in some are extended down on the abdomen and in others almost obsolete. Young specimens are often more or less stained. The few taken in Uruguay appear to have slightly larger bills than those from the westward. An adult female taken January 29 is molting wing and tail feathers while the body plumage is worn.

In the present paper I have followed current usage in adopting the generic term *Siptornis* for a broad group of species that without question may be divided into two or more restricted genera. At this time only a little more than half of the species known in this assemblage are available to me, sufficient to demonstrate that while Mr. Cory's treatment of the group [38] has certain merits, it is far from conclusive, and in certain respects requires modification as regards the allotment of species in the genera recognized. Further material is needed before an intelligent discussion of the matter may be made.

The present species, one of more active habit than many of its relatives, frequents groves and thickets often near the borders of openings and clearings, where it works about from 2 to 10 meters or more from the ground. In ordinary circumstances the birds are suggestive of titmice as they clamber and hop about among the smaller limbs, and the long tail, with its pointed, rather stiffened feathers, seems almost an encumbrance to their movements, as it is often held in awkward positions. It may not be used even when the bird swings around on the underside of a horizontal limb, so that one comes to wonder at the possible function of the stiffened tip, when suddenly one of the little birds may start up a tree trunk in orthodox woodpecker fashion. Should any return be necessary they are not averse to whirling around and sidling down head first as acrobatically as any nuthatch, though this is not a usual habit.

The birds are gregarious, and, in addition to traveling two or three together, have a predilection for association with other small brush-haunting birds, and form in a way the guides for little traveling companies of *Thamnophilus gilvigaster*, *Picumnus*, etc., as chickadees do in similar bands in more northern regions.

When excited *S. pyrrhophius* comes about with rapid, sputtering explosive call notes that may be represented as *spee-ee-ee-ee* or *tsee-ee-ee-a.* In the breeding season they utter a low trill, not unpleasing in sound, that is suggestive of the song of *Synallaxis f. frontalis*, and is given from the cover of branches.

[38] Proc. Biol. Soc. Washington, vol. 32, Sept. 30, 1919, pp. 149–160.

At Victorica, Pampa, at the end of December young were out of the nest, and though fully grown were still fed by their parents. Young birds, fully grown, were secured in Uruguay late in January, and one was taken at Tapia in April. In the Chaco this species was known as *Alonzo ca-a guëpe*.

SIPTORNIS D'ORBIGNYII CRASSIROSTRIS (Landbeck)

Synallaxis crassirostris LANDBECK, in Leybold, Anales Univ. Chile, vol. 26, no. 6, June, 1865, p. 713. (Between Melocoton and the Rio Tunuyan, Mendoza.)

A female *Siptornis* secured on the arid flats above the city of Mendoza, on March 13, 1921, seems to agree with Hellmayr's observations regarding this form.[39] The specimen in question has renewed the tail feathers and the wing feathers, save for the primary coverts, while feathers of the head are still in molt. The bird has a distinct rufous throat patch, the outer rectrix is wholly cinnamon, the second one is cinnamon, save for the end of the shaft and an elongate patch on the inner web near the tip which are blackish. The third rectrix has nearly half of the web at the distal end blackish, while the fourth rectrix is black, except for the basal half of the outer web. The fifth has the cinnamon color still more restricted, while on the sixth there is a mere wash of cinnamon near the base of the outer web. The lower rump is cinnamon like the upper tail coverts. The wings are distinctly washed with the same color.

From *Siptornis steinbachi*, which was found near Mendoza in the same area, *S. d. crassirostris* differs in paler brown lower mandible, rusty throat patch, whiter under surface, lighter dorsal region, browner crown, and in differently marked rectrices as indicated under *steinbachi*.

As no typical specimens of *d'orbignyii* are at hand for comparison, the individual from Mendoza is allocated subspecifically solely on its agreement with Hellmayr's statement concerning *crassirostris*. A skin presented to the United States National Museum by Mr. B. H. Swales, collected by L. Dinelli near Colalao del Valle, Tucuman, at an altitude of 2,500 meters, seems to stand intermediate between *d'orbignyii* and *crassirostris*, as the upper parts are more rufescent than in the bird from Mendoza, though the tail is the same.

These birds usually were found in pairs that ran or hopped about on the ground beneath the scattered bushes or clambered swifty through the branches. When excited they chattered and called, and worked rapidly away through the brush. They are characterized by large size and distinctly reddish-brown coloration.

---

[39] Verh. Ornith. Ges. Bayern, vol. 13, Feb. 25, 1917, p. 116.

The one taken has the following measurements: Wing, 64.2; tail, 73.3; culmen from base, 12.3; tarsus, 22.1 mm.

SIPTORNIS STEINBACHI Hartert

Siptornis steinbachi, HARTERT, Nov. Zool., vol. 16, December, 1909, p. 213. (Cachi, Salta, Argentina, altitude 2,500 meters.)

The skin of a female *Siptornis* secured March 13, 1921, near the city of Mendoza, is referred to the present species with some reservation. The bird under discussion was taken in the same type of country as that inhabited by the one allotted as *S. d. crassirostris* and resembles that bird in general appearance, differing in lack of a rufous throat patch, in blackish bill, and deeper rufous on wings, under tail coverts, and flanks. The throat has a faint yellowish tinge in the center, with obscure blackish tips on the feathers. The specimen, apparently adult, is in full molt, and has only three rectrices, the external ones on the right side. The two outermost are cinnamon, the third is cinnamon save for a blackish stripe, that extends along the middle of the distal half of the outer web and spreads to the shaft at the tip. The skin agrees substantially with the original diagnosis except for the dusky streaks on the lesser wing coverts. The specimen has the following measurements: Wing, 64; tail imperfect; culmen from base, 14.4; tarsus, 22 mm. These correspond closely to the measurements given by Hartert for his type, namely, wing, 66; culmen, 14; and tarsus, 22 mm.

The bird does not seem to have been reported before save from the type-locality from which no specimens are available for comparison.

SIPTORNIS BAERI Berlepsch

Siptornis baeri, BERLEPSCH, Bull. Brit. Orn. Club, vol. 16, May 28, 1906, p. 99. (Cosquin, Cordoba.)

Five specimens of this species were taken at Victorica, Pampa, on the following dates: December 24, 1920, adult male; December 27, immature male; and December 28, two adult males, one immature female. The species is easily distinguished from *S. sordida*, which it resembles superficially, by its heavier more robust bill and the darker median rectrices. Skins in juvenal plumage differ from adults in lighter ventral surface, more indistinct throat patch, and fine vermiculations of dusky on breast.

The species inhabits the same type of country as that frequented by *Synallaxis a. albescens*, a species quite similar in general appearance. Apparently *baeri* rears two broods during the season, as adults taken in December were in breeding condition, while a first brood was already on the wing.

Victorica marks a slight extension in the previously known range.

ARID HILLS GROWN WITH SCATTERED BRUSH, HAUNT OF CRESTED TINAMOU
(CALOPEZUS ELEGANS MORENOI)

North of General Roca, Rio Negro, December 2, 1920

A QUIET CHANNEL OF THE RIO NEGRO. THE BIRDS IN THE WATER
ARE COOTS (FULICA ARMILLATA AND F. LEUCOPTERA)

South of General Roca, Rio Negro, December 3, 1920

ELEVATED PLAIN SURROUNDING THE TOWN OF ZAPALA, NEUQUEN. HAUNT
OF ENICORNIS PHOENICURUS

Taken December 8, 1920

FOOTHILLS OF THE PRE-CORDILLERA, SHOWING FLATS GROWN WITH CREO-
SOTE BUSH

Kilometer 32, Ferrocarril Trasandino, Mendoza, March 15, 1921

In addition in the United States National Museum there is a specimen taken by Capt. T. J. Page, labeled as secured in Uruguay in July, 1860.

SIPTORNIS SORDIDA FLAVOGULARIS (Gould)

Synallaxis flavogularis GOULD, Zool. Voy. Beagle, pt. 3, Birds, November, 1839, p. 78, pl. 24. (Bahia Blanca and Santa Cruz.)

The present spinetail, of wide distribution in Argentina, was recorded at the following points: General Roca, Rio Negro, November 24 to December 3, 1920; Zapala, Neuquen, December 9; Ingeniero White (Bahia Blanca), Buenos Aires, December 13; Guamini, Buenos Aires, March 8, 1921; Potrerillos, Mendoza, March 17 and 19; Tunuyan, Mendoza, March 23 to 27; and at Formosa, Formosa, August 24, 1920.

The series from Ingeniero White, the port of Bahia Blanca, includes two males and seven females, taken December 13, all breeding birds, in worn, abraded plumage. These may be considered topotypical skins, as Darwin states that he collected his specimens at Bahia Blanca and Santa Cruz. The series of females shows considerable variation in the form of the throat patch. In one the throat spot is fairly large (though smaller than in males) and tawny in color; in three the patch, while similar in hue, is restricted and more or less obscured by a mixture of white; in another the area in question is faintly washed with yellowish; and in two the throat is plain white. In one male the throat is tawny and in the other cinnamon buff. Five specimens from General Roca, Rio Negro, include four adult males shot November 24 (two), 27, and December 3, and two adult females killed November 25 and December 9. These do not differ appreciably in color from those from Bahia Blanca when allowance is made for the fact that they are in better feather. All have the colored throat patch with the usual range in depth of color, and show more distinct blackish points on the tips of the feathers than in case of the more worn specimens from the coast. One adult female has a patch of white feathers in the center of the nape. An adult female taken at Zapala, Neuquen, on December 9 is badly worn and is renewing the central rectrices. The single specimen from Guamini, in western Buenos Aires, similar in general appearance to those described above, is an immature female in molt into adult plumage.

At Potrerillos, Mendoza, an immature female and another immature bird whose sex was not known were taken at an altitude of over 1,500 meters on March 17; and at Tunuyan, Mendoza, two immature males and an adult female were taken March 23 and two immature females on March 27. The adult female, in full molt, is somewhat deeper brownish gray, less whitish below than others that have been noted thus far. Immature birds in juvenal plumage lack the colored throat patch and have the breast and foreneck rather faintly vermiculated with dusky. The throat patch is assumed early in the post-juvenal molt, save in those birds that do not develop it. Its lack in certain individuals is an apparent retention of a character of immaturity, as it may be held that the throat patch has been a secondary acquirement. A single adult female secured at Formosa, Formosa, on August 24, while somewhat duller and darker on the breast than the average, is matched closely by some individuals in the considerable series at hand from the south, and is similar to an old specimen in the United States National Museum collection from Santa Fe. I have not seen *S. s. affinis* Berlepsch from Tucuman, but, on the basis of Berlepsch's original description, consider the Formosan specimen nearer to *flavogularis*. It is possible that the one taken at Formosa represents a northern migrant, as it was shot in winter.

Though recorded from northern Argentina, this spinetail was most abundant in the semiarid regions of the south and west, where it found a congenial home in low growths of bushes, particularly of piquillín (*Condalia lineata*), atriplex, and a sort of grease wood (*Grahamia bracteata*) whose dense branches offered it safe cover. In places, as near the coast at Bahia Blanca, the birds were abundant and formed the dominant element among passerine species; at other localities they were found sparingly or only in limited areas.

As they clamber around among the thorny twigs of dense bushes these spinetails seem rather clumsy in their movements, and when touched by thorns often fall with seeming awkwardness to one side. In reality, however, they are expert in progression, and, in spite of their seeming lack of skill, are working rapidly and surely through difficult passages. At intervals one may pause, often on a concealed perch near the ground, to sing a low, double-noted trill, *tsee-ee-ee-ee tsee-ee-ee-ee*, that is given by both males and females. Occasionally one may come up to sing from a more pretentious perch on the top of a bush where it has a wider view of its chosen world. The habit of song among females seems common in this group as I have noted it among others of related genera. Flight, tilting and fairly rapid, is practiced for short distances only, as the birds drop into safe cover as soon as possible. In addition to their song they uttered chattering scolding calls, especially during the breeding season and were considerably excited by squeaking. It was not unusual to see them on the ground, particularly under heavy cover.

A nest found near General Roca, on December 3, 1920, was placed in a spiny bush among heavy branches about a meter from the ground where it was difficult to reach it. This nest was composed of firmly interlaced thorny twigs that made a ball 200 mm. in diameter. At one side near the top was an entrance opening, located among thorny limbs of the shrub, and further protected by small spiny twigs grouped about it. After some difficulty I opened the structure to find that the interior cavity was firmly and closely felted with a lining of fur from the introduced hare, a device that not only made a safe cushion for the eggs and young, but also gave protection from the severe winds of this region, that otherwise had free passage amid interstices in the nest material. This nest contained two white eggs, that when first seen were beautifully tinted by the yolk within through the somewhat translucent shell. When blown they became dull white in color. These eggs measure as follows, in millimeters: 18.4 by 14.4, and 18 by 14.6. In form the eggs are rather bluntly pointed with little distinction between large and small ends.

It is probable that two broods are reared in a season, as young only recently grown were taken in March.

SIPTORNIS PATAGONICA (d'Orbigny)

Synallaxis patagonica d'Orbigny, Voy. Amér. Mérid., Ois., 1835–1844, p. 249. (Banks of the Rio Negro.)

Near General Roca, Rio Negro, from November 23 to December 2, 1920, the present species was fairly common; adult females were collected on November 23 and 24 and a male on November 25. The skins preserved do not differ appreciably from two taken near San Antonio del Oeste, not far from the type locality. The species has been recorded west to the Rio Limay in Neuquen,[40] a short distance beyond Roca. The throat in *patagonica* of both sexes has the feathers slate at the base with the tips white, forming a distinct dark throat patch spotted rather irregularly with white, as prominent as the throat patch of *Synallaxis f. frontalis*. No mention of this is made in the description in the British Museum Catalog (vol. 15, p. 69), and in the key (p. 65) the throat is said to be unspotted, though the original description by d'Orbigny describes the throat as distinctly marked. The species differs notably in structural characters from the stiff-tailed forms of *Siptornis* and in some ways seems quite aberrant.

This bird was found in the semiarid region that bordered the Rio Negro, where it frequented the denser, taller stands of *Atriplex* and other shrubs that grew in the river bottom or occurred more sparingly in the smaller, more scattered growth that clothed the gravel hills above the flood plain. Individuals hopped about among the twigs or walked slowly around on the ground, always under protec-

---

[40] Hellmayr, Nov. Zool., vol. 28, September, 1921, p. 268.

tion of brush. Their flight was rapid and tilting with the black tail showing prominently as they darted away. The song is a musical, even trill, resembling the syllables *tree-ee-ee-ee-ee*. On November 24 I found a nest of this species, one of the prominent structures made of sticks so abundant in the brush of this region. The nest, placed in the top of a bush 3 feet from the ground, without concealment, was an irregular ball, approximately 400 mm. in outside diameter, constructed of thorny twigs from 100 to 300 mm. long, ranging in size from fine sticks to those as large in diameter as a lead pencil. A tubular entrance tunnel, made of small, very thorny twigs, closely and firmly interlaced, led out at one side for a distance of 400 mm., supported by a limb that grew out beneath the nest. The nest ball was so compactly made that it required some time and trouble to open it. The inner cavity was 125 mm. in diameter and had in the bottom a firmly felted cup of plant down, fur of the introduced hare (common in this region), and feathers. Three eggs that lay on this soft bed, dull white in color, were on the point of hatching and were badly broken in preparation. Two that are more or less entire offer the following measurements, in millimeters: 20.9 by 15.1 and 20.1 by 14.5.

SIPTORNIS HUMICOLA (Kittlitz)

Synnalaxis humicola KITTLITZ, Mém. Acad. Imp. Sci. St.- Pétersbourg, Div. Sav., vol. 1, 1831, p. 185. (Valparaiso.).

Near Concon, Chile, where *Siptornis humicola* was fairly common, skins of two males were preserved on April 26 and 27 and a female on April 28, 1921. The birds frequented dense thickets of low brush that grew over the slopes of rolling hills, where they worked slowly about among the limbs or occasionally on the ground, where it was open but heavily protected above. In actions they suggested *Synallaxis f. frontalis*. Usually they were silent and so were difficult to find, but on one encounter one burst out in a clear, trilled song like that of some wren. The muscular part of the stomach in this species is large and strong, heavier, in fact, in proportion to the size of the bird than in some seed-eating finches.

A male, when first taken, had the maxilla and tip of the mandible black; base of mandible gray number 8; tarsus, deep olive gray; toes, tea green; iris, natal brown.

SIPTORNIS LILLOI Oustalet

Siptornis Lilloi, OUSTALET, Bull. Mus. Hist. Nat. (Paris), vol. 10, 1904, p. 44. (Lagunita, Tucuman.)

An immature female, shot at an altitude of 2,300 meters on the Sierra San Xavier, above Tafi Viejo, Tucuman, was the only one of these birds seen. The specimen is in the immature stage described by Lillo as *S. dinellii.*[41] The tips of the tail feathers are rounded and do not display the acuminate points found in *hudsoni* and *anthoides*, a condition that may change with age. Hellmayr[42] has indicated that *S. lilloi* is the bird described by Chapman[43] from above Tafi del Valle, Tucuman, as *Siptornis punensis rufala.*

The individual taken was flushed among tussock grass that covered an open slope at the summit of the cumbre, and was killed on the wing. It appeared dark in color when in the air, with a distinct reddish-brown band in the wings. A bird of similar appearance that I flushed but did not secure in tussock grass on the highest points near El Salto above Potrerillos, Mendoza, during March was probably *Siptornis anthoides.*

SIPTORNIS HUDSONI (Sclater)

Synallaxis hudsoni, P. L. SCLATER, Proc. Zool. Soc. London, 1874, p. 25. (Conchitas, Buenos Aires.)

Hudson's spinetail in general coloration suggests a pipit, while in attitude and habits it is strongly suggestive of the allied *Anumbius*. It was encountered at only two localities in the Province of Buenos Aires, first near Lavalle, Buenos Aires, where it was fairly common in marshes on October 23 and November 6 and 9, 1920, and a second time near Guamini, where, on March 8, 1921, I found three in a dense patch of thistles and other weeds on the shore of the Laguna del Monte. Near Lavalle the bird was partial to marshes grown heavily with *Juncus acutus*, a sharply pointed rush that grew in clumps with little runways between. The birds were shy and secretive and were seldom seen until they darted out and flew rather swiftly with undulating flight to new cover. When once under shelter they crept rapidly away so that it was difficult in many instances to flush them a second time. The light outer margins of the tail showed prominently in flight. Occasionally one perched quietly among dead rushes at the border of some lagoon.

On the evening of November 6, while setting a line of mouse traps among hollows between the dunes south of Cape San Antonio, I flushed one of these birds from the base of a clump of *Juncus* and after careful search discovered a nest, a domed structure placed directly on the ground in the base of the tussock of rush, with a runway like that of some mouse leading into it. Never have I seen the nest of a bird more completely concealed from any possible view, and save for the chance that directed my hands to the base of the clump in question, so that the female flew out almost in my

---

[41] Rev. Letr. Cienc. Soc. Tucuman, vol. 3, July, 1905, p. 53.
[42] Arch. Naturg., vol. 85, November, 1920, p. 72.
[43] Bull. Amer. Mus. Nat. Hist., vol. 41, Sept. 1, 1919, p. 328.

grasp, I should never have suspected its presence. The nest was composed of bits of grass, and was warmly lined with many feathers. The three eggs, with incubation partly begun, are dull white with a very faint tinge of cream. The shell is slightly roughened. The eggs measure, in millimeters, as follows: 22.2 by 16.8, 21.5 by 16.7, and 21.4 by 16.5.

An adult female, shot October 23, when first taken, had the maxilla and tip of mandible fuscous black; base of mandible drab gray; iris light seal brown; tarsus and toes ecru drab.

Adult females, preserved as skins, were secured near Lavalle on October 23 and November 9, and an immature female was shot at Guamini on March 8. The two adults, though fully grown, differ strikingly in color of throat patch, as in one it is yellow and in the other cinnamon buff. The immature bird has the breast heavily and the abdomen more lightly streaked with dusky. Doctor Hartert[44] has treated *S. hudsoni* as a subspecies of *S. anthoides*, with which decision I find that I can not agree, as *anthoides* is distinguished by its somewhat broader, less narrowly pointed rectrices, and more suffused, less definite color pattern on the dorsal surface, in addition to its darker coloration.

CORYPHISTERA ALAUDINA ALAUDINA Burmeister

Coryphistera alaudina BURMEISTER, Journ. für Ornith., 1860, p. 251. (La Plata States.)

Near Tapia, Tucuman, the present bird was common from April 6 to 13, 1921; four of the six taken were preserved as skins. These comprise adult males, shot April 6 and 12, and an immature male and female, taken April 11. The immature specimens resemble adults, but have the streaks of the undersurface narrower and less sharply defined. The skin in this species is thick and very tough.

An adult male, when killed, had the maxilla sayal brown, shading to deep mouse gray at the tip and at the base of the culmen; mandible dull light-grayish vinaceous; tarsus and toes cinnamon; iris natal brown.

I have not seen specimens of *C. a. campicola* Todd,[45] a northern subspecies said to differ from the typical form in paler, more buffy coloration above, with the underparts less heavily streaked.

These curious birds inhabited dry, open scrub growing over rolling hills, where they were found in parties numbering from three to six, apparently families, as adults and young were taken from the same flock. Though the birds were common, they were silent un-

---

[44] Nov. Zool., vol. 16, December, 1909, p. 214.
[45] Proc. Biol. Soc. Washington, vol. 28, November 29, 1915, p. 170. (Guanacos, Bolivia.)

less alarmed, and in the dense scrub were easily overlooked. It was not unusual to see them feeding on open ground among thorny bushes where, like an ovenbird, they walked about with long strides. When startled they rise at once into the limbs of the trees where they clamber quickly away until they are concealed behind twigs or leaves. Frequently they work rapidly along, flying from tree to tree, until they are lost to view. In fact, when thoroughly alarmed it is difficult to keep near them, so artful are they in seeking a screen behind which they may move rapidly away without being observed. Though not breeding at this season, they spent considerable time about nests, constructed of sticks, placed at low elevations in the trees. The little bands frequently rested in close proximity to these structures, or when not too much alarmed hopped or climbed rapidly to shelter behind them, where they rested in a crouching attitude with crest erect. and head turning quickly from side to side. Their need for. protection from sight was obvious since the light, streaked color pattern and the erect crest made them very conspicuous. When alarmed they gave a sputtering metallic rattle that was very peculiar.

The stick nests that they frequented were 300 mm. in diameter, globular in form, with an entrance through a small tunnel that led into one side. The nests were strongly made with thorny, crooked twigs so interwoven that it was difficult to open them for examination. The twigs used were often 300 mm. or over long and as large around as a pencil. The birds delighted in resting in the entrance tunnel or in clambering about over the top. On April 12 I observed three busy with the arrangement of a few small twigs about the entrance to one of these domiciles, a labor that was accompanied by odd chattering and trilling notes. These changed to the sputtering alarm note as soon as I was sighted and the whole party moved rapidly away.

In the collections of the United States National Museum there is a specimen secured in February, 1860, on the Rio Bermejo.

ANUMBIUS ANNUMBI (Vieillot)

Furnarius annumbi VIEILLOT, Nouv. Dict. Hist. Nat., vol. 12, 1817, p. 117. (Paraguay.)

The present species, the one usually indicated by the name *leñatero*, though that designation is applied to all of the tracheophones that build stick nests, was recorded and skins were collected as follows: Kilometer 182, Formosa, August 21, 1920; Formosa, Formosa, August 23 (adult male) and 24 (adult female); Puerto Pinasco, Paraguay, September 3 (adult male); Kilometer 80, west of Puerto Pinasco, September 6 to 18 (adult male, September 7); Dolores, Buenos Aires, October 21; Lavalle, Buenos Aires, October 29 to November 13 (adult male, November 13); Victorica, Pampa, December 26 (adult male); Carrasco, Uruguay, January 9 and 16, 1921; La Paloma, Uruguay, January 23; San Vicente, Uruguay, January 25 to February 2; Lazcano, Rocha, February 3 to 9 (two adult males, February 3); Rio Negro, Uruguay, February 17 to 19; Guamini, Buenos Aires, March 3 to 8 (four taken March 6 include adult and immature males and females). .

The material secured indicates that there are two forms of this curious and interesting bird, but as the ranges to be assigned and the names to be applied are uncertain my notes are given under the specific name. Two skins from the vicinity of Puerto Pinasco, Paraguay, in the Chaco, are somewhat duller in color and seem more heavily streaked above than those from the pampas. Two skins from Formosa are somewhat intermediate in appearance between these two northern birds and the paler, buffier, less heavily streaked birds secured in the Province of Buenos Aires and Uruguay. Vieillot takes his *Furnarius annumbi* from Azara, but makes no mention of locality. Azara (vol. 2, p. 226) remarks that the *anumbi* is fairly common, without being abundant in the countries of the Rio de la Plata, and that it seems most common in Paraguay. Paraguay is commonly accepted as the type locality. *Anthus acuticaudatus* Lesson [46] is described with no locality indicated. *Anumbius anthoides* d'Orbigny and Lafresnaye [47] was said to have come from Corrientes, which is near the boundary of Paraguay.

This species ranges through areas of open brush, often scanty in character, or may even penetrate far into the open where heavy growths of thistles afford low coverts. During the entire year it seems to remain paired, so that only after the close of the breeding season, when grown young still accompanied their parents, were more than two found in company. The birds were tame, inconspicuous, and usually quiet in actions. They fed on the ground where they walked about on open grass sward or in shelter of low tussocks or clumps of weeds. Often they rested on fence posts or wires at a considerable distance from cover. When alarmed or at rest it was usual for them to seek refuge near their stick nests, or in spiny clumps of thistles, and when frightened they passed from one such covert to another, dropping down to fly with a direct or slightly undulating flight just above the ground. Save for an occasional flash of white from the tail they were wholly plain in appearance.

---

[46] Traité Ornith., 1831, p. 424.   [47] Mag. Zool., 1838, Cl. II, p. 17.

The stick nests of the present species were seen everywhere in their range and were of remarkable construction. Thorny twigs were woven in an irregular form, entirely inclosed, with an entrance opening frequently in the form of a short tunnel. These structures were placed in low, often thorny, trees, or occasionally in thistles. It was not unusual to see them built about the wires on the cross arms of telephone and telegraph poles, and occasionally one was built in the head of a railroad semaphore. In one such instance that came under my observation the birds had filled a space between iron uprights 4 feet long and were still busily engaged in carrying sticks up to 10 meters in the air, though lower nesting sites abounded. (Pl. 10.)

On one occasion I observed a pair that had evidently just chosen the site for a new home. The birds had selected a slight opening among more or less horizontal limbs of a thorny tree, where they hopped about a few inches apart as they examined the space critically, or rested near together and pecked and pulled at near-by twigs. At intervals the male sang in a low tone. Nest construction is apparently a prolonged process and the birds seem to work at it when not in breeding condition. As the nests are firmly woven and durable they last for several years. Though ordinarily peaceful, males fight savagely in defense of their chosen territory, battling with intruders until exhausted.

The peculiar song of this species, given at times by the female as well as the male, may be represented as *chick chick chick chee-ee-ee-ee-ee*, uttered in a rapid monotone, with an effort that shakes the whole body. It is repeated frequently and is often heard from a nest. The call note is a sharp *tschick*. In the Chaco the *guira anumbi* of Azara was known in Guaraní as *buituitui* in imitation of the song while the Angueté Indians knew it as *kas mis ka now ah.*

The development of the skull with age in the Tracheophone mesomyodi seems less rapid than in the Oscines, so that the usual age criterion for our smaller Passeriformes of the extent of cancellation between the plates of the cranium may not be trusted. On March 6, I killed four *Anumbius*, an adult pair and two fully grown immature birds. In the adult male the top of the skull was entirely covered with cancellations. In its mate, an adult female, the top of the skull was still open, as it was in the two young, the offspring of the adult pair.

An adult male, taken August 23, had the maxilla and tip of mandible cinnamon drab; base of mandible pale drab gray; iris liver brown; tarsus drab gray; toes smoke gray. An immature male, taken March 6, had the maxilla and tip of the mandible slightly darker than hair brown; extreme tip of culmen shading to chaetura black; rest of mandible puritan gray; iris cinnamon drab; tarsus and toes tea green. In an adult female, also taken on March 6, the bill resembled that of the immature male; the iris was Hay's russet; tarsus slightly duller than deep olive buff; toes tea green.

PHACELLODOMUS STRIATICOLLIS STRIATICOLLIS (d'Orbigny and Lafresnaye)

Anumbius striaticollis d'ORBIGNY and LAFRESNAYE, Mag. Zool., 1838, Cl. II, p. 18. (Buenos Aires.)

When compared with *P. rufifrons* alone the present species seems generically distinct, as it carries to extreme development the characters of strongly graduated tail and broadened shafts on the breast feathers found to a less degree in *rufifrons*. *Phacellodomus ruber*, however, offers such an appearance of transition between the two that in my opinion *Phaceloscenus* Ridgway (for *Anumbius striaticollis* d'Orbigny and Lafresnaye) may not be maintained as a distinct genus.

The subspecies *maculipectus* Cabanis, as represented by a skin in the Field Museum from Cuesta de Angama, Tucuman, is distinctly darker above than true *striaticollis*.

On October 23, 1920, I found several of the present species in a partly dry, grass grown tidal marsh bordering the Rio Ajo at Lavalle, Buenos Aires, and collected a pair of adults. Three or four fed together on the ground, and as I approached climbed into a low, thorny bush, while others were encountered in thick, long grass where they were seen only as they flushed and flew with an undulating flight for a meter or more. At San Vicente, Uruguay, young birds were fairly common in dense brush on rocky hillsides, where I collected an immature female on January 25, 1921, and another (preserved in alcohol) on January 28. The birds scolded vigorously with rattling call notes but kept concealed behind leaf-grown branches. An immature male was shot in a growth of saw grass in a marsh near Lazcano, Uruguay, on February 5.

The adult birds taken have the plumage worn so that it is harsh and hard to the touch from the prominence of the shaft tips of the feathers. Immature specimens in juvenal plumage are less rufescent on the back and breast, and, though the broadened shafts are evident on the feathers of head and breast, have the plumage softer than adults. The rectrices in *striaticollis* are slightly broader than in *rufifrons*.

An adult, when first killed, had the maxilla blackish; mandible and maxillar tomia, below nostril, gray number seven; iris cream buff; tarsus and toes neutral gray.

PHACELLODOMUS RUBER RUBER (Vieillot)

Furnarius ruber VIEILLOT, Nouv. Dict. Hist. Nat., vol. 12, 1817, p. 118. (Paraguay.)

This species was recorded at Resistencia, Chaco, on July 8, 1920, when two were seen, and an adult male was taken, at the Riacho Pilaga, Formosa, August 16, when an adult female was secured, at Formosa, Formosa, where an adult male was taken August 23, and others seen on the day following, and at Puerto Pinasco, Paraguay, on September 3. The three specimens taken do not differ appreciably from one another in color. Cherrie [48] has described a northern form (that I have not seen) as *P. r. rubicula* on the basis of more rufous coloration on the dorsal surface.

These birds were found in swamps grown with saw grass, and in the cat-tails and other vegetation that bordered lagoons, particularly in the areas known as *palmares*, where low palms grew in scattered groves over marshy ground. They were frequently shy, especially when feeding in dense marsh vegetation from which they refused to be called by unusual noises. In fact, they were more often seen when I remained perfectly quiet and waited for them to dart out into sight for a few seconds. When the wind was quiet their noisy rustling among the dead cat-tail stalks was plainly audible though the birds themselves were entirely hidden. Rarely one came out for a few seconds with a sharp, scolding *check check*, and jerked nervously up and down on its perch with the body inclined well forward. The light eyes showed plainly when the birds were not too far distant. Where they were encountered amid palms they often came up into the palm tops, where on cool mornings they rested in the sun to preen their plumage.

An adult male, taken August 23, had the maxilla and extreme tip of the mandible dull sooty black; rest of mandible dawn gray; iris primuline yellow, shading to mustard yellow at outer margin; tarsus and toes storm gray. Another male, shot July 8, had the iris apricot orange.

PHACELLODOMUS RUFIFRONS SINCIPITALIS Cabanis

Phacellodomus sincipitalis CABANIS, Journ. für Ornith., 1883, p. 109. (Tucuman, Tucuman.)

An immature female taken near Tapia, Tucuman, on April 12, 1921, differs from *P. r. rufifrons* in darker coloration on the dorsal surface. The specimen mentioned was shot from a small flock in dense brush where the birds remained in close concealment.

---

[48] Bull. Amer. Mus. Nat. Hist., vol. 35, May 20, 1916, p. 186.

PHACELLODOMUS SIBILATRIX Sclater

Phacellodomus sibilatrix "Döring, MS.," SCLATER, Proc. Zool. Soc. London, 1879, p. 461. (Cordoba.)

An adult male taken July 15, 1920, at Las Palmas, Chaco, near the Riacho Quia, was the only one seen.

PSEUDOSEISURA GUTTURALIS (d'Orbigny and Lafresnaye)

Anabates gutturalis d'ORBIGNY and LAFRESNAYE, Mag. Zool., 1838, cl. 2, p. 15. (Rio Negro, Patagonia.)

The present species of shy, secretive habit, an inhabitant of arid regions, was encountered near General Roca, Rio Negro, November 24 to December 2, 1920; at Zapala, Neuquen, from December 7 to 9; and near the city of Mendoza, in western Argentina, on March 13. These birds frequent the denser growths of low, thorny brush that grows over the dry, barren slopes of stony, sandy hills in these regions, where, though they may be noted at a distance, it is difficult to approach them. They are seen occasionally running swiftly on the ground, with the tail erect, taking advantage of any cover that may offer. When excited they may come out to rest on some low branch with jerking wings and tail, while they peck nervously at the branches near at hand. They are found in pairs or family groups that remain near their huge stick nests placed in the low bushes. Their song is a loud, rattling call given in chorus by male and female, similar in a way to that of *Furnarius rufus*, but less loud and perhaps slightly more metallic in sound. In addition they give a low, clucking call. Though their laughing calls may be heard frequently at a distance, the birds slip away rapidly through the brush, so that it is difficult to approach within gunshot. Their undulating flight carries them barely above the ground.

An adult female when first taken had the maxilla and tip of the mandible dark mouse gray; base of mandible pale Russian blue; iris cream buff; tarsus gray number 6.

Adult females were taken at General Roca, November 24 and 29, and an adult female and three immature birds (two females and one male) at Zapala, on December 9. The adult birds are in exceedingly worn plumage and have begun molt on the head. Immature birds of both sexes are alike, and differ from the adults in having faint, dusky cross bars on the breast. The white throat feathers in this species are soft and silky to the touch.

PSEUDOSEISURA LOPHOTES (Reichenbach)

Homorus lophotes REICHENBACH, Handb. Spec. Ornith., August, 1853, p. 172. (Bolivia [49] ?)

The present species probably may be separated generically from *Pseudoseisura gutturalis*, from which it differs to a considerable extent. The two are here associated pending further study.

Though reported of fairly wide range, the present bird was encountered only near Victorica, Pampa, from December 23 to 29, and at Rio Negro, Uruguay, from February 16 to 18. It is an inhabitant of open groves of low trees where it feeds on the ground, often in company with *Drymornis bridgesi*. At the slightest alarm the crested *Pseudoseisura* flies up with chattering calls, and hops about in the shelter of the limbs as alertly as a jay, pausing to peer out or to peck nervously at the limbs. The flight is strongly undulating.

These birds build huge nests of sticks, as large in diameter as a bushel measure, somewhat flattened, with an entrance at one side, that are placed in the tops of low trees from 4 to 6 meters from the ground. The conspicuous nests are seen frequently, but the birds are usually so shy and retiring that it may be difficult to find them. However, on occasion they may come familiarly into dooryard trees, and at Victorica, where they were known as *chorloco*, they were accused of stealing the eggs of domestic fowls. The birds are found in pairs save for the period when adults are accompanied by young. (Pl. 10.)

The paired birds shriek in chorus with nasal, laughing calls that close with rattling notes suggestive of those of some melanerpine woodpecker. When excited they utter in a low tone a note resembling the syllable *cuck cuck cuck*. In an immature male, barely grown, the bill was black; iris ecru drab; tarsus and toes dark grayish olive.

Two immature specimens taken at Victorica on December 23, though fully feathered, had the borders of the gape soft and the bill shorter than in adults. These, in juvenal plumage, differ from older individuals in having indistinct, narrow, dusky bars on the breast and sides of the head.

XENICOPSIS RUFO-SUPERCILIATUS OLEAGINUS (Sclater)

Anabazenops oleaginus P. L. SCLATER, Proc. Zool. Soc. London, 1883, p. 654. (Sierra de Totoral, Catamarca, Argentina.[50])

When Sclater described the present form of *Xenicopsis* he did so on the basis of skins secured by White in the Sierra de Totoral, Catamarca, and on others (in the United States National Museum)

---

[49] Reichenbach remarks that the locality on the label of his specimen, given as Bolivia, is probably incorrect.
[50] For citation of type specimen see Cat. Birds Brit. Mus., vol. 15, 1890, p. 106.

taken by the Page expedition at Parana. The type-locality was subsequently designated as Catamarca. Two immature specimens (fully grown) that I secured April 17, 1921, on the lower slopes of the Sierra San Xavier, above Tafi Viejo, Tucuman, are distinctly more yellowish olive both above and below than specimens from the Chaco, Parana (the Page specimens), and Uruguay, and indicate that there are two forms of the bird under discussion in Argentina. It is assumed that the Tucuman specimens are similar to those from Catamarca so that they are given the name *oleaginus*. The somewhat duller-colored bird from farther east must take the name *acritus* (Oberholser) described from Sapucay, Paraguay.

On the lower forested slopes of the Cumbre above Tafi Viejo these birds were fairly common in dense growths of bushes and herbaceous vegetation, but worked about under the cover of large nettlelike plants where it was difficult to secure them.

XENICOPSIS RUFO-SUPERCILIATUS ACRITUS (Oberholser)

, *Anabazenops acritus* OBERHOLSER, Proc. Biol. Soc. Washington, vol. 14, Dec. 12, 1901, p. 187. (Sapucay, Paraguay.)

The type of *X. r. acritus* is an immature bird in the somewhat brighter more yellowish olivaceous plumage that distinguishes immature from adult individuals in this species, a fact that seems to have led to its separation originally. With *oleaginus* restricted to a more western range in Catamarca and Tucuman (probably north into Bolivia), the name *acritus* becomes available for the form of northeastern Argentina and Paraguay, distinguished from *oleaginus* by duller more grayish coloration. Typical *X. r. rufosuperciliatus* is brighter, more rufescent on the dorsal surface, especially on the wings and has the markings on the under surface somewhat less sharply defined.

An adult male taken at Las Palmas, Chaco, on July 21, 1920, was the only one seen until I reached Lazcano, Uruguay, where two were seen and an immature female was taken near the Rio Cebollati on February 8. At Rio Negro, Uruguay, where the birds were fairly common from February 15 to 18, an immature male was preserved February 17, and a female of the same age on February 18. The adult male from Las Palmas, in full winter plumage, though a trifle brighter than two specimens from the Page expedition, is duller colored than the type of the subspecies. Two from Rio Negro, Uruguay, resemble the type save that they have not quite completed the molt into full plumage. The one from Lazcano in the same stage of molt is darker.

These birds were found in dense thickets, usually in lowlands, where it was more or less wet and swampy. They worked deliberately in the dense plant growth, often somewhat awkwardly, clambering about like large titmice, always well under cover. On one occasion one uttered a high-pitched note that may be written *chee-a;* otherwise they were silent. An adult male, taken July 21, had the maxilla, sides, and tip of the mandible dull blackish brown; rest of mandible Rood's lavender, with the extreme base tinged with yellow; iris fuscous; tarsus and toes dark ivy green; lower surface of toes yellowish.

CULICIVORA STENURA (Temminck)

Muscicapa stenura TEMMINCK, Nouv. Rec. Planch. Col. Ois., vol. 3, November, 1822, pl. 167, fig. 3. (Brazil.)

The single individual observed, an adult female shot at the Riacho Pilaga, Formosa, August 14, 1920, was found in a dense stand of a tall grass (*Andropogon condensatus*) near the bank of a sluggish stream. It flew for short distances with a tilting flight and when at rest clung to the upright grass stems. The specimen is in full plumage, but has only eight rectrices, so that two must have been lost, as the species is supposed to possess 10. In Temminck's original plate it is figured with 12, probably through error.

The specimen secured in Formosa had the bill dull black; iris Rood's brown; tarsus and toes black.

In Argentina the species has been recorded previously from Mocovi and Ocampo, Santa Fe, and from Itapua, Misiones.

Mr. Ridgway[51] calls attention to the fact that this species, with nonexaspidean tarsi and only 10 rectrices, must be removed from the Tyrannidae, and suggests tentatively that it be put in the Furnariidae.

Family FORMICARIIDAE

TARABA MAJOR MAJOR (Vieillot)

Thamnophilus major VIEILLOT, Nouv. Dict. Hist. Nat., vol. 3, 1818, p. 313. (Paraguay.)

The present species was recorded at the following points: Resistencia, Chaco, July 9, 1920 (adult male taken); Las Palmas, Chaco, July 19 to 31 (male secured July 19, and a second one on July 31); Riacho Pilaga, Formosa, August 8 to 18 (two pairs, shot August 8 and 18, respectively); Formosa, Formosa, August 23 and 24; Kilometer 80, west of Puerto Pinasco, Paraguay, September 8 and 11; Kilometer 200 (in the same region), September 25; Tapia, Tucuman, April 10 (a female shot) to 13 (male taken), 1921.

The bill in the pair from Tapia is slightly heavier than in others, but otherwise the series is very similar, and all are referred to true

[51] U. S. Nat. Mus., Bull. 50, vol. 4, 1907, p. 340.

major. The two specimens from Tucuman have molted and renewed a part of the flight feathers.

The present species is an inhabitant of thickets or dense growths of weeds, and, though as retiring in habit as a chat (*Icteria*), has more curiosity and may be decoyed more readily into view. It impresses one as a bird of character that will repay observation with curious and interesting traits. When it chooses to appear it is alert, certain in every movement, and jaunty in carriage, but often one merely has sight of a brilliant red eye glaring with a positively evil expression through some crevice between leaves, with only a suggestion of a darker body behind. The bird is more frequently heard than seen, even where it is common. Its calls, usually uttered in a complaining tone, are considerably varied, a common one being a low *pruh-h-h-h*, another *tur-r-r te tuh*, while others do not lend themselves readily to representation. The song, strange and deceptive in volume and tone, may be written as *heh heh heh heh heh heh heh-h-h-h-h-h quo-ah.* It begins slowly, becomes increasingly rapid until it changes to a rattle, and then, after a slight pause, terminates in a squall exactly like that uttered by a gray squirrel (*Sciurus carolinensis*). During August the birds were usually in pairs, and males sang constantly in early morning in sunny weather.

The Toba Indians called this species *soo loo likh,* while the Angueté knew the female as *al lakh tik tik,* and the male as *yum a oukh.*

An adult male, taken July 9, when fresh, had the maxilla and line of the gonys blackish slate; rest of mandible, basal tomia of maxilla and tarsus gray number 6; iris spectrum red. The colors in females were similar, though both sexes varied slightly in the depth of red of the eye.

THAMNOPHILUS RUFICAPILLUS RUFICAPILLUS Vieillot

Thamnophilus ruficapillus VIEILLOT, Nouv. Dict. Hist. Nat., vol. 3, 1816, p. 318. (Corrientes, Argentina.)[52]

The present species seems to be rare or local, as it was seen on only two occasions. At the Paso Alamo, on the Arroyo Sarandi, an adult male in rather worn plumage was taken on February 2, 1921. Another specimen, preserved in alcohol, was secured near Rio Negro, Uruguay, on February 16. The birds were found in dense brush near water, where they worked slowly about under heavy cover, in habit suggesting *Thamnophilus gilvigaster.* Their call note was a low whistle.

Hellmayr[53] gives *Thamnophilus subfasciatus* Sclater and Salvin and *T. marcapatae* Hellmayr as subspecies of *T. ruficapillus.* Mr.

---

[52] Type locality selected by Hellmayr, Field Mus. Nat. Hist., Zool. Ser., vol. 13, pt. 3, Nov. 20, 1924, p. 108.
[53] Arch. Naturg., vol. 85, 1919 (November, 1920), p. 85.

Ridgway has recognized the genus Rhopochares of Cabanis and Heine for the present group, but I have preferred to use the broader generic term for it.

THAMNOPHILUS GILVIGASTER GILVIGASTER Pelzeln

Thamnophilus gilvigaster "TEMMINCK" PELZELN, Ornith. Brasiliens, 1868, p. 76. (Curytiba, Parana, Brazil.)

The male of the common ant bird of Uruguay differs from that of northern Argentina in having the under tail coverts and flanks darker buff, and the darker gray of the breast extended back to the upper abdomen. Females likewise share the darker color of the underparts so that both sexes may be distinguished at a glance from dinellii. Doctor Hellmayr[54] has considered the buff-bellied ant birds of southern distribution as subspecies of Thamnophilus caerulescens of Paraguay. With all due respect for the weight of Doctor Hellmayr's opinion in such matters, I am not now prepared to accept this in view of the greater difference in color between the sexes in caerulescens. Although gilvigaster belongs in the same group as caerulescens, and apparently occupies a contiguous range, intergradation between the two does not seem to have been proven. Thamnophilus ochrus Oberholser is the female of T. caerulescens, as is shown by examination of the type. In the specimens that I have seen, gilvigaster is readily told from Thamnophilus caerulescens Vieillot by the buff on the posterior portion of the body and by its slightly larger bill.

Males were collected along the Rio Cebollati, near Lazcano, Uruguay, on February 5 and 8, 1921, and an immature female on February 5. An immature male was shot at Rio Negro, Uruguay, on February 17, and an adult female on February 19. The latter specimens are fully as dark as those from eastern Uruguay and show no intergradation toward the paler Argentine form. Aplin[55] has recorded a bird of this genus from the thickets along the Arroyo Grande (a tributary of the Rio Negro) near Santa Elena, but I failed to note it at San Vicente near the coast of southeastern Uruguay. In its distribution in Uruguay this bird from its thicket-haunting habit would of necessity follow the courses of streams.

These birds were common near Lazcano, from February 5 to 8, and in the vicinity of Rio Negro, from February 14 to 19. On both the Rio Cebollati and Rio Negro they inhabited dense, heavy growth near the streams where they moved about with jerking tails, or occasionally perched quietly, twitching the tail at intervals. They showed considerable curiosity at strange sounds, and came about to peer at me, sometimes within 3 or 4 feet of my face,

---

[54] Nov. Zool., vol. 28, May, 1921, p. 199, and Field Mus. Nat. Hist., Zool. ser., vol. 13, pt. 3, Nov. 20, 1914, p. 102.

[55] Ibis, 1894, p. 185.

jerking their tails continually. They have several soft mewing or nasal notes, but do not become as excited at intrusion as do many small birds. A male taken February 7 (preserved as a skeleton) was in breeding condition and from the appearance of the abdomen had been engaged in incubation.

THAMNOPHILUS GILVIGASTER DINELLII Berlepsch

Thamnophilus dinellii BERLEPSCH, Bull. Brit. Orn. Club, vol. 16, May 28, 1906, p. 99. (Estancia Isca Yacu, Santiago del Estero, Argentina.)[56]

The bird here treated has been known variously as *maculatus*, *caerulescens*, and *gilvigaster* until *dinellii* was described by Berlepsch. The present combination seems to have been made first by Hartert.[57]

Of this form I collected the following skins: Resistencia, Chaco, July 9, 1920, male and female; Las Palmas, Chaco, July 13, 14, 20, 21, and 30, two males and three females; Riacho Pilaga, Formosa, August 8 and 11, two females; Tapia, Tucuman, April 12, 1921, immature female. These differ from *gilvigaster* from Uruguay in uniformly paler coloration on the ventral surface, with the grayish wash of the breast lighter and less extensive. Birds from the Chaco agree in general with the single skin from Tapia, Tucuman, though this bird from Tucuman, and two from the Riacho Pilaga, Formosa, are slightly grayer on the dorsal surface than the series from the Territory of Chaco.

The present ant bird, in the Chaco, was common in dense undergrowth under heavy trees, but at times was found in more open groves scattered over the savannas. In general it had the motions of a titmouse, save that it did not cling to limbs, but with this mannerism combined the jerking of the tail and twitching of the wings of a flycatcher. It fed at times on the ground, where it hopped slowly about, pausing to peer around, but more often worked through limbs rising 3 or 4 meters from the ground. The birds had much curiosity and were easily called out from the heavier coverts. Their notes were somewhat varied, the usual one resembling *pruh pruh pruh-h*, given in a mewing tone. In spring, males mounted into the top of some bush or low tree where, concealed in the leaves, they sang a pleasing whistled repetition of notes suggestive of the song of the white-breasted nuthatch (*Sitta carolinensis*).

The Toba Indians called them *kwo o likh*.

A male, shot July 9, had the maxilla and extreme tip of the mandible black; base of mandible gray number 5, with an indistinct, grayish line along the cutting edge; tarsus and toes slate gray, underside of toes yellowish; iris dark brown. A female was similar but had the colors somewhat duller.

---

[56] Hellmayr, Field Mus. Nat. Hist., Zool. ser., vol. 13, pt. 3, Nov. 20, 1924, p. 103, states that Berlepsch's type specimen comes from Santiago del Estero, though in the original description it is listed from Tucuman.

[57] Nov. Zool., vol. 16, 1909, p. 221.

This ant bird was recorded at the following localities: Resistencia, Chaco, July 9, 1920; Las Palmas, Chaco, July 13 to 30; Riacho Pilaga, Formosa, August 7 to 18; Formosa, August 24; Kilometer 25, Puerto Pinasco, Paraguay, September 1; Kilometer 80 (Puerto Pinasco), September 9, 11, and 15; Kilometer 200 (in the same region), September 25; Tapia, Tucuman, April 12, 1921. The species was uncommon both at Tapia and west of Puerto Pinasco. No specimens were taken in the Paraguayan Chaco, so that it is possible that birds from that region belong to another form.

MYRMORCHILUS STRIGILATUS SUSPICAX Wetmore

Myrmorchilus strigilatus suspicax WETMORE, Journ. Washington Acad. Sci., vol. 12, Aug. 19, 1922, p. 327. (Riacho Pilaga, near Kilometer 182, Ferrocarril del Estado, Gobernacion de Formosa, Argentina.)

The present form, described from specimens from the Riacho Pilaga, Formosa, and the Rio Bermejo, in Argentina, differs from typical *strigilatus* from Bahia, Brazil (the type locality) in more buffy superciliary stripe, browner post-ocular mark, and buffy wash on the sides, flanks, and under tail coverts. Mr. Ridgway[58] in including *Myrmorchilus* in a section containing birds with 10 rectrices was misled by an imperfect specimen as, in a small series of skins, I find 12 tail feathers present.

At the Riacho Pilaga, Formosa, the present form was fairly common as it was recorded on August 11, 13, and 18; eight specimens including males and females were secured on the first and last of these dates. Individuals recorded but not taken near Kilometer 80, west of Puerto Pinasco, Paraguay, on September 8 and 9, belong perhaps to this southern subspecies.

These birds range in pairs in dense undergrowth where it is difficult to see them, and it was several days after my arrival at the Riacho Pilaga before I succeeded in obtaining specimens, though they had been heard frequently. The call note of the male is a loud, shrill whistle that is repeated several times, and may be represented as *chee-ah chee-ah chee-ah chee-ah.* This is answered by the female in a lower tone. On the ground they walk about easily, usually under such heavy cover that I had mere glimpses of their dark forms. When excited they bobbed up near at hand with a loud *thrut thrut* of the wings, at times on open limbs, where they rested, at intervals swinging the tail over the back like the handle of a pump, frequently through an angle of 90°.

The Angueté Indians called them *keh yow.*

An adult male (taken August 11), when first killed, had the bill black, becoming neutral gray at the base of the mandible; tarsus mouse gray; iris dull brown.

---

[58] Birds of North and Middle America, vol. 5, 1911, p. 13.

STIGMATURA BUDYTOIDES INZONATA Wetmore and Peters

Stigmatura budytoides inzonata WETMORE and PETERS, Proc. Biol. Soc. Washington, vol. 36, May 1, 1923, p. 143.   (Tapia, Tucuman.)

The present race differs from *S. b. flavo-cinerea* in the presence of a white spot on the inner web of the outer tail feather, and by the bright yellow of superciliary and underparts. *S. b. budytoides* has the white marking in the tail more extensive. It was fairly common at Tapia, Tucuman, from April 9 to 13, 1921, where seven skins preserved were collected on the following dates: Two males, April 9 and 10, a female and one of unknown sex on April 11, a male on April 12, a male and one with sex not determined on. April 13. The birds frequented dry, rather open forest of low trees, with frequent clumps of thorny bushes, where they ranged well under cover in pairs or little bands of three or four individuals. As they hopped about in search for food they jerked and twitched the tail, frequently throwing it above the back, while the wings were drooped, a mannerism that with their slender forms gave them the appearance of gnatcatchers. In fact, as *Polioptila dumicola* was found in the same situations it was at times difficult to distinguish readily between the two, when the birds were partly concealed behind screens of branches. At intervals *Stigmatura* emitted a series of sharp, explosive call notes in which two or more joined, a medley that suggested the explosive calls of *Tyrannus verticalis*. Save for these the birds would often have passed unnoted in the scrub. Their flight was weak and tilting, and was seldom pursued for any great distance. Specimens taken were in various stages of molt.

Mr. Ridgway[59] calls attention to the resemblance of *Stigmatura* (usually considered a tyrannid) to certain Formicariidae and suggests that it may belong in that family.

STIGMATURA BUDYTOIDES FLAVO-CINEREA Burmeister

Phylloscartes flavo-cinereus BURMEISTER, Reise La Plata-Staaten, vol. 2, 1861, p. 455.   (Valleys of Sierra Uspallata, Mendoza, Argentina.)

S. b. flavo-cinerea is distinguished from the northern forms of the species by the duller yellow of the undersurface and the white superciliary. The two specimens that I have seen have no indication of a white spot on the inner web of the outer rectrix. A specimen secured by the Page expedition on the Rio Bermejo in March, 1860, is intermediate between *flavo-cinerea* and *inzonata* of Tucuman, as it has the dull breast of *flavo-cinerea* and the tail of *inzonata*, while the superciliary is very dull yellow.

---

[59] U. S. Nat. Mus., Bull. 50, vol. 4, 1907, p. 339.

An adult male, secured near Victorica, Pampa, December 27, 1920, hopped about in the bushes at the border of a thicket with the tail held like that of a gnatcatcher at a jaunty angle above the back. The bill in this specimen was black; iris hessian brown; tarsus and toes dark neutral gray.

EUSCARTHMUS MELORYPHUS MELORYPHUS Wied

Euscarthmus meloryphus WIED, Beitr. Nat. Brasilien, vol. 3, 1831, p. 947. ("Campo geral" and the border line between the Provinces of Minas Geraes and Bahia, Brazil.)

An adult male secured at Las Palmas, Chaco, on July 23, 1920, does not seem to differ markedly from a specimen of this bird in the Field Museum from Macaco Secco, near Andarahy, Brazil. In the skin from Las Palmas dull olive-green tips on the central crown feathers almost entirely obscure the ochraceous-tawny color of the central crown stripe.

This small bird was encountered in swampy woods in a heavy growth of caraguatá (*Aechmea distichantha*), a spiny-leaved plant that covered the forest floor, where it worked about a few inches from the ground, hopping slowly over the broad plant leaves or fluttering feebly from perch to perch in its search for food.

Oberholser [60] has indicated that, through a type fixation by Gray in 1840 (List Gen. Birds, p. 32), the genus *Euscarthmus* Wied, 1831, is applicable to the present species, replacing *Hapalocercus* Cubanis, 1847. Mr. Ridgway [61] considers this genus as possibly a member of the Formicariidae. It is certainly not a true flycatcher, and is included tentatively at this point.

Family RHINOCRYPTIDAE [62]

SCYTALOPUS FUSCUS Gould

Scytalopus fuscus, GOULD, Proc. Zool. Soc. London, February, 1837, p. 89. (Chile.)

An immature male in full plumage, secured April 27, 1921, near Concon, Chile, was the only bird of this group encountered. While crossing a deep gulch with a small stream at the bottom, heavily shaded by a dense growth of trees, the individual in question, in its dull plumage barely visible in the somber shadows, came silently into

---

[60] Auk, 1923, p. 327.

[61] U. S. Nat. Mus., Bull. 50, pt. 4, p. 339.

[62] Apparently the first family designation for the tapaculos is that of Lafresnaye, who, in an *Essai de l'Ordre des Passereaux* (the first part of which seems to have been published at Falaise in 1838, though doubt attaches to the date of succeeding sections), has as his third family (p. 13) the Rhinomidae. This, Lafresnaye continues, has for its type the genus *Rhinomye* Geoffroy, established in 1832, an evident emendation of *Rhinomya*. With *Rhinocrypta* replacing *Rhinomya* as a generic term the family name for the group becomes Rhinocryptidae instead of Pteroptochidae or Hylactidae, two terms that have been in common use.

a dense mat of tangled branches 3 meters above the stream, attracted by a squeak, and bobbed about in a wrenlike manner. The bill, in life, was black; base of mandible light neutral gray; tarsus fuscous black: toes smoke gray; iris carob brown.

The bird measures as follows: Wing, 52; tail, 44; exposed culmen, 12.5; tarsus, 18.5 mm.

RHINOCRYPTA LANCEOLATA (Is. Geoffroy and d'Orbigny)

Rhinomya lanceolata, ISIDORE GEOFFROY and d'ORBIGNY, Mag. Zool., 1832, cl. 2, pl. 3. (Carmen and Salina d'Andres Paz, Rio Negro.[63])

In the valley of the Rio Negro, below General Roca, Rio Negro, the present species was common from November 23 to December 3, 1920; a female was taken November 23 and males on November 23, 27, and December 3. The birds were encountered usually in rather heavy growths of open brush that clothed the arid flood plain of the stream, and few seemed to range inland through the still drier, gravelly hills that formed the northern border of the valley. Though common, *R. lanceolata* was shy, and was seen or secured only at the expense of considerable effort. The birds normally ran about on the ground with crest erect and tail cocked at an angle above the back. As I traversed their haunts I was greeted by a low *prut prut prut*, or a musical *tullock*, from the brush on either hand, and occasionally had a glimpse of one of the elusive birds as it darted across some little opening. Occasionally, when safe behind a protective screen of low weeds or a drooping branch, one stopped to peer back at me, or less frequently with a running jump one sprang into the branches of a bush and clambered up for a better outlook. It is doubtful if *Rhinocrypta* has occasion to fly a hundred meters in the course of a month, a circumstance that has given rise to the appropriate local name of *corre corre que no vuela*. The slight breast muscles were pale in color, indicating a poor blood supply, and loose and flabby in substance in contract to the strong, heavy leg muscles. The heavy operculum that overhung either nostril was movable, and was so developed that it may serve as a protective device that aids breathing during the constant heavy windstorms of these regions, during which the loose earth forms a dense dust cloud in the air. On December 3 I found a nest placed more than a meter from the ground amid heavy branches in a dense, thorny bush. The somewhat bulky structure was an untidy affair made of weed stems, bits of bark, and grasses, lined with finer material. The top was covered in an arch, and the entrance was a large irregular opening in one side. The nest contained two white eggs and two of the spotted eggs of the glossy cowbird (*Molothrus*

---

[63] d'Orbigny, Voy. Amér. Mérid., Ois., vol. 4, pt. 3, 1835–44, p. 195.

bonariensis). The two white eggs measure, respectively, in millimeters, 24.5 by 18.3 and 27.3 by 20.5. The smaller egg has a glossier, more irregular surface than the large one, which is smoother and dull in color. The difference in size and appearance between the two is so striking that I believe that the larger egg alone is that of *Rhinocrypta*, and that the smaller one was deposited in the nest by some *Synallaxis*. The parent (*R. lanceolata*) slipped down from the nest through the branches to the ground; when secured, to my surprise, it was a male.

Near Mendoza, Mendoza, this species was fairly common on March 13, 1921, but was very shy. At Potrerillos, Mendoza, it was recorded on March 17, and several were seen and a female taken at El Salto, at an elevation of 1,800 meters. Others were recorded on March 27 below Tunuyan, Mendoza, through a low range of sandy hills east of the Rio Tunuyan. In fall the birds were much shyer than during the breeding season, and, though their notes were heard, it was only by chance that one was seen as it ran across some little opening in the brush. Usually they took care to edge aside without exposing themselves. The fall specimen secured is brighter, buffier brown above and on the flanks and under tail coverts than summer birds, but is otherwise similar. The difference in color is attributed to the fact that this bird is in fresh fall plumage, while the others are somewhat worn.

An adult male, taken November 27, had the maxilla and tip of mandible dull black; base of mandible, neutral gray; iris, Vandyke brown; tarsus and toes, black.

TELEDROMAS FUSCUS (Sclater and Salvin)

Rhinocrypta fusca SCLATER and SALVIN, Nom. Av. Neotr., 1873, p. 161. (Mendoza.)

Near General Roca, Rio Negro, the barrancolino, as this tapaculo is known, was fairly common from November 25 to December 2, 1920. Adult males, preserved as skins, were taken on November 25 and 26, and a female on November 26. The species was found in low, open brush over the dry gravel hills that bordered the level valley of the Rio Negro, in a region wholly without permanent water. As I passed along the shallow, steep-walled barrancas that traversed this area, I had an occasional glimpse of a wrenlike bird of moderate size, as, with tail cocked over its back, it sprinted away over open ground, swerving constantly behind low clumps of vegetation for protection. Like snowy plover the birds ran on and on and on long after one might expect them to stop, and hard running was required on the part of the collector in order to keep up with them. At other times as I cared for specimens that I had shot one would run out from the shelter of bushes with tail erect and wings drooped to watch me curiously.

The song of this species is loud, and, though simple in its nature, was rather pleasing in contrast to the harsh surroundings among which it was heard. It may be represented by the syllables *took took took took took*, repeated rapidly. In running *Teledromas* takes very long steps. In a series of tracks that I measured I found the steps to average 150 mm. apart, a remarkable distance when it is considered that the length of the bird from bill to tail is less than 175 mm.

On the evening of November 25, while setting a line of mouse-traps along a small barranca, I flushed a female from the entrance of a nest tunnel, an opening near the top of perpendicular cut bank in loose soil, 3 feet above the bottom of the dry channel. The round tunnel, from 60 to 75 mm. in diameter, led back for a distance of 375 mm. to end in a chamber 150 mm. in diameter, with the bottom lined with bits of grass that formed a roughly, cup-shaped nest. The two eggs, in which incubation had begun, were white in color, without markings. One was broken in transportation. The other measures, in millimeters, as follows: 26.2 by 20.5.

The genus *Teledromas* Wetmore and Peters[64] for the present species is distinguished from *Rhinocrypta* Gray by its smooth, uncrested head, relatively stronger, heavier bill; short under tail coverts; and relatively longer hind toe and claw. In addition, it will be noted that while *T. fuscus* made its nest in a tunnel in the side of a cut bank, *R. lanceolata*, as I have noted elsewhere, deposited its eggs in a stick nest in a low shrub.

MELANOPAREIA MAXIMILIANI ARGENTINA (Hellmayr)

Synallaxis maximiliani argentina HELLMAYR, Bull. Brit. Orn. Club, vol. 19, Apr. 29, 1907, p. 74. (Ñorco, Tucuman, Argentina.)

Four specimens of this bird secured (three skins and one in alcohol) from the localities listed below, in the absence of comparative material, are referred on the basis of range to the subspecies *argentina*. W. D. Miller[65] has with reason removed the present genus from the Formicariidae to the Rhinocryptidae, in part on information from the bodies of the present specimens (preserved in alcohol) as when, at his request, I examined these I found that they possessed four notches on the posterior border of the metasternum. It may be added that the habits of these birds are not radically different from those of other tapaculos. *Melanopareia*[66] was first taken

---

[64] Proc. Biol. Soc. Washington, vol. 35, Mar. 20, 1922, p. 41.
[65] In notes not yet published.
[66] Reichenbach, Handb. Spec. Ornith., August, 1853, p. 164.

at Las Palmas, Chaco, on July 10, 1920, when an adult male was secured.

At the Riacho Pilaga, Formosa, two were taken (a female and one other put in alcohol) on August 14, and others were seen. On the Sierra San Xavier near Tafi Viejo, Tucuman, they were fairly common on the open slopes just above the forest on April 17, 1921, and a female was secured. In the region of the Chaco the species was found in heavy saw grass near the borders of low thickets. The birds flew or climbed into the bushes where they worked away to safety or remained at rest, often giving a curious tilting jerk to the tail. They seemed to feed mainly on the ground. Above Tafi Viejo they clambered slowly about among dense growths of weeds. The call note was a rapid *chit tuck.*

Family COTINGIDAE

PACHYRAMPHUS VIRIDIS VIRIDIS (Vieillot)

Tityra viridis VIEILLOT, Nouv. Dict. Hist. Nat., vol. 3, 1816, p. 348. (Paraguay.)

A female, July 13 and a male July 14, 1920, shot near Las Palmas, Chaco, were the only ones seen. The birds were encountered in heavy forest near a stream, where they worked through the tops of the trees in search of insects, in movement suggesting vireos, as they frequently flew to a perch and remained for several seconds while peering about. A large insect was beaten heavily on a limb. The female uttered a low note that may be rendered as *preer.* The male, when first killed, had the bill clear green-blue gray, except the tip of the maxilla which was dusky; tarsus and toes deep glaucous gray; iris dull brown.

The two taken are similar to a specimen from Sapucay, Paraguay, and agree with it in being larger than *P. v. cuvierii* from Bahia. The wing measurement of the skins from Las Palmas is as follows: Male, 80.4 mm.; female, 76 mm.

PACHYRAMPHUS POLYCHOPTERUS NOTIUS Brewster and Bangs [67]

Pachyrhamphus notius BREWSTER and BANGS, Proc. New England Zool. Club, vol. 2, Feb. 15, 1901, p. 53. (Concepcion del Uruguay.)

The only one observed was an adult male that was shot January 31, 1921, near San Vicente, Uruguay, in a small tract of low forest near the Laguna Castillos, where the bird perched like a flycatcher on a dead limb in a small, well-shaded opening among the trees. This specimen, with a wing measurement of 86.5 mm., presents in a

[67] For use of the name *notius,* see Bangs and Penard, Proc. Biol. Soc. Washington, vol. 35, Oct. 17, 1922, p. 225.

marked degree the characters of large size and dark coloration that distinguish the southern form.

XENOPSARIS ALBINUCHA (Burmeister)

Pachyrhamphus albinucha BURMEISTER, Proc. Zool. Soc. London, 1868, p. 635. (Rio de la Plata, near Buenos Aires.)

Near Laguna Wall, 200 kilometers west of Puerto Pinasco, an adult male was taken September 25, 1920, as it watched alertly for insects from a low perch at the border of a thicket of vinal. The bill in this specimen is decidedly larger than in a topotype of the species examined in the collections of the United States National Museum, and may perhaps represent a distinct form. The culmen from the base measures 12.3 mm., while in the bird from Argentina it is 10.4 mm. Difference between the two in heaviness of bill is noticeable. A second specimen from Monteagudo, Tucuman, a male, in the collection of the Museum of Comparative Zoology, agrees in size with the bird from Buenos Aires (culmen from base 11 mm.).

CASIORNIS RUFA (Vieillot)

Thamnophilus rufus VIEILLOT, Nouv. Dict. Hist. Nat., vol. 3, 1816, p. 316. (Paraguay.)

Near Las Palmas, Chaco, a female of this cotinga was shot July 13, 1920, and another, placed in alcohol, was taken July 21. Males were secured at Kilometer 80, west of Puerto Pinasco, Paraguay, September 11 and 16, and others were seen at Kilometer 25, September 1, and on the eastern bank of the Rio Paraguay, opposite Puerto Pinasco, on September 30. The female from Las Palmas is deeper rufous above than others examined. A specimen shot July 21 had the tip of the bill dull black; base of mandible tilleul buff; base of maxilla avellaneous; iris natal brown; tarsus and toes deep purplish gray. Hellmayr [68] considers *C. fusca* Sclater and Salvin, of which I have seen only one skin, a race of *rufa* Vieillot.

A retiring species, the present form was found singly or in pairs only in dense undergrowth in heavy woods, where it hopped about from perch to perch or remained motionless among green leaves in the pose of a flycatcher. The call note, heard rarely, resembled *tsa ah* given in a high-pitched tone.

HABRURA PECTORALIS PECTORALIS (Vieillot)

Sylvia pectoralis VIEILLOT, Nouv. Dict. Hist. Nat., vol. 11, 1817, p. 210. (Paraguay.)

An adult male secured at Las Palmas, Chaco, on July 17, 1920, agrees in size and color with another taken (with an adult female)

---

[68] Nov. Zool., vol. 15, June, 1908, p. 56.

near the Rio Paraguay at Puerto Pinasco, Paraguay, on September 3, so that both are assumed to represent the typical form, which would seem then to range south through the Chaco of northern Argentina. A female in the Page collection, taken at Parana, has the underparts browner and the bill larger than the female from Puerto Pinasco, and is supposed to be *H. p. minimus* (Gould).[69] *Habrura bogotensis* Chapman,[70] of which there is an adult male, taken by Hermano Apolinar Maria at the type-locality, in the United States National Museum, differs from *pectoralis* in much more rufescent coloration and in wholly black bill, so that it seems to represent a distinct species.

On the two occasions that I encountered this tiny bird, it was found among weeds and low bushes in pastures not far distant from water. The few individuals seen, rather wild and difficult to approach, fluttered about a few inches from the ground until flushed, when they flew with a rapid, quickly undulating flight to low perches on the sides or tops of small weeds. When at rest they occasionally twitched the tail quickly. In general appearance and mannerisms they suggested small flycatchers.

Mr. Ridgway [71] considers this bird out of place in the Tyrannidae and suggests that it may be better located among the Cotingas.

Family TYRANNIDAE

AGRIORNIS LIVIDA LIVIDA (Kittlitz)

Tamnophilus lividus, KITTLITZ, Mém. Acad. Imp. Sci. Saint Pétersbourg, vol. 2, 1835, p. 465, pl. 1. (Valparaiso and Concepcion Bay, Chile.)

Three specimens of this large species, all in full winter plumage, were secured at Concon, Chile, a male on April 24, 1921, and females on April 25 and 28. Wing measurements of these three are as follows: Male, 131.5 mm.; females, 124 and 124.5 mm. *Agriornis l. fortis* Berlepsch [72] is distinguished by larger size and paler coloration. Two skins of *fortis* secured from E. Budín, taken April 17 and 27, 1918, at Puesto Burro, Maiten, Chubut, have wing measurements of 144.6 (male) and 134.5 mm. (female). Two immature birds (U. S. National Museum coll.), secured February 24, 1897, at the head of the Rio Chico, Santa Cruz, and March 7, 1897, on the Pacific slope of the Cordillera, beyond the locality first mentioned, have the head, back, and upper breast indistinctly streaked with dusky, but

---

[69] *Pachyramphus minimus* Gould, Zool. Beagle, pt. 3, Birds, July, 1839, p. 51. (Montevideo, Uruguay.)

[70] *Habrura pectoralis bogotensis* Chapman, Bull. Amer. Mus. Nat. Hist., vol. 34, Dec. 30, 1915, p. 646. (Suba, Bogota Savanna, Colombia.)

[71] U. S. Nat. Mus., Bull. 50, pt. 4, 1907, p. 339.

[72] *Agriornis livida fortis* Berlepsch, Proc. Fourth Int. Ornith. Cong., February, 1907, p. 352. (Valle del Lago, Chubut, Argentina.)

otherwise resemble adults. These birds are in molt from juvenal to first winter plumage.

Near Concon, Chile, the large *Agriornis* was common from April 24 to 28, 1921. The birds were found on open flats near the Rio Aconcagua, or in pastures dotted with bushes on the hill slopes above, where they rested quietly on the top of some bush that offered outlook. Occasionally one dropped down to the ground where, like a robin, it ran rapidly along for a few steps and then paused abruptly with head thrown up and body erect. The long, heavy bill marked them from other birds of similar size, even at a distance. All were silent.

In the adult male of this species the tenth primary is abruptly narrowed at the tip for 10 to 16 mm., while the tip of the ninth is narrowed for about half the amount of the tenth. In females the primaries are normal. In the male taken at Concon the primaries resemble those of the female. Apparently this sexual distinction does not develop until the primaries have been molted once, so that the wing of males in their first winter is like that of females.

The male shot April 24 had the maxilla dull black; mandible light drab, shaded with deep quaker drab toward the tip; iris army brown; tarsus and toes, dull black.

AGRIORNIS STRIATA STRIATA Gould

Agriornis striatus GOULD, Zool. Voy. Beagle, pt. 3, Birds, 1839, p. 56. (Santa Cruz, Argentina.)

Specimens of the present species secured number three, an adult male from General Roca, Rio Negro, taken November 29, 1920, an adult female from Zapala, Neuquen, shot December 7, and a male from Tunuyan, Mendoza, collected March 27, 1921. The two summer birds are in somewhat worn breeding dress, while the fall skin is in full winter plumage. The female has the two outermost primaries very slightly sinuated on the outer margin. In adult males the ninth and tenth primaries are narrowed distally, and are incised deeply for 12 to 16 mm. at the tip, this incision being only slightly less on the ninth than on the outermost primary. The form *Agriornis s. andecola,*[73] of which I have seen no specimens, is said by Berlepsch[74] to differ from the typical bird in having fainter brownish black throat stripes and a stronger buffy wash on the lower surface. It ranges in the higher Andes of western Bolivia.

Near General Roca, Rio Negro, these flycatchers were encountered on November 29, 1920, and again on December 2, in a region of arid gravel hills covered with an open growth of low brush. They often

---

[73] *Pepoaza andecola* d'Orbigny, Voy. Amer. Merid., vol. 4, pt. 3, Oiseaux, 1835–1844, p. 351. (5,000 meters above the sea, in Bolivia.)

[74] Proc. Fourth Int. Ornith. Cong., February, 1907, p. 464.

rested in usual flycatcher fashion on the top of a bush, but frequently dropped down to run rapidly about on the ground. They were wary and when approached flew away barely above the ground, many times traveling for long distances before a pause, passing so low among the bushes that it was difficult to follow their course. Others were seen Dec.mber 7 and 9 in similar territory near Zapala, Neuquen, four being observed together on one occasion, when they pursued one another with high-pitched, petulant calls. On March 27, 1921, several were seen in brush-grown areas east of the Rio Tunuyan, near Tunuyan, Mendoza, where they sought the tops of bushes that offered a commanding outlook over the surrounding ground. At rest their erect position, large head, and long bill are marked characters, while when flying their clay brown coloration and long wings are displayed.

A male, taken November 29, had the maxilla dull black; mandible pale drab gray; iris natal brown; tarsus and toes black.

AGRIORNIS MONTANA (d'Orbigny and Lafresnaye)

Pepoaza montana d'Orbigny and Lafresnaye, Mag. Zool., 1837, Cl. 2, p. 64. (Chuquisaca, Bolivia.)

A male was shot at an altitude of 1,500 meters above Potrerillos, Mendoza, on March 17, 1921, and another (preserved in alcohol) was secured March 19, near El Salto, at 1,800 meters. The skin secured is of a bird in molt into first-winter plumage that has the primaries normal as in females. In the fully adult male the ninth and tenth primaries are slender and are narrowed for a distance of 14 or 15 mm. at the tip. Berlepsch[75] has found that *Agriornis maritima*[76] (d'Orbigny and Lafresnaye) is based on an adult male of *A. montana* of the same authors. As the name *montana* occurs on the page preceding the one where *maritima* is found, it has priority and must be used for the species.

Though it is probable that the bird from western and southern Argentina should be distinguished as the subspecies *leucura* Gould,[77] material at hand does not include specimens from Bolivia, so that adequate comparisons may not be made.

The two examples of *Agriornis montana* observed were found on the ground or on low bushes near streams. They did not differ in actions from *striata* or *livida*, but were readily distinguished by the white in the tail.

The male taken had the bill dull black; iris natal brown; tarsus and toes black.

[75] Proc. Fourth Int. Ornith. Congr., February, 1907, p. 464-465.
[76] *Pepoaza maritima* d'Orbigny and Lafresnaye, Mag. Zool., Cl. 2, 1837, p. 65. (Cobija, Bolivia.)
[77] *Agriornis leucurus* Gould, Zool. Voy. Beagle, pt. 3, Birds, 1839, pl. 13. (Patagonia.)

TAENIOPTERA CINEREA CINEREA (Vieillot)

Tyrannus cinereus VIEILLOT, Anal. Nouv. Ornith. Élém., 1816, p. 68. (Argentina.[78])

Examination of the type and three other specimens of *T. c. obscura* Cory[79] indicates that this form is slightly darker than true *cinerea*.

This flycatcher was recorded at the following points: Las Palmas, Chaco, July 13 to August 1; Riacho Pilaga, Formosa, August 8 to 21; Puerto Pinasco, Paraguay, September 3; Kilometer 80, west of Puerto Pinasco, September 8; San Vicente, Uruguay, January 26 to February 2; Lazcano, Uruguay, February 3 to 9 (noted on February 9 as far as Corrales). Three adult males taken at Las Palmas, Chaco, July 28, Riacho Pilaga, Formosa, August 8 and San Vicente, Uruguay, January 26 were preserved as skins. The series of this species at hand is far from adequate for comparison, but it may be noted that a female from Matto Grosso appears paler above and has a broader light tip on the tail than the specimens listed above. The two skins from Las Palmas and Formosa are in full winter plumage. The one from Uruguay is in very worn breeding feather, but has not yet begun the molt.

The iris is martius yellow, lined heavily toward the inner margin with peach red, so that at casual inspection it appears wholly red; bill and tarsus dull black.

The present species ranges through the warmer areas of the region visited, as none were seen in the pampas of Buenos Aires. It is an inhabitant of open country, but chooses localities where trees or bushes are not far distant. The birds were encountered frequently about houses, or in the outskirts of little towns, or were observed in numbers along roads that wound through the open country. They chose commanding perches on posts, tops of bushes, telegraph wires, or, failing these, on elevated clods or little eminences on the ground, where they rested with heads drawn down in usual flycatcher attitude in watch for prey. When food was sighted they darted down to the ground to seize it, alighted perhaps to run along for a few steps, and then returned to a higher perch. Their wings were long and pointed and their flight, accompanied by a flashing of the black and white wing markings, light and graceful. As they alight the wings are often raised above the back for a second until they gain a secure hold on the perch. They are alert, active, and graceful in all their movements. In coloration they are strongly suggestive of a mocking bird (*Mimus*), and even close at hand give this impression, a likeness that is at once belied when the bright red eye, pointed

---

[78] See Brabourne and Chubb, Birds South America, 1912, p. 259.
[79] Field Mus. Nat. Hist., Orn. Ser., vol. 1, Aug. 30, 1916, p. 341. (Jua, Ceara.)

wing, and short tail are observed. During sharp, frosty winter mornings in the Chaco, as the first rays of the sun stretched with pleasing warmth across the open prairies, these flycatchers often uttered a little whistled song that could undoubtedly be readily set to the scales used in human music by one versed in musical annotation.

In Uruguay they were especially common along the country roads, seeming in hot weather more sluggish than during the winter season. In the warmer part of the day they chose perches on fence wires, where their heads were in the shadow of a post. At rest they were so inconspicuous as to be often overlooked. Near San Vicente they were common in an extensive forest of palms where they appeared to be nesting, though no nests were found. In the summer time I heard them utter a faint *swee*.

TAENIOPTERA CORONATA (Vieillot)

Tyrannus Coronatus VIEILLOT, Tabl. Enc. Méth. Orn., vol. 2, 1823, p. 855. (Paraguay.)

Two were secured at Victorica, Pampa, on December 26, 1920, an adult male in worn breeding plumage, and another preserved in alcohol. The birds were encountered on this date in fair numbers scattered through rolling pampa, where low bushes or small trees were spread at intervals. They rested in the tops of bushes or occasionally among open limbs in a tree, at intervals jerking the tail. The flight was slow and direct and was performed with rapid beats of the partly opened wings. The birds were silent.

The adult male taken had the bill and tarsus black; iris natal brown.

TAENIOPTERA IRUPERO (Vieillot)

Tyrannus Irupero VIEILLOT, Tabl. Enc. Méth. Orn., vol. 2, 1823, p. 856. (Paraguay.)

Though Hudson has recorded this flycatcher as common throughout the Argentina of his time, the species now seems restricted in its range, as I did not find it in the open pampas. It was noted at the following points: Los Amores, Santa Fe, to Charadrai, Chaco, July 5, 1920 (seen from train at frequent intervals); Resistencia, Chaco, July 9; Las Palmas, Chaco, July 13 to August 1; Riacho Pilaga, Formosa, August 8, 14, and 21; Formosa, Formosa, August 23 and 24; Puerto Pinasco, Paraguay, September 1 and 3; Kilometer 80, west of Puerto Pinasco, September 6 to 21; Victorica, Pampa, December 23 to 29; San Vicente, Uruguay, January 25 to February 2, 1921; Lazcano, Uruguay, February 3 to 9; Modesto Acuña, Cordoba, March 31 (seen from train); Tapia, Tucuman, April 7 to 13. It will be noted that the species was not recorded during field work in the Province of Buenos Aires, nor was it seen on the numerous occasions that I crossed the Province in trains. In its present distribution the species seems common from central Pampa, southern Cordoba, central Santa Fe, Entre Rios, and southern Uruguay northward into Paraguay and southeastern Brazil.

Three skins were preserved, an adult male shot at Las Palmas, Chaco, July 14, an adult female from Kilometer 80, Puerto Pinasco, Paraguay, September 18, and an adult female at Victorica, Pampa, December 24, 1920. The last named is in worn soiled breeding plumage. Comparisons of small series do not show differences between birds from distant localities. Females as well as males have the tail tipped with black, so that Sclater's statement [80] that the female has no black band on the tail is incorrect. In the male the narrowed tip of the outer primary measures 8 mm. or more, in females it is less than 6 mm. in length. The bill, tarsus, and toes in this species are black, the iris vandyke brown.

This beautiful bird, known as *blanca flor*, *viudita*, or *irupero*, though a flycatcher, has the habits and mannerisms of a bluebird (*Sialia*), so much so that as it flits its wings from some fence post or bush one is almost surprised that it does not break into warbling song. The birds frequent open country, where posts, low trees, or bushes offer convenient stations from which to watch for food, which seems to consist largely of insects secured from the ground. The *viudita* rests quietly, eyeing the ground intently, until food is observed, when it flies gracefully down with rapid movement of its long pointed wings to rest and look about for a few seconds before returning to a higher perch. The pure white body plumage with black primaries and black-tipped tail make it a prominent and beautiful figure in the landscape, especially since it invariably seeks an open perch, in spite of which it is tame and unsuspicious.

The flight is quick, nervous, and undulating, but the birds seem sedentary and seldom fly for great distances. The bird seemed wholly silent. A nest discovered near Victorica, Pampa, on December 24 was placed in a hollow in the crotch of a large caldén tree (*Prosopis nigra*) that stood somewhat separated from its fellows. The chamber that concealed the nest was an irregular hollow 200 or 250 mm. in diameter, with an entrance through a slight crevice at one side. The only lining of this domicile consisted of a few feathers arranged carelessly on the loose rubbish in the bottom of the cavity. The one young bird that this nest contained, when compared to its beautiful parents, with their clean, contrasted colors, was an ugly duckling, indeed, since its dark skin was scantily covered with dull gray down. (Pl. 9.)

---

[80] Cat. Birds Brit. Mus., vol. 14, 1888, p. 14.

TAENIOPTERA PYROPE PYROPE (Kittlitz)

Muscicapa Pyrope KITTLITZ, Mém. Acad. Imp. Sci. St.-Pétersbourg, Div. Sav., vol. 1, 1831, p. 191. (Tome, Concepcion, Chile.)

A sufficient series from central Chile indicates that the bird of that region is decidedly paler than that from the wooded region of the southern Andes, or from the vicinity of the Straits of Magellan. From April 24 to 28, 1921, these birds were recorded at Concon, Chile, where five skins were preserved, two males on April 25 and 28, and three females on April 24, 26, and 27. In this species the females have the two external primaries narrowed at the tip, but lack the marked incision and attenuation found in the males. The bill, tarsus, and toes are shining black in both sexes. The iris in general is maize yellow, clouded with xanthine yellow or English red about the pupil. In some specimens the extent of the clouding of red varies in the right and left eyes so that one eye may be much brighter than the other.

Though, when at rest this species in attitude suggests a *Myiarchus* or a *Sayornis*, its flight, characterized by alert dash with sudden turns and whirls, is like that of others of the genus. The birds were found among openings in brush over rolling hillsides or along hedge rows and small streams, where they rested on commanding perches. In flight they are graceful and active as they swing out after some passing insect and then with an abrupt loop drop to another perch. Their call was a low *tick tick* given infrequently and barely audible at 80 yards.

On the morning of April 27 there was a pronounced migration among them and the number present was greatly increased.

TAENIOPTERA MURINA (d'Orbigny and Lafresnaye)

Pepoaza Murina d'ORBIGNY and LAFRESNAYE, Mag. Zool., 1837, Cl. 2, p. 63. (Rio Negro, Argentina.)

An adult male was shot November 23, 1920, near General Roca, Rio Negro, from three that were encountered among bushes. The birds ran rapidly on the ground or flew up to perch on the bushes. Others were noted November 30. The birds pursued one another with sharp squeaky notes through the bushes. On December 13 several were noted among greasewood bushes near Ingeniero White, Buenos Aires. The tail in flight appears dead black.

TAENIOPTERA RUBETRA Burmeister

Taenioptera Rubetra BURMEISTER, Journ. für Ornith., 1860, p. 247. (Sierra de Mendoza.)

The present species was encountered first near General Roca, where from November 23 to December 2 it was recorded as fairly common on the plains that bordered the Rio Negro. Females were prepared as skins on November 23 and 24, and two additional specimens, one as a skeleton and one in alcohol, were preserved.

They ran swiftly along on the ground to pause and stand with head erect, or perched in alert attitudes on fence wires or the tops of bushes. In flight they traveled for long distances barely above the ground to rise finally to a perch on a low bush. Males at intervals flew up to make a metallic rattle with their wings as they turned abruptly and dropped to the ground.

Near Zapala, Neuquen, from December 7 to 9, these flycatchers were observed in little valleys where the grass had been closely cropped by stock. An adult male was taken December 7. Near Ingeniero White, the port of Bahia Blanca, Buenos Aires, probably near the northern border of the breeding range, two were recorded on December 13.

The male in the present species has the two outermost primaries decidedly attenuate, while in the female these two feathers are normal. In addition, the back of the male is more rufescent than in the opposite sex. An adult male, when first taken, had the base of gonys pallid brownish drab; rest of bill, tarsus, and toes black; iris Rood's brown.

LICHENOPS PERSPICILLATA PERSPICILLATA (Gmelin)

Motacilla perspicillata GMELIN, Syst. Nat., vol. 1, pt. 2, 1789, p. 969. (Rio de la Plata.)

Specimens of the widely distributed silverbill were secured at the following localities: San Vicente, Uruguay, January 27, 1921, adult male; Las Palmas, Chaco, July 23, 1920, adult female; Resistencia, Chaco, July 10, adult male; Berazategui, Buenos Aires, June 29; Dolores, Buenos Aires, October 21, adult male and female; Ingeniero White, Buenos Aires, December 13, immature male; Tunuyan, Mendoza, March 21, 1921, adult male, March 24 and 28, females; General Roca, Rio Negro, November 30, 1920, adult male. Adult males examined from the northern part of the Province of Buenos Aires into Paraguay and southern Brazil have the white patch in the wing at its maximum extent and may be considered as typical of true *perspicillata*. The black on the outer webs of the primaries extends only 2 or 3 mm. beyond the level of the primary coverts, the median portions of the shafts of the ninth and tenth primaries are white, and the dark distal tip is restricted. The white area forms a prominent streak along the side of the closed wing. The wing varies from 86.2 to 91.5 mm. (average of 12 specimens, 88.8 mm.). In skins from near Bahia Blanca, from the valley of the Rio Negro, and from Mendoza the white wing patch becomes somewhat more restricted, so that instead of seven outer primaries that are nearly white the number is reduced to six, as the outer web of the fourth primary is extensively black. The black on the others is also increased both on the outer web and at the tip. The wing varies from 92 to 93.6 mm. (average of three specimens, 92.7 mm.). These show a distinct approach to *andina*, but are nearer *perspicillata*. Finally, at Zapala and from the region south of the Rio Negro come specimens in which the black on the outer webs of the primaries may extend 5 or 6 mm. beyond the primary coverts, in which the shafts are black and the distal dark patch extensive. The wing in these ranges from 90.1 to 96.2 mm. (average of four specimens, 92.9 mm.). Though the white may be somewhat more extensive than in some from Chile, these seem best referred to *andina*. In these the white in the closed wing appeared streaked with black owing to the extent of the dark markings on the outer webs of the primaries.

A male taken at San Vicente, Uruguay, January 27, is molting the body feathers, while in three from Tunuyan, Mendoza (March 21 to 28), the outer primaries are being renewed.

Following are the localities and dates when this flycatcher was recorded: Berazategui, Buenos Aires, June 29, 1920; Resistencia, Chaco, July 9 and 10; Las Palmas, Chaco, July 13 to August 1; Riacho Pilaga, Formosa, August 8 to 20; Formosa, Formosa, August 23; Puerto Pinasco, Paraguay, September 3; Kilometer 80, west of Puerto Pinasco, September 17; Dolores, Buenos Aires, October 21; Lavalle, Buenos Aires, October 23 to November 15; General Roca, Rio Negro, November 23 to December 3; Ingeniero White, Buenos Aires, December 13; Carrasco, Uruguay, January 9 and 16, 1921; La Paloma, Rocha, January 23; San Vicente, Rocha, January 27 to February 2; Lazcano, Rocha, February 5 to 9; Potrerillos, Mendoza, March 16; Tunuyan, Mendoza, March 22 to 29.

The silverbill is restricted in its haunts to the vicinity of water. Though common in the Chaco north to the Rio Pilcomayo, it seemed rare or local beyond. In the pampas it was locally common in northern Buenos Aires, but was not detected in the regions that I visited near Guamini or Carhue. In northern Rio Negro it frequented the vicinity of streams, but was also common in the irrigated alfalfa fields, haunts that will enable the bird to extend its range, as cultivation, through enlarged irrigation projects, increases in the arid sections of northern Patagonia and western Argentina. In the Province of Mendoza, where part of the birds noted may have been migrant from the south, the species was found along streams and irrigating ditches, on one occasion at an altitude of 1,500 meters in the valley at Potrerillos.

These flycatchers run about freely on the ground, stopping ab--ruptly to throw up the head, so that they frequently suggest small thrushes. At other times they rest on clods of earth, bushes, or fence posts, from which they dart out at passing insects. Like *Taenioptera irupero*, they suggest in many of their mannerisms the bluebirds (*Sialia*) of North America. The crenulated lobe encircling the eye is easily seen in females, while in males, in which it is larger, its extent and light color produce an effect that is almost uncanny. During the breeding season males frequently rise 3 or 4 meters in the air, to whirl over and descend head first, with rapidly vibrating wings that produce a white halo about the body. Occasionally one in the same display describes erratic parabolas in the air, that reveal its contrasted colors to the utmost. Not content with these conspicuous displays, it attempts song, a squeaky effort barely audible at 50 meters. At other seasons the birds are wholly silent.

On November 24 a female was seen near Roca carrying material for nest lining, while at Ingeniero White, the port of Bahia Blanca, two or three broods of fully grown young were seen December 13. These last uttered low, squeaky calls.

The species is known locally as *pico plato*, or more rarely *ojo plato*, misnomers both since bill and eye are yellow. An adult male, taken July 10, had the bill straw yellow, tipped faintly with dusky; rosette about eye baryta yellow; iris barium yellow; tarsus and toes black. A female, shot July 23, had the maxilla and tip of mandible bone brown, becoming blackish at extreme tip; sides of maxilla, behind and below nostril, and base of mandible chartreuse yellow; iris vinaceous buff, with spots and mottlings of a darker color; rosette about eye deep olive-buff; tarsus and toes black.

LICHENOPS PERSPICILLATA ANDINA Ridgway

Lichenops perspicillatus andinus RIDGWAY, Proc. U. S. Nat. Mus., vol. 2, May 22, 1879, p. 483. (Santiago, Chile.)

An adult male silverbill shot at Zapala, Neuquen, on December 9, 1920, is representative of the present race, as the white wing patch is restricted by encroachment of black, especially on the outer webs of the primaries, and the wing has a measurement of 96.2 mm. Females of the two races of *Lichenops* appear indistinguishable in color, though in *andina* they average somewhat larger than in true *perspicillata*. The difference is slight and measurements overlap, so that many specimens of this sex, taken alone, may not be certainly identified.

On December 8 and 9, 1920, these birds were fairly common in areas where water offered them a suitable haunt. Extensive tracts in this arid region were not suited to them, so that their number was not great.

MACHETORNIS RIXOSA RIXOSA (Vieillot)

Tyrannus rixosus VIEILLOT, Nouv. Dict. Hist. Nat., vol. 35, 1819, p. 85. (Paraguay.)

Common and widely distributed through the pampas and the Chaco, the present bird was noted at the following points: Berazate-gui, Buenos Aires, June 29, 1920; Santa Fe, Santa Fe, July 4; Resistencia, Chaco, July 9 and 10; Las Palmas, Chaco, July 14 to 23; Formosa, Formosa, August 5, 23, and 24; Riacho Pilaga, Formosa, August 21; Puerto Pinasco, Paraguay, September 1 and 30; Kilometer 80, west of Puerto Pinasco, September 6 to 21; Lavalle, Buenos Aires, October 23 and November 9; Montevideo, Uruguay, January 14, 1921; La Paloma, Uruguay, January 23; San Vicente, Uruguay, January 27 to February 2; Lazcano, Uruguay, February 5 to 9. The species was most abundant in the Chaco, and was not recorded in the arid interior sections of central Argentina. An adult female taken at Formosa, August 24, and a pair secured at the Estancia Los Yngleses, Lavalle, Buenos Aires, on November 9, resemble other skins examined from Paraguay and southern Brazil. *M. r. flavogularis* Todd,[31] named from Venezuela, of which I have seen no specimens, is said to be brighter below than *rixosa*, with the throat but little paler than the abdomen, while the gray crown is duller, contrasting less strongly with the back. This form is said to occupy all the northern portion of South America.

These flycatchers inhabit wet localities in open regions, where occasional trees offer suitable night roosting places, a predilection that explains their greater abundance in the Chaco, where wet savannas with scattered trees are the characteristic feature of the country. *Machetornis*, though it roosts at night among leafy branches, spends most of its day on the ground, preferably near or among horses, cattle, or sheep that it follows as assiduously as do cowbirds for the sake of insects frightened up or attracted by the feeding stock. It is common to find *Molothrus* and *Machetornis* in company in suitable situations, and the flycatcher flies up to perch on the back of an ox or a horse as fearlessly as on a log or a post. In fact, the birds showed preference for such perches and frequently alighted on an animal when frightened from the ground. The ordinary method of progression of this bird was peculiar. A long, hesitating step made with bobbing head, was followed by a run for four or five steps, then another long step with the run repeated. In pursuit of insects or to evade too familiar approach it often ran

---

[31] Ann. Carnegie Mus., vol. 8, May 20, 1912, p. 210. (Tocuyo, Estado Lara, Venezuela.)

rapidly for a considerable distance. With its evident predilection for the vicinity of large animals, one may wonder if, hundreds and thousands of years ago, *Machetornis* was as familiar with the giant ground sloths and glyptodons that ranged then through these same regions as it is now with the stock introduced by man.

When not in company with cattle, *Machetornis* frequently ran about on the aquatic plants that covered the surface of small lagoons, where the long legs came in play in enabling the birds to step over interstices between the leaves of the plants. With the coming of September mating activities began, and the flycatchers pursued one another with snap and vigor, uttering high-pitched, squeaky calls and rattling their wings. Occasionally one rested in a tree top to utter a soft song *swee see dee*, a low effort that, though simple, was pleasing.

The Angueté Indians called this bird *yeht tin a bas gookh*.

MUSCISAXICOLA MACULIROSTRIS d'Orbigny and Lafresnaye

Muscisaxicola maculirostris d'ORBIGNY and LAFRESNAYE, Mag. Zool., 1837, cl. 2, p. 66. (La Paz, Bolivia.)

The 10 specimens taken were secured as follows: General Roca, Rio Negro, November 29, 1920, adult female; Zapala, Neuquen, December 8, adult female; Mendoza, Mendoza (altitude, 850 meters), March 13, 1921, male and female; Potrerillos, Mendoza (1,500 to 1,800 meters), March 16, 17, 18, and 19, four males and two females. The two adults secured in northern Patagonia in summer are breeding birds in slightly worn plumage. The series secured in the Province of Mendoza are all in fresh fall dress, and may be migrants come from the south. Three specimens seen from Calca, Peru are duller on the abdomen, and grayer on the sides of the head and neck than birds from Argentina, a difference that though slight seems distinct. With specimens from the type locality for comparison, it is probable that two forms may be recognized.

Following are the dates on which *M. macᵘlᵢrostris* was recorded: General Roca, Rio Negro, November 29, 1920; Zapala, Neuquen, December 8; Mendoza, Mendoza, March 13, 1921; Potrerillos, Mendoza, March 15 to 19; Tunuyan, Mendoza, March 25.

Those seen on their breeding grounds were found among low bushes on sandy or gravelly hillsides. Those noted in winter chose similar haunts, usually on sloping flats near streams, or on hillsides above water, where scattered bushes offered cover but left the ground bare in between. At this season they were found in little parties of two or three that ran alertly about on the ground, or rested for a few minutes in the tops of bushes. Passing insects were secured by a quick spring in the air, while others were picked up in the scanty herbage. Short, low flights, near the ground, revealed the long, pointed wings, gray back, and black tail with its light border that form the characters by which the bird may be recognized in the field. The flight is strong and certain, and the birds alighted alertly with head erect. All noted were silent. On March 17, along the Rio Mendoza below Potrerillos, they were in passage downstream in small parties that appeared to be in migration from the higher altitudes.

A female, taken November 29, had the base of the mandible cream buff; remainder of the bill black; iris Hay's brown; tarsus and toes black.

LESSONIA RUFA RUFA (Gmelin)

Alauda rufa GMELIN, Syst. Nat., vol. 1, pt. 2, 1789, p. 792. (Buenos Aires.)

Mathews [82] shows that *Alauda nigra* Boddaert for this species is antedated by *Alauda nigra* of the same author for another bird so that the specific name becomes *rufa* of Gmelin.

This ground-inhabiting flycatcher was recorded at the following localities: Berazategui, Buenos Aires, June 29, 1920 (adult male taken); Santa Fe, Santa Fe, July 4; Zapala, Neuquen, December 8 and 9 (two adult males shot); Carrasco, Uruguay, January 16, 1921; Guamini, Buenos Aires, March 3 to 8 (four males, three females taken); Tunuyan, Mendoza, March 22, 23, and 27 (adult male shot); and Concon, Chile, April 23 (adult male taken). Twelve skins were secured in all. Birds from Zapala, shot in summer, were in full breeding plumage. Four males, shot in March at Guamini, were all in immature plumage, in which they are similar to females, save that the back is more rufescent. One shows distinct signs of molt, apparently from a juvenal plumage. On this basis the young males molt from a juvenal plumage into a first winter plumage that is similar to the dress of the female. A specimen taken in September at Conchitas, Buenos Aires (in the United States National Museum), is in molt from the dull winter dress into the black adult plumage. On this slender evidence it may be supposed that the young males assume adult dress by a prenuptial molt in spring. An adult male, shot at Tunuyan March 27, in full adult plumage, is renewing the outer primaries. A skin in the United States National Museum, taken in April at Conchitas, has the throat white and the lower surface mottled with whitish. Other winter taken adult males do not differ from breeding specimens, save for an occasional specimen with very faint whitish tips on the feathers of the lower surface. It is common usage in recent years to consider *Lessonia oreas*,[83] which dif-fers from *nigra* in larger size and whitish edgings on the inner webs

---

[82] Austr. Av. Rec., vol. 3, Nov. 19, 1915.
[83] *Centrites oreas* Sclater and Salvin, Proc. Zool. Soc. London, 1869, p. 154. (Tinta, Peru.)

of the primaries, as a subspecies of *nigra* (in which the inner webs of the primaries are black in the adult male, and cinnamon in the female and immature male). The only specimens of *oreas* that I have seen are from Peru, and show no evidence of intergradation.

Adult males have the eighth and ninth primaries narrowed distally, with the seventh and tenth of normal width. As the narrowed feathers are concealed beneath the external primary they seem not to have been noted previously. In the male in first winter plumage, and the female at all seasons, all of the flight feathers are normal.

At all seasons of the year these interesting flycatchers frequent open ground, preferably near water, where they hop or run about on the ground, pausing to peck at the turf or to throw the head up and flit the wings rapidly. They are almost as terrestrial as pipits, a fact that may account for the elongated pipitlike claw on the hallux; like birds of that group they often seek elevated perches on little mounds of earth. They also fly up to rest on fence posts, or low bushes. Their flight is tilting and usually carries them only a foot or two above the ground. During the breeding season, near Zapala, males were common in the close-cropped grass of the lowland pastures, often in the vicinity of barrancas. As *rufa* has been supposed to breed only in Patagonia, it is unfortunate that a fine male that I watched for some time on January 16, 1921, near Carrasco, Uruguay, was not secured.

During the winter season *rufa* comes north to winter in abundance on the open pampas, but does not seem to penetrate beyond the limit of the plains. By March 3 I found the birds common on the level flats bordering the Laguna del Monte near Guamini, Buenos Aires, where they associated in little flocks. Others continued to arrive from the southward, driven up by the encroachment of cold in their summer homes. The birds now had the full lax plumage that protects them in winter, and ran about on the open flats unmindful of the heavy wind, as they made no effort to seek shelter from its blasts. In early morning members of the little scattered flocks pursued one another or chivied passing pipits vivaciously. On March 23, near Tunuyan, Mendoza, a flock of 15 arrived suddenly on the flats bordering the river, evidently a migrant flock from the south.

Like many other pampas birds, during winter they were entirely silent; and as my experience with them in summer was limited, I heard no calls from them whatever.

FLUVICOLA ALBIVENTER (Spix)

Muscicapa albiventer SPIX, Av. Spec. Nov. Brasiliam, vol. 2, 1825, p. 21, pl. 30, fig. 1. (Brazil.)

An adult female shot at Formosa, Formosa, on August 23, 1920, was found among open brush and saw grass at the border of a marsh, where it ran about on the ground or perched with rapidly jerking tail in the bushes. Another was seen, but not secured, at the border of a flooded estero near Puerto Pinasco, Paraguay, on September 3.

YETAPA RISORUS (Vieillot)

Muscicapa risora Vieillot, Gal. Ois., vol. 1, 1825, p. 209, pl. 131. (Brazil.)

The type locality of the present species is cited in most references as Paraguay, on the supposition that it is based on the *cola rara pardo y blanco* of Azara. Vieillot, however, described the species from an actual specimen and remarks "Le nom latinisé sous lequel nous décrivons cette espèce est celui qu'elle porte au Brésil," and discusses further certain differences between his specimen and the description of Azara. The specific name, usually given *risorius*, is spelled as above in the original publication. The form of the tail in both sexes of the present bird, taken in connection with its long, slender claws, are so different from the condition found in *Alectrurus tricolor* as to warrant generic separation. Hence *risorus* is placed in the genus *Yetapa* Lesson.[84] In form *Yetapa* is more similar to the large *Gubernetes* than to *Alectrurus*.

An adult female, the only one taken, was shot August 18, 1920, at the Riacho Pilaga, Formosa, in a heavy growth of tall grass that covered a small prairie. The bird perched on the side of the tall-stemmed seed heads or flew for short distances with a tilting flight. Another was seen but not secured near Carhue, western Buenos Aires, in December.

The Angueté Indians, in Paraguay, recognized the specimen that I had and called it *uh yuh ka bi ba oi koh.*

GUBERNETES YETAPA (Vieillot)

Muscicapa yetapa Vieillot, Nouv. Dict. Hist. Nat., vol. 21, 1818, p. 460. (Paraguay.)

An adult female was taken near Las Palmas, Chaco, on July 27, 1920, from a flock of three that passed me on an open savanna with direct, slightly tilting flight and undulating tails. Their call was a harsh note that may be represented as *rut rut*. The bird secured is in full plumage and offers no peculiarities worthy of remark. The claws in this species, while less developed than in *Yetapa risorus*, are elongate and rather slender.

SISOPYGIS ICTEROPHRYS (Vieillot)

Muscicapa icterophrys Vieillot, Nouv. Dict. Hist. Nat., vol. 21, 1818, p. 458. (Paraguay.)

Adult males were secured at the Estancia Los Yugleses, near Lavalle, Buenos Aires, on October 30 and November 1, 1920, and an

---

[84] Trait. Ornith., 1831, p. 387.

immature female was shot near Tapia, Tucuman, on April 11, 1921. The two adults are in full breeding plumage. The third specimen, in partial juvenal plumage, is duller, less yellow below and on the superciliary, is grayer above, and has two broad, light streaks on the wing, formed by whitish tips on greater and middle wing coverts.

At Los Yngleses, on October 30, a pair of these flycatchers had begun a nest in a fruit tree in the yard, and the female was busily engaged in carrying nest material to arrange it in a cuplike form in a convenient crotch. Other birds were recorded on November 1 and 9, and one was seen near Lavalle November 13. In appearance and actions the birds were typical flycatchers. They chose resting perches among leafy branches, and when their backs were turned to the observer were inconspicuous. One captured a large cater-pillar, killed it, and swallowed it. At Tapia, Tucuman, two were observed occupying low perches in open trees in dry forest. They were in company with a band of other passerines that appeared to be in migration as they moved rapidly through the scrub.

ARUNDINICOLA LEUCOCEPHALA (Linnaeus)

Pipra leucocephala LINNAEUS, Mus. Ad. Frid. Reg., vol. 2, 1764, p. 33. (Surinam.[85])

At the Riacho Pilaga, Formosa, an adult male was secured on August 8, 1920, and a male in immature dress on the day following. The birds were recorded on August 16 and 17, and one shot on the latter day was preserved in alcohol. The immature specimen has the black of the adult replaced by white on the breast and abdo-men, and by brownish gray on the back, wings, and sides. The adult specimen seems to have a larger bill (culmen from base, 18.5 mm.) than the few examined from Bahia, Santarem, Demerara, and other points in northern South America.

These odd flycatchers were fairly common in the outer growths of tall cat-tails that fringed lagoons, where it was difficult to secure them as one might work about the shore line for days without catch-ing sight of one save by chance. When I paddled out across the water in a clumsy boat hewn from the trunk of a silk-cotton tree, or on a crude raft made of a bundle of cat-tails lashed together, these birds were more in evidence, but it was difficult at that to retrieve specimens that were shot. Ordinarily these flycatchers rested quietly on low perches among the rushes, with the tail twitching quickly, while at short intervals they sallied out after passing insects. In flight a rattling sound was often produced by the wings. Their only call was a thin high-pitched note that may be represented as *tseet*.

---

[85] See Berlepsch and Hartert, Nov. Zool., vol. 9, April, 1902, p. 34.

KNIPOLEGUS CYANIROSTRIS (Vieillot)

Muscicapa cyanirostris VIEILLOT, Nouv. Dict. Hist. Nat., vol. 21, 1818, p. 447. (Paraguay.)

The present species was found at Las Palmas, Chaco, July 14 (male taken), and 30, 1920 (female secured), at San Vicente, Uruguay, January 28 (two males shot), 30 (female), and 31, 1921 (female), at Lazcano, Uruguay, February 6 (female), and at Rio Negro, Uruguay, February 14 and 18. The seven skins preserved offer no differences from others examined from Paraguay and southern Brazil. The male of this species resembles *K. aterrimus* in color, but has the white on the wing restricted to a narrow margin on the inner webs of the primaries. The female differs from that of *aterrimus* in being heavily streaked. The form of the primaries in both sexes is normal. Specimens secured at Las Palmas are in full winter plumage, while those shot in Uruguay, in midsummer, are in molt. An immature female that has not quite completed the molt into fall plumage is more rufescent above and less heavily streaked below than adult females secured in winter.

In an adult male, taken July 14, the tip of the bill was black; remainder pale Medici blue; iris coral red; tarsus and toes black. An adult female, taken July 30, had the maxilla dull black; mandible light Payne's gray, slightly darker toward tip; iris Rood's brown; tarsus and toes black.

In the Chaco these birds frequented dense growths of heavy forest, while in Uruguay they were found in heavy thickets near water. They were especially common in the low growth along the Rio Cebollati, near Lazcano. They were silent and, save for the twitching of the tail, were rather quiet, though alert and active in the pursuit of insects on the wing. Their general appearance was that of a phoebe (*Sayornis*). The streaked females are so different in color from the black males that they may easily be mistaken for another species.

KNIPOLEGUS ATERRIMUS ATERRIMUS Kaup

Cnipolegus aterrimus, KAUP, Journ. für Ornith., 1853, p. 29. (Cochabamba, Yungas, Moxos, Chiquitos, Bolivia.[86])

According to Berlepsch[87] *Knipolegus anthracinus* Heine,[88] described from Bolivia and in current use for birds from northern Argentina, is identical with *K. aterrimus* of Kaup, since examination of the type specimen did not bear out the supposed character of smaller size in *aterrimus*. I have seen no specimens from Bolivia, but Berlepsch states that skins from Argentina and Bolivia are

---

[86] From d'Orbigny and Lafresnaye, Mag. Zool., 1837, p. 59.
[87] Proc. Fourth Int. Orn. Congr., 1907, p. 471.
[88] *Cnipolegus anthracinus* Heine, Journ. für Ornith., 1859, p. 334. (Bolivia.)

identical, so that without advantage of comparative material my notes are given herewith under the typical subspecies. Doctor Chapman's decision [89] that *ockendeni* of Hartert from Peru should rank as a subspecies of *heterogyna*, which he considers specifically distinct from *aterrimus*, in my opinion, after examination of five specimens (including a male from Carabaya, the type locality) is erroneous. Though *ockendeni*, in addition to smaller size, and darker coloration in the female, has a somewhat heavier bill, it appears that it is a form of *aterrimus*.

The male of the present species is distinguished from *K. cyanirostris* by the broad white band across the underside of the wing, while the female is plain brown, unstreaked. An adult male of *aterrimus* in the United States National Museum, collected at Chilecito, La Rioja, Argentina, an abnormal specimen, has scattered white feathers on the sides and abdomen.

This form was encountered first at General Roca, Rio Negro, where the birds were fairly common from November 25 to December 2, 1920. An adult male was shot here on November 29, and females on November 25 and 29. Part of the birds observed frequented willow thickets along the Rio Negro, where they were probably on their breeding grounds, while others were found in the open brush through the arid, gravelly hills to the north of Roca. The number of these last varied from day to day, and it was my opinion that the individuals in these areas were still in migration. The birds rested on low perches, flirting the tail constantly, at intervals darting out after small insects, or dropping down to run along for a few feet on the ground. The flash of white from the wings of the somberly clad males, as they took flight, was almost startling, while the reddish brown color in the tail and rump of females in the glaring desert sun appeared almost red. Their only note was a faint *tseet*.

Later, at Tapia, Tucuman, the species was recorded from April 7 to 12, 1921, and two specimens were secured, a male April 12 and a female April 11. Here the birds were encountered along deep barrancas in the open forests, apparently in fall migration. The male taken at Tapia is in full plumage, while the female is just completing a fall moult. The two appear identical with specimens from Rio Negro.

In an adult male, taken November 29, the tip of the bill was black; base glaucous gray; iris Rood's brown; tarsus and toes black. A female shot on the same date had the tip of the bill blackish; base all around glaucous gray, much duller on the maxilla; iris Rood's brown; tarsus and toes black.

---

[89] U. S. Nat. Mus., Bull. 117, 1921, p. 89.

ENTOTRICCUS STRIATICEPS (d'Orbigny and Lafresnaye)

Muscisaxicola striaticeps d'ORBIGNY and LAFRESNAYE, Mag. Zool., 1837, cl. 2, p. 66. (Chiquitos, Bolivia.)

Dr. Hellmayr [90] has determined that this species, known for many years as *cinereus* Sclater,[91] should bear the name of *striaticeps* as indicated above. He writes that the type, in the Paris Museum, is labeled as taken at Chiquitos, Bolivia, though in the original description the species is said to come from La Paz. The greatly narrowed primaries in this species distinguish it not only from *Knipolegus*, but also from all other flycatchers. Mr. Ridgway described the peculiarities of this bird when he erected the genus *Phaeotriccus*, but as through inadvertence he designated *Cnipolegus hudsoni* as type, the name *Phaeotriccus* must be used for *hudsoni*. As *striaticeps* is undoubtedly peculiar, the present writer and Peters [92] have proposed that it be called *Entotriccus*. It is characterized by greatly narrowed primaries with the sixth to the tenth (outermost) distinctly falcate; seventh primary longest; tenth shorter than the first.

This flycatcher was recorded at the following points: Riacho Pilaga, Formosa, August 8 (adult female taken), 13 (two females and a male) and 18; Puerto Pinasco, Paraguay, September 3; Kilometer 80, west of Puerto Pinasco, September 9, 10 (a female shot) and 15; Tapia, Tucuman, April 7 to 13, 1921 (two males and a female taken April 7 and 9). The female taken at Tapia is less heavily streaked on throat and breast than skins from the Chaco, so that it is paler below, a difference due in part perhaps to the fact that the specimen is in fresh fall plumage. Females are identical with males in wing formula and in the narrowed form of the primaries.

In the Chaco, during the winter season, these alert little flycatchers sought low perches on the sheltered sides of dense groves of forest, where they were protected from cold winds. In the warmer, more open scrub near Tapia, Tucuman, they were scattered at random through little valleys, though more frequent perhaps along deep-cut barrancas that were common in this region. When at rest the tail twitched constantly, heightening their superficial resemblance to small *Empidonax*. During warm forenoons, in pleasant weather, males, from a perch at the top of a low tree or a dead limb, frequently shot straight up for a distance of 20 feet, turned and descended head first, with closed wings until just above the former perch, when the velocity of their fall was checked with a sudden rattle of wings, and the bird once more was at rest, as nonchalant and jaunty as though

[90] Nov. Zool., July, 1906, pp. 318–319.

[91] *Cnipolegus cinereus* Sclater, Proc. Zool. Soc. London, 1870, p. 58. (Corumba, Matto Grosso.)

[92] Proc. Biol. Soc. Washington, vol. 36, May 1, 1923, p. 144. Type, by original designation, *Muscisaxicola striaticeps* d'Orbigny and Lafresnaye.

it had not moved. The directness with which they rose and descended gave the same impression as a ball that is snapped into the air to fall back to the hand that had tossed it. This odd action was witnessed frequently and was probably a mating display intended for the season of spring. The only call heard from these little birds was a low *tsu wip*.

PHAEOTRICCUS HUDSONI (Sclater)

Cnipolegus hudsoni SCLATER, Proc. Zool. Soc. London, 1872, p. 541, pl. 31.
(Rio Negro, eastern Rio Negro, Argentina.)

When Mr. Ridgway [93] characterized the genus *Phaeotriccus* as new he evidently did so, as shown in his diagnosis, from *Entotriccus striaticeps*, but for some reason designated *Cnipolegus hudsoni* Sclater as type. Though the structural characters cited can cover *striaticeps* alone, the term *Phaeotriccus* may be applied only to *hudsoni*. As it happens *Cnipolegus hudsoni* Sclater is sufficiently distinct from the typical forms of *Knipolegus* Boie to warrant its separation so that *Phaeotriccus* comes into use for a valid generic group. It may be characterized as follows: Similar to *Knipolegus* Boie, but the three outermost primaries (eight to ten) narrow, tapering gradually from beyond center to tips; seventh primary broader but still narrower than normal; sixth and seventh primaries about equal; first longer than fourth, shorter than fifth. The wing is illustrated in the original description of the species.[94] The male, in addition to a narrow band of white across the primaries, has a white spot on the flanks that is concealed beneath the wing. The female of this species does not seem to have been described.

Males of Hudson's flycatcher were seen near Victorica, Pampa, on several occasions from December 23 to 29, 1920. An adult male was shot December 27, and another December 29, but I have no record of the female. The species is an alert, aggressive little bird that frequents openings in dense scrub, where it selects a low perch from which to watch for food. At times it gives a sharp explosive note followed by a loud popping of its bill.

The species, according to published notes, has been previously known from eastern Rio Negro (the type-locality) and eastern Mendoza (according to Fontana).

MECOCERCULUS LEUCOPHRYS LEUCOPHRYS (d'Orbigny and Lafresnaye)

Muscicapa leucophrys d'ORBIGNY and LAFRESNAYE, Mag. Zool., 1837, cl. 2, p. 53. (Yanacache, Yungas, Bolivia.[95])

Three specimens, a male and two females, all immature individuals in molt from juvenal to first winter plumage, were shot

---

[93] Proc. Biol. Soc. Washington, vol. 18, Sept. 2, 1905, p. 209.
[94] Proc. Zool. Soc. London, 1872, p. 542.
[95] See d'Orbigny, Voy. Amer. Merid., vol. 4, pt. 3, Oiseaux, 1835–1844, p. 327.

April 17, 1921, at an altitude of more than 1,800 meters on the Sierra San Xavier, above Tafi Viejo, Tucuman. These three specimens are distinctly more olivaceous, less brown above and on the sides of the breast than a series of *M. l. setophagoides* from Peru and Colombia. I have not had the advantage of specimens from Bolivia in comparison and so only assume that the Tucuman birds are typical.

In the groves and low thickets that were scattered over the open slopes of the Cumbre above the heavy rain forest, these small flycatchers were common. The majority ranged between 1,800 and 2,000 meters, though a few were found in alders just below the summit, 150 meters higher. In appearance they resembled other small flycatchers as they moved about under cover of leaves. In general aspect and coloration they were also suggestive of *Stigmatura budytoides*. They gave a low trilling song.

RHYNCHOCYCLUS SULPHURESCENS (Spix)

Platyrhynchus sulphurescens SPIX, Av. Spec. Nov. Brasiliam, vol. 2, 1825, p. 10, pl. 12. (Rio de Janeiro, Piauhy, and River Amazons.)

Through lack of a sufficient series for comparison, it is not practicable to identify subspecifically the specimens of this species that I secured in Paraguay and northern Argentina. They are brighter colored than the type of Oberholser's *Rhynchocyclus scotius* [96] from an unknown locality in Brazil. *Rhynchocyclus grisescens* Chubb [97] may be a distinct species, as it is said to be olive gray above instead of green, though it is possible that the type specimen, a female, may represent an individual phase of *sulphurescens*, in which case the name would apply to the subspecies found in the lower half of the Paraguay River Valley.

At Las Palmas, Chaco, I shot a female of this flycatcher on July 13, 1920, the only one seen in Argentina. The species has been recorded previously within the limits of the Republic only in Misiones and at Ledesma, Jujuy. [98] In the vicinity of Puerto Pinasco, Paraguay, the species was more common, as an immature male was secured at Kilometer 25 West on September 1, and a pair were shot near Kilometer 80 on September 8. The bird was common near Kilometer 80 through September, but was not seen in the drier areas farther west. On September 30 it was recorded on the Cerro Lorito, on the eastern bank of the Paraguay River. The birds were encountered in heavy forest, where they frequented the dense tops of low trees. Though they sallied out frequently to capture insects

---

[96] *Rhynchocyclus scotius* Oberholser, Proc. U. S. Nat. Mus., vol. 25, 1902, p. 63. (Brazil.)

[97] *Rhynchocyclus grisescens* Chubb, Ibis, 1910, p. 588. (Sapucay, Paraguay.)

[98] Dabbene, Orn. Argentina, An. Mus. Nac. Buenos Aires, vol. 18, 1910, p. 324.

on the wing, they often suggested vireos as they searched alertly but rather slowly among small limbs, a simulation that was especially notable when one hopped to a perch to remain quietly peering about without movement of the body for a short period. Their song was a curious effort that may be written as *sweet swees swee-ees*, given slowly, with every syllable uttered distinctly and separately. A male shot September 8 was nearly in breeding condition.

EUSCARTHMORNIS MARGARITACEIVENTER MARGARITACEIVENTER (d'Orbigny and Lafresnaye)

Todirostrum margaritacei venter d'ORBIGNY and LAFRESNAYE, Mag. Zool., 1837, cl. 2, p. 46. (Chiquitos, Bolivia.)

This tody flycatcher was fairly common in the Chaco from the vicinity of Resistencia north to northern Paraguay. At Resistencia, males were preserved on July 8 and 9, 1920, and others were seen July 10. At Las Palmas, Chaco, the species seemed less abundant since the only ones observed were two females shot July 19 and 27. Near the Riacho Pilaga, where the forest as a rule was drier than nearer the Rio Paraguay, *Euscarthmornis* was recorded in a particularly heavy stand of timber known as the *Monte Ingles*, where several were seen and a female taken on August 18. Near Kilometer 80, west of Puerto Pinasco, Paraguary, they were fairly common from September 9 to 20; a male and female were taken September 10, and a second male September 20. Another male was shot September 30 on the Cerro Lorito, a wooded hill on the eastern bank of the Rio Paraguay opposite the town of Puerto Pinasco. The series secured, all in full plumage, vary in color individually in the definiteness of streaking on the undersurface. No specimens from Bolivia are available for comparison.

These small birds frequent the lower brush at the border of heavy forest, where they hop about among the twigs in search for food, always near cover. Their light eyes give them an odd appearance. Their movements, while active, are somewhat heavy, entirely different from the sprightly actions of warblers. At times they utter a low call that resembles *tsu tsu.*

A male, shot July 8, had the maxilla dull brown; mandible a trifle paler; tarsus and toes russet vinaceous; iris yellowish white, suffused near pupil with dull buff.

Oberholser[99] has shown grounds for transfer of the generic name *Euscarthmus,* to what has been known as *Hapalocercus,* and has proposed *Euscarthmornis* for birds of the present group.

---

[99] Auk, 1923, p. 327.

MYIOSYMPOTES FLAVIVENTRIS (d'Orbigny and Lafresnaye)

Alectrurus flaviventris d'ORBIGNY and LAFRESNAYE, Mag. Zool., 1837, Cl. 2, p. 55. (Corrientes, Argentina.)

This small flycatcher, an inhabitant of marshes and the borders of swamps, was recorded and collected at the following points: Puerto Pinasco, Paraguay, September 3, 1920 (a pair of adults taken); Las Palmas, Chaco, July 22 (adult female); Dolores, Buenos Aires, October 21 (adult male); Lavalle, Buenos Aires, October 25 and November 9 (adult male taken October 25); General Roca, Rio Negro, November 24 to December 3 (a pair of adults); San Vicente, Uruguay, January 31 (adult male taken at Laguna Castillos) and February 2 (seen at Paso Alamo on the Arroyo Sarrandi); Lazcano, Uruguay, February 5 to 9; Tunuyan, Mendoza, March 22 and 26 (two males, three females). An adult female, taken November 27, had the bill black save at the base where it was tinged with tilleul buff; inside of mouth dull antimony yellow; tarsus and toes black. In an adult male, shot November 24, the inside of the mouth and the tongue were jet black.

Specimens from the Province of Buenos Aires north into Paraguay have slightly shorter wings than those from Mendoza, Patagonia, and Chile, but the difference seems too slight to warrant a name. Eight skins (there is no appreciable sexual difference in size) from Paraguay (Puerto Pinasco), Chaco (Las Palmas), and Buenos Aires (Conchitas, Dolores, and Lavalle) have a wing measurement ranging from 45.6 to 48.4 mm. Seven others from Chile (vicinity of Santiago), Mendoza (Tunuyan), Rio Negro (General Roca), and Chubut (Rio Chubut, below Leleque) have the wing from 48.9 to 50.4 mm. (Several from Tunuyan are molting primaries and do not offer true measurements.) If the difference indicated proves valid in further series, the large southern and western form will be known as *Myiosympotes flaviventris citreola* (Landbeck).[1]

During winter, in the saw grass marshes of the Chaco, these little birds worked about so quietly among weeds and low bushes over the water that it was a distinct surprise to find them more alert and active in willow thickets on the Rio Negro in the breeding season. At this period they came out within a few feet of me, apparently through curiosity, and males often rested in the sun on the tops of low willows from which they made short sallies for flying insects. Their song, heard frequently in early summer, was peculiar. It began with a low, clicking sound, like that made by striking two rounded pebbles together lightly, that was repeated slowly, then

---

[1] *Arundinicola citreola* Landbeck, An. Univ. Chile, vol. 24, no. 4, April, 1864, p. 338. (Mapocho, above Santiago, Chile.)

with increasing rapidity while the bill was thrown up perpendic-
ularly, and terminated in an abrupt note with which the bill was
jerked down suddenly to its usual position, *tick tick tick tick-tick-
tick-you*. In March this species was common about swales, weed
patches, and cornfields near the Rio Tunuyan, in Mendoza, and was
apparently in migration. Specimens in fresh fall plumage are
brighter yellow than those secured in summer.

PSEUDOCOLOPTERYX SCLATERI (Oustalet)

Anaeretes sclateri OUSTALET, Nouv. Arch. Mus. Paris, ser. 3, vol. 4, 1892,
p. 217. ("Chili.")

An adult female was taken at Las Palmas, Chaco, on July 22, 1920,
a male (skeleton) July 28, and a second male at the Riacho Pilaga,
Formosa, August 16. The sixth and seventh primaries in the female,
though not minute as in the male, are noticeably shorter than the
fifth and eighth. There is no apparent reason for not recognizing
Pseudocolopteryx of Lillo[2] as a valid genus.

These odd little birds were found among sedges and other low
growth at the borders of lagoons, often above shallow water covered
with floating vegetation. As they worked about through such
growth they were so well concealed that it was difficult to locate them.
Occasionally one flew for a meter or perhaps a little more with feeble
flight. Before alighting, males at times produced a sudden *whir*, a
sound caused by the attenuate sixth and seventh primaries.

SERPOPHAGA SUBCRISTATA (Vieillot)

Sylvia subcristata VIEILLOT, Nouv. Dict. Hist. Nat., vol. 11, 1817, p. 229.
(Paraguay.)

Serpophaga subcristata is an inhabitant of eastern Argentina,
Uruguay, Paraguay, and southern Brazil, where it is common in
forests and bush-grown pastures. It is migratory in Buenos Aires,
but remains through the winter in the Chaco. Thirteen specimens
were secured as follows (localities arranged in geographic sequence):
Kilometer 80, west of Puerto Pinasco, Paraguay, September 15,
1920, male; Las Palmas, Chaco, July 15 and 19, male and female;
Resistencia, Chaco, July 8 and 9, male and female; Rio Negro, Uruguay, February 14 and 19, 1921, male and female; San Vicente,
Uruguay, January 31, immature female; Lavalle, Buenos Aires, November 1, 1920, female; Victorica, Pampa, December 26 and 29, male
and female; General Roca, Rio Negro, November 27, male and
female. It is supposed that *Serpophaga* seen at the Riacho Pilaga,
Formosa, August 12 to 18, 1920; at Formosa, Formosa, August 23
and 24; at La Poloma, Uruguay, January 23, 1921, and at Lazcano,

---

[2] Rev. let. cienc. soc., (Tucuman), vol. 3, July, 1905, p. 48.

Uruguay, February 7 and 8, belonged to the present species, but no specimens were collected for identification.

It is probable that there are two forms of *subcristata* in the range as outlined in the opening paragraph above. The single specimen secured from west of Puerto Pinasco and a few others seen from the Chaco are paler below than birds from the Province of Buenos Aires. Birds that in bright coloration resemble those of the south were also secured at Las Palmas and Resistencia. It is possible that the pale birds represent a resident form of the Chaco and that the brighter ones are winter migrants from the south.

S. munda is so similar to *S. subcristata* that it would appear that the two should stand as subspecies of one form. However, west of Puerto Pinasco, Paraguay, on the ranch known as Kilometer 80, I found *subcristata* and *munda* ranging in the same forests without evidence of intergradation, so that they must be considered specifically distinct. As *S. inornata* was taken there also, three distinct species of *Serpophaga* were found at this point.

Serpophaga subcristata is one of the familiar species of the forested regions and brushy areas in the Chaco, that on the pampas inhabits groves about the estancias, and in the more arid south is found in heavy stands of *Baccharis* and *Salix* along the larger streams. Though undoubtedly a flycatcher, it is so sprightly and vivacious in its movements that in life it gives little suggestion of its tyrannine affinities. In fact, as the birds flit and hop about among the twigs, often calling or singing excitedly, they bear a striking resemblance to warblers. During winter they were found constantly with little bands of other little birds that ranged the forests and came around without fear to inspect me, often hopping out almost within reach. Both sexes sang frequently, a fact that I established by collecting specimens, but the notes of males were louder than those of the opposite sex. The entire song may be repre-sented as *chois chois chee chee chee chee-ee-ee-ee*, a few twittering notes followed by a hard trill, of which the first part was frequently omitted.

The first young bird seen, a male not quite grown, was taken December 29, near Victorica, Pampa. Other juvenile birds were recorded at San Vicente, Uruguay, January 31, and Rio Negro, Uruguay, February 14. In juvenal plumage *subcristata* is washed with brown above, especially on the upper tail coverts, and has the wing bars light-pinkish cinnamon instead of whitish. Beneath the birds are whitish, and the black and white markings found in adults in the crown are lacking. An adult female shot at Rio Negro, Uruguay, February 19, is in fall molt.

The bill and tarsus in this species are black.

SERPOPHAGA MUNDA Berlepsch

Serpophaga munda BERLEPSCH, Ornith. Monats., vol. 1, 1893, p. 12. (Samaipata, Valle Grande, Bolivia.)

The present species seems identical with *S. subcristata*, save that the lower breast and abdomen are white instead of yellow, and the dorsal surface usually is grayer. The species inhabits western and northwestern Argentina, and extends eastward in the Chaco into Paraguay. It is said to occur also in the Argentine Chaco. The following specimens referred to *munda* were collected: Kilometers 25 and 80, west of Puerto Pinasco, September 1 and 9, 1920, two males; Mendoza, Mendoza (altitude, 850 meters), March 13, 1921, male; Potrerillos, Mendoza, March 16, 17, and 21, one male and two females; and Tapia, Tucuman, April 9, 1921. Birds of this genus recorded March 27, near Tunuyan, Mendoza, were supposed to be this species.

One of the skins taken at Puerto Pinasco was an immature bird, though fully grown, with a slight olive wash on the lower back that is absent in the adult. Those shot in Mendoza and Tucuman are in fall molt. In juvenal plumage the two light wing bars are distinctly buff, while in the succeeding plumages these bars are much lighter to nearly white. In immature birds in first winter plumage the lower abdomen is very faintly washed with yellow, suggesting the condition found in *subcristata*, where this color in deeper hue extends over the abdomen and lower breast. At first glance this wash of yellow in *munda* is confusing, but specimens are easily distinguished when compared in series as *subcristata* is told at once by the much yellower color. Careful comparison of an adequate series of the two fails to indicate differences that may separate them other than those that have been noted.

West of Puerto Pinasco, *S. munda* was encountered in fair numbers, in heavy timber where it worked actively about in the smaller branches like some warbler. During fall in the Province of Mendoza the birds were found in low scrub that clothed the dry slopes above small valleys or in better watered sections in growths of weeds. They were fairly common and from their movements appeared to be in migration. In early morning, especially, they were recorded as moving actively through the thickets or weed patches, often uttering a low *tseet*, like the fall calls of some of our warblers. Near Tapia, Tucuman, they were found occasionally in the scrubby forest.

SERPOPHAGA INORNATA Salvadori

Serpophaga inornata SALVADORI, Boll. Mus. Zool. Anat. Comp. Univ. Torino, vol. 12, no. 292, May 12, 1897, p. 13. (San Francisco, Chaco of Bolivia.)

Near Kilometer 80, west of Puerto Pinasco, Paraguay, on September 20, 1920, two were taken in heavy forest, as they worked actively about through the trees. Their song was a low trill that may be represented as *chee-ee-ee-ee*. No others were recorded. The present species differs from *S. subcristata* and *S. munda*, which it resembles superficially, in the lack of black and white markings in the crown. The bill in addition is longer than usual in the other two species. The abdomen is white centrally, while the sides and lower tail coverts are washed with yellowish. It is distinguished at a glance from its allies.

SERPOPHAGA NIGRICANS (Vieillot)

Sylvia nigricans VIEILLOT, Nouv. Dict. Hist. Nat., vol. 11, 1817, p. 204. (Paraguay and shores of the Rio de la Plata.)

The present species is somewhat rare at the present time, and was found in few localities. At Berazategui, Buenos Aires, June 29, 1920, an adult female was taken on low ground near a ditch. The bird was active in pursuit of insects and when quiet rested indifferently on low grass stems, twigs, lumps of mud, or level ground. The feet were bedaubed with mud. In Uruguay the species was found on three occasions, each time in lowland marshes where dense thickets of low willows and other water-loving shrubs stood in shallow water. One was observed February 3, 1921, at the Paso Alamo on the Arroyo Sarandi. An adult male was taken February 7, and another seen on the day following near the Rio Cebollati below Lazcano. A third, an immature female, was taken at Rio Negro, Uruguay, on February 18. The birds hop rather actively about in their dense cover, jerking the broad black tail or nervously spreading it like a fan even when at rest.

The adult male taken February 7 had the bill and tarsus black; iris warm sepia; inside of mouth, including tongue, warm chrome. The bird taken at Berazategui in June is in full winter plumage. The one shot near Lazcano, February 7, is badly worn and is molting on the body. It appears much darker than the winter bird. The immature specimen taken, still in juvenal plumage, is browner above and on the lower abdomen and under tail coverts than adults and has no concealed white spot in the crown.

COLORHAMPHUS PARVIROSTRIS (Gould)

Myiobius parvirostris GOULD, Zool. Voy. Beagle, pt. 3, Birds, July, 1839, p. 48. (Santa Cruz, Patagonia.)

Near Concon, Chile, a male was secured on April 26, 1921, and another on the day following. The first mentioned, when first killed, had the maxilla and tip of mandible black; base of mandible hair brown; iris chestnut brown; tarsus and toes black. The birds were found near small streams where they sought low perches on weeds or bushes in little open spaces, whence they made sallies for passing insects. In appearance and actions they resembled the usual type of small flycatchers.

SPIZITORNIS FLAVIROSTRIS FLAVIROSTRIS (Sclater and Salvin)

Anaeretes flavirostris SCLATER and SALVIN, Proc. Zool. Soc. London, 1876, p. 355. (Tilotilo, Yungas, Bolivia.)

Five adult males secured near General Roca, Rio Negro, on November 25 and 29, and December 2, 1920 (one prepared as a skeleton), mark a considerable extension in range for this species, since it has been recorded by Dabbene [3] only south to the Sierra de Cordoba. It is possible that it has been overlooked through its similarity to *Spizitornis parulus*. *S. f. flavirostris* was found with *S. p. patagonicus*, but was readily distinguished by the yellowish base of the mandible, by the heavier black streaks on the underparts, and, when in the flesh, by its dark eye. It was fairly common in the low bushes that dotted the sides of little valleys in the arid gravel hills north of the flood plain of the Rio Negro. In general appearance, aside from its crest, it suggested a gnatcatcher, as it hopped about in the tops of the low bushes or occasionally darted up to secure some insect in the air. The resemblance was heightened when it threw the tail at a jaunty angle over the back, though the slender, recurved crest of a few black feathers broke the illusion at first glance. The birds were active and alert and often difficult to approach since they flew with tilting flight from bush to bush at the slighest suggestion of danger.

Males were practically in breeding condition and were singing constantly, a low buzzing, squeaky effort, barely audible above the wind, that I wrote as *seet zwee-ee seeta seeta seeta*.

The inside of the mouth and base of the mandible were zinc orange; rest of bill black; iris Hay's brown; tarsus black.

Chapman [4] has named two subspecies of *flavirostris* from Peru.

SPIZITORNIS PARULUS PARULUS (Kittlitz)

Muscicapa Parulus KITTLITZ, Mém. Acad. Imp. Sci. St.-Pétersbourg, vol. 1, 1831, p. 190. (Concepcion and Valparaiso, Chile.)

On the grounds that *Anairetes* of Reichenbach, 1850,[5] is preoccupied by *Anaeretes* Dejean, 1837,[6] Oberholser [7] has proposed the generic name *Spizitornis* for this bird.

[3] Orn. Argentina, An. Mus. Nac. Buenos Aires, vol. 18, 1910, p. 331.
[4] Amer. Mus. Nov., no. 118, June 20, 1924, p. 8.
[5] *Anairetes* Reichenbach, Av. Syst. Nat., 1850, pl. 66.
[6] *Anaeretes* Dejean, Cat. Col., ed. 3, 1837, p. 181. E. A. Schwarz informs me that this work, though marked as the third edition and universally so cited, is in reality a fourth print since, when the third revision of Dejean had been printed, it was destroyed by fire before more than a few copies had been distributed. It was set up again, and this reprint was still marked as the third edition though in reality it was the fourth.
[7] Auk, 1920, p. 453.

A series of three males and two females secured April 24, 26, and 27, 1921, near Concon, Chile, with five more in the United States National Museum collection (Tome and other localities in Chile not specified), serve to demonstrate the characters of the typical form. From these it appears that true *parulus* is marked by grayish coloration, somewhat limited streaking of the undersurface, and the reduction or absence of white wing bars. Hellmayr[8] found this to be true in 10 skins from Valparaiso and Valdivia. *A. p. curatus* Wetmore and Peters, from Argentina, which is yellowish below like *parulus*, is somewhat more broadly streaked, on the average, is lighter, more grayish above, and has two broad white wing bars. One of the females from Concon has the greater and middle coverts faintly tipped with buffy white and is somewhat paler above than four others (all in fresh fall plumage). It indicates a near approach to *curatus*, and may possibly be a migrant from some higher region where there is a tendency toward intergradation between the two forms. Barros[9] records the birds as resident at an altitude of 1,700 meters on the upper Rio Aconcagua. This one specimen was the cause of some uncertainty as to the validity of *curatus* from east of the mountains; nine supposedly typical examples of *parulus*, in which there was a mere trace at most of a pale edging to the coverts, and Hellmayr's account of 10 more, in which the condition is similar, seem to indicate that this one specimen represents an intergrade.

These tiny birds frequented the dense brush on the hill slopes above the Rio Aconcagua, where they traveled actively about through the bushes. In general appearance and actions they suggested king- lets, as they flitted the wings constantly, an appearance that was belied by the jaunty black crest that came into view when the birds were seen clearly. They were frequently aggressive and drove one another about petulantly. At this season they were in full fall plumage.

In a male the upper third of the iris was raisin black, the remainder marguerite yellow. The dark and light areas were sharply defined and the unusual pattern with two distinct colors gave the eye an appearance that was exceedingly strange.[10]

SPIZITORNIS PARULUS CURATUS Wetmore and Peters

Spizitornis parulus curatus WETMORE and PETERS, Auk, 1924, p. 145. (Rio Colorado, Gobernación de Rio Negro, Argentina.)

The present form is represented by a female shot at Potrerillos, Mendoza, on March 15, 1921 (another seen but not taken March 17),

---

[8] Arch. für Naturg., vol. 85, November, 1920, p. 51.
[9] Rev. Chilena Hist. Nat., vol. 25, 1921, p. 185.
[10] This condition has been figured accurately by Barros, Rev. Chilena Nat. Hist., vol. 25, 1921, p. 185, fig. 26.

and a male killed March 24 on an arid brush-grown flat 15 miles west of Tunuyan in the same Province. An adult male shot December 24, 1920, near Victorica, Pampa, in an open forest of caldén and algarroba, is distinctly intermediate between *curatus* and *patagonicus*, and marks a point near the dividing line between the two. The abdomen in this intermediate skin is very faintly yellowish, and the back only faintly darker than in *patagonicus*. It might, with equal propriety be identified with either of the forms concerned.

The geographic races at present known for the species *parulus* are indicated below:

SPIZITORNIS PARULUS PARULUS (Kittlitz).

> *Muscicapa Parulus* KITTLITZ, Mém. Acad. Imp. Sci. St.-Pétersbourg, vol. 1, 1831, p. 190. (Concepcion and Valparaiso, Chile.)

Above dark, wing bars absent or faint, auricular dark patch sharply defined.

Central Chile. Skins from Bariloche, Rio Negro, while not typical, have been assigned to this race.[11]

SPIZITORNIS PARULUS AEQUATORIALIS (Berlepsch and Taczanowski).

> *Anaeretes parulus aequatorialis* BERLEPSCH and TACZANOWSKI, Proc. Zool. Soc. London, 1884, p. 296. (Cechce, western Ecuador.)

Similar to *parulus*, but breast more heavily streaked, darker above, crown with white markings restricted.

Ecuador and Peru.

SPIZITORNIS PARULUS CURATUS Wetmore and Peters.

> *Spizitornis parulus curatus* WETMORE and PETERS, Auk, 1924, p. 145. (Rio Colorado, Gobernación de Rio Negro, Argentina.)

Similar to *parulus*, but paler above, with two distinct wing bars, dark auricular patch less sharply defined and upper breast whiter, less yellowish.

Eastern Rio Negro through Pampa to Cordoba and the foothills of the Andes in Mendoza. Eastern Chubut?

SPIZITORNIS PARULUS PATAGONICUS Hellmayr.

> *Spizitornis parulus patagonicus* HELLMAYR, Arch. f. Naturg., vol. 85, November, 1920, p. 51. (Neuquen, Gobernacion de Neuquen, Argentina.)

Similar to *curatus* but paler above with abdomen white.

Eastern Neuquen and northwestern Rio Negro, intergrading with *curatus* in central Pampa.

---

[11] See Wetmore and Peters, Auk, 1924, p. 145.

VALLEY OF RIO BLANCO, WITH SIERRA DEL PLATA IN THE BACKGROUND
Above Potreiillos, Mendoza, March 18, 1921

A MUDDY CHANNEL, OR CIENAGA, HAUNT OF PAINTED SNIPE (NYCTI-
CRYPHES SEMI-COLLARIS) AND JACKSNIPE (CAPELLA PARAGUAIAE)
Near Tunuyan, Mendoza, March 28, 1921

RIO TUNUYAN, NEAR TUNUYAN, MENDOZA

Taken March 22, 1921

DRY FOREST NEAR TAPIA, TUCUMAN

Taken April 14, 1921

SPIZITORNIS PARULUS LIPPUS Wetmore.

Spizitornis parulus lippus WETMORE, Univ. California Publ. Zoöl., vol. 21, 1923, p. 336. (Mayne Harbor, Evans Island, Owens Islands, Chile.)

Similar to *parulus* but darker above, black of head duller, and breast more abundantly streaked.

Straits of Magellan.

SPIZITORNIS PARULUS PATAGONICUS Hellmayr

Spizitornis parulus patagonicus, HELLMAYR Arch. für Naturg., vol. 85, November, 1920, p. 51. (Neuquen, Gobernacion de Neuquen, Argentina.)

Two adult males secured at General Roca, Rio Negro, November 29 and December 3, 1920, have the underparts white, rather heavily streaked, two well-marked white wing bars, and the dorsal surface gray and may be considered typical of this form since they were taken only a short distance east of the type locality.

Near Roca this bird often was found in low bushes in the same areas that were occupied by *A. flavirostris.* However, on December 3, I encountered *patagonicus* among growths of *Atriplex* and similar shrubs on the low flats near the Rio Negro, where *flavirostris* was not seen, so that when the birds are settled for the summer at their breeding stations, the two species may affect different ecological associations. The two were similar in actions but had slightly different notes and were easily distinguished by color.

A male, shot November 29, had the bill black; inside of mouth zinc orange; iris pale olive bluff save for a purplish area that covered a segment on the upper side; tarsus black.

TACHURIS RUBRIGASTRA RUBRIGASTRA (Vieillot)

Sylvia rubrigrastra VIEILLOT, Nouv. Dict. Hist. Nat., vol. 11, 1817, p. 277. (Paraguay and Buenos Aires.)

Four specimens of this bright-colored marsh flycatcher include an adult male from Dolores, Buenos Aires, October 21, 1920, a male and a female from Tunuyan, Mendoza, March 26, 1921, and a female from Concon, Chile, April 29. Comparison of a series of 19 skins from Buenos Aires, Rio Negro, Mendoza, and Chile, indicates that birds from the southern part of the range of the species do not differ sectionally in spite of the wide range included. The two skins from Tunuyan and the one from Concon are immature individuals in first winter plumage, distinguished from older individuals by a yellow spot on the rump.

True to their reputation these handsome mites of the feathered world were found among rushes in marshes, usually where the water was something less than a meter in depth. They were local in occurrence and appeared to gather in little scattered colonies, as many extensive areas of suitable growth were not inhabited by them. They were shy and apprehensive, so that it was often difficult to approach them. Their custom was to clamber about among the rush stalks, where their long legs fitted them for progress, or occasionally to fly across little openings with slightly tilting, but direct flight, performed with head erect and rapidly flitting wings. The white in wing and tail are prominent in flight. Occasionally they descended to run about on little mud bars at the bases of clumps of cat-tails.

They were first recorded at Dolores, Buenos Aires, on October 21, 1920, when two were seen. Near Lavalle, in the same Province, they were found casually on October 30, November 2 and 9, but were not common. On March 26, and 28, 1921, a number were recorded near Tunuyan, Mendoza, in the rush-grown marshes known as cienagas. They were found here in little family parties, and, though shy, were tolled out by squeaking from concealment among the cattails. Near Concon, Chile, April 28, one was seen, and on the day following one was brought by a boy as, in company with Dr. E. P. Reed, I was about to leave for Valparaiso.

LEPTOPOGON AMAUROCEPHALUS Cabanis

Leptopogon amaurocephalus CABANIS, Arch. Naturg., vol. 1, 1847, p. 251. (Brazil.)

On July 21, 1920, at Las Palmas, Chaco, an adult female *Leptopogon* was killed in dense brush near the Rio Quia. The bird hopped about actively under cover of the branches, or paused to rest for considerable intervals on hidden perches. In this specimen, when first taken, the extreme base of the mandible was tilleul buff; rest of bill black; iris natal brown; tarsus and toes fuscous.

With only three specimens of *L. amaurocephalus* at hand, I do not care to express an opinion as to the forms into which this species may be divided. The specimen from Las Palmas, which has the wing 65.2 mm. long, is slightly deeper and richer in color throughout than a skin from Victorica, Sao Paulo, or the type of *icastus* Oberholser [12] from Sapucay, Paraguay. Chubb [13] has indicated that specimens from Sapucay do not differ from others from Brazil.

CAMPTOSTOMA OBSOLETUM OBSOLETUM (Temminck)

Muscicapa obsoleta TEMMINCK, Nouv. Rec. Planch. Col. Oiseaux, vol. 3, 1838, pl. 275, fig. 1. (Curytiba, Parana, Brazil.[14])

Five males, one shot at Resistencia, Chaco, July 8, 1920, one from Laguna Wall, 200 kilometers west of Puerto Pinasco, Paraguay,

[12] Proc. Biol. Soc. Washington, vol. 14, Dec. 12, 1901, p. 187.
[13] Ibis, 1910, p. 582.
[14] Nov. Zool., vol. 15, June, 1908, p. 43.

taken September 25, 1920, two from the Cerro Lorito on the east bank of the Rio Paraguay opposite Puerto Pinasco, secured September 30, and one from Tapia, Tucuman, collected April 8, 1921, are referred to the typical form, as they agree in color with three seen from Taquara do Mundo Novo, Rio Grande do Sul, and have the measurements assigned by Hellmayr[15] to that form. The wing in these birds, in the order cited, measures as follows: 53, 55.4, 55.8, 55.2, and 54.3 mm. As these represent the chord of the folded wing taken with dividers, they are comparable with Doctor Hellmayr's figures, in which it is supposed that the wing was measured flat.

Mr. Ridgway[16] has removed *Ornithion inerme*, the type of *Ornithion* Hartlaub to the Pipridae, as it has a pycnaspidean tarsus, leaving the species with exaspidean tarsi, that have been associated with it, in the Tyrannidae under the name *Camptostoma* of Sclater.

This small bird was of local occurrence and was seldom seen. At Resistencia the one taken was shot from a little flock of three that came flitting actively through some low trees in dense growth, occasionally uttering a vireolike, scolding note. Near Laguna Wall, in the Chaco, 200 kilometers west of Puerto Pinasco, the birds were fairly common in the dense growths of thorny vinal that covered large areas. On September 30, near the Rio Paraguay, I found a number in heavy timber over a wet area where the forest was open. The birds were seen high in the tops of the still leafless trees, where they perched quietly except when they darted out to secure passing insects. Their song was a rattling, laughing *chee chee chee chee chee* that was almost swiftlike in its tones. In fall, near Tapia, Tucuman, the species was encountered again in low scrub in company with other small, brush-haunting birds.

One taken July 8 had the tip of the mandible blackish; base of culmen dull slate; rest of mandible pinkish white, becoming dull orange at gape; inside of mouth, orange; iris, brown; tarsus, slate black.

ELAENIA ALBICEPS ALBICEPS (d'Orbigny and Lafresnaye)

Muscipeta albiceps d'ORBIGNY and LAFRESNAYE, Mag. Zool., 1837, cl. 2, p. 47. (Yungas, Bolivia.)

The present species seems to be one that ranges throughout Patagonia, north at least to the Rio Negro, and that northward extends through the foothills of the Andes in Chile and Argentina to southern Peru. In Argentina it is reported from the isolated Sierra de Cordoba. From examination of a considerable series it seems that *albiceps* differs from *E. parvirostris*, which superficially appears identical in narrower bill, browner, less greenish dorsal coloration,

---

[15] Nov. Zool., vol. 15, June, 1908, p. 44.
[16] Proc. Biol. Soc. Washington, vol. 19, Jan. 29, 1906, p. 14.

grayer breast, more extensive white in crown stripe, with feathers of occiput somewhat longer and fuller, and lack of whitish margins on the lesser wing coverts. The young of *albiceps* in juvenal plumage are duller brown than those of *parvirostris*. Specimens assigned to *albiceps* include an adult male shot in willows near the Rio Negro, south of General Roca, Rio Negro, December 3, 1920; immature male and female from an altitude of 1,500 meters at Potrerillos, Mendoza, March 17, 1921; and an immature female from 700 meters' elevation at Tapia, Tucuman, April 7, 1921. The last three are in juvenal plumage. The specimen from Tapia is considerably paler above than those from Potrerillos, but is duller in color than *parvirostris*, from the same locality, while in addition it has a narrower bill and no white on the lesser wing coverts. As it was taken in late fall it may be migrant from a higher elevation.

Near General Roca, Rio Negro, *Elaenia a. albiceps* was fairly common among thickets of willows near the river, but was so shy and retiring that it was difficult to secure. The call note of males, a rapidly given *wheur*, was audible at some distance, but in the dense growths of small willows the bird was difficult to see, while in more open groves they were too wary to permit easy approach. Elsewhere the species was recorded in the Andean foothills near Potrerillos, Mendoza, from March 17 to 21, where the birds were found in growths of creosote bush (*Covillea cuneifolius*) along streams. These, as well as one found in dry brush near Tapia, Tucuman, on April 7, were apparently in migration.

ELAENIA PARVIROSTRIS Pelzeln

Elainea parvirostris PELZELN, Orn. Bras., pt. 2, 1868, p. 178. (Curytiba. Parana, Brazil.)

This form of *Elaenia* ranges from northern Buenos Aires (Lavalle, Conchitas, etc.) northward through northern Argentina, Uruguay, Paraguay, and Brazil. In general it extends eastward of the area inhabited by *albiceps*, and seems to frequent lower altitudes, since it is not found in the Andes nor is it known in the colder area of Patagonia on the south. Though similar in general appearance to *albiceps*, *E. parvirostris* is distinguished by broader bill, more greenish dorsal coloration, lighter breast, shorter occipital feathers, and less amount of white in crown. The lesser wing coverts in *parvirostris* frequently are tipped with white, forming an indistinct third band on the wing. The young in juvenal plumage are brighter, more greenish in coloration than those of *albiceps*.

The five specimens secured include an adult male taken November 9, 1920, at the Estancia Los Yngleses, near Lavalle, Buenos Aires, adult male and female from San Vicente, Uruguay, January 25 and 28, 1921, an immature female, in juvenal plumage, shot at Rio Negro,

Uruguay, February 14, and a male from Tapia, Tucuman, April 13. The species was common in thickets near streams in southern Uruguay (La Paloma, January 23, San Vicente, January 25 to 31, Rio Negro, February 14), but in Argentina was seen only on the two occasions when specimens were taken, as noted above. The birds were found on low perches among rather dense growth, where they made short flights from twig to twig or rested quietly with twitching tail. At the end of January they were breeding, and young were on the wing by the middle of February. Males in the nesting season gave an emphatic little song like that of some *Empidonax*, uttered from secure retreat among willows or similar shrubs. When alarmed about their nests they uttered a low *tsip*.

MYIOPAGIS VIRIDICATA VIRIDICATA (Vieillot)

Sylvia viridicata VIEILLOT, Nouv. Dict. Hist. Nat., vol. 11, 1817, p. 171. (Paraguay.)

On September 30, 1920, at the base of the Cerro Lorito on the eastern bank of the Rio Paraguay, opposite Puerto Pinasco, Paraguay, attention was attracted to this flycatcher by its sharp explosive note *chur esp*. The bird perched in heavy cover in the tops of trees and shrubs, in dense forest, occasionally flying out after insects. Only through its peculiar call was I able to follow it and secure it. It proved to be an adult male in breeding condition. This bird has the following measurements: Wing, 63.7; tail, 59.8; culmen from base, 10.6; tarsus, 15.7 mm.[17]

SUIRIRI SUIRIRI (Vieillot)

Muscicapa suiriri VIEILLOT, Nouv. Dict. Hist. Nat., vol. 21, 1818, p. 487. (Paraguay.)

The suiriri flycatcher was common through the Chaco from northern Argentina into Paraguay, and in the wooded country in the territory of Pampa and northern Tucuman. Dates and localities (with specimens preserved indicated in parentheses) at which it was recorded are as follows: Las Palmas, Chaco, July 13 to 31, 1920 (male taken July 13, female July 18); Riacho Pilaga, Formosa, August 7 to 18; Formosa, Formosa, August 23; Kilometer 80, west of Puerto Pinasco, Paraguay, September 7 to 20 (females taken September 7 and 11); Kilometer 200, west of the same point, September 25; Victorica, Pampa, December 23 to 29 (adult male taken December 23, two adult females December 24 and 28); Tapia, Tucuman, April 6 to 13) males taken April 6, 11, and 13, one juvenile with sex indeterminate September 11). Birds secured in December were in worn plumage, those shot in April were in molt. In the juvenal plumage (shown by one individual) described previously

---

[17] For discussion of the forms of this species, see Berlepsch, Proc. Fourth Int. Ornith. Congr., February, 1907, pp. 425–431.

by Hartert [18] this bird presents a curious appearance for a species of this group as the entire dorsal surface from forehead to upper tail coverts, including the lesser wing coverts, is marked with triangular spots of white.

I am uncertain as to the validity of *Suiriri suiriri albescens* (Gould) [19] separated by Oberholser [20] on supposed grayer dorsal surface and whiter wing bars. With a fair series I find these characters somewhat variable in birds from Paraguay and from points farther south, so that I can not make a definite separation with the material at hand.

. : Though these birds frequented forest or brush-grown areas, they were conspicuous and easily seen, as they were usually encountered among open branches where there was little concealment of twigs 'or foliage. It was usual to find two or three together. The species had several notes that served to advertise its presence, one that resembled *chee-ee-ee-ee-ee-ee*, a rolling whinny, being most common. The ordinary call note was a low *chee chee*, and in the breeding season they uttered a musical song in a low tone. Their movements were slow and rather methodical, so that at times they gave some suggestion of vireos.

SUIRIRI IMPROVISA Wetmore

Suiriri improvisa WETMORE, Auk, 1924, p. 595. (Tapia, Province of Tucuman, Argentina.)

The type and only specimen seen of this species was shot near Tapia, Tucuman, on April 9, 1921, as it worked slowly through the tops of trees in dry, open forest. In general appearance the bird suggests *Suiriri suiriri* except that it has a longer, heavier bill, but with this structural resemblance is combined a type of coloration resembling that of *Sublegatus fasciatus*. In a way *improvisa* is representative of *Suiriri affinis* (Burmeister) (long considered an *Elaenia*, but placed in *Suiriri* by Berlepsch),[21] but is distinctly different in its darker color, distinct grayish band across the chest, and the lack of yellowish at the bases of the rectrices. It is surprising to discover so distinct species in a locality so well worked as Tapia.

SUBLEGATUS FASCIATUS (Thunberg)

Pipra fasciata THUNBERG, Mém. Acad. Imp. Sci. St. Pétersbourg, vol. 8, 1822, pp. 283, 285. (Brazil.) (Reference from Brabourne and Chubb.)

This flycatcher was first recorded at Las Palmas, Chaco, where specimens were collected July 13, 27, and 30, 1920. Others were

---

[18] Nov. Zool., vol. 16, December, 1909, p. 200.
[19] *Pachyrhamphus albescens* Gould, Zool. Voy. Beagle, pt. 3, Birds, July, 1839, p. 50, pl. 14. (Buenos Aires.)
[20] Proc. U. S. Nat. Mus., vol. 25, 1902, p. 136.
.: [21] Proc. Fourth Int. Ornith. Congr., 1907, p. 442.

noted at Formosa, Formosa, August 23 and 24, and in September in the area west of Puerto Pinasco, Paraguay, they were observed at Kilometer 80, on September 10, 15, 16 (a male taken), and 20, and at Kilometer 200 (near Laguna Wall) on September 25. These birds frequented the dense branches of algarrobas and other low, thorny trees that were scattered across small, open prairies, where they hopped about under cover, avoiding the open save when flying with a tilting flight from tree to tree. At intervals the tail was jerked quickly two or three times, and occasionally one uttered a soft note, *whit*, or a more explosive sound, *whit sfee*.

A specimen taken September 30 when fresh had the bill blackish brown number 3; iris Hay's brown; tarsus and toes black.

Berlepsch and Hellmayr [22] have considered *Sublegatus glaber* Sclater and Salvin a geographic form of *fasciatus*, a contention not borne out by the scanty material at hand, since *glaber* has a distinctly broader, heavier bill that is black in color and is more distinctly uncinate at the tip. A skin seen from Santa Ana, in the Urubamba Valley, Peru (male, July 15, 1916), one of the five recorded by Chapman [23] as *Sublegatus fasciatus fasciatus*, that has a much smaller bill and is paler in color than my skins from the Chaco, is probably *Sublegatus griseocularis* Sclater and Salvin,[24] described from Maranura, a short distance below Santa Ana.

PITANGUS SULPHURATUS BOLIVIANUS (Lafresnaye)

Saurophagus bolivianus LAFRESNAYE, Rev. Mag. Zool., 1852, p. 463. (Chuquisaca, Bolivia.)

The eight specimens secured of this common bird are as follows: Female, Formosa, Formosa, August 24, 1920; female, Kilometer 80, west of Puerto Pinasco, Paraguay, September 7; a pair, Lavalle, Buenos Aires, November 13; adult and juvenile males and immature female, San Vicente, Uruguay, January 25, 26, and 28, 1921; female, Lazcano, Uruguay, February 8. The species ranged through the humid eastern pampas of Argentina and Uruguay, along the base of the Andes, and in the Chaco, north into Paraguay, but was not seen in the semiarid interior. Points at which it was recorded are as follows: Puerto Pinasco, Paraguay, September 3 and 30, 1920; Kilometer 80, September 6 to 20; Kilometer 110, September 23; Kilometer 200, September 25; Formosa, Formosa, August 5, 23, and 24; Riacho Pilaga, Formosa, August 8 to 21; Las Palmas, Chaco, July 13 to August 1; Resistencia. Chaco, July 8 to 10; Santa Fe, Santa Fe, July 4; Carrasco, Uruguay, January 9 and 16, 1921; La Paloma, Uruguay, January 23;

---

[22] Journ. für Ornith., 1905, pp. 4–5.
[23] U. S. Nat. Mus., Bull. 117, 1921, p. 96.
[24] Proc. Zool. Soc. London, 1876, p. 17. (Maranura, Peru.)

San Vicente, Uruguay, January 25 to February 2; Lazcano, Uruguay, February 3 to 9; Rio Negro, Uruguay, February 14 to 19; Quilmes, Buenos Aires, June 27, 1920; Berazategui, Buenos Aires, June 29; Dolores, Buenos Aires, October 21 and 22; Lavalle, Buenos Aires, October 27 to November 15; Carhue, Buenos Aires, December 17; Guamini, Buenos Aires, March 4, 1921; Tunuyan, Mendoza, March 23 to 29; Tapia, Tucuman, April 6 to 14; Tafi Viejo, Tucuman, April 17.

This large flycatcher, though found on the pampas, was more common at the border of forests, along fence rows, or in open brush. It was especially partial to water and was observed frequently perched on limbs or rushes that overhung small streams or marshes. It was observed often in parks in cities where tree growth was extensive, and is one of the first of the native birds to attract the attention of the traveler, its presence constantly advertised by its querulous notes. The birds usually watched for prey from some open post, and on occasion, from their intent gaze at the water, I suspected that they were on the lookout for small fishes. Occasionally they hovered in the air like kingfishers or small hawks, with the body suspended at an angle of 45° and rapidly beating wings. They were solitary save during the breeding season, when they congregated in pairs.

Their mating display, observed occasionally, was peculiar. The individual giving it, stood bolt erect with the neck perpendicular, threw the point of the bill down and exposed the flaring, colored crest directly in front, while it shook the wings rapidly and made a loud cracking sound with its bill. Nest building began in October, and during November their large nests, often with crudely formed roofs, were seen on several occasions. Domed nests, common among pampas inhabiting birds, apparently give protection from predatory animals and shelter from the heavy storms of spring. A young bird, recently from the nest, was shot at San Vicente, Uruguay, January 25.

Local names for this well-known species are given almost universally in imitation of its notes, as witness *bien-te-veo* in Spanish, *pit-o-güe* in Guaraní, and *heht aow pah* in Anguetè.

MYIODYNASTES SOLITARIUS (Vieillot)

Tyrannus solitarius VIEILLOT, Nouv. Dict. Hist. Nat., vol. 35, 1819, p. 88. (Paraguay.)

The present species, seen only in Paraguay, in the region near Puerto Pinasco, was migratory, and did not appear during spring until September 20, when a male was shot near Kilometer 80. Another was recorded near the same point on September 21, one near

Kilometer 110 on September 23, and a number at the base of the Cerro Lorito, on the eastern bank of the Rio Paraguay, September 30. The birds frequented heavy woods where they sought sheltered perches 6 to 10 meters from the ground. Though their streaked plumage made them conspicuous they were difficult to see, as they watched intruders alertly, and at any suspicious movement flew to a safer distance.

The male taken had the base of the mandible pale olive buff; remainder of bill black; iris bone brown; tarsus and toes slate gray; claws black.

MYIOPHOBUS FASCIATUS AURICEPS (Gould)

Myiobius auriceps GOULD, Zool. Voy. Beagle, pt. 3, Birds, July, 1839, p. 47. (Buenos Aires.)

At Lazcano, Uruguay, an adult male was shot February 5, 1921, and at Rio Negro, in the same country, a female (with two birds in juvenal plumage, one a female, the other with sex unknown) was taken February 15, and another female on February 17. The crown spot in the male is ochraceous orange, and in the females yellow (nearly absent in some specimens). The young are more rufescent brown above and on the wing bars than adults, washed more with brownish below, and lack the coronal patch. Though Hellmayr [25] and Dabbene [26] state that Argentine specimens of this species are not distinguishable from those of Brazil, the birds at hand substantiate Ridgway's recognition [27] of *auriceps* as a race distinct from *fasciatus* of Brazil and northern South America. The wing in five specimens from Venezuela, with sex not marked, measures from 53 to 60 mm., and in a male from Para, 56.5 mm. Three males from Buenos Aires and eastern Uruguay have the wing from 64 to 64.2 mm., and three females (two in much worn plumage) from the same region 58 to 62 mm. There seems to be a recognizable difference in size between series from Argentina and Uruguay and elsewhere.

These small flycatchers, with the habits of *Empidonax*, were found in dense, lowland thickets near streams. They chose resting places hidden among leaves, where they remained quietly on watch for insects, at short intervals twitching the tail. When food appeared it was pursued with snap and vigor, its capture announced by a click of the bill. Occasionally one uttered a plaintive note that may be rendered as *tsi bur*.

---

[25] Nov. Zool., vol. 15, June, 1908, p. 52.
[26] An. Mus. Nac. Buenos Aires, vol. 18, July 16, 1910, p. 343.
[27] Birds North Middle Amer., vol. 4, 1907, p. 543.

PYROCEPHALUS RUBINUS RUBINUS (Boddaert)

Muscicapa rubinus Boddaert, Tabl. Planch. Enl., 1783, p. 42. (Brazil.)[28]

This handsome flycatcher, common in central and northern Argentina, was recorded and collected as follows: Riacho Pilaga, Formosa, August 8 and 14, 1920 (two males taken); Formosa, Formosa, August 23 and 24 (male taken on 24th); Puerto Pinasco, Paraguay, September 3; Kilometer 80, west of Puerto Pinasco, September 9 to 21 (male September 21); Lavalle, Buenos Aires, October 24 to November 13 (pair November 13); General Roca, Rio Negro, November 27 to December 3; Victorica, Pampa, December 23 to 29 (male December 28, adult female December 24, immature female December 28); Carrasco, Uruguay, January 9 and 16, 1921; La Paloma, Uruguay, January 23; San Vicente, Uruguay, January 26 to February 2 (female January 27); Rio Negro, Uruguay, February 18.

This bird inhabited regions similar to those in which its northern representative, *mexicanus*, is found in the southwestern United States, namely, open thickets and groves of low trees, often in the vicinity of dry watercourses, where it chose low perches, frequently where a network of small limbs protected it from the sudden onslaught of bird-eating hawks. The young males carry the streaked, immature plumage until late winter following the season in which they were hatched, as is shown by three males from the Territory of Formosa, killed in August, which are in transition from this streaked phase to the brilliant plumage of the adult. The period from September, when they had attained full feather, until December constituted the mating season during which their beautiful display, in which they flew out with head erect and crest raised, and supported themselves in air with rapid beats of the wings, vivid burning spots of red that instantly attracted the eye, was seen frequently. At a distance of a few yards, during this action, a thin, steely note *tsit-tsur-ee-ee* was faintly audible. Occasionally one gave a low, crackling note like the sound made by breaking dry twigs. An immature specimen secured December 28 had only recently left the nest.

In some localities the brilliant display of the male had given this species the name of *brazita de fuego;* elsewhere it was called *churinche.*

EMPIDONAX TRAILLII TRAILLII (Audubon)

Muscicapa traillii Audubon, Birds Amer. (folio), vol. 1, 1828, pl. 45. (Woods along the prairie lands of the Arkansas River.)

An adult female, very fat, was found on the deck of the steamship *Santa Elisa* at daybreak on the morning of May 11, 1921. The bird, which belongs to the eastern race (formerly known as *alnorum*

---

[28] See Brabourne and Chubb, Birds of South America, vol. 1, 1912, p. 298.

Brewster), was supposed to have come aboard when we were opposite Cape Mala, Panama, in the Gulf of Panama.

EMPIDONAX EULERI (Cabanis)

Empidochanes Euleri CABANIS, Journ. für Ornith., 1868, p. 195. (Cantagallo, Rio de Janeiro, Brazil.)

Of six skins attributed to this species, two males were taken at a low hill 25 kilometers west of Puerto Pinasco, Paraguay, September 1, 1920, a female at San Vicente, Uruguay, January 31, 1921, another female at Lazcano, February 8, and a male at Rio Negro February 14. All are adult. Specimens from Uruguay appear darker and browner than those from the Paraguayan Chaco, due perhaps in part to their more worn condition of plumage. The status and range of this species are in considerable confusion in current literature, as formerly *euleri* was supposed to occur only in Brazil. Lillo[29] has said that, according to Hellmayr, the bird ranges into Argentina, but it does not appear to have been recorded before from Uruguay. *Empidonax argentinus* (Cabanis), of which *Empidonax brunneus* Ridgway appears to be a synonym, is said to be smaller than *euleri*, and though sometimes considered a geographic race, on the basis of one specimen seen (the type of *brunneus*), appears specifically distinct. The wing in *euleri* has the following measurements: Males (4 specimens), 64.4–66.6 mm.; females (2 specimens), 60.5–62.8 mm.

These small flycatchers frequented low brush in heavy forest, where the ground was densely shaded. Perches were chosen under shadow of growths of leaves, where the birds remained motionless, save for the rapidly twitching tail. In Uruguay they were fairly common, especially in swampy localities. In addition to those recorded above, one preserved in alcohol was taken on September 30, 1920, on the eastern shore of the Paraguay River, opposite Puerto Pinasco.

MYIARCHUS TYRANNULUS TYRANNULUS (Müller)

Muscicapa Tyrannulus MÜLLER, Natursyst., Suppl., 1776, p. 169. (Cayenne.)

The present species, marked by broad, rufescent margins on the inner webs of the rectrices, was collected only at Kilometer 25 and Kilometer 80, west of Puerto Pinasco, Paraguay, where adult males were secured September 1 and 11, 1920. In this region they were fairly common in heavy forest, where they watched for insects or hopped slowly about in the outer branches of tall trees. On September 15 several were found together, calling excitedly, apparently engaged in mating. Their usual call resembles one of the notes of *Nuttallornis borealis*, and in addition, they have a song, a rattling

---

[29] Apunt. Hist. Nat., vol. 1, No. 3, Mar. 1, 1909, p. 42.

whit whir-r-r-r whit. The species was recorded west to Kilometer 200.

MYIARCHUS SORDIDUS Todd

Myiarchus sordidus TODD, Proc. Biol. Soc. Washington, vol. 29, June 6, 1916, p. 96. (El Trompillo, Carabobo, Venezuela.)

An adult male secured near San Vicente, in extreme eastern Uruguay, January 31, 1921, is listed under *sordidus* with reservation. Todd, in his review of the genus *Myiarchus*,[30] examined this bird and marked it "*Myiarchus* sp. (near *sordidus*)." After study and comparison with Todd's revision of the genus it appears that the specimen shows the characters of darker dorsal surface that distinguish *sordidus* from *pelzelni*, and under present understanding of the group it can be listed only as *sordidus*. It may be noted that Todd[31] records *sordidus* from Rio Grande do Sul, so that eastern Uruguay is not a remarkable extension of range, especially since other south Brazilian species were obtained at the same point.

The bird was shot in heavy brush bordering the Laguna Castillos.

MYIARCHUS PELZELNI Berlepsch

Myiarchus pelzelni BERLEPSCH, Ibis, 1883, p. 139. (Bahia, Brazil.)

The present species was found near Victorica, Pampa, where an adult female was secured December 24, 1920, and an adult male and a young female December 27. The adult male is much grayer above than the female and has little yellow below. The young bird, not yet fully feathered, is yellower on the abdomen, and has rectrices and remiges margined with cinnamon.

The birds were found in heavy growth of the semiarid, open forest of low, thick-trunked trees characteristic of this region, where they were located through their low-pitched mournful whistled calls.

MYIARCHUS FEROX SWAINSONI Cabanis and Heine

Myiarchus swainsoni CABANIS and HEINE, Mus. Hein., pt. 2, 1859, p. 72. (Brazil.)

The present bird resembles *M. t. tyrannulus* superficially, but has a smaller bill and lacks the rufescent coloring in the tail. The bird was recorded at Resistencia, Chaco, July 8, 1920 (male taken); Las Palmas, Chaco, July 17 to 31 (two males July 17 to 21); and Tapia, Tucuman, April 8 to 13, 1921 (male shot April 8). They were found in open woods in fair numbers. In feeding they hopped easily about among twigs and leaves, snatching at insects, and occasionally resting stationary for a time.

---

[30] See Proc. Biol. Soc. Washington, vol. 35, Oct. 17, 1922, pp. 181–218.
[31] Idem, p. 197.

EMPIDONOMUS AURANTIO-ATROCRISTATUS (d'Orbigny and Lafresnaye)

Tyrannus aurantio-atrocristatus d'ORBIGNY and LAFRESNAYE, Mag. Zool., 1837, cl. 2, p. 45. (Valle Grande, Bolivia.)

The present species appears to be migrant in the southern part of its range, since it was not recorded until September 15, 1920, when three were found and two males taken near Kilometer 80, west of Puerto Pinasco, Paraguay. Others were noted there September 20 and 21, and the birds were seen in fair numbers September 23 at Kilometer 170, and September 25 at Kilometer 200. From December 24 to 29 the species was fairly common near Victorica, Pampa, where two males in rather worn breeding plumage were taken. One other male was shot at Rio Negro, Uruguay, on February 17, 1921. The birds frequented open, brushy areas, and where the forest was thick were encountered only at the borders of the groves. In actions they were somewhat similar to kingbirds, as they always chose perches at the tips of low branches, or at the top of small trees where they might watch for prey. Their flight, as they darted or turned swiftly in the air after insects, and then alighted with an expert flirt of their long wings, was alert and graceful. The call note of males was a low, whistling *pree-ee-ee-er*, that may be likened to the noise produced in flight by the wings of *Nothura maculosa*. At other times they uttered a series of squeaky calls that might pass for a song.

The Lengua Indians in the Paraguayan Chaco called them *snak pi tik*.

The bill, tarsi, and toes in fresh specimens were black; iris Vandyke brown.

TYRANNUS MELANCHOLICUS MELANCHOLICUS Vieillot

Tyrannus melancholicus VIEILLOT, Nouv. Dict. Hist. Nat., vol. 35, 1919, p. 84. (Paraguay.)

As in the time of Azara, this kingbird arrived in Paraguay in September, since the first one taken, a male, was secured on September 23, at Kilometer 110, west of Puerto Pinasco. Others were seen here September 26, and at Kilometer 80, September 28, while a male was taken from a perch above the Rio Paraguay, opposite Puerto Pinasco, on September 30. Near General Roca, Rio Negro, a few were noted December 3, in willows along the Rio Negro, and at Victoria, Pampa, on December 23 and 24, the species was common. Two breeding males (one prepared as a skeleton) were taken there December 23. Near San Vicente, Uruguay, from January 27 to 31, 1921, the birds frequented groves of palms, where an adult female was shot January 27. The species was found in small numbers at Lazcano, Uruguay, from February 5 to 8, and was recorded near

Tunuyan, Mendoza, March 26 to 28. It is migrant and retreats northward in winter.

These birds were found at the borders of groves where they sought commanding perches and watched for passing insects. They were stolid and inactive save when in alert pursuit of prey. Occasionally one uttered a high-pitched, trilling call, and a wing-tipped bird gave staccato cries like those of other kingbirds, but ordinarily they were silent. When not hurried, their flight was of the fluttering type, common to other kingbirds, performed with short, rapid vibrations of the partly opened wings.

The two summer skins preserved, in worn breeding plumage, in appearance are much darker than those secured in spring. One from Victorica in particular shows little greenish wash on the upper surface. A male, shot December 23, had the bill, tarsus, and toes black; iris natal brown.

The Angueté Indians in the Paraguayan Chaco called this species *T'a pah.*

MUSCIVORA TYRANNUS (Linnaeus)

Muscicapa Tyrannus LINNAEUS, Syst. Nat., ed. 12, vol. 1, 1766, p. 325. (Cayenne.)

Specimens of the fork-tailed flycatcher from Argentina appear somewhat darker on the back than the average of those from northern South America. Northern and southern forms may not be separated with certainty on the basis of material at hand, as dark birds occur in the north in the small series seen. The question is complicated by the extensive northward migration of the species from temperate areas into the Tropics.

. The species was widespread from spring until fall, and was noted as follows: Kilometer 80, west of Puerto Pinasco, Paraguay, September 6 to 27, 1920; Dolores, Buenos Aires, October 21; Lavalle, Buenos Aires, October 25 to November 15; Santo Domingo, Buenos Aires, November 16; General Roca, Rio Negro, November 30; Carhue, Buenos Aires, December 17; Victorica, Pampa, December 23 to 28; Carrasco, Uruguay, January 9 and 16, 1921; La Paloma, Uruguay, January 23; San Vicente, Uruguay, January 25 to February 2; Lazcano, Uruguay, February 3 to 9; Rio Negro, Uruguay, February 14 to 19; Franklin, Buenos Aires, March 11. The three adult males and one female taken (Kilometer 80, west of Puerto Pinasco, Paraguay, September 9, Carhue, Buenos Aires, December 17, Victorica, Pampa, December 28, and San Vicente, Uruguay, January 30) offer no striking peculiarities. Two females in juvenal plumage shot near Victorica, Pampa, December 28, are recently from the nest and have no suggestion of the long tail found in adults.

The fork-tailed flycatcher was frequent throughout the open pampas, but was most abundant where there was scattered tree growth. In habits these birds resembled kingbirds. They invariably sought perches in the open on fences, low bushes, or the tops of small trees, where they rested quietly. Though in appearance they suggested *Muscivora forficata*, they were less noisy and active. Most of their notes were flat, with little carrying power against the force of the pampan winds. The call of young recently from the nest was a low *tsip* that suggested a note of *Brachyspiza*, while adults uttered an explosive call note, somewhat flat in tone, varied by a staccato rattle when tilting among themselves or in pursuit of other birds. Hawks and other large birds were attacked viciously, and the flycatchers frequently darted out at any bird that passed too near.

Fall migration among fork-tailed flycatchers began by the first of February. On February 2, while passing through a region of rolling hills north of San Vicente, Uruguay, I recorded at least 2,000, many of them young with partly grown tails. The birds were found in small flocks, and were spread along wire fences for a distance of several miles. By February 9 they had lessened in abundance but continued common in Uruguay until February 19. Throughout this period they were obviously traveling northward. Shortly after daybreak on February 18, near Lazcano, a band of 16 individuals paused to rest for a few minutes in bushes bordering a lagoon, and then, in straggling formation, passed on to the northeast. On March 2 occasional individuals were recorded from a train, from the suburbs of Buenos Aires as far south as 25 de Mayo in the Province of Buenos Aires, but none were seen beyond that point. The last one recorded was observed from a train near Franklin, Buenos Aires, on March 11.

The species is known universally as *tijerita*.

Family PHYTOTOMIDAE

PHYTOTOMA RUTILA Vieillot

Phytotoma rutila VIEILLOT, Nouv. Dict. Hist. Nat., vol. 26, 1818, p. 64. (Paraguay.)

During winter, near Las Palmas, Chaco (July 26 and 31, 1920), this strange bird was found in small flocks in bush-grown pastures, and occasional individuals were recorded at the Riacho Pilaga, Formosa (August 14, 18, and 21). Near Las Palmas a pair of adult birds was secured July 26. The species seemed irregular in its occurrence in the Chaco region, and may have been only a winter visitant. The birds were found quietly at rest on the tops of low bushes, with crest erect, and were usually difficult to approach. The light eye was prominent at a considerable distance. The birds passed from perch to perch, with an undulating flight not far above the ground, and as they checked their momentum in order to alight spread the tail, displaying its prominent white markings. At times they hopped down among the branches to conceal themselves among dense leaves.

In the vicinity of Victorica, Pampa, from December 23 to 29, *Phytotoma rutila* was encountered in abundance on its breeding grounds, and six preserved as skins include three adult males, an adult and an immature female, and a fledgling. At this season the species ranged in wooded areas of low, heavy-trunked caldén, algarroba, and similar trees where there was abundant dense undergrowth of chañar, piquillín, and a variety of low shrubs, many with thorn-protected branches. On entering these tracts for the first time my attention was drawn at once by one of the most bizarre bird notes that has as yet come to my ears, and, hastening to trace it, I found to my astonishment that it was the song of this plant cutter. Adult males in full, handsome plumage rested on open limbs, often at the tops of low trees, and with great earnestness gave a succession of low notes that may be likened only to the drawn-out squeaking produced when two tree limbs, moved by winds, rub slowly across one another. Occasionally a more strenuous effort produced a sound like the squeaking of leather, that might terminate in a froglike croak. The whole, while ranking as a musical performance of the highest rank in the opinion of the one responsible for it, and without doubt the sweetest of music to his mate, was to human understanding ridiculous to an extreme. At intervals they flew from perch to perch with slow affected flight, performed with rapid beats of the wings, different entirely from their usual method of progression.

It is possible that two broods are raised, as one fledgling and one fully grown immature bird were taken, while adult males were in full breeding condition. The fledgling, though as yet unable to fly, retreated precipitately among the thorny limbs of a piquillín, where it was secured only with difficulty. The birds were feeding on green drupes of various sorts, and all those taken, here and elsewhere, had the sides of the bill covered with gum from plant juices.

At Potrerillos, Mendoza, an immature female was shot March 19, 1921, at an elevation of 1,800 meters.

In the specimens from the areas listed, there are no differences apparent other than those that may be considered individual. The fledgling taken, a young male, suggests the female in color pattern, save that the coloration is softer and darker, and the streaks on both dorsal and ventral surfaces more obscure. The under tail coverts are tawny olive, with obscure shaft streaks of black. Immature females in full juvenal plumage have the streaking narrower. One has a buffy suffusion on the undersurface. An adult female in winter plumage is more buffy than one shot in the breeding season. A female, taken July 26, had the iris ochraceous buff; bill andover green, shading to vetiver green at base of mandible; tarsus and toes deep neutral gray.

Family HIRUNDINIDAE

IRIDOPROCNE MEYENI (Cabanis)

Petrochelidon meyeni CABANIS, Mus. Hein., pt. 1, 1850, p. 48. (Santiago, Chile.)

At Guamini, Buenos Aires, several were seen and an immature female was taken on March 5, 1921. A dozen were recorded in company with *Pygochelidon* on March 7. Near Concon, Chile, the species was fairly common on April 27 and 28, and an adult male was collected on the date first mentioned.

As Cabanis's name for this swallow is a substitute for *Hirundo leucopyga* "Lichtenstein" of Meyen,[32] the type locality for *meyeni* must be the same as that for Meyen's *leucopyga*, that is, the city of Santiago.

IRIDOPROCNE LEUCORRHOA (Vieillot)

Hirundo leucorrhoa VIEILLOT, Nouv. Dict. Hist. Nat., vol. 14, 1817, p. 519. (Paraguay.)

Seven skins preserved of this swallow include the following: Two adult males, Kilometer 80, west of Puerto Pinasco, Paraguay, September 6 and 9, 1920; two immature males, San Vicente, Uruguay, January 27, 1921; adult male and female and immature male, Bañado de la India Muerta, 12 miles south of Lazcano, Uruguay, February 3. Specimens in the United States National Museum from the Province of Buenos Aires (collected in the sixties) have a decidedly more greenish cast above than two from Paraguay, a character in which skins from Uruguay in worn plumage seem somewhat intermediate. A definite difference possibly may be established with better series.

A male, taken September 6, had the bill black; tarsus and toes, blackish brown number 1. Two juvenile specimens recently from the nest had the gape and base of the bill yellowish.

In the Paraguayan Chaco, from September 6 to 23, these little swallows, like *Iridoprocne bicolor* in habits and appearance, were common in areas where cavities in broken palms that stood near or in shallow lagoons offered suitable nest sites. Males perched at or above old woodpecker holes or other openings, calling and lifting

---

[32] *Hirundo leucopyga* "Lichtenstein" Meyen, Nov. Act. Acad. Caes. Leop.-Carol. Nat. Curios., vol. 16, Suppl., 1834, p. 73, pl. 10, fig. 2. (In der Stadt Santiago sehr haufig.)

their wings in an endeavor to entice females to examine the proposed nest site or circled about near by. Their broken, warbling song was heard continually. Though apparently mating, the two taken were not yet in breeding condition, and two nest holes that I examined September 9 were empty. The species was recorded west to Kilometer 110. The white rump is a field mark prominent in flight. At San Vicente, Uruguay, from January 27 to February 3, families of young recently from the nest were recorded, while near Lazcano, Uruguay, from February 3 to 9, the birds had gathered in flocks which contained from 10 to 100 individuals that rested on fence wires or circled low over the fields. The birds were so abundant here that hundreds frequently were observed during a day.

HIRUNDO ERYTHROGASTRA Boddaert

Hirundo erythrogaster BODDAERT, Tabl. Planch. Enl., 1783, p. 45. (Cayenne.)

On the evening of September 24, 1920, when near the Laguna Wall, in the Chaco, 200 kilometers west of Puerto Pinasco, Paraguay, I was delighted to distinguish the graceful forms of several barn swallows among the members of a flock of *Pygochelidon* that came circling over the marshes in search of a secure resort to spend the night. The Lengua Indian who was with me distinguished the specimen that I collected, an adult female, from other swallows as *mem a sakh sa heht kil wa nah*. In the southern part of their winter range, barn swallows were not common as subsequently I saw them on only three occasions, at Puerto Pinasco, near the Rio Paraguay, September 30; near Lezama, Buenos Aires (from a train), October 19, and near Lavalle, Buenos Aires, November 15. Two were seen on each of these dates.

ALOPOCHELIDON FUCATA (Temminck)

Hirundo fucata TEMMINCK, Nouv. Rec. Planch. Col. Ois., vol. 4, 1838, pl. 161. (Brazil.)

This species was observed on August 23 and 24 near my hotel in Formosa, Formosa, under such conditions that it could not be collected. On one occasion a pair alighted on the ground to waddle about with sidling steps, picking up bits of sand. A female shot March 28, 1921, from a flock of *Pygochelidon* at Tunuyan, Mendoza, was the only one taken. On the wing this bird resembles a rough-winged swallow (*Stelgidopteryx*).

Chubb[33] has described a subspecies of this bird from Mount Roraima, British Guiana, as *A. f. roraimœ*, on the basis of brighter coloration of head and throat, paler dorsal surface, and smaller size

---

[33] Bull. Brit. Orn. Club, vol. 40, June 30, 1920, p. 155.

(wing, 96 mm.). The specimen from Mendoza has a wing measurement of 96 mm., while another from the same Province, in the United States National Museum collections (taken by Weisshaupt), measures 92.5 mm., figures that cast some doubt upon supposed difference in size in birds from southern localities.

PYGOCHELIDON CYANOLEUCA (Vieillot)

Hirundo cyanoleuca VIEILLOT, Nouv. Dict. Hist. Nat., vol. 14, 1817, p. 509. (Paraguay.)

Near Lazcano, Uruguay, the present species was recorded from February 5 to 8, 1921, and two, an immature female and an adult male, were taken February 7 and 8. The immature bird, only recently from the nest, has breast and flanks washed with buffy brown. It is possible that swallows recorded at La Paloma January 23, San Vicente January 26 and 31, and Rio Negro February 17 and 18 (all in Uruguay) were also this species.

There is considerable doubt in my mind as to which of the two closely allied *Pygochelidon* is intended by Vieillot's *Hirundo cyano-leuca*. The description in Azara of the *Golondrina de la timoneles negros* on which Vieillot's name is based may apply with almost equal propriety to either of the birds at present known as *cyanoleuca* or *patagonica*, as the present quotation from Azara [34] will show. "Longitud 4–11/12 pulgadas, y las demas medidas á proporción de la anterior [the barn swallow]. Del pico á la cola, todo el resto encima y el costado de la cabeza, son turqui. La cola, remos y cobijas, son lo mismo que en la precedente, aunque sin gotas blancas en la cola. La tira que en la mencionada se adelanta desde la raiz del ala, en la presente es parda, y termina al fin de la garganta con una manchita obscura. De la horqueta á la cola blanco, con los timoneles inferiores negros, ó casi como el lomo. La cola y remos debaxo pardos, como las tapadas; aunque las de junto al encuentro tienen ribetillos blancos." Small size (*patagonica* measures 125 mm. or more in length) is the only absolute character in the above that indicates *cyanoleuca*, as the colors described may apply to either species. It must be stated that *patagonica* is the only species that I heve seen from Paraguay since the specimen (in the United States National Museum) cited by Chapman [35] as from Paraguay in reality comes from the Rio Paraguay, in southern Brazil. The only *Pygo-chelidon* that I collected in Paraguay (taken 200 kilometers west of Puerto Pinasco) was *patagonica*. Chapman has called attention to the rarity of records for the species with dark underwing coverts in the interior lowlands of South America, and it is my impression

[34] Hist. Nat. Pax. Paraguay, vol. 2, 1805, p. 508.
[35] Am. Mus. Nov., No. 30, Feb. 28, 1922, pp. 1, 3, and 11.

that *Hirundo cyanoleuca* Vieillot will eventually be transferred to the bird now known as *P. patagonica*, while *P. cyanoleuca* auct. will become *P. minuta* (Maximilian).[36] The matter, however, is best left in abeyance pending more extended collecting in Corrientes and Paraguay.

PYGOCHELIDON PATAGONICA PATAGONICA (d'Orbigny and Lafresnaye)

Hirundo patagonica d'ORBIGNY and LAFRESNAYE, Mag. Zool., 1837, cl. 2, p. 69. (Patagonia.)

The present swallow is the most common species of the present family in Argentina. Records, based on specimens, of the occurrence of this bird are as follows: Kilometer 200, west of Puerto Pinasco, Paraguay, September 24 (male taken) and 25, 1920 (the most northern record for the species in the interior); Carhue, Buenos Aires, December 15 to 18 (immature male, December 18); Guamini, Buenos Aires, March 4 to 8, 1921 (four skins, adults and immature of both sexes); General Roca, Rio Negro, November 27 to December 3 (adult female, November 27); Zapala, Neuquen. December 8 (adult female); Potrerillos, Mendoza, March 17 to 21 (adult female, March 19).

The following records are assumed to represent this species, but identification was not checked by the collection of skins; Puerto Pinasco, Paraguay, September 3; Formosa, Formosa, August 24; Riacho Pilaga, Formosa, August 9 to 21; Las Palmas, Chaco, July 14 to 31; Dolores, Buenos Aires, October 21, Lavalle, Buenos Aires, October 23 to November 13; Ingeniero White (near Bahia Blanca), Buenos Aires, December 13; Tunuyan, Mendoza, March 22 to 29. A northern subspecies of this bird from Peru, *P. p. peruviana* Chapman, differs from the typical form in smaller size, and paler under wing coverts. It is possible that *P. flavipes* Chapman,[37] based on a single skin from Maraynioc, Peru, represents merely an immature phase of *cyanoleuca*. An immature female, *cyanoleuca*, from Matchu Picchu, Peru, shot June 25, 1915, (Cat. No. 273,300 U. S. N. M.,) has the feet and tarsi decidedly yellowish brown.

Though it has been stated that the light external margin of the outer rectrices is found in *patagonica* alone, I have noted it in slight degree in skins of *cyanoleuca* from Costa Rica, Colombia, Peru, and Uruguay.

During winter, in the Chaco, these little swallows were found in flocks about lagoons where they often rested on tufts of grass in the water, or after hawking about for insects in the dusk sought roosts among the rushes. One shot, September 24, was fat and in

---

[36] *Hirundo minuta* Maximilian, Reis. Brasilien, vol. 2, p. 336.   (Rio de Janeiro, Brazil.)
[37] Am. Mus., Nov., no. 30, Feb. 28, 1922, p. 8.

good plumage in marked contrast to the worn feathers and thin body of a newly arrived migrant barn swallow from North America, taken at the same time. During summer, *P. patagonica* was found through the pampas in pairs about cut banks near water, and was especially common along the Rio Negro in northern Patagonia. Though small and light in body, so that the birds blew about in the wind, they were able at will to breast the strongest blasts. Their nests were placed in little tunnels excavated in the sides of banks. Males sang a low and rather squeaky song at intervals, but on the whole the species was silent. On March 5 little flocks were common on the open pampa at Guamini, and by March 7 the number of those present was greatly increased, apparently by migrants driven by colder weather from the south. The birds, many of them immature, hawked for food along the lake shore, and, when tired of buffeting the constant wind, settled in little open spaces on the ground. Occasionally a few joined *Iridiprocne meyeni* at rest on the wires of a fence, but seldom did one pause on the higher telephone wires frequented by their companions of larger size.

PHAEOPROGNE TAPERA TAPERA (Linnaeus)

Hirundo Tapera LINNAEUS, Syst. Nat., ed. 12, vol. 1, 1766, p. 345. (Brazil.)

This martin, in color and marking a larger counterpart of the bank swallow, is migrant in the southern part of its range, as it was not seen until September 17, 1920, when it was recorded at Kilometer 80, west of Puerto Pinasco, Paraguay. It was observed from then until September 30, and on October 21 was already present at Dolores, Buenos Aires. The species was noted at Lavalle, Buenos Aires, from October 25 to November 13, and at the following points in Uruguay; Carrasco, January 9 and 16; Montevideo (in the Prado), January 14; La Paloma, January 23; San Vicente, January 27 to 31; Lazcano, February 5 to 9, and Rio Negro, February 15 to 18. The five skins taken include two males from Kilometer 80, Puerto Pinasco, a pair from Lavalle, and an adult female from San Vicente. These are similar to one another and all possess the dark spots on the median undersurface that are lacking in *P. t. immaculata* Chapman [38] from Colombia and Venezuela.

A male, shot September 18, had the bill, tarsus, and toes black; iris bone brown.

This martin was encountered among dead trees in open woods or groves, and in the north was especially common among groves of palms. The birds appear weaker in flight than most swallows, and pause frequently to rest on dead limbs after short circling flights.

---

[38] Bull. Amer. Mus. Nat. Hist., vol. 31, July 23, 1912, p. 156. (Chicoral, near Giradot, Alt. 550 meters, Tolima, Colombia.)

Their call is a flat *chu chu chup* that has little carrying power. During the breeding season males often circled about with stiffly held, decurved wings that formed an inverted Λ. They. nested frequently in the mud nests of the *hornero* (*Furnarius rufus*).

PROGNE CHALYBEA DOMESTICA (Vieillot)

Hirundo domestica VIEILLOT, Nouv. Dict. Hist. Nat., vol. 14, 1817, p. 520. (Paraguay and Rio de la Plata.)

In winter the present species was found in the central and northern portion of the Chaco, while in summer it was more widely spread. It was recorded as follows: Formosa, Formosa, August 23 to 24, 1920; Puerto Pinasco, Paraguay, September 1 to 3, (adult male taken September 3); Kilometer 80, west of Puerto Pinasco, September 6 to 20 (female taken September 19); Buenos Aires, Argentina (on the Avenida de Mayo in the heart of the city), October 17; Dolores, Buenos Aires, October 21; Lavalle, Buenos Aires, October 26 to November 10 (male shot October 31); Santo Domingo, Buenos Aires, November 17 (male taken); Carrasco, Uruguay, January 9 and 16, 1921; La Paloma, Uruguay, January 23; Lazcano, Uruguay, February 5 to 8; Corrales, Uruguay, February 10; Rio Negro, Uruguay, February 15; Mendoza, Mendoza, March 14. This form differs from typical *chalybea* in larger size and more extensive whitish tips on the feathers of throat and upper breast.

In notes and actions similar to *Progne subis*, these martins were fairly common about many of the towns that I visited. In the pampan region they nested in crevices and openings about roofs and cornices of houses, while in the north it was common to see them about openings in palm trees. In Paraguay, birds were observed examining nest sites on September 3, while a male taken at Lavalle, Buenos Aires, on October 31, was in full breeding condition. On February 10, at Corrales, eastern Uruguay, 50, including many young, had gathered in a flock that rested on telephone wires. Males during the breeding season were active in the pursuit of carranchos (*Polyborus*) and other hawks.

PROGNE ELEGANS Baird

Progne elegans BAIRD, Rev. Amer. Birds, May, 1865, p. 275. (Bermejo River, Argentina.)

The present species was found at General Roca, Rio Negro, on November 24, 1920, when two adult males were taken along the Rio Negro, and again on November 27, when it was found in town as well as in the country. An adult female that I shot fell in the river and was swept away in the swift current. On December 19, at Carhue, Buenos Aires, a pair examined crevices among the rafters of a covered passage at the hotel where I was stopping, and on the following morning the male martin threw two young *Passer domesticus* from a nest near his chosen site, in spite of the protests of the adult sparrows. As there were many openings suited for nesting still unoccupied, this act must be attributed to a wanton meanness of disposition. At Victorica, Pampa, the fork-tailed martin was fairly common in town from December 23 to 29. The last seen were three noted on March 31, 1921, as my train stopped at the station of Monte Ralo, Cordoba.

In general appearance, action, and calls, the male of this species is similar to that of *Progne subis*, but is marked when on the wing by its longer, more deeply forked tail. The female is entirely dark underneath.

Mr. Todd[39] has called attention to the fact that *Progne elegans* Baird (based on an immature male) is the same as *P. furcata* Baird, the name current for this martin for many years, and must replace it, as the name *elegans* occurs on an earlier page of the same work in which Baird described *furcata*.

Family TROGLODYTIDAE

TROGLODYTES MUSCULUS MUSCULUS Naumann

Troglodytes musculus NAUMANN, Vög. Deutschl., vol. 3, 1823, p. 724, (Bahia.)

The series of house wrens taken during my work in South America has been studied by Chapman and Griscom during their revision of *Troglodytes musculus* and identifications of specimens are theirs. The only skin of typical *musculus* is an adult male taken at Kilometer 80, west of Puerto Pinasco, Paraguay, on September 11, 1920. The birds at that point were common at the borders of forest and came familiarly about the ranch buildings, where they sang from the doorways and searched for food among the split palm trunks that formed the roofs. It is assumed that this form was the one recorded on the Rio Paraguay, at Puerto Pinasco, and possibly the one seen in the Chaco near Laguna Wall, 200 kilometers west.

TROGLODYTES MUSCULUS REX (Berlepsch and Leverkühn)

Troglodytes furvus var. *rex* BERLEPSCH and LEVERKÜHN, Ornis, vol. 6, 1890, p. 6. (Samaipata, Bolivia.)

A small series of wrens from the Chaco have been identified as intermediate between *rex* and *musculus*, but nearer *rex*. These include the following localities: Resistencia, Chaco, July 9 and 10, 1920 (adult female taken); Las Palmas, Chaco, July 13 to 31 (two

---

[39] Auk, 1925, pp. 276–277.

pairs collected) ; and Riacho Pilaga, Formosa, August 7 to 18 (adult male shot). It is supposed that those recorded at Formosa, Formosa, August 23; Tapia, Tucuman, April 6 to 13; and Tafi Viejo, April 17, were of this same race.

The southern house wren, of whatever race, in action and general appearance is the same busy bird, full of life and energy, that greets us in our northern dooryards, and is one of the first species to be recognized on arrival in unfamiliar southern scenes. Notes and actions are unmistakably those of a house wren, and even the bubbling song is not noticeably different. In the Chaco they were found in tangles of brush bordering thickets or in clumps of grass at the borders of savannas from which they darted back into the brush when alarmed. They were in full song in June and July, and sang occasionally during April when they were in molt. They were seen on the Sierra San Xavier, above Tafi Viejo, Tucuman, from the base to an elevation of 1,800 meters.

TROGLODYTES MUSCULUS BONARIAE Hellmayr

Troglodytes musculus bonariae HELLMAYR, Anz. Ornith. Ges. Bayern, no. 1, Feb. 25, 1919, p. 2. (La Plata, Buenos Aires.)

The pampan house wren was found at Berazategui, Buenos Aires, June 29, 1920 (immature female taken); Dolores, Buenos Aires, October 21; Lavalle, Buenos Aires, October 23 to November 13 (adult male taken); Montevideo, Uruguay, January 9, 1921 (seen in the city); Carrasco, Uruguay, January 16; La Paloma, January 23; San Vicente, January 25 to 31 (adult male collected); Lazcano, Uruguay, February 5 to 7 (a pair); Rio Negro, Uruguay, February 14 to 19 (four males, adult and immature shot). One that I saw March 3 near Guamini, Buenos Aires, was perhaps this same subspecies but may have been *chilensis*.

A nest found February 7, 1921, near Lazcano, Rocha, was concealed in a hollow in a fence post standing near a thicket. The birds used a crack in one side as an entrance, and had constructed a slight cup of feathers and fine grasses lined with horsehair that contained four eggs with incubation far advanced. These eggs have the ground color grayish white, finely and uniformly spotted throughout with vinaceous, corinthian red and brick red. They measure 17 by 13.8; 16.8 by 13.8; 16.4 by 13.8; and 16.1 by 13.4 mm.

TROGLODYTES MUSCULUS CHILENSIS Lesson

Troglodytes chilensis LESSON, Voy. Coquille, Zool., vol. 1, pt. 2, April, 1830, p. 665. (Concepcion, Chile.)

Under this name is grouped a series of specimens from the following localities; Victorica, Pampa, December 23 to 29 (a pair taken); Mendoza, Mendoza, March 13; Potrerillos, Mendoza, March 15 to 21 (five immature specimens); Tunuyan, Mendoza, March 22 to 29 (immature female); Concon, Chile, April 24 to 27 (two). House wrens seen at General Roca, Rio Negro, November 23 to December 3, may also have been this form.

CISTOTHORUS PLATENSIS PLATENSIS (Latham)

Sylvia platensis LATHAM, Index Orn., vol. 2, 1790, p. 548.   (Buenos Aires.)

At Lavalle, Buenos Aires, four adult males were taken October 23, 1920, while at Tunuyan, Mendoza, a series of 10, all immature (both sexes represented), was secured between March 23 and 28, 1921. Hellmayr,[40] in a review of this species, considers that typical *platensis* ranges from Bahia Blanca north to Santa Elena, Entre Rios, and west to Mendoza. Of the present series the skins from near the mouth of the Rio Ajo, at Lavalle, may be considered as topotypical of Latham's *platensis* from "Bonaria." The skins from the Province of Mendoza are different, but as all are young must be allotted to *platensis* until adult specimens may be examined. All of the birds from Tunuyan appear slightly more heavily streaked above, and have the dark tail bars broader than skins from Lavalle. Half of those seen have the rump plain brownish and the streaks on the crown nearly obsolete. In the remainder the head is distinctly lineated; the back markings are heavier and extend down over the rump. In a way these specimens appear intermediate between *C. p. hornensis*, which is very heavily marked above from neck to upper tail coverts, and is strongly rufescent, and *platensis* but are nearer the latter. It is probable that *Cistothorus fasciolatus* Burmeister[41] may prove a valid subspecies.

At Lavalle, Buenos Aires, I found a small colony of these marsh wrens in low growths of dead rushes at the border of a tidal marsh, where attention was attracted by their tinkling songs *tu-tu-tu tee-tee-tee ter-ter-ter tsee-ee-ee-ee*, each triplet being pitched in a slightly different key, while the whole terminated in a metallic music-box rattle. As the birds sang from the tops of rushes they were easily located by their light breasts, but as I approached they dropped down into heavy cover which they often refused to leave in spite of various attractive noises, from the usual squeak to the low clinking of a brass shell against a gun barrel, made to excite their curiosity. Their flight was undulating, often at a height of 1½ to 2 meters above the marsh. Near Tunuyan, Mendoza marsh wrens were en-

---

[40] Nov. Zool., vol. 28, September, 1921, p. 250.
[41] Journ. für Ornith., 1860, p. 252.   (Mendoza.)

countered in growths of heavy weeds along irrigation ditches, in small marshes, or in corn fields, where they sang and chattered or worked quietly about. At this season it was easy to draw them into sight, as at a squeak they came clambering and hopping out through the dense growth that sheltered them until often they were scarcely a meter away. All were in molt into first fall plumage. Several of those secured were badly infested about the anus with larvae of a parasitic fly, which, however, seemed to cause them no inconvenience.

An adult male, when fresh, had the maxilla dull black; base of mandible vinaceous buff, with the tip washed with quaker drab; iris natal brown; tarsus and toes wood brown; claws fuscous.

Family MIMIDAE

MIMUS TRIURUS (Vieillot)

Turdus triurus VIEILLOT, Nouv. Dict. Hist. Nat., vol. 20, 1818, p. 275. (Paraguay.)

In the series of 11 skins preserved there seems to be no constant difference between specimens from northern Patagonia, Pampa, Mendoza, and Paraguay. The banded mocking-bird was collected and observed as follows: Santa Fe, Santa Fe, July 4, 1920; Resistencia, Chaco, July 10; Las Palmas, Chaco, July 13 to August 1 (male July 15, female July 27); Riacho Pilaga, Formosa, August 13 to 21 (female August 18); Formosa, Formosa, August 23 and 24; Puerto Pinasco, Paraguay, September 3; Kilometer 80, west of Puerto Pinasco, September 6 to 20 (males September 7 and 15); Kilometer 110, September 23, and Kilometer 200, September 25 (west of the same point); General Roca, Rio Negro, November 23 to December 3 (female November 24); Carhue, Buenos Aires, December 19; Victorica, Pampa, December 23 to 29 (adult male and two females December 26, immature male, December 29); Tunuyan, Mendoza, March 24 and 27, 1921 (immature female, March 27); Tapia, Tucuman, April 12.

Birds taken at the end of December were in much worn plumage. A young male in juvenal plumage, taken December 29, has the breast obscurely mottled with grayish brown, but is otherwise similar to adults. An immature female, shot March 27, has not quite completed the post-juvenal molt, and in fresh plumage is darker and richer in color than others examined. Some skins from Victorica, Pampa, and the one from Mendoza, have the bill slightly heavier than in specimens seen from more eastern localities.

In general appearance and habits the banded mocker is similar to *Mimus polyglottos*. It inhabits dense growths of low brush, though attracted frequently about farmhouses or courtyards. It was common in the Chaco, was less numerous on the pampas because of lack of suitable coverts, and was found in fair numbers on the Rio Negro of Patagonia, where it finds its southern limit. It reached its greatest abundance in the dry, thorny scrubs of the Pampa Central, where, near Victorica, it was one of the common birds. It appears to be sedentary in habits, and individuals may be observed in the same vicinity day after day perched above some dense tangle of brush. During winter it indulged in occasional snatches of song, but with the coming of spring gave utterance to the melody that caused such pleasure and admiration in Hudson and occasioned his eulogy of its vocal powers. The song strongly resembles that of *Mimus polyglottos*, and is accompanied frequently by aerial gyrations, in which the birds spring into the air and support themselves with slow beats of the widely opened wings, that, with the spread tail, display their contrasted markings to the utmost. They are excellent mimics, a frequent imitation being that of the low *chur-r-r-ri* of the vermilion flycatcher. In ordinary flight the extensive white markings of the tail, with a flash of white in the wings, are prominent characteristics, while in proper light the brown coloration of the back may be distinguished.

The Angueté Indians called this species *pihn mukh*.

A nest found December 26 near Victorica, Pampa, was placed 2 meters from the ground in a small shrub near the center of an open thicket. The structure was composed of thorny branches, lined with various soft materials, while a bulwark of coarse, very spiny twigs, erected to a height of 50 mm., protected the rim on all sides. The nest contained six eggs, heavily incubated, of which one is that of *Molothrus bonariensis*. Four of the others, unquestionably those of *Mimus triurus*, are pale Niagara green, spotted throughout with a color varying from walnut to Rood's brown, varied with occasional markings of dull lavender. The spots are most numerous at the larger end. These four measure: 26.2 by 18.3; 25.7 by 18.6; 25.5 by 18.2; and 25.2 by 18.7 mm. The sixth egg has the background dull white, with a faint greenish tinge, and is blotched boldly with hazel and chestnut brown, with a few markings of light plumbago gray. It is distinctly different in type from the four described above but may be an aberrant egg of *triurus*. It measures 24.6 by 19.1 mm.

Young mockers, seen at the end of December, rested quietly on open shaded perches, but hopped alertly to cover among heavy branches when at all alarmed.

MIMUS SATURNINUS MODULATOR (Gould)

Orpheus modulator GOULD, Proc. Zool. Soc. London, 1836, p. 6. (Mouth of Rio de la Plata, Montevideo, and Maldonado, Uruguay.)[42]

Six skins preserved include a pair from Lavalle, Buenos Aires, shot November 13, 1920; a pair from San Vicente, Uruguay, secured January 25 and 27, 1921; and immature male and adult female from Rio Negro, February 18 and 19. The two from Lavalle (wing, male, 120 mm.; female 117.8 mm.) are in breeding plumage. Adults from San Vicente (wing, male, 121 mm.; female 110 mm.) that, from their locality, may be considered typical, are considerably worn but resemble in color those from Buenos Aires. An adult female from Rio Negro is in full molt, as is an immature male from the same locality (wing, 121.5 mm.).

The present form was common in tala woods near the town of Lavalle, but was local as it was not noted at the Estancia Los Yngleses, a few miles distant. In Uruguay it was common in brushy regions. Young recently from the nest were seen at La Paloma, January 23.

MIMUS SATURNINUS CALANDRIA (d'Orbigny and Lafresnaye)

Orpheus calandria D'ORBIGNY and LAFRESNAYE, Mag. Zool., 1837, cl. 2, p. 17. (Corrientes, Argentina.)

Two adult males from the Riacho Pilaga, Formosa, shot August 7, 1920, have a wing measurement, respectively, of 105.2 and 109.6 mm. These are dark in color on the back, like *M. s. modulator*, and differ from that form only in smaller size. A male taken September 25 at Kilometer 200, west of Puerto Pinasco, Paraguay (wing 109 mm.), is somewhat browner above, indicating perhaps an approach to *M. s. frater*,[43] but is nearer to *calandria*.

The present form was a bird of the open that was recorded in the Chaco upon comparatively few occasions. Near the Riacho Pilaga I found half a dozen feeding in an old cornfield, or on another occasion noted several running about on open ground. When flushed they uttered a sharp, scolding *check check*. Occasionally they were heard giving a mockerlike song, though the season was winter. At times, in singing, one sprang into the air to hover for a few seconds with slowly vibrating wings. Several observed in company at the border of a marsh in the Paraguayan Chaco flew immediately to heavy cover.

---

[42] See Hellmayr, Nov. Zool., vol. 21, February, 1914, p. 159. Type-locality given erroneously by Gould as Straits of Magellan.

[43] *Mimus saturninus frater* Hellmayr, Verh. K. K. Zool.-bot. Ges. Vienna, May 22, 1903, p. 220. (Ypanema, Sao Paulo, Brazil.)

MIMUS PATAGONICUS PATAGONICUS (d'Orbigny and Lafresnaye)

Orpheus patagonicus d'ORBIGNY and LAFRESNAYE, Mag. Zool., 1837, cl. 2, p. 19. (Patagonia.)

Two males and a female of the Patagonian mocking bird, in slightly worn breeding dress, secured at General Roca, Rio Negro, November 24 and 29, 1920, show the normal development of grayish-brown ventral surface and cinnamon-buff flanks that characterize the species. One when first killed had the bill and tarsus black; iris buffy brown.

This mocker was common in growths of atriplex (*Atriplex lampa* and *A. crenatifolia*), creosote bush, and greasewoods on the flood plain of the Rio Negro, near General Roca, and proved to be a true desert form since it spread out through the arid, gravel hills north of the railroad, where water was wholly lacking. On December 6 it was recorded from the train near Challaco, Neuquen (approximately 100 kilometers west of the town of Neuquen), but was not found at Zapala.

These mockers watched alertly from the tops of bushes, or flew ahead of me, showing a band of white, interrupted in the middle, at the end of the tail. Their song, mockerlike in type, delivered from some low perch, consisting of many broken phrases interspersed with frequent imitations of the notes of other brush birds, was similar to that of *M. triurus*, though the performers were less flamboyant in actions during delivery.

MIMUS PATAGONICUS TRICOSUS Wetmore and Peters

Mimus patagonicus tricosus WETMORE and PETERS, Proc. Biol. Soc. Washington, vol. 36, May 1, 1923, p. 145. (Lujan de Cuyo, Province of Mendoza, Mendoza.)

The present form is separated from typical *patagonicus* by its grayer dorsal surface, a difference most evident in birds in fall and winter plumage.

Two were observed near an old puesto above the city of Mendoza, on March 15; and in the foothills of the Andes near Potrerillos, Mendoza, the birds were common from March 15 to 19. An immature male, taken at El Salto, at an elevation of 1,800 meters, on March 19, is in post-juvenal molt.

MIMUS THENCA (Molina)

Turdus Thenca MOLINA, Sagg. Stor. Nat. Chili, 1782, p. 250. (Chile.)

Near Concon, Chile, *M. thenca* was common from April 24 to 28, 1921, and two specimens were taken. This species seems closely allied to *M. longicaudatus* Tschudi [44] of Peru, which may prove sub-

---

[44] Arch. für Naturg., 1844, p. 280. (Peru.)

specifically related to it. A male, when fresh, had the bill, tarsus, and toes black, the iris buffy olive. Known locally as *thenca*, this mocker was common through open scrub that covered large areas of rolling hills. In general the bird is reminiscent of *Mimus patagoni-cus*. It delights in resting quietly on the tops of low trees where it has a commanding outlook, but at any danger may dive into safe cover below. The black moustache and broad white superciliary stripe are characteristic markings, while in flight the white tips of the rectrices show plainly. Though the season was fall these mockers sang more or less frequently, at intervals imitating other birds, but giving many notes that appeared to be their own, that, while distinctly mockerlike, differed from any that I had heard previously.

DONACOBIUS ATRICAPILLUS ATRICAPILLUS (Linnaeus)

Turdus atricapillus LINNAEUS, Syst. Nat., ed. 12, vol. 1, 1766, p. 295. (Brazil.)

Two females, killed September 30, 1920, at Puerto Pinasco, Paraguay, were preserved as skins, and a third bird, shot at the same time, was placed in alcohol.

Griscom [45] has recognized a northern race of this bird as *D. a. brachypterus* Madarász, with a range from north central and northern Colombia to eastern Panama, on the basis of lighter coloration above.

A female, when first killed, had the bill black, except for a spot of pale Medici blue at base of gonys; iris apricot yellow; tarsus and toes fuscous; bare skin on side of neck light cadmium.

While passing in a narrow dugout canoe through the masses of floating water hyacinth and grass, called *camalote*, that covered great areas in the less rapid stretches of the Rio Paraguay, a bird showing considerable white in the tail, dark above and buff below, flew up, with a curious scolding note, to a perch on a grass stem. I shot it at long range and after pushing in to it, with some trouble, was astonished to find a *Donacobius*, a species that I had associated mentally with the brushy haunts of thrashers and catbirds. At once I told my Indian—who knew the bird as *guira pecobá* (pecobá being banana) and said that they were found in banana plantations—that we must secure others. After some search another popped out to rest with hanging tail on a low perch, and as I shot this one half a dozen more came into sight around us, flying for short distances with tilting flight or perching near by with twitching wings and tail. The plant growth in which they were found was luxuriant and extensive, and the birds lived with all the seclusion of marsh wrens.

---

[45] Auk, 1923, pp. 215–217.

Family TURDIDAE

TURDUS RUFIVENTRIS RUFIVENTRIS Vieillot

Turdus rufiventris, VIEILLOT, Nouv. Dict. Hist. Nat., vol. 20, 1818, p. 226. (Brazil.[46])

The rufous-bellied robin was recorded as follows: Kilometer 25, west of Puerto Pinasco, Paraguay, September 1, 1920; Kilometer 80, west of Puerto Pinasto, Paraguay, September 13 (adult male); Villa Concepcion, Paraguay, October 3; Formosa, Formosa, August 23 and 24; Las Palmas, Chaco, July 15 to 30 (adult male, July 30); Resistencia, Chaco, July 8 to 10 (adult male, July 8); Tafi Viejo, Tucuman, April 17, 1921 (female); Lazcano, Uruguay, February 5 to 8; San Vicente, Uruguay, January 25 to 31 (young males, adult females); La Paloma, Uruguay, January 23; Berazategui, Buenos Aires, January 29, 1920; Lavalle, Buenos Aires, October 27 to November 13 (pair, October 30).

The nine skins preserved offer no differences save those due to wear or change in plumage. A male killed January 30 is in fully developed juvenal plumage. Adult females taken January 25 are badly worn and have initiated the post breeding molt in wing and tail, while from that date on birds were in the ragged condition common the world over to robins in fall.

Cory has described a subspecies *juensis*[47] on the basis of paler dorsal and posterior ventral regions. I have seen no skins from Ceara, the type-locality, but a specimen ·from Bahia seems somewhat paler than those from farther south.

A male, shot July 30, had the tomia olive ocher, becoming light yellowish olive on mandible and yellowish olive at base of maxilla; bare eyelid yellow ocher, becoming light yellowish olive toward outer margin, forming a prominent light eyering; iris May's brown; tarsus and toes neutral gray.

The *zorzal colorado* (or red thrush) was an inhabitant of thickets or semiopen forests often in the vicinity of water, though in northern Buenos Aires it was encountered in the dry, open groves on the larger estancias. The birds were seen frequently on the ground in forest, or occasionally at the open borders of lagoons, but at any alarm darted to cover. On large estates, as at Los Yngleses, the Gibson estancia near Lavalle, they came out familiarly on the lawns. In notes, appearance, and habits they were strongly suggestive of the robins of North America. Their song was a sweet, broken warble, given from a resting place behind leafy branches in the top of

---

[46] Brabourne and Chubb, Birds South Amer., vol. 1, December, 1912, p. 344, cite " Brazil=Rio."

[47] *Planestious rufiventer juensis* Cory, Field Mus. Nat. Hist., Orn. ser., vol. 1, no. 10, Aug. 30, 1916, p. 344. (Jua, near Iguatu, Ceara, Brazil.)

a low tree. The song period began, in Paraguay, the first of October, and extended to the time of molt in February. Their usual call note was a low *pup pup* varied with several louder, laughing calls that were heard especially toward dusk. On a fall evening in April, as J. L. Peters and I descended a trail on the slopes of the Sierra San Xavier, in Tucuman, robins called and answered on every side from the heavily forested slopes. In spite of their retiring habits the birds were curious, so that it was easy to call them out and shoot what specimens were wanted. On several occasions they were observed feeding on wild fruits, once (at Resistencia, Chaco) on the berries of *Rapanea laetevirens.* The species is one that is hunted to some extent and will need protection to maintain it in its present abundance.

A nest, found October 30 (at Los Yngleses), was placed in a tala tree 4 feet from the ground, where several small shoots projecting from the side of the trunk (which was 12 inches in diameter) furnished a firm support. The nest was made of the dried stalks of weeds mixed with a small quantity of fresh green material, and was lined with rootlets. A rim made of cow dung ran part around, but there was no complete cup of such material. This nest contained three eggs of the thrush, and three of those of the common cowbird (*Molothrus b. bonariensis*). The thrushes' eggs have the ground color much paler than pale Niagara green, blotched and spotted with brick red and chestnut brown. Two of the eggs are boldly marked over the entire surface. The other has small, scattered spots throughout, with heavy markings at the larger pole. These eggs measure 28.7 by 21.4; 28.7 by 4; and 28.5 by 20.9 mm. Another nest, examined November 10, that contained one egg, had a solid cup of hardened earth that contained the nest lining.

TURDUS MAGELLANICUS PEMBERTONI Wetmore

Turdus magellanicus pembertoni WETMORE, Univ. California Publ. Zool., vol. 21, no. 12, June 16, 1923, p. 335. (Cerro Anecon Grande, Rio Negro, Argentina.)

The present form is distinguished from *T. m. magellanicus* by grayer coloration both above and below, a distinction easily evident when series are compared. Hellmayr [48] considers *magellanicus* as subspecifically allied to *T. falcklandii* from the Falkland Islands, but, though the two are evidently of the same stock, difference between them, in my opinion, is sufficiently great to warrant their specific separation.

This bird, which might with propriety be known as the willow robin, was fairly common in the groves of large willows bordering

[48] Nov. Zool., vol. 28, September, 1921, p. 238.

the Rio Negro, below General Roca, Rio Negro, where an adult male was shot November 27 and others were seen November 30. It resembled other related species in actions and in high-pitched call notes, but was shy and retiring. A fully grown young in spotted plumage was recorded November 27.

The present species, in addition to the black crown, is distinguished from *T. rufiventris* by the pale color of tarsi and feet.

TURDUS ALBICOLLIS Vieillot

Turdus albicollis VIEILLOT, Nouv. Dict. Hist. Nat., vol. 20, 1818, p. 227. (Brazil.)

An adult male was taken September 30, 1920, opposite Puerto Pinasco on the eastern bank of the Rio Paraguay, where the birds were fairly common on the wooded slopes of the Cerro Lorito. At any alarm they flew in to view the cause, and perched with jerking wings and tail, while they called with a robinlike *pimp*. When one chanced to spy me, however, it dropped in the dense, low cover at once and was lost to view. The one taken had the maxilla and tip of mandible dull black; base of mandible primuline yellow; iris bone brown; bare eyelid and gape yellow ocher; tarsus and toes fuscous. The wing measures 105.2 mm.

Chubb [49] has described a Paraguayan form of this thrush as *Merula albicollis paraguayensis*, from skins secured at Sapucay, Paraguay, which is said to "differ from the true *M. albicollis* Vieill. in being olive brown above instead of rufous brown, while the gray band across the throat is paler and narrower, and the white on the middle of the abdomen more extended, imparting a whiter appearance." My specimen when compared with a single skin from Brazil of uncertain locality does not bear out these alleged differences, though they may prove evident when suitable material is examined.

TURDUS AMAUROCHALINUS Cabanis

Turdus amaurochalinus CABANIS, Mus. Hein., pt. 1, 1850, p. 5. (Brazil.)

This robin was recorded and collected as follows: Resistencia, Chaco, July 10 (two males); Las Palmas, Chaco, July 13 to 31 (adult female, July 30); Riacho Pilaga, Formosa, August 6 to 18 (adult female, August 18); Formosa, Formosa, August 23 and 24; Cerro Lorito, opposite Puerto Pinasco, Paraguay, September 30 (adult female); San Vicente, Uruguay, January 30, 1921 (juvenile male); Lazcano, Uruguay, February 3 to 8 (juvenile male, February 3); Rio Negro, Uruguay, February 17 (adult female); Tapia, Tucuman, April 13 (immature male and adult female). Specimens in juvenal plumage resemble adults in general appearance, but are

---

[49] Ibis, 1910, p. 608.

heavily spotted below and more or less streaked with whitish and buff above. Immature specimens in first winter plumage sometimes have small cinnamon tips on some of the greater coverts. An adult female, taken February 17, is in full molt. The bill, in adult males at least, becomes yellow at the approach of spring, but in young males is dark. Females acquire yellow on part of the lower mandible, but none were recorded with the bill entirely yellow. An immature male, taken July 10, has the bill fuscous tinged with lighter brown, becoming blacker at the base of the culmen; iris Hay's brown; tarsus slate gray. An adult male, killed on the same date, had the bill chamois, slightly darker about the nostril; tarsus light drab. The present species does not have the colored skin about the eye found in *T. albicollis*, which also differs in having a rufescent wash on the sides.

These robins frequented thickets and growths of heavy timber where they remained well hidden save when human or other dangerous intruders were not known to be near. In heavy forest I found them working about deadfalls, or less frequently found a little flock running about on some grass plot where surrounding bushes afforded a screen that gave them some sense of security. Flocks also gathered to feed on the drupes of fruit-bearing trees, such as *Rapanea laetevirens* and others. At the slightest alarm all dived precipitately into the brush and were lost to view. When not afraid they ran about with wings and tail jerking jauntily like common robins, but when startled disappeared with all the furtiveness of the smaller thrushes. Their flight is direct and fairly rapid.

Specimens from Tapia, Tucuman, appear whiter below than birds from farther east and south. In the series at hand successful division into geographic races may not be made.

TURDUS ANTHRACINUS Burmeister

Turdus anthracinus BURMEISTER, Journ. für Ornith., 1858, p. 159. (Mendoza.)

Semimerula Sclater proposed for this and allied species does not seem sufficiently distinct to merit generic rank.[50] As Selby[51] has designated the European blackbird as the type of *Turdus*, this name must supplant *Planesticus*, an observation made first by Dr. C. W. Richmond and recorded by Oberholser.[52] Since Hellmayr[53] has shown that *Turdus fuscater* d'Orbigny and Lafresnaye, the name previously applied to the species under discussion here, is in reality the large northern bird formerly called *gigas*, old friends of the

[50] See Ridgway, Birds North and Middle America, vol. 4, 1907, p. 90.
[51] Ill. British Orn., vol. 1, pt. 1, Land Birds, 1825, p. xxix.
[52] Proc. Biol. Soc. Washington, vol. 34, June 30, 1921, p. 105.
[53] Nov. Zool., vol. 28, September, 1921, p. 236.

zorzal negro of the western Argentine will recognize it under its present appellation only with difficulty. These black robins were encountered in the vicinity of Potrerillos, Mendoza, from March 15 to 21, 1921, where seven were taken, all, except one adult female, immature. The species was recorded to an elevation of 1,800 meters at an estancia known as El Salto.

A male in the spotted and streaked juvenal plumage, shot March 15, had the bill, in general, dull black, shading to onionskin pink at tip; gape between ochraceous buff and ochraceous orange; eye ring chamois; iris light drab; tarsus ochraceous buff, with a tinge of gray; nails dull black. A male partly molted into first fall plumage had the bill including the gape zinc orange; eye ring mustard yellow; bare eyelids water green; iris avellaneous; tarsus and toes yellow ochre; nails dark neutral gray. The change from black to yellow bill is coincident with the post-juvenal molt. An adult female taken March 19 is in full molt.

The present species was met in thickets in the vicinity of water, either near irrigation ditches or along streams. At intervals they flew from covert to covert, and after alighting might pause for an instant in the open with jerking wings and tail, but at the slightest alarm dropped into heavy cover. The taking of specimens was difficult as it was restricted to what snapshots might offer. The birds were observed feeding on the berries of the piquillín (*Condalia lineata*).

Family SYLVIIDAE

POLIOPTILA DUMICOLA (Vieillot)

Sylvia dumicola VIEILLOT, Nouv. Dict. Hist. Nat., vol. 11, 1817, p. 170. (Uruguay.)[54]

The gnatcatcher was recorded as follows: Resistencia, Chaco, July 8 to 10, 1920 (a pair); Las Palmas, Chaco, July 13 to 31 (female, July 20); Riacho Pilaga, Formosa, August 7 to 18 (male, August 7); Formosa, Formosa, August 23 and 24; Kilometer 80, west of Puerto Pinasco, Paraguay, September 11, 15, and 20 (male, September 11); Kilometer 200, west of same point, September 25; Lavalle, Buenos Aires, October 27 to November 13 (male, October 30); San Vicente, Uruguay, January 31, 1921 (male); Lazcano, Uruguay, February 6 (immature female) and 7; Rio Negro, Uruguay, February 19 (two males, two females); Tapia, Tucuman, April 6 to 13 (two males April 8 and 9). A specimen from Kilometer 80, Puerto Pinasco, is slightly smaller than others, otherwise the series exhibits only the slight differences due to age, sex, or seasonal wear. April specimens are completing a fall molt.

[54] See Dabbene, El Hornero, vol. 1, 1919, p. 240.

This handsome gnatcatcher was found in thorny trees at the border of forest, in semiopen scrub, and lowland thickets or, at the Estancia Los Yngleses, near Lavalle, Buenos Aires, in groves of tala (*Celtis tala*). The birds hopped jauntily about among the branches with drooping wings and elevated tail, in a manner similar to other *Polioptila*, but the songs and call notes, as loud as those of a small warbler, were surprisingly strong for a bird of this group. The ordinary scolding note resembled *zhree* or *pree-ee* or a low *chit-it*. The usual song, given in a melodious tone, was *whit see wheety wheety wheety* or *tee tee tee tee wheety wheety wheety*. Occasionally they indulged in a more varied warble. They were found at times in mixed flocks with other small brush birds.

An adult male taken on October 27, near Lavalle, was in breeding condition, and an immature female was secured at Lazcano, February 6. The birds were encountered in pairs throughout the year, and remained attentively near one another at all times. In the autumn month of April, near Tapia, Tucuman, young males of the year were in full song and were active and demonstrative toward the females, so that pairing seemed to be in progress, though the nesting season was several months in the future. The case is analagous to that of *Thryothorus ludovicianus* and *Sitta carolinensis*, where young birds pair in their first fall and remain mated through the winter.

Family MOTACILLIDAE

ANTHUS FURCATUS FURCATUS d'Orbigny and Lafresnaye

Anthus furcatus d'ORBIGNY and LAFRESNAYE, Mag. Zool., 1837, cl. 2, p. 27. (Near Carmen, on the Rio Negro, Patagonia.)[55]

A series of 16 skins of this pipit afford the following distributional data: Rio Negro, Uruguay, February 21, 1921, immature male; Bañado de la India Muerta, 12 miles south of Lazcano, Uruguay, February 3, three males, two females, adult and immature; San Vicente, Uruguay, January 31, adult male; Berazategui, Buenos Aires, June 29, 1920, female immature; 15 miles south of Cape San Antonio, Buenos Aires, November 6, adult female; Carhue, Buenos Aires, December 15 to 17, two adult and one young male, three adult females; Victorica Pampa, December 26, adult male. In addition I have examined a male taken September 27, 1919, by E. G. Holt at Rio Grande, Rio Grande do Sul, Brazil, and two males and two females shot near Bahia Blanca, Buenos Aires, by S. A. Adams, from October 31 to November 9, 1903. These records in part somewhat extend the boundaries of the range assigned by Hellmayr.[56]

---

[55] See Hellmayr, El Hornero, vol. 2, Aug. 2, 1921, p. 181.
[56] El Hornero, vol. 2, Aug. 2, 1921, pp. 181–182.

This pipit was locally common in some of the areas visited, and though closely similar to *A. c. correndera*, with which it was often associated, was readily told by its grayer, less distinctly streaked dorsal surface, and by the fact that in walking it did not tilt the tail. Near Berazategui, Buenos Aires, I secured my first specimen on June 29, 1920, on low, wet ground near the Rio de la Plata. At Carhue, Buenos Aires, from December 15 to 18, the birds were common over rolling, open country covered with low tufts of grass. At this season they were in pairs and were nesting. As I crossed the plains it was common for a pair to rise to circle about with strongly undulating flight and utter chirping calls of alarm until I had passed beyond their limits. Often males alone rose to accompany me for a short distance, darting down frequently to pass near the female when she remained upon the ground. On December 16, as I walked rapidly across the open prairie, a male pipit suddenly rose behind me with a sharp alarm call that brought his mate fluttering out from a nest concealed beneath a clump of grass almost at my feet. The nest was a thin-walled cup of grasses, lined with material of a finer texture than the exterior, placed in a slight depression, so that the rim was flush with the surface. The two hard-set eggs that it contained have a buffy white ground color, almost concealed by obscure spots and blotches of pale ecru drab and snuff brown. They measure 19 by 14.6 and 19 by 14.4 mm. A young bird only recently from the nest, taken December 15, is dull blackish above, with each feather margined with pinkish buff, producing a mottled appearance. The hind claw already is well developed, though the tail has not yet attained its full length.

At Victorica, Pampa, three were found in a little opening surrounded by bushes, and a male, which I shot, flew up to alight on a twig. At San Vicente, Uruguay, January 31, a breeding male was taken on the open shore of a lagoon in the same area where a breeding male of *A. c. correndera* was secured. On February 3 adult and juvenile individuals in molt into fall plumage were found south of Lazcano, Uruguay, and the birds were common until February 9 as far as Corrales. Scattered flocks frequented rolling uplands near Rio Negro on February 21. At Guamini, Buenos Aires, scattered flocks were found through fields and along alkaline shores near the Laguna del Monte.

ANTHUS CORRENDERA CORRENDERA Vieillot

Anthus correndera VIEILLOT, Nouv. Dict. Hist. Nat., vol. 26, 1818, p. 491. (Paraguay and Rio de la Plata.)

Specimens taken of this pipit include the following: San Vicente, Uruguay, January 31, 1921, adult male; Dolores, Buenos Aires, October 21, 1920, a pair; Lavalle, Buenos Aires, adult male, November

2, female, October 25; Guamini, Buenos Aires, adult male, adult and immature females March 7, 1921; Zapala, Neuquen, immature male (juvenal dress) December 8, 1920; Tunuyan, Mendoza, adult male, March 28, 1921. The present species may be distinguished from *A. furcatus*, with which it is often associated, by the nearly straight, much elongated hind claw, longer bill, and more distinctly streaked plumage. Streaks on the flanks are blackish and plainly marked, instead of indistinct as in *furcatus*.

Limits of subspecies in this pipit[57] are difficult to draw. A skin from Tunuyan, Mendoza, is clearly intermediate toward *chilensis*, but seems best allotted to *correndera*. A bird in juvenal plumage from Zapala, Neuquen, seems to belong to the typical form.

These pipits were abundant on the eastern pampas near Dolores and Lavalle, but were not noted in numbers elsewhere. During the breeding season, in October and November, males rose constantly to sing on the wing, circling with direct flight and rapidly flitting wings, a flight distinct from their undulating movement at other times. When tired the wings were set and the bird dropped slowly into the grass to rise and continue its evolutions in a short time. On October 21, near Dolores, a female flushed from a nest at my feet, ran rapidly away, and after a short flight joined her mate. The nest, placed in a mat of dead grass stems, was a cup composed of grass stems sunk in a little hollow so that its margin was flush with the surface. It was entirely concealed from above except for the opening that led into the cavity. Bottom and sides were damp from moisture that exuded from the soil so that the eggs were wet. The three slightly incubated eggs have a dull white ground color and are covered heavily with an irregular wash and spotting of natal and bone brown. The shell was very delicate. They measure 21.7 by 15.4; 21.6 by 14.9; and 21.1 by 14.9 mm.

Apparently two broods may be reared, as though young were fully grown at Zapala, Neuquen, on December 8 and 9, males were still in song. On January 31 a breeding male (the only one of the species seen in Uruguay) was taken on the open shore of the Laguna Castillos, near San Vicente.

Near Guamini, Buenos Aires, the birds were common in grass, in little scattered flocks, March 7 and 8, and adults and young taken were in molt into fall plumage. The tongue in these individuals was blackish, being almost jet black in juveniles and paler in adults. One was shot in the act of eating a butterfly (*Colias lesbia* Fabricius) that was captured where it had sought shelter from cold and wind in the grass. Near Tunuyan, Mendoza, these pipits were noted at times in weed-grown fields.

---

[57] See Hellmayr, El Hornero, vol. 2, August, 1921, pp. 185–188.

ANTHUS CORRENDERA CHILENSIS (Lesson)

Corydalla chilensis LESSON, Rev. Zool., vol. 2, 1839, p. 101. (Chile.)

Near Concon, Chile, on April 24 and 25, 1921, several of these pipits were seen. A male taken on the first date mentioned shows the pronounced yellow wash above and below that distinguishes this form from typical *correndera*. An immature male, shot near Guamini, in southwestern Buenos Aires, on March 7, 1921, is identical in coloration with the Chilian form, and must be designated as that race under our present understanding of the forms involved. Specimens of *chilensis* in the United States National Museum include birds from near Santiago, and a small series from Gregory Bay and Elizabeth Island, in the Straits of Magellan. Skins from Lago San Martin, Santa Cruz (Museum of Vertebrate Zoology collection), while not typical, are nearer *chilensis* than *correndera*. It is assumed that the specimen from Guamini is a migrant from the south or southwest. Patagonian skins seem more or less intermediate and specimens from Rio Negro are not wholly typical of *correndera*. The Guamini skin may possibly represent an extreme variant toward *chilensis* from some region in Patagonia where the two forms intergrade.

ANTHUS LUTESCENS LUTESCENS Pucheran

Anthus lutescens PUCHERAN, Arch. Mus. Hist. Nat. Paris, vol. 7, 1855, p. 343. (Rio de Janeiro, Brazil.)

Eleven skins of this small pipit come from the following localities : Las Palmas, Chaco, July 15 and 22, 1920, one male, two females; Riacho Pilaga, Formosa, August 9, female; Puerto Pinasco, Paraguay, September 3, two females; Kilometer 80, west of Puerto Pinasco, September 8, 9, and 21, three males, two females. Though single birds were seen occasionally, it was usual to encounter this species in flocks that contained as many as 50 individuals. The birds frequented wet meadows or the borders of lagoons where low, scattered clumps of bunch grass furnished a certain amount of shelter, or less often were found on open spaces at the borders of ponds, or even on mats of vegetation floating on shallow water. When first alarmed they crouched motionless in little depressions or under slight cover, where they entirely escaped the eye, or if too closely pressed took to wing with a curious, hesitant flight, in which the body was held at a vertical angle of 45° and the bird progressed in a series of jerking undulations. Though at times flocks rose to wheel about in the air, they usually dropped back to the ground in a short space to remain quiet until danger seemed past. When feeding in cover they walked slowly about in a crouching position, creeping under wisps of grass and seeking any slight protection that offered. At such times they bore some resemblance in mannerism to some of the grass finches. When in the open they stood more erect and seemed bolder. This species belies the common name of its family in that it does not wag the tail in walking, a modesty of action that was verified on several occasions.

On October 3, near Villa Concepcion, Paraguay, a little colony of these pipits, in pairs, was found in the short grass of pasture land behind the town. Males sang a drawn-out song that resembled *tsee-ee-ee-a yuh-h-h* in a high, thin tone. The ordinary call note given in flight resembled *chees chees, tsu* or *tsea*.

Family CORVIDAE

CYANOCORAX CYANOMELAS (Vieillot)

Pica cyanomelas VIEILLOT, Nouv. Dict. Hist. Nat., vol. 26, 1818, p. 127. (Paraguay.)

The present species is represented by skins of three adult males, two from Las Palmas, Chaco, July 13, 1920, and one from Kilometer 80, west of Puerto Pinasco, September 15. The latter specimen has a slightly more slender bill than the others and a somewhat shorter wing. These jays were recorded at Las Palmas, Chaco, from July 13 to 30, 1920; Riacho Pilaga, Formosa, August 8 to 21; Kilometer 25, west of Puerto Pinasco, September 1; Kilometer 80, west of the same point, September 6 to 20; and the Cerro Lorito opposite Puerto Pinasco, September 30. Old skins of this jay vary much in color through fading, some appearing so different as to suggest another species.

The birds were found in little bands of five or six, probably family parties from the previous season, that ranged, often in company with *Cyanocorax c. chrysops*, through stands of tall trees bordering streams or the groves that dotted the prairies of the Chaco. At any curious sound they glided in on set wings to perch familiarly near at hand and peer about, while if one of their number was killed the others gathered above it for a vociferous wake, their remarks punctuated by vigorous jerks of wings and tail. Their flight, when traveling for any distance, was peculiar. While straight and direct like that of other jays, it was accomplished by a number of slow beats of the wings followed by perhaps half a dozen quicker strokes, and every effort at flying ended in a long, upward glide that carried the bird to the desired perch. Their ordinary call is a loud *car-r-r* decidedly crowlike in sound, while at other times they called *chah chah* or *quaw*. At times they descended to hop about on the ground in search for food. Occasionally one was encountered that was bold to impudence, as when, in a wild, uninhabited region, in the Formosan Chaco, one came for scraps of cooked meat from my lunch, at the same time keeping a sharp eye on the dog who accompanied me. The cooked flesh of the muscovy duck that I threw to it must have been strange fare, yet the bird held the fragments between its toes and ate with relish. Once or twice jays of this species stole skeletons of small birds hung out to dry near camp.

CYANOCORAX CHRYSOPS CHRYSOPS (Vieillot)

Pica chrysops VIEILLOT, Nouv. Dict. Hist. Nat., vol. 26, 1819, p. 124. (Paraguay.)

In the wooded area of the southern section of the Chaco this strikingly marked jay was locally common where extensive tracts of forest still existed. It was recorded at the following points: Las Palmas, Chaco, July 14 to 27, 1920 (adult male July 14); Riacho Pilaga, Formosa, August 8 to 18; Kilometer 80, west of Puerto Pinasco, Paraguay, September 8 (two seen). A skeleton and a specimen in alcohol were preserved at Las Palmas in addition to the skin already mentioned.

The present species, during winter and early spring, was encountered in little bands of five or six (probably families of the previous season) that ranged in the heaviest *monte*, or occasionally in groves scattered over open prairies, usually in company with the larger *Cyanocorax cyanomelas*. Both species exhibited great curiosity and were easily decoyed up within a distance of 5 or 10 meters. When they were within hearing, at any squeaking note they came sailing in with spread wings and crest fully erect to perch on some open limb and eye me with no semblance of fear. They uttered a number of jaylike calls, and on one occasion one suddenly jerked up and down on its perch, rising to the full length of its legs and then dropping back, while it called *kuk kuk kuk kuk* loudly. On the whole, *chrysops* was more noisy than *cyanomelas*.

CYANOCORAX CHRYSOPS TUCUMANUS Cabanis

Cyanocorax tucumanus CABANIS, Journ. für Ornith., 1883. p. 216. (Tucuman.)

Three specimens of the present form, a male, a female, and one with sex not determined, were secured near Tapia, Tucuman, on April 12, 1921. These, compared with typical skins, exhibit the characters of heavier bill, more strongly arched culmen, and darker, blacker dorsum that characterize this subspecies. The differences in crest and color of abdomen that have been alleged are not apparent. On the date in question two flocks, each numbering five or six individuals, were encountered in rather heavy forest near the Rio Tapia. They came hopping out rather curiously when they first saw me, but retreated at once to heavy cover, and worked away, their presence indicated only by their jaylike calls. April 17 another small band was encountered in one of the dense groves on the upper slopes of Sierra San Xavier, above Tafi Viejo, Tucuman, at an altitude of about 2,100 meters. These remained concealed among heavy branches and slipped away down the steep slopes to more distant quarters. The habits in general are similar to those of the typical form.

Family CYCLARHIDAE

CYCLARHIS GUJANENSIS VIRIDIS (Vieillot)

Saltator viridis, VIEILLOT, Tabl. Enc. Méth., vol. 2, 1823, p. 793. (Paraguay.)

The three skins of this species preserved include an adult male from Las Palmas, Chaco, July 31, 1920; adult male, Riacho Pilaga, Formosa, August 11; and immature female, Tapia, Tucuman, April 12, 1921. The two males have a wing measurement of 82.5 and 79.5 mm., respectively, and the female, with wing not quite grown, 78.5 mm. The southern form, to which these birds belong, is distinguished from *C. g. cearensis* Baird by larger size.

An adult male, taken July 31, had the maxilla and tip of mandible cinnamon drab, changing to neutral gray at tip of culmen; base of mandible deep green-blue gray; iris ochraceous buff with a tinge of ochraceous orange; tarsus and toes gray number 7.

These birds inhabited low trees in brush-grown pastures or at the borders of barrancas, where they hopped slowly and deliberately about among the dense branches with erect carriage, examining twigs and leaves for food. That the strong, heavy bill was of service was shown when one tore and pulled at a strip of insect-infested bark, using much strength in its efforts. The song, heard August 23 near Formosa, Formosa, was a pleasant warble, somewhat accented, so that it did not seem monotonous though constantly repeated. From its tone I had supposed that it came from some finch and was astonished to trace it to a pepper shrike.

The Toba Indians in Formosa called this species *si trih.*

CYCLARHIS OCHROCEPHALA Tschudi

Cyclarhis ochrocephala TSCHUDI, Arch. für Naturg., 1845, pt. 1, p. 362. (Southern Brazil and Buenos Aires.)

The first of these birds observed was an adult female taken at Berazategui, Buenos Aires, on June 20, 1920, in a thicket near the Rio de la Plata. In southern Uruguay the species was common, as two adult males and one immature bird of the same sex were shot at San Vicente on January 28 and 30, 1921, and adult and immature males on February 6 and 8. The species was observed at Rio Negro, Uruguay, from February 17 to 19. The adult female from

Berazategui is peculiar in having the chestnut superciliary extended behind the eye thus suggesting *viridis*. An adult male, taken January 30, had the tip of the culmen fuscous; rest of maxilla benzo brown; base of mandible light brownish drab; rest pallid quaker drab becoming fuscous at tip; iris brick red; tarsus and toes deep green-blue gray.

Like its congener, this pepper shrike frequented brush where it hopped slowly about among the dense limbs with all the assurance of a tyrant flycatcher.

Attention was often drawn to the bird by its rollicking, warbling song that carried for a considerable distance. A second song given with bill pointed toward the sky resembled *too too too wheur*. In addition to its songs, the species has several peculiar calls uttered in a loud tone. Those seen in January were accompanied by grown young.

Where color may not be distinguished, the strong, heavy bill of this bird is a prominent field mark.

Family VIREONIDAE

VIREO CHIVI CHIVI (Vieillot)

Sylvia chivi VIEILLOT, Nouv. Dict. Hist. Nat., vol. 11, 1817, p. 174. (Paraguay.)

Eight skins preserved offer certain differences in coloration, but may be referred to typical *chivi*. An adult male shot September 30, 1920, on the Cerro Lorito opposite Puerto Pinasco, Paraguay, may represent the usual form of the type race. Four adult males from San Vicente, Uruguay (January 28, 29, and 30, 1921), and a pair from the Rio Cebollati, near Lazcano (February 6), are somewhat duller in coloration. Deeper coloration also characterizes an adult male from Tapia, Tucuman, shot April 9, 1921.

On September 30 the species was common in the forests near the Rio Paraguay, apparently newly come in spring migration, since none had been seen previously. One was recorded at Asuncion, Paraguay, on October 6. In eastern Uruguay, the species bred commonly on brush-grown slopes of canyons in the rocky hills near San Vicente (January 28 to 30, 1921), and was fairly common in the dense thickets along the Rio Cebollati, below Lazcano (February 6 and 8). Spring and summer birds sang as persistently as does *V. olivaceus* in the north, a species of which *chivi* is so much a counterpart in appearance, actions, and notes that it is recognized at first glance. Their smaller size and yellow-green sides and flanks are apparent on close scrutiny, while in the hand it is found that the iris is duller, as it varies from Rood's to Vandyke brown. The birds work quietly through the limbs, pausing frequently to peer about, in the slow manner usual to vireos. At a squeak they came down to peer at me with the crown feathers raised. The song, given without interruption during search for food, is a series of phrases, similar to but possibly slightly less emphatic than that of the northern red-eye.

At Tapia, Tucuman, from April 10 to 13, the species was fairly common in dense scrub, where it traveled in company with parula warblers and other small bush and tree haunting birds. The birds had ceased singing then, though their complaining call note was heard at intervals. An adult male taken was extremely fat.

An adult male, shot January 30, had the maxilla dusky neutral gray; base of gonys washed with pallid brownish drab; rest of mandible clear green-blue gray; iris Vandyke brown; tarsus and toes light Medici blue.

Family COMPSOTHLYPIDAE

BASILEUTERUS HYPOLEUCUS Bonaparte

Basileuterus hypoleucus BONAPARTE, Consp. Av., vol. 1, 1850, p. 313. (Brazil.)

An adult female, shot September 1, 1920, at Kilometer 25, west of Puerto Pinasco, Paraguay, the only specimen collected, differs from skins of *hypoleucus* from Matto Grosso in the yellow under tail coverts and the yellowish wash on the lower breast and abdomen. It is possible that this bird represents a distinct race.

The specimen was one of several found in heavy woods on a low hill, where the birds fed actively through the tops of the lower growth. They were observed in little parties of three or four, apparently families since a part were fully grown young.

BASILEUTERUS AURICAPILLUS AURICAPILLUS (Swainson)

Setophaga auricapilla SWAINSON. Anim. in Menag., 1838, p. 293. (Brazil.)

Near Las Palmas, Chaco, this *Basileuterus* was fairly common on July 13, 14, 17, and 21, 1920. Two were taken, an adult female, July 13, and one with sex not marked, July 21. The birds frequented dense thickets and heavy woods, where they hopped actively about among the smaller twigs with flitting wings and jerking tail. On February 5, 1921, I secured another, an immature male, found in dense growth along the Rio Cebollati, near Lazcano, Uruguay, in company with a mixed flock of *Serpophaga* and *Thamnophilus*. The *Basileuterus* worked through the lower limbs between 1 and 2 meters from the ground with the tail wagging in a characteristic manner.

The female shot July 13 had the maxilla and tip of mandible bone brown; base of mandible paler; tarsus tawny olive; feet slightly yellowish; iris very dark brown.

The single bird from Uruguay is much darker on the flanks and dorsum than those from the Argentine Chaco, differences that may be due to immaturity or freshness of plumage, or may prove to be of subspecific value. The species has not been recorded previously from Uruguay.

BASILEUTERUS LEUCOBLEPHARIDES LEUCOBLEPHARIDES (Vieillot)

Sylvia leucoblepharides VIEILLOT, Nouv. Dict. Hist. Nat., vol. 11, 1817, p. 206. (Paraguay.)

The present warbler was encountered in the following localities: Resistencia, Chaco, July 9 and 10, 1920 (one of unknown sex taken July 9); Las Palmas, Chaco, July 13 to 30 (adult female, July 13); Riacho Pilaga, Formosa, August 7 and 18; San Vicente, Uruguay, January 30 (adult female taken); Lazcano, Uruguay, February 5 to 8 (adult female, February 5); Rio Negro, Uruguay, February 14 to 18. Specimens secured in Uruguay, in full post-breeding molt, appear darker than skins taken in winter in the Chaco, a condition due perhaps to their new plumage. All are assigned to the typical form, though the single skin in United States National Museum that is supposed to represent *B. l. superciliosus* (Swainson)[58] is not in satisfactory condition for comparison.

Though in the Tableau Encyclopedique et Methodique,[59] Vieillot calls this species *S. Leucoblephara*, in the original description the specific name is spelled *leucoblepharides*.

Basileuterus l. calus Oberholser[60] is a synonym of *B. l. leucoblepharides*, since the typical form comes from Paraguay.[61]

This bird inhabited dense thickets or the heavy growth that often borders clearings, where it frequented low growth or, with constantly wagging tail, walked about on the ground. The birds were inquisitive and came very near to me when suitable cover offered, a proximity that intensified the ear-piercing *tsee* that served them for call note. They were found in pairs. The song of the male was made up of a repetition of a single, clear, whistled note repeated several times in a slowly descending scale that in sound and cadence suggested the song of a canyon wren (*Catherpes mexicanus*) but lacked the carrying power of the notes of that bird. A juvenile individual was recorded February 5.

---

[58] *Trichas superciliosus* Swainson, Anim. in Menag., 1838, p. 295. (Brazil.)
[59] Vol. 2, 1823, p. 459.
[60] *Basileuterus leucoblepharus calus* Oberholser, Proc. Biol. Soc. Washington, vol. 14 Dec. 12, 1901, p. 188. (Sapucay, Paraguay.)
[61] See Tabl. Encyc. Meth., vol. 2, 1823, p. 460.

One taken July 9 had the bill blackish slate; tarsus and toes honey yellow; iris very dark brown.

BASILEUTERUS FLAVEOLUS (Baird)

Myiothlypis flaveolus BAIRD, Rev. Amer. Birds, May, 1865, p. 252. (Paraguay.)

This warbler was found only in the region west of Puerto Pinasco, Paraguay, where it was seen September 1, 1920, near Kilometer 25, and September 9 to 20, near Kilometer 80. Adult males, taken September 1 and 10, were preserved as skins. The species ranged in pairs in dense forest growth, feeding on or near the ground. The birds were shy but were occasionally seen walking or hopping about with constantly jerking tail. Males sang a sweet, warbling song, and the call note was a sharp *chip*.

MYIOBORUS BRUNNICEPS (d'Orbigny and Lafresnaye)

Setophaga brunniceps d'ORBIGNY and LAFRESNAYE, Mag. Zool., 1837, p. 50. (Yungas, Bolivia.)

On April 17, 1921, the handsome brown-capped redstart was common on the slopes of the Sierra San Xavier above Tafí Viejo, Tucuman, between 1,800 and 2,100 meters, where it ranged in thickets of low, rather dense undergrowth scattered over rolling slopes above the forest, or occasionally came into more open areas among the groves of tree alders. The birds, alert and active in every movement, flew from perch to perch with a flirt of the tail that displayed the prominent white of the outer feathers.

The specimen preserved is an immature male in fresh fall plumage.

GEOTHLYPIS AEQUINOCTIALIS VELATA (Vieillot)

Sylvia velata VIEILLOT, Hist. Nat. Ois. Amer. Sept., vol. 2, 1807, p. 22, pl. 74. (No locality. "De la collection de M. Dufresne.")

The present yellowthroat was so local in its distribution and so sedentary that it was probably overlooked in many localities. It was recorded as follows: Las Palmas, Chaco, July 20, 22 (adult male taken), and 28 (male and female shot); Riacho Pilaga, Formosa, August 9 and 17 (a male taken on each of the dates mentioned); Formosa, Formosa, August 24 (immature male secured); Lazcano, Uruguay, February 5 (immature male); Rio Negro, Uruguay, February 17 (immature female) and 18 (immature female, adult male); Tapia, Tucuman, April 11 and 12 (adult females on the two dates given). Immature birds are somewhat browner than others, while adults shot in winter are more richly colored than those secured in summer. Immature birds were common in February, and adults taken in Uruguay in February and in Tucuman in April were in full molt. Several seen had lost all of the rectrices and all were in ragged condition. No definite differences are apparent in birds from the localities mentioned.[62]

This yellowthroat frequented cat-tails or other aquatic growth standing in water, or dense tangles of herbaceous vegetation bordering wet swales, or other low localities. In this safe cover they crept about cautiously, at times flying for short distances with quick, tilting flight to some safe retreat among the grasses. Though often common it was difficult to catch sight of them. Their call note was a harsh *tseep tseep*, quite different from the scolding call of *Geothlypis trichas*. I did not identify their song.

COMPSOTHLYPIS PITIAYUMI PITIAYUMI (Vieillot)

Sylvia pitiayumi VIEILLOT, Nouv. Dict. Hist. Nat., vol. 11, 1817, p. 276. (Paraguay.)

The present species, with a broad distribution through humid wooded areas in the northern half of Argentina, was recorded at the following points: Resistencia, Chaco, July 8 to 10, 1920 (male, July 8); Las Palmas, Chaco, July 13 to 30; Riacho Pilaga, Formosa, August 18; Formosa, Formosa, August 23 and 24; Kilometer 25, west of Puerto Pinasco, Paraguay, September 1 (male); Kilometer 80, west of Puerto Pinasco, September 6 to 21 (male, September 8); Cerro Lorito, opposite Puerto Pinasco, September 30; San Vicente, Uruguay, January 28 to 31, 1921 (adult male, January 28); Lazcano, Uruguay, February 5 to 8 (immature male and female, February 5); Rio Negro, Rio Negro, February 15 (one with sex not determined); Tapia, Tucuman, April 6 to 13 (immature female, April 9; Tafi Viejo, Tucuman, April 17 (male).

The series of specimens taken is fairly uniform with exception of a male in fresh plumage shot April 17 at an altitude of 1,800 meters on the slopes of the Sierra San Xavier. This bird is faintly darker above than typical birds from Paraguay and indicates an approach to the coloration found in *C. p. elegans* Todd,[63] though lighter than the average of that form as shown in a series examined from Colombia to southern Peru. More recently Todd has described an additional form, *Compsothlypis p. melanogenys*,[64] from Yungas de Cochabamba, Bolivia (elevation 1,500 meters), which is said to be much deeper in color, particularly above, than *elegans*. This I have not seen. It is possible that the bird from the Sierra San Xavier represents an approach toward *melanogenys*. Further collections

[62] For the use of the name *velata* see Hellmayr, Nov. Zool., vol. 28, September, 1921, pp. 243–244.
[63] *Compsothlypis pitiayumi elegans* Todd, Ann. Carnegie Mus., vol. 8, May 20, 1912, p. 204. (Anzoategui, Estado Lara, Venezuela.)
[64] Proc. Biol. Soc. Washington, vol. 37, July 8, 1924, p. 123.

from the mountains of northwest Argentina may record a darker mountain form as an addition to the Argentine list.

The present warbler was one the most widely distributed of forest-haunting birds through the Chaco, in suitable areas in Uruguay, and in the wooded areas of Tucuman. During winter it joined little roving bands of birds of similar habits and was found everywhere in groves and thickets. The pitiayumi warbler is characterized by active, agile motions that carry it rapidly through the smaller branches. As spring approached males sang a song, similar to that of the northern species of the genus, that may be represented as *swois swois swois see-ee-ce zee-ee-ee-ee-up*. Young that had not finished the post-juvenal molt were taken at Lazcano, Uruguay, on February 5, and the birds here were seen in little flocks of 20 to 25 individuals. The usual call was a sharp *tsip*.

A male, taken July 8, had the upper mandible dusky black; lower mandible and extreme edge of upper for basal half ivory yellow, shading toward dusky at tip; inside surface of lower mandible ivory yellow; of upper mandible dull olive yellow; tongue flesh color, becoming dusky at tip; tarsus dusky brown; toes yellowish; iris dark brown.

ATELEODACNIS SPECIOSA SPECIOSA (Temminck)

Sylvia speciosa TEMMINCK, Nouv. Rec. Planch. Col. (vol. 3), livr. 49, pl. 293, fig. 2. (Rio de Janeiro, Brazil.)

On July 27, 1920, an adult male was taken near Las Palmas, Chaco, among the thorny bushes of an open scrub scattered across a broad savanna. Another was recorded on September 1, at Kilometer 25, west of Puerto Pinasco, Paraguay, in company with other small brush birds in the tree tops, and a pair was seen September 21 near Kilometer 80, in the same region. The birds were active, but at the same time deliberate and certain in their movements as they passed among the branches or bent forward and down to examine the underside of twigs.

The male taken has a wing measurement of 59 mm. and is slightly larger than two seen from Bahia. The only previous Argentine record for the species seem to be one taken at Ledesma, Jujuy, on July 13, 1906, by L. Dinelli.[65]

Family ICTERIDAE

DOLICHONYX ORYZIVORUS (Linnaeus)

Fringilla oryzivora LINNAEUS, Syst. Nat., ed. 10, vol. 1, 1758, p. 179. (Cuba.)

Field observations in the Chaco, where the bobolink is found during the northern winter, were carried on during the period when

---

[65] Lillo, M., Apuntes Hist. Nat. (Buenos Aires), vol. 1, Mar. 1, 1909, p. 43.

the species was still in the north. My only records, therefore, are those of captives offered for sale as cage birds in the market. Four males seen in the city of Mendoza in the principal market on March 30, 1921, were almost in full plumage though molting the tails. Another was seen in a bird store in Tucuman, Tucuman, on April 2. The bobolink was known in Spanish as *charlatan* and in the western Argentina was in vogue as a cage bird.

TRUPIALIS MILITARIS MILITARIS (Linnaeus)

Sturnus militaris LINNAEUS, Mant. Plant., app., 1771, p. 527. ("Terra Magellanica.")

The following records were made for the military blackbird: Coast of the Province of Buenos Aires, 24 kilometers south of Cape San Antonio, November 3 to 8 (male shot November 4) ; Lavalle, Buenos Aires, November 13; General Roca, Rio Negro, November 23 to December 2 (male and female, November 23 and 24) ; Zapala, Neuquen, December 8 and 9 (male December 9) ; Ingeniero White, Buenos Aires, December 13; Carhue, Buenos Aires, December 16 (male) ; Victorica, Pampa, December 23 to 28; Potrerillos, Mendoza, March 19 to 21 (male, March 19 at El Salto, altitude 1,800 meters) ; Tunuyan, Mendoza, March 26 and 28; Concon, Chile, April 25 and 28 (female).

The present species seems to reach its maximum size in the Falkland Islands, where it is characterized as *Trupialis militaris falklandica* Leverkühn [66] (one specimen seen, male, wing, 133.5; tail, 97.1; tarsus, 40.5; culmen from base, 37.4 mm.). The bill in this insular form is especially long and heavy.

Males from the Straits of Magellan (Gregory and Laredo Bays), Santa Cruz (Rio Gallegos, and near Rio Coy), central Neuquen (Zapala) north into Mendoza (Potrerillos and "Mendoza"), seem to represent typical *militaris militaris*, with the wing ranging from 125 to 139.1 mm. Skins from Rio Negro (Paja Alta and General Roca), Buenos Aires (Bahia Blanca, Carhue, and the coast south of Cape San Antonio), and Rio Grande do Sul, Brazil (one skin), are somewhat smaller, as in these the wing runs from 118.6 to 124 mm. With extensive series it may prove expedient to recognize these last as a distinct form for which the name *Pezites brevirostris* Cabanis [67] will be available.

On the coast of Buenos Aires this bird ranged among sand dunes partly grown with vegetation, where there was more or less shelter from the winds. In the arid sections of northern Patagonia, as at General Roca, Rio Negro, it was partial to the vicinity of water, but

---

[66] Journ. für Ornith., 1889, p. 108. (Falkland Islands.)
[67] *Pezites brevirostris* Cabanis, Mus. Hein., pt. 1, 1851, p. 191. (Brazil.)

was not averse to penetrating inland among the scanty bushes that clothed the slopes of arid gravel hills. Through the pampas country it was found amid clumps of rough bunch grass that covered extensive rolling pastures.

. The birds are inhabitants of the ground, where they walk about like meadowlarks (*Sturnella*) amid the grass. When at all alarmed they usually presented their obscurely marked backs to the observer, and when one chanced to turn about the flash of brilliant red on the breast came as a pleasant surprise from a bird apparently plain in coloration. Their flight is straight and direct, and is accompanied by a flash of white from beneath the wing. Their call note was a low *pimp*, while from the ground or some low perch males sang a wheezy song.

TRUPIALIS DEFILIPPII (Bonaparte)

Sturnus defilippii BONAPARTE, Consp. Gen. Av., vol. 1, 1850, p. 429. (Brazil, Paraguay, and Montevideo.)

Cultivation and grazing appears to be restricting the numbers of this species and of *T. m. militaris* in the Province of Buenos Aires. Personally I found *T. defilippii* only at Carhue, from December 15 to 18, 1920, where four males and two females, with two additional birds as skeletons, were collected, and at Guamini, Buenos Aires, on March 3, 1921. Between Empalme Lobos and Bolivar on the same date flocks of 100 or more were recorded from a train on the date last mentioned.

While *Trupialis m. militaris* is suggestive of the meadowlarks (*Sturnella*) the present bird, in the form of a study skin so closely similar in appearance, is more like an *Agelaius* in actions. In fact *defilippii* resembles *Leistes* in habits more than it does *T. militaris*.

T. defilippii seems abundant now in the area between Saavedra and the foothills of the Sierra de la Ventana. Near Carhue the birds were scattered over rolling hills and prairies south of town, in a region grown with clumps of a rough, harsh bunch grass. The birds were gregarious and were found in flocks that fed in company. In collecting specimens on one occasion I witnessed a curious example of the value of sentinels in flock feeding. A flock of these birds was feeding on the ground among open bunch grass, while one of their number, a fine plumaged male, remained on guard on the top of a clump of grass. I made several attempts to approach the flock, but each time the alarm was given by the sentinel and all rose and flew before I was within range. Finally I killed the sentinel by a very long shot, and though the others heard the discharge of the gun, as no alarm was given, they remained motionless in the grass with heads erect, where they could look about. I approached nearer then and killed another with a short-range shell. Two now arose and circled uncertainly for a meter or so, undecided where to go. I shot one of these and the other immediately dropped into the grass beside it to share its fate a moment later. Loss of the sentinel thus broke up for a few minutes the entire flock organization.

Early in morning long, straggling flocks were observed in flight across country to favored grounds. In feeding, the birds remained hidden in cover of the grass, and were difficult to see until they rose. In flight they often mount for 30 meters in the air and start as though bound for some distant point, but suddenly pitch down into the grass perhaps not more than a hundred meters from the point where they flushed.

In certain areas they seemed to be on their breeding grounds, though no nests were found. In these localities males rested on clumps of grass or on fence posts, where they displayed their brilliantly marked breasts. At short intervals they rose from 1 to 2 meters in the air to give a high-pitched song and then with spread wings set stiffly above the back dropped rapidly to the ground with a shrill, rattling call. Females remained under cover, and when flushed hid at once in the grass. The call note of the males was a note like *chep*, while females uttered a low, chattering call. At times males pursued females swiftly over the prairies.

The black under wing surface of the present species is easily seen when the birds are in flight and distinguishes it readily from *T. militaris* in which the under wing coverts are white.

LEISTES SUPERCILIARIS (Bonaparte)

Trupialis superciliaris BONAPARTE, Consp. Gen. Av., vol. 1, 1850, p. 430. (Matto Grosso, Brazil.[68])

An adult female shot at Formosa, Formosa, August 24, 1920, was the only skin preserved, as a male killed September 25 at Laguna Wall, 200 kilometers west of Puerto Pinasco, Paraguay, was through force of circumstances made into a skeleton. Bangs[69] has named a southern subspecies of this bird as *petilus* (type locality Concepcion del Uruguay, Entre Rios), but with the scanty material of this species at hand I am unable to make out geographic forms. The female recorded above has a wing measurement of 88 mm. Hellmayr[70] indicates *superciliaris* as a race of *Leistes militaris*. The presence of a distinct superciliary in the male of the southern bird seems to indicate specific distinction between the two.

The present species was seen on comparatively few occasions. The first one was recorded near Santa Fe, Santa Fe, on July 4, 1920; one was seen later in July at Las Palmas, Chaco (date uncertain) ; a half

---

[68] See Berlepsch, Nov. Zool., vol. 15, June, 1908, p. 123.
[69] Proc. Biol. Soc. Washington, vol. 24, June 23, 1911, p. 190.
[70] Arch. für Naturg., vol. 85, 1919 (November, 1920), p. 34.

dozen were found near Formosa, Formosa, August 24; two were recorded at Puerto Pinasco, Paraguay, September 3; and at Laguna Wall, 200 kilometers west of Puerto Pinasco, 20 were found on September 25. At Lavalle, Buenos Aires, November 2, a boy brought me a male, too badly shot to preserve. The birds inhabited pastures, wet meadows, or recently burned stubble, where they worked about in as inconspicuous a manner as possible. As a crouching attitude usually concealed the brilliant red of the breast they were difficult to make out. When startled they rose with uncertain, undulating flight like that of pipits, and, though they might alight for a brief space in bushes, soon dropped to the ground. As the primaries are little longer than the secondaries, and the tail is short, they present a curious appearance on the wing. After alighting in the grass the wings are flitted several times, and as the bird walks about the tail is frequently opened and shut in a nervous manner. Their call note is a low *chuck*. As many as 20 were encountered in one scattered flock.

PSEUDOLEISTES GUIRAHURO (Vieillot)

Agelaius guirahuro VIEILLOT, Nouv. Dict. Hist. Nat., vol. 34, 1819, p. 545. (Paraguay and Rio de la Plata.)

The present species, distinguished by its yellow rump, is noticeably more yellow below than *virescens* when seen in the field. Near Lazcano, Uruguay, February 5, 1921, I saw a flock of six feeding on the ground in an open pasture, where I was not able to approach within gunshot. Near Rio Negro, Uruguay, on February 14, I collected an adult male from two found in a stretch of open camp dotted with bushes. From my limited experience, *P. guirahuro* seemed similar in habits and general appearance to *virescens*, but its calls appeared clearer and louder and the song stronger. The birds may have been breeding near Rio Negro, as they rested in low tree tops and scolded me with twitching tails. The one taken is in worn plumage and has begun to molt the wing coverts.

PSEUDOLEISTES VIRESCENS (Vieillot)

Agelaius virescens, VIEILLOT, Nouv. Dict. Hist. Nat., vol. 34, 1819, p. 543. (Paraguay and Buenos Aires.)

Pseudoleistes virescens, marked from its congener *guirahuro* by its plain back and the restriction of yellow on the flanks, was encountered in greater abundance than in case of the related species. It was seen during July, 1920, at Las Palmas, Chaco, but was not collected until I reached Lavalle, Buenos Aires. The species was common here from October 23 to November 15, and adult males were collected October 23, November 6 and 13. In Uruguay the bird was recorded at San Vicente, January 24 to February 2; Lazcano, February 5 to 9 (adult male, February 5); and Rio Negro, February 15. Specimens from Uruguay appear to have slightly shorter bills·than those from Buenos Aires, but the species is somewhat variable in length of culmen.

These blackbirds were found in little scattered flocks at the borders of marshes or in wet localities on the open pampa, where they walked about like grackles in preoccupied search for food. They were common near the beach below Cape San Antonio, Buenos Aires, and many were noted about the cañadones farther inland. During October and November the birds often scolded me as I traveled through the rushes, and on November 13 I saw young recently from the nest. In fall and winter it was usual for them to rest in clumps of saw grass to warm themselves in the early morning sun. Their calls, given usually on the wing, were musical and pleasant to the ear.

An adult male killed October 23 had the bill and tarsus black; iris sayal brown.

AMBLYRAMPHUS HOLOSERICEUS (Scopoli)

Xanthornus holosericeus SCOPOLI, Del. Flor. Faun. Ins., pt. 2, 1786, p. 88. (Islands of the Parana Delta.[71])

This brilliant inhabitant of rush-grown marshes was fairly common in central and northern Argentina. At Las Palmas, Chaco, from July 14 to 28, 1920, it was observed occasionally in flight overhead, in usual blackbird fashion, and my acquaintances among the people of the little village were not satisfied with my daily bag of specimens until, on July 28, I had added a beautiful male to my collection. An immature male with only scattered orange feathers on the otherwise black head and chest was frowned upon as hardly worth preservation, but to me was as interesting as the adult. This bird is in slow molt into the adult plumage.

These blackbirds were recorded frequently about the lagoons near Kilometer 182, Formosa, August 9 to 21, and on the latter day were noted in numbers in the esteros between Fontana and Formosa. A few were found with flocks of *Agelaius ruficapillus* September 17, near Kilometer 80, west of Puerto Pinasco, Paraguay, where a male taken had nearly completed the molt into the adult stage. The species was seen at Dolores, Buenos Aires, October 21, and a female about to breed was shot at Lavalle, Buenos Aires, October 29. Another was noted near here November 16. One was seen at the Paso Alamo on the Arroyo Sarandi north of San Vicente, Uruguay, February 2, 1921.

These blackbirds are found universally in marshes where often they rest concealed among the rushes and seem rather shy. They

---

[71] See Dabbene, An. Mus. Nac. Hist. Nat. Buenos Aires, vol. 23, Dec. 26, 1912, p. 372. The original locality cited by Scopoli "Antigua, Panay," was in error. Brabourne and Chubb (Birds South America, December, 1912, p. 436) cite Brazil.

congregate in small bands. Their call is a rapid *check check*, and the song a clear flutelike whistle that may be represented as *tu tee-ee-ee-te*. Though on August 16 males in song were seen in pursuit of females as though mating, no nests were found. The broad, flattened bill is used as a pry by thrusting it in the ground and then spreading the mandibles apart as described in *Amblycercus*.

The Toba Indians called them *kwus to ta*.

A male, taken September 17, had the bill and tarsus black, and the iris Vandyke brown.

NOTIOPSAR CURAEUS (Molina)

Turdus Curaeus MOLINA, Sagg. Stor. Nat. Chili, 1782, p. 252. (Chile.)

Oberholser [72] has proposed the name *Notiopsar* for *Curaeus* Sclater (1862) because of an earlier *Cureus* Boie (1831) for a genus of cuckoos.

The Chilian blackbird was seen at Concon, Chile, from April 25 to 28, 1921, where two were preserved as skins. The birds were found in small flocks in open brush on hill slopes where their slender forms and long tails at first sight suggested thrushes. This impression was dispelled at once by their more or less agelaiine songs and their clucking calls, and on closer acquaintance they proved quite similar in habits to the *Pseudoleistes* from east of the Andes. Their usual call is a high pitched *whee whee* followed by a low *chuck a lah*.

The bill, tarsus, and feet are black, the iris fuscous.

GNORIMOPSAR CHOPI CHOPI (Vieillot)

Agelaius chopi VIEILLOT, Nouv. Dict. Hist. Nat., vol. 34, 1819, p. 537. (Paraguay.)

The chopi was common through the Chaco, where it was recorded at Las Palmas, Chaco, July 16 to 30, 1920 (female taken July 16); Riacho Pilaga, Formosa, August 7 to 20 (adult male shot August 17); Formosa, Formosa, August 24; Puerto Pinasco, Paraguay, September 1 to 29 (adult male taken at Kilometer 80, September 18); and San Vicente, Uruguay, January 26 and 27, 1921 (adult male secured January 26).

Three specimens secured at Las Palmas, Riacho Pilaga, and Puerto Pinasco, with wing measurements ranging from 116.5 mm (female) to 127 mm. (male), refer to the typical subspecies without difficulty. The fourth skin, an adult male secured near San Vicente, in the palm groves that spread over the lowlands near the eastern frontier of Uruguay, is much larger (wing, 136; tail, 94.5; culmen from base, 26; tarsus, 32.7 mm.) and must represent another form. Leverkühn

---

[72] Proc. Biol. Soc. Washington, vol. 34, p. 136.

has described a large bird from Bolivia,[73] which must stand as *Gnorimopsar c. megistus* (one skin seen in United States National Museum), but which can have no relation to the bird of almost equal size from far distant Uruguay. *Gnorimopsar sulcirostris* (Spix) said doubtfully to come from Minas Geraes, according to Hellmayr,[74] is a distinct species with a wing measurement of 155 mm. The Uruguayan skin is listed here with these comments, though it can not be considered typical *chopi*.

The *chopi* or *tordo* (the latter a name shared with the cowbird) was partial to marshy meadows where it fed in little bands that flew up at intervals with musical calls to rest in the sun in scattered clumps of trees. It was not unusual to see them at dusk in flight to a roost in some reed bed, and it was reported that bands gathered in late summer in such numbers as to damage fields of corn. Though their ordinary call was a harsh *chuck*, when several were gathered they uttered slow whistled calls that formed a pleasing medley. They much resembled grackles in their direct flight and general habits. Near Puerto Pinasco they came about dwellings and ranch buildings to investigate garbage cast out from the kitchen, varying their search for food by whistled concerts from near-by shade trees.

By the middle of August males were seen circling with set wings in short spirals before females, and by the end of September they had scattered to breeding grounds. In January, in eastern Uruguay, I found them among groves of palms where they seemed to be nesting as they scolded with rattling calls, while they walked about on the palm fronds overhead or occasionally gave a burst of song.

An adult male, taken September 18, had the bill, tarsus, and toes black, and the iris Hay's brown. The genus is remarkable for the sharply rounded ridge that passes obliquely across the base of the mandible.

AGELAIUS THILIUS CHRYSOPTERUS Vieillot

Agelaius chrysopterus VIEILLOT, Nouv. Dict. Hist. Nat., vol. 34, 1819, p. 539. (Paraguay.)

The yellow-shouldered blackbirds of the west coast and the Andean region from Chile (the type-locality of *Turdus thilius* of Molina [75]) north to Calca, Peru, have the wing distinctly longer than those from east of the Cordillera. Average measurements of four males from Chile are, wing, 94; tail, 63.5 mm.; of three females, wing, 86; tail, 67.8 mm. Eleven males from Argentina (Buenos Aires and Santa Fe) measure, wing, 85.1; tail, 65.6 mm.; seven females, wing, 79.6; tail, 60.4 mm. These differences are sufficient to warrant the

---

[73] *Aphobus megistus* Leverkühn, Journ. für Orn., 1889, p. 104. (Santa Cruz and San Miguel, Bolivia.)

[74] Abh. Kön. Bayerischen Akad. Wiss., Kl. 2, vol. 22, Abt. 3, 1906, pp. 614–615.

[75] *Turdus Thilius*, Sagg. Stor. Nat. Chili, 1782, p. 250.

recognition of two races under this species as I have done here. The earliest name apparent for the eastern race is *Agelaius chrysopterus* Vieillot from the reference cited above. The male of Vieillot's description is a hopeless composite drawn apparently from *Agelaius xanthomus* of Porto Rico, and whatever species of *Icterus* is indicated by Latham's *Oriolus cayanensis*. The female, however, is taken from Azara's account of the *Tordo negro cobijas amarillos* and refers to the present bird. The name may thus be identified definitely with the eastern form of the yellow-shouldered blackbird.

An adult male taken at Berazategui, Buenos Aires, on June 29, 1920, is in full black plumage save for a few faint, paler margins that remain on the lower breast, abdomen, and lower tail coverts (where they are most pronounced) and on the dorsal surface. A pair in breeding plumage, somewhat worn, were taken at Lavalle, Buenos Aires, November 6 (female) and November 15 (male). At the Laguna Castillos, near San Vicente, Uruguay, I killed a pair in very worn breeding plumage and an immature female not quite fully grown. The adult female is very dark, and both of the older birds are smaller than others I have seen, as the wing in the male measures 80 mm. and in the female 75.4 mm., a condition due perhaps in part to wear. The juvenile female is dark in tone and much browner than others slightly older from other regions. At Tunuyan, Mendoza, on March 23 I secured an adult male, a female of equal age, and an immature female. The male has just completed the postnuptial molt and has the black obscured in a peculiar way. The crown is bone brown with a central streak of deep olive buff and a superciliary streak of pale olive buff. The hind neck is darker than buffy brown, the back feathers are margined with deep olive buff and natal brown, the greater and median wing coverts tipped with pale olive buff, and the feathers of rump and upper tail coverts bordered with deep olive buff. The entire undersurface has the feathers margined with pale olive buff. These lighter markings are lost through wear, so that by spring the bird is black save for the yellow shoulders. Pale markings persist on the under tail coverts longer than elsewhere.

The adult female from Tunuyan has about completed the body molt, but has all of the rectrices and a few wing feathers growing in anew. Females from Mendoza are distinctly paler than birds from Buenos Aires, and it is probable that they represent a pale form with a range covering marshes in the arid western regions of Argentina. Present material is not sufficient to demonstrate this satisfactorily.

The yellow-shouldered blackbird is an inhabitant of rush-grown fresh-water marshes, though ranging near the coast where salt water is tempered by springs or streams. At Berazategui, Buenos Aires, on June 29, 1920, the yellow-shouldered blackbird was fairly common in marshy spots near the Rio de la Plata in small flocks or scattered singly through the rushes. On July 15 a flock of a dozen was noted near a small lagoon at Las Palmas, Chaco, ·the farthest north at which the birds were recorded. On October 21, near Dolores, Buenos Aires, they were in pairs and males were singing. Near Lavalle from October 23 to November 15 they were common in the rush grown cañadones. Where shrubs or low trees offered perches, the males rested in them, though elsewhere they were content to cling to the side of a reed.

Near General Roca, Rio Negro, from November 23 to December 3, a few were recorded mainly in irrigated alfalfa fields. At Car-hue, Buenos Aires, a pair was seen in a small marsh from December 15 to 18. On January 16, 1921, birds in molt were noted along th' Arroyo Carrasco, east of Montevideo. At San Vicente, Uruguay, flocks containing grown young were seen January 31, and at Lazcano, Uruguay, small numbers were found in marshes grown with saw grass from February 5 to 9. At Tunuyan, Mendoza, from March 22 to 28, they gathered with cowbirds in cornfields bordering the cienagas where they did considerable damage to corn still in the milk. There was a small but regular evening flight here to some roost in the swamps south of town.

AGELAIUS CYANOPUS Vieillot

Agelaius cyanopus VIEILLOT, Nouv. Dict. Hist. Nat., vol. 34, 1819, p. 552. (Paraguay.)

The form here under discussion was fairly common in the Chaco, but was not seen elsewhere. It was noted at Las Palmas, Chaco, July 22 to 31, 1920 (adult females, July 22 and 28); Riacho Pilaga, Formosa, August 9 to 17 (immature male and female, August 16); and Kilometer 80, west of Puerto Pinasco, Paraguay, September 10 (male). The wholly black adult males of this *Agelaius* were so wary that none in that plumage were taken. A male shot August 16 in color is like the adult female but has a larger bill and longer wing. The immature female is duller below than the adult of the same sex. A male shot September 10 is in molt into adult feather as black is appearing on the head and sides of the breast.

These blackbirds were found in marshy localities, near lagoons, or about cornfields or other small cultivated tracts on low ground. It was usual to see them resting on rush stems or walking about on masses of vegetation that floated on shallow water. They were frequently rather wild, and when startled flew out in disorder with undulating flight. Among growths of weeds on the ground they walked about rapidly, clambering over considerable obstructions.

54207—26——25

Their usual call notes were closely similar to those of *Passer domesticus*, while another note resembled *check check*. The song of the male may be represented as *chee-ee-ee-ee*, a rather hard, rattling note that was followed by some twittering calls. The Angueté Indians called this species *mah ho*.

AGELAIUS RUFICAPILLUS Vieillot

Agelaius ruficapillus VIEILLOT, Nouv. Dict. Hist. Nat., vol. 34, 1819, p. 536. (Paraguay.)

The present species was another that I encountered only in the Chaco. It was recorded at Las Palmas, Chaco, July 23 and 28, 1920, when two adult males were taken from small flocks at the borders of lagoons. Other flocks were observed near Puerto Pinasco, Paraguay, September 3, and in the vicinity of Kilometer 80, west of Puerto Pinasco, from September 9 to 17. They were especially common here on the banks of esteros and lagoons flooded by heavy rains. At the latter point a male was collected September 10 and a female September 17. Often the birds walked about on floating vegetation among rushes where they were entirely concealed, but when startled flew up to alight in a close flock in some clump of grass, rushes, or low tree. Their call was a low *chick*. Their flight was undulating.

The Angueté Indians called this bird *gwas gookh*, apparently a group name for several species of blackbirds.

ICTERUS PYRRHOPTERUS PYRRHOPTERUS (Vieillot)

Agelaius pyrrhopterus VIEILLOT, Nouv. Dict. Hist. Nat., vol. 34, 1819, p. 543. (Paraguay.[76])

This curious oriole was recorded as follows: Resistencia, Chaco, July 10, 1920; Las Palmas, Chaco, July 27 (adult male taken); Riacho Pilaga, Formosa, August 14 (female); Formosa, Formosa, August 24 (female); Puerto Pinasco, Paraguay, September 1 (male); Kilometer 80, west of Puerto Pinasco, Paraguay, September 7 to 20 (female, September 7; male prepared as skeleton, September 15); Tapia, Tucuman, April 11 and 13, 1921. The specimens secured have slightly smaller bills and shorter tails than three (including the type) from the Province of Buenos Aires, which represent *Icterus p. argoptilus* Oberholser. It is possible, however, that with sufficient material for adequate judgment the two supposed forms will prove inseparable.

These orioles were found in groves and gardens in the Chaco, and were occasionally noted in the dry forests of northern Tucu-

---

[76] Azara, from whom Vieillot's account of this bird is taken, mentions no definite locality in connection with this species.

man. They were sprightly and active in movement and jerked the tail rapidly and violently as they moved about, especially when alarmed at the presence of a snake or some other enemy. In spring they were found about the handsome blossoms of such trees as the lapacho (*Tecoma obtusata*) and at all seasons were partial to growths of vines or creepers. Their active search for food often led them to swing head down from small twigs. The call note of the male was a sharp *spick spick*, and the song was clear and whistled. The female uttered a mewing note or a harsh rattling call.

An adult female had the bill black; iris Kaiser brown; tarsus neutral gray; claws fuscous.

MOLOTHRUS BONARIENSIS BONARIENSIS (Gmelin)

Tanagra bonariensis GMELIN, Syst. Nat., vol. 1, pt. 2, 1789, p. 898. (Buenos Aires.)

The series of 15 skins secured seems uniform in size and color except that specimens from northern Rio Negro and Mendoza appear very slightly larger and have heavier bills than those from Paraguay. All are considered as representative of the typical form. Records for the species are as follows: Santa Fe, Santa Fe, July 4, 1920; Resistencia, Chaco, July 8 to 10; Las Palmas, Chaco, July 12 to 31 (female taken July 19; male, July 31); Riacho Pilaga, Formosa, August 7 to 20 (males collected August 14 and 15); Formosa, Formosa, August 23 and 24; Puerto Pinasco, Paraguay, September 2 and 3; Kilometer 80, west of Puerto Pinasco, September 6 to 25 (male, September 10); Dolores, Buenos Aires, October 21; Lavalle, Buenos Aires, October 23 to November 15 (male, October 29; females, October 29 and November 13); General Roca, Rio Negro, November 23 to December 3 (male, November 30); Zapala, Neuquen, December 9; Carhue, Buenos Aires, December 15 to 18; Victorica, Pampa, December 23 to 29; Carrasco, Uruguay, January 16, 1921; La Paloma, Uruguay, January 23; San Vicente, Uruguay, January 25 to February 2 (female, January 27; juvenile male, January 31); Lazcano, Uruguay, February 3 to 9; Rio Negro, Uruguay, February 14 to 18; Potrerillos, Mendoza, March 16 and 18; Tunuyan, Mendoza, March 22 to 28 (adult female, March 26; adult and immature males, March 26 and 28).

Some of the skins from Formosa and Paraguay, taken in August and September, are in very worn plumage at this early season, so that males show only traces of the gloss that normally covers the entire body plumage. One adult female from Las Palmas, Chaco, is so much darker than others that it was supposed to be *brevirostris* until carefully compared. Adults from Tunuyan, Mendoza, are in full molt so that two have lost all of the rectrices. Young birds in juvenal plumage are similar to females but are streaked with whitish below. A skin in this stage from Mendoza is much browner both above and below than one from eastern Uruguay.

The naturalist from North America finds little in the habits and appearance of the shining cowbird that is not reminiscent of *Molothrus ater* of his native continent. *Molothrus b. bonariensis* was found in small flocks during the most of the year, the largest bands being noted in fall when at times several hundred flocked together to feed in fields of ripened corn. Bands of from 20 to 50 frequented corrals and plazas at the larger ranches, and became so familiar that they often walked about on dirt floors beneath the porticos of the dwelling houses. At Las Palmas, Chaco, during cold weather a little flock came about the *fonda* where I had living quarters; but when it turned warm and pleasant for a day or two flew off to less sheltered grounds. The birds fed in sociable groups on the ground but at any alarm flew up suddenly to low perches in trees. In pastures and fields these cowbirds gathered about grazing animals and followed along in their company. They treated domestic stock with the greatest familiarity. It was common on cold days to see a little flock whirl into a yard and alight on the back of some horse or ox which paid no attention to their coming. I have seen as many as 15 birds resting on the back of one horse and have no doubt that the warmth of its skin was grateful to its smaller companions who seemed to find especial delight in burying their toes in the hair of its back. Cowbirds were always attracted when grain was fed to horses. They worked over sandy areas by scratching in the usual blackbird fashion by jumping forward and then back, dragging their claws in the dust on the return. In early morning flocks frequently resorted to the open shores of lagoons to bathe. (Pl. 20.)

The first indication of pairing was noted at the end of October when three, taken on October 29 near Lavalle, Buenos Aires, were all breeding; a female contained an egg nearly ready for the shell. At this period only small flocks were seen and it was not unusual to find the birds in pairs. At Los Yngleses, the Gibson estate, I was interested one morning in observing the maneuvers of a male cowbird about the newly completed nest of a *Sisopygis icterophrys*. The female flycatcher was on the nest, as though brooding, though in reality she was merely resting as her home was not yet completed. The cowbird hopped about among the limbs approaching nearer and nearer until the other bird finally darted out at him, though without apparent animosity. The cowbird hopped up then to a point above the nest, peered down into it for a few seconds, after which he flew away apparently satisfied.

SHINING COWBIRDS (MOLOTHRUS B. BONARIENSIS) WARMING THEIR TOES
ON A COLD WINTER MORNING. THESE BIRDS ALIGHT FEARLESSLY ON
DOMESTIC STOCK

Las Palmas, Chaco, July 13, 1920

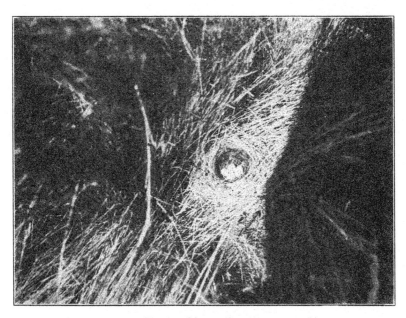

NEST AND EGGS OF MISTO (SICALIS A. ARVENSIS)

Carhue, Buenos Aires, December 18, 1920

At this season when one male joined a little band of others it was common for all to point the bill directly up and peer about, posturing thus for several seconds. In display before the female, males bent the head down, expanded the neck feathers, and spread the wings. One observed in this display in a low bush was so demonstrative that he toppled off his perch backward at short intervals, but regained an upright position instantly. This performance was varied by short circles on the wing about his mate's perch. Males uttered a bubbling, gurgling song.

Eggs ascribed to this cowbird were found in five sets of eggs that were collected. They exhibit much variation in size and coloration. Two taken from the nest of *Rhinocrypta lanceolata* at General Roca, Rio Negro, December 3, 1920, are dull white in color, finely, uniformly, and somewhat sparsely dotted with vinaceous tawny; they measure 21.6 by 16.9 and 21.6 by 17 mm. One secured with a set of *Furnarius rufus rufus* is heavily spotted throughout with cinnamon rufous and hazel, the spots confluent at the larger pole, with a scattering of purplish spots due to concealment of part of the brown beneath shell structure; this egg, which is much larger than the first two, measures 24.1 by 20.1 mm. Another, found in a nest of *Mimus triurus* at Victorica, Pampa, on December 26, 1920, is white with a very faint bluish tinge, entirely unmarked. It measures 21.8 by 19 mm. Three eggs from a nest of *Turdus rufiventris* collected at Lavalle, Buenos Aires, October 30, 1920, are variable. One is pure white with a very few fine widely scattered dots of cinnamon rufous, another is dull white evenly and sharply spotted with hazel, cinnamon rufous, and chestnut brown, with a few spots of a purplish hue, and the third has the ground color tinted distinctly with pinkish brown with the entire surface spotted with suffused markings of cinnamon rufous, hazel, and purplish. Measurements of these three are as follows: 21.5 by 18.6, 23 by 17.6, and 22.6 by 17.8 mm. The fifth and last parasitized nest found was that of a *Diuca minor*, taken at Victorica, Pampa, December 26, 1920, which contained one egg similar to the one collected from the nest of *Furnarius*, but less heavily marked; it measures 23.4 by 18.9 mm.

It is my opinion that the well-known variation in color among eggs of this cowbird is due to a mimicry similar to that so widely discussed and debated in the case of certain parasitic cuckoos of the Old World. The greater number of the tracheophone species which form so conspicuous an element among the smaller birds that breed in the area frequented by this cowbird lay white, unmarked eggs. As it is obviously of advantage for a parasitic egg to resemble that of the foster parent, it may be supposed that certain groups or individuals among the cowbirds that persistently parasitize these tracheophone species have developed an unmarked egg, while others who foist their product on oscines, whose eggshells in the main are boldly marked, have maintained a heavily spotted egg. The parasitic instinct among many of the cowbirds, however, seems poorly regulated, so that at the present time many deposit in nests other than those in which they should if their eggs are to have a maximum opportunity to escape detection through their inconspicuousness.

In addition to the short list of parasitized species, recorded above, I killed a juvenile cowbird that had a *Poospiza personata* as foster parent. A young bird secured at Tunuyan, Mendoza, had the side of the mandible near the nostril greatly swollen, apparently from the presence of a larval dipteran.

In an adult male *M. b. bonariensis* the bill, tarsus, and feet were black, the iris dull brown.

MOLOTHRUS BREVIROSTRIS (d'Orbigny and Lafresnaye)

Icterus brevirostris d'ORBIGNY and LAFRESNAYE, Mag. Zool., 1838, cl. 2, p. 7. (Maldonado, Uruguay, and Corrientes, Argentina.)

The screaming cowbird, known with the common cowbird as *tordo*, was local in distribution so that in Argentina the species was noted in few localities. An adult female in old worn plumage was shot at the Riacho Pilaga, Formosa, August 8, 1920, and on November 13, near Lavalle, Buenos Aires, two males were seen and one was taken. In Uruguay the species was more widespread but was far from common. At San Vicente one or two were seen in vacant lots within the town limits on January 26 and 28, 1921, and near Lazcano several were seen and three were taken in an area of open pastures on February 8. A juvenile male was shot at Rio Negro, Uruguay, on February 14, and another was recorded on February 17.

The notes of this cowbird were characteristic and entirely different from those of the allied species. In February they were breeding and males were markedly attentive to females. When flushed it was usual for little flocks to alight on wire fences, when males sidled over toward their mates with bowed heads and elevated neck ruffs while they uttered a harsh note that may be rendered as *zhree-ah*. Their general attitude at such times was reminiscent of *Tangavius aeneus*. At times they uttered a sweetly whistled song.

Adult males were distinguished from the common cowbird even when afield by the shorter, heavier bill, while in the hand they are marked by blacker coloration and shorter tarsus. The adult female is much blacker and the juvenile male darker than the corresponding stages of *bonariensis*. The young bird, however, is not easily separated. In addition to the differences noted the present species seems to have the base of the maxillar tomium heavier, and the rictus more abruptly deflected than *bonariensis*.

MOLOTHRUS BADIUS BADIUS (Vieillot)

Agelaius badius Vieillot,˙ Nouv. Dict. Hist. Nat., vol. 34, 1819, p. 535.
(Paraguay and Rio de la Plata.)

The bay-winged cowbird was widely distributed, as the following records indicate: Las Palmas, Chaco, July 17 to 30, 1920 (two males July 17); Kilometer 110, west of Puerto Pinasco, Paraguay, September 23; Lavalle, Buenos Aires, October 23 to November 1 (female, October 25, male, November 1); Victorica, Pampa, December 23 to 29 (male, December 24); San Vicente, Uruguay, January 27 to 31, 1921 (juvenile male, January 27, juvenile female, January 31); Lazcano, Uruguay, February 5 to 8; Rio Negro, Uruguay, February 17 (juvenile female) to 19; Potrerillos, Mendoza, March 21; Tunuyan, Mendoza, March 25 to 28; Tapia, Tucuman, April 6 to 14. Specimens taken do not offer any marked variation.

Hellmayr [77] has named a form of this species from northern and central Bolivia (type locality, Chuquisaca) as *Molothrus badius bolivianus*, distinguished by larger size and somewhat browner shade above. A skin in the United States National Musuem from Mendoza is somewhat browner above than others and may represent an approach to *bolivianus* though it is no larger than others from the Province of Buenos Aires. A male taken at Jujuy, July 29, 1922, by D. S. Bullock, is also somewhat browner above than the average, but has a wing length of only 91 mm. I have seen skins from northern Buenos Aires with the wing 95 mm. long, a greater size than indicated by Hellmayr in the paper mentioned for the typical form. The shortened wing tip of *Molothrus badius* distinguishes it structurally from other cowbirds and warrants the use of *Agelaioides* Cassin [78] as a subgenus at least. Further investigation may reveal other differences that will warrant generic separation. The question has been discussed briefly by Mr. Ridgway.[79]

Stempelmann and Schulz [80] in a list of the birds found in Cordoba have included this species as *Demelioteucus badius*, without comment as to the source of the generic name used. *Demelioteucus* thus appears to be new here since I have not yet found it at any other place. At this point it is a synonym of *Agelaioides* Cassin.

The breeding season for this bird apparently extends from December to January. At the end of December, in the stunted open forest about Victorica, Pampa, these cowbirds were found in twos and threes about large, deserted stick nests of various tracheophones and showed considerable solicitude as I approached. At the end of

---

[77] Verb. Orn. Ges. Bayern, vol. 13, Feb. 25, 1917, p. 108.
[78] *Agelaioides* Cassin, Proc. Acad. Nat. Sci. Philadelphia, vol. 18, 1866, p. 15. Type *Agelaius badius* Vieillot.
[79] Birds North and Middle America, vol. 2, 1902, pp. 205, 206, and 207.
[80] Bol. Acad. Nac. Cienc. Cordoba, vol. 10, 1890, p. 399.

January young still in juvenal plumage were found near San Vicente, Uruguay, while at Rio Negro other young were seen during the middle of February. A fledgling was taken here on February 17 from an old nest of *Pseudoseisura lophotes*. The large domed structures constructed by this species are very durable so that they resist the weather for a considerable time after the tracheophones are through with them. Though I searched a number at various times I did not succeed in finding eggs of the cowbirds.

The call notes of this cowbird are harsh and emphatic, *chrut* or *check* repeated incessantly when the birds are anxious about some nest site. The males utter a sweet warbling song, a pleasant melody that has earned the species the sobriquet of *musico*, a name well warranted. Even during the breeding season little bands of bay-winged cowbirds were found in company often voicing their sweetly modulated whistled song to the accompaniment of the more prosaic bubbling of *Molothrus bonariensis*. During winter bay-winged cowbirds gathered in little bands of 20 to 50 members that frequented the vicinity of ranch buildings or little open savannas in the Chaco, where they fed on the ground or rested in close companies in the tops of low trees. Their brown wings distinguished them easily from other blackbirds.

Young in juvenal plumage are similar to adults, but are faintly and indistinctly streaked with whitish below, and spotted obscurely with dusky on crown and back.

ARCHIPLANUS ALBIROSTRIS (Vieillot)

Cassicus albirostris VIEILLOT, Nouv. Dict. Hist. Nat., vol. 5, 1816, p. 364. (Paraguay.)

The genus *Archiplanus* Cabanis, with the present species as its type, has been separated from *Cacicus* by Miller [81] on smaller, more wedge-shaped bill, with culmen and commissure nearly straight, shorter, more rounded wing tip, with ninth primary shorter than third, and (usually) better developed aftershaft. Miller has included in *Archiplanus* the species current as *Cacius chrysopterus*, *C. chrysonotus*, *C. leucoramphus*, and the bird described by Dubois as *Agelaius sclateri*. Todd [82] has also recognized *Archiplanus* as distinct and has added to it the species previously known under the name *Amblycercus solitarius*.

With removal of the present species, known in recent years as *Cacicus chrysopterus* (Vigors), to *Archiplanus* its name will become *Archiplanus albirostris* Vieillot, a specific name not available

---

[81] Auk, 1924, pp. 463–465.
[82] Proc. Biol. Soc. Washington, vol. 37, July 8, 1924, pp. 114–115.

in *Cacicus* as it was antedated by *Tanagra albirostris* Linnaeus, 1764, a synonym of *Cacicus cela* (Linnaeus), 1758.

Vieillot's description of this small cacique was based on Azara who gave no locality in connection with his notes on the *yapu negro y amarillo*. The type-locality may be assumed to be Paraguay.

The small, yellow-rumped cacique was common in the Chaco, but was not recorded elsewhere. It was seen frequently at Resistencia, Chaco, from July 8 to 10 (adult female, taken July 8), and at Las Palmas, July 13 to 31. Near the Riacho Pilaga, Formosa, it was found from August 7 to 18, and was seen near Formosa, Formosa, August 23. In the vicinity of Puerto Pinasco it was noted from September 1 to 30, from the Rio Paraguay, west for 200 kilometers.

This cacique frequented forest trees in much the same manner as orioles, though occasionally it came into bushes growing among saw grass at the borders of savannas. Though oriolelike in form and color in habits these birds differed, as, in addition to searching the smaller branches, they crept about on the larger limbs or examined dead stubs by pushing and prying in crevices or bark with the pointed mandibles separated at the tip. This was done forcefully while the birds clung with their feet or braced with their tails. On September 9 they were in pairs but were not breeding, and nesting was not noted until September 30, when new nests under construction were found. These were woven of a black rootlet and hung as pendant bags from the tips of slender limbs. Some were 2 or 3 feet long, with a globular bag at the end to contain the nest proper.

The song was a sweet, broken warble and the note a mewing call that may be represented as *char*. They were known commonly as *pajaro sergente* from the yellow shoulder epaulettes, while the Toba Indians designated them as *ve da lla koh*.

A female, shot July 8, had the bill sky gray; iris pale ochraceous salmon; tarsus and toes blackish brown. There was variation among them as to color of bill, from light to dark. One taken has the crown and nape mottled with yellow.

ARCHIPLANUS SOLITARIUS (Vieillot)

Cassicus solitarius VIEILLOT, Nouv. Dict. Hist. Nat., vol. 5, 1816, p. 364. (Paraguay.)

Hellmayr [83] has considered *Cassicus nigerrimus* of Spix [84] distinct from *Cassicus solitarius* Vieillot, but reference to Azara's original description of the *yapu negro* [85] indicates that the two are identical. Azara remarks of the bill "del color y materia que en el precedente."

---

[83] Abh. Kön. Bayerischen Akad. Wiss., Kl. 2, vol. 23, Abt. 3, May 20, 1906, p. 613.
[84] Av. Spec. Nov. Brasiliam, vol. 7, 1824, p. 66, pl. 63, fig. 1. (Banks of the Amazon.)
[85] Apunt. Hist. Nat. Pax. Paraguay, vol. 1, 1802, pp. 268–269.

El precedente I take to refer to the preceding species, the *yapu* (*Ostinops decumanus*) that has a light colored bill which controverts Vieillot's statement that in *solitarius* the bill is black.

Though formerly placed in the genus *Amblycercus* the species *solitarius* is apparently a cacique. *Amblycercus holosericeus* has a strongly operculate nostril, no crest, and the eighth primary shorter than the second, while *Archiplanus solitarius* has the nostril nonoperculate, the eighth primary longer than the third, the wing longer, and a decumbent crest.

The differences between *solitarius* and *holosericeus* have been discussed recently by Miller,[86] but without making a change in their current status. Todd[87] includes *solitarius* in the genus *Archiplanus* Cabanis, a group segregated by Miller in the paper cited above. I had recognized also that *solitarius* must be removed from *Amblycercus*, and concur in Todd's allocation of it in the genus *Archiplanus*. The bill in *solitarius* is broader at the tip than in *Archiplanus albirostris*, type of Cabanis's genus, but the form of the decumbent crest and of nostril, wing, and tail are closely similar in the two birds and indicate clearly their relationship. Todd is somewhat confused in the structural characters assigned to *holosericeus* and *solitarius*, since it is *holosericeus* that has nostrils linear and overhung by a membrance, not *solitarius* as stated.

The species here discussed was found at only three localities: Las Palmas, Chaco, July 23 (adult male taken) and 26; Riacho Pilaga, Formosa, August 14 (adult male); and Kilometer 200, west of Puerto Pinasco, Paraguay, where three were seen September 25. The birds were encountered in heavy brush where attention was called to them by their harsh notes, *quay quay*. They were alert and active and peered out with tail thrown over the back, but at any alarm disappeared in dense scrub and were lost to view. The light colored bill showed prominently, even when they were seen in tilting flight across openings in the thickets. Like *Archiplanus chrysopterus* when feeding they poked and pried at leaves or loose bark with open bill. On examining them I was struck by the utility of a development noted frequently in some orioles, *Amblyrhamphus*, other Icterids, and in the genus *Sturnus* among other birds, where the angle of the lower jaw (the *processus angularis posterior*) is prolonged behind the articulation as a slender bar. Contraction of the bands of muscle that pass from this bar of bone to the side of the skull force the tip of the lower jaw away from the upper, with the articulation of the lower jaw with the quadrate acting as a fulcrum. The bill is thrust into or under bits of bark, a rolled leaf,

---

[86] Auk. July, 1924, pp. 463–467 (received in Washington July 5, 1924).
[87] Proc. Biol. Soc. Washington, vol. 37, July 8, 1924, p. 114.

or into the ground and then a cavity opened by thrusting apart the tips of the mandibles so that any food concealed may be seized. .

In one male taken the bill at the base was dark neutral gray shading through light grape green and water green to olive buff at tip; iris Rood's brown; tarsus benzo brown; toes dark neutral gray; claws mouse gray. Another male had the bill deep sea-foam green at center, shading to number 7 gray at base and olive buff at tip; iris maroon.

In the Chaco the bird was known locally as *que ve* in imitation of its note. The Toba Indian name was *kom kom*.

Family THRAUPIDAE

TANAGRA CHLORITICA VIOLACEICOLLIS (Cabanis)

Acroleptes violaceicollis CABANIS, Journ. für Ornith., 1865, p. 409. (Brazil.)

Two male euphonias shot at Las Palmas, Chaco, July 23 and 26, 1920, are identified under this name in accordance with the treatment of Hellmayr.[88] Berlepsch[89] considered the bird from this region as representative of *Tanagra serrirostris* (d'Orbigny and Lafresnaye). The skins preserved have a wing measurement of 58.5 and 59 mm. respectively, and are somewhat larger than two skins from Bahia. They are barely larger and have faintly broader bills than three of the same type of coloration from Chapada, Matto Grosso.

The species was recorded frequently at Las Palmas from July 23 to 31, and was seen at Formosa, Formosa, August 23 and 24. Attention was drawn to it by its high-pitched whistled call *tee tee tee whee-ee* (identical, so far as my memory served, with that of *Tanagra sclateri* of Porto Rico); and on looking about the birds were found near some clump of mistletoe, the berries of which seemed to form their food. At times they hopped about actively among the twigs, singing a warbling song, but usually perched quietly, uttering their whistle at intervals.

TACHYPHONUS RUFUS (Boddaert)

Tangara rufa BODDAERT, Tabl. Planch. Enl., 1783, p. 44. (Cayenne.)

The white-shouldered tanager was found at Resistencia, Chaco, July 8 to 10 (male and female taken); Las Palmas, Chaco, July 13 to 30 (male skinned, July 26); Riacho Pilaga, Formosa, August 9 to 18 (male and female, August 15); Kilometer 25, west of Puerto Pinasco, Paraguay, September 1, and Kilometer 80, west of the

[88] Nov. Zool., vol. 30, October, 1923, pp. 232–235.
[89] Verh. V. Int. Ornith.-Kongr. Berlin, 1911, pp. 1014, 1125.

same point, September 21 (female). In Chaco and Formosa the bird was common. It was a ground-haunting species that ranged with other small birds in the undergrowth of forests, or, in morning and evening, ventured out into the more open cover of scattered bushes at the borders of the savannas. The birds were curious and were readily enticed to view from dense coverts. The striking jet-black males, that occasionally showed a flash of white from the shoulder in flight, seemed more shy than the cinnamon-colored females.

In a male taken July 9 the maxilla and tip of the mandible were black; base of mandible pale neutral gray; iris dull brown; tarsus brownish black; underside of toes washed with yellow.

HEMITHRAUPIS GUIRA GUIRA (Linnaeus)

Motacilla guira LINNAEUS, Syst. Nat., ed. 12, vol. 1, 1766, p. 335. (Eastern Brazil.[90])

An adult male was taken at Kilometer 25, west of Puerto Pinasco, Paraguay, September 1, 1920, and a female on the Cerro Lorito, on the eastern bank of the Rio Paraguay, opposite Puerto Pinasco, on September 30. These two agree in. appearance with skins from Matto Grosso and Rio de Janeiro. Male specimens from southern Paraguay (Sapucay), representing *H. g. fosteri* Sharpe, have the rump and breast slightly darker, but only one has the broadened yellow markings on the superciliary and forehead supposed to distinguish this form. This subspecies is apparently poorly marked.

Though the generic term *Hemithraupis* is here used for this tanager, I see little save difference in color pattern to differentiate it from *Nemosia* of Vieillot.

These birds frequented the tops of tall trees in heavy forest. They were observed gathering nesting material September 30. The female had the maxilla and tip of mandible blackish brown number 3; rest of mandible cinnamon buff; iris bone brown; tarsus deep green-blue gray.

PIRANGA FLAVA (Vieillot)

Saltator Flavus VIEILLOT, Encyc. Méth., vol. 2, 1822, p. 791. (Paraguay.)

This tanager was fairly common in forested areas. It was observed at Resistencia, Chaco, July 8 and 10, 1920; Las Palmas, Chaco, July 30 and 31; Riacho Pilaga, Formosa, August 18; Formosa, Formosa, August 24; Kilometer 200, west of Puerto Pinasco, Paraguay, September 25; and Tapia, Tucuman, April 9, 1921. The four skins preserved include male and female from Resistencia, July 8 and 10; female from the Laguna Wall, 200 kilometers west of Puerto

---

[90] See Hellmayr, Nov. Zool., 1908, p. 30.

Pinasco, September 25; and male from Tapia, Tucuman, April 9. The bird last mentioned is just completing a molt from a worn, yellowish-green plumage to the red of the adult. It appears then that the immature male wears a livery similar to that of the female until the second fall when it acquires adult dress.

Piranga flava was found with bands of other small birds in forest growths, usually in the taller trees. The birds were marked when excited by a slow wagging motion of the tail, that was almost as regular as that of a *Cinclodes*. Their call note was a loud *chip* or *chu chu*. Their flight was undulating.

THRAUPIS BONARIENSIS (Gmelin)

Loxia bonariensis, GMELIN, Syst. Nat., vol. 1, pt. 2, 1789, p. 850. (Buenos Aires.)

The tanager known as *siete colores*, a name applied to other species of bright-colored birds as well, was fairly common in wooded areas, and was encountered in groves on the pampas of northern Buenos Aires. It was noted as follows: Riacho Pilaga, Formosa, August 8 and 21, 1920; Lavalle, Buenos Aires, November 1 to 14; Victorica, Pampa, December 23 to 29; La Paloma (near Rocha), Uruguay, January 23, 1921; San Vicente, Uruguay, January 25 to 31; Potrerillos, Mendoza, March 15 and 18; Tapia, Tucuman, April 6 to 13. Seven specimens were prepared as skins. Birds from the Chaco seem to have smaller bills on the average than those from farther south, but this is a highly variable character. Two in immature dress taken at Tapia, April 7 and 9, are more rufescent on breast and rump than adults.

This tanager ranges alone or in little flocks of three or four individuals that feed restlessly through the tops of the trees or fly with swift, undulating flight to fresh hunting grounds. They are active and seemed shy and easily alarmed.

The song, heard frequently during December, was sibilant, with little carrying power, and was uttered with much effort. It was given from a perch on the highest twig of low trees. In January, in Uruguay, the/birds were feeding young among low, dense thickets, and scolded with sharp notes at possible enemies.

THRAUPIS SAYACA OBSCURA Naumburg

Thraupis sayaca obscura NAUMBURG, Auk, 1924, p. 111. (Parotani, Dept. Cochabamba, Bolivia.)

The blue tanager was common in the broken woodland of the Chaco, but was not found elsewhere. It was recorded at Resistencia, Chaco, July 8 to 10, 1920; Las Palmas, Chaco, July 13 to 28 (immature male taken July 13, adult female July 17); Formosa,

Formosa, August 23 and 24, and Kilometer 80, west of Puerto
Pinasco, Paraguay, September 16 (two seen and a female taken).

Mrs. E. M. B. Naumburg [91] has recently separated the blue tanager
of Bolivia and Argentina (except Misiones) as a race that may be
distinguished by darker coloration especially above, somewhat
longer wing, and usually smaller bill than typical *sayaca* from northeastern Brazil (Bahia and Ceara). Skins from Las Palmas, Chaco,
are representative of *obscura*, while one from Kilometer 80, west of
Puerto Pinasco, while intermediate, is so dark that it is best identified with the western race. In the rather limited series that I have
seen skins from Matto Grosso, Paraguay, Sao Paulo, and Rio Grande
do Sul are distinctly intermediate and appear on the whole nearer
obscura than *sayaca*. This statement is made with reservation as it
is based on a limited number of skins.

During the winter period these birds were found in bands of three
or four to a dozen that worked quickly through the tops of the trees,
or that flew with undulating flight for long distances across country.
They were shy and difficult to see among the dense leaves of the
trees. Little flocks frequented orange trees where they cut open ripe
fruit to eat the juicy pulp.

Hartert is apparently correct in his statement [92] that there are no
Argentine records for *T. cyanoptera*, as in a fair series of Argentine
skins in the United States National Museum that species is not represented.

STEPHANOPHORUS DIADEMATUS (Temminck)

Tanagra diademata TEMMINCK, Nouv. Rec. Planch. Col. Ois., vol. 3, livr.
41, pl. 243, Dec. 27, 1823. (Brazil.)

The name *Tanagra diademata* is ordinarily cited from Mikan's
Delectus Florae et Faunae Brasiliensis as "pl. 2, 1820–25." On
consulting this rare work it is found that the name in question occurs
in the fourth fascicle which did not appear until 1825, so that Temminck's publication at the end of 1823 has priority.

This handsome species was common in Uruguay, where it was
noted at La Paloma, the port of Rocha, January 23 (adult and immature males taken); San Vicente, January 27 to 31 (male and
female collected); Lazcano, February 6 to 8 (male taken), and Rio
Negro, February 15 to 19. The birds inhabited dense thickets.
They rested often on the tops of bushes or trees, and when alarmed
dropped into heavy cover beneath. They decoyed readily and were
collected without difficulty. The song is a pleasant, rapidly uttered
warble, finchlike in tone, and the call a soft *chewp chewp*. The
flight is tilting. Unless light conditions are favorable they appear

---

[91] Auk, 1924, pp. 105–116.
[92] Nov. Zool., December, 1909, p. 171.

dull black so that their brilliant colors when in the hand come as a distinct surprise.

The immature male, in juvenal plumage, in general is deep mouse gray with a wash of blue on the back, rump, sides of the head and neck, and crown, with a grayish wash at the sides of the crown.

The bill in this species suggests that of *Phytotoma*, as it is evidently designed for cutting and crushing. The margin of the upper mandible is faintly crenulate. Near the tip there are some slightly projecting serrations just inside the cutting edge. Still farther inward is a row of distinct corneous projections that extend along either side of the palate, and converge to meet near the tip of the bill. On the inner margin of the lower mandible are slight projections that meet those above. The tongue is broad, strong, and bifurcate at the tip.

Family FRINGILLIDAE [93]

SALTATOR SIMILIS SIMILIS d'Orbigny and Lafresnaye

Saltator similis d'ORBIGNY and LAFRESNAYE, Mag. Zool., 1837, cl. 2, p. 36. (Corrientes, Argentina.)

This saltator was first seen near Resistencia, Chaco, when an adult female was taken July 9, 1920. At Las Palmas, Chaco, it was frequent from July 19 to 31 and two immature males were prepared as skins (July 19 and 26). Saltators that I assumed to be this species were recorded at the Riacho Pilaga, August 7, 11, and 14, and at Formosa, Formosa, August 24. Two males were secured at Kilometer 25, west of Puerto Pinasco, on September 1 and 3. These are all fairly uniform in dimensions and in color, the only differences noted being between adult and immature individuals, the latter being greener above and buffier on the breast than older birds. Berlepsch [94] has described *S. s. ochraceiventris* from Santa Catherina and Rio Grande do Sul (type locality Taquara) as deeper buff below, a character present in one skin in the United States National Museum from Rio Grande.

Like other saltators the present species inhabited thickets and groves where it fed either on the ground or among the tree tops. Those taken usually had the bill gummed with plant juices. Near Las Palmas they were seen in orange groves which they appeared to visit for the fruit. One brought to me alive was caught in a snare baited with corn.

Their usual call note was a low *pree-ee* or *prut prut*. They sang a cheerful whistled song from amid leaves in the tops of low trees,

---

[93] The family name Fringillidae is here used in the broader sense pending an allocation of South American forms in the logical scheme proposed by Doctor Sushkin (Auk, 1925, p. 260).

[94] Verh. V. Int. Ornith.-Kongr., 1911, p. 1146.

persistent notes that were heard constantly during the breeding season.

The Angueté Indians called this species *yum a pow ookh.*

SALTATOR CAERULESCENS CAERULESCENS Vieillot

Saltator caerulescens VIEILLOT, Nouv. Dict. Hist. Nat., vol. 14, 1817, p. 105. (Paraguay.)

On July 17, 1920, an adult male was taken at Las Palmas, Chaco. Several were seen in brush-grown areas, where they were shy and difficult to approach. Their call was a sharp *tsip.* It is possible that they were seen on other occasions, but all others taken were *Saltator s. similis.*

SALTATOR AURANTIIROSTRIS AURANTIIROSTRIS Vieillot

Saltator aurantii rostris VIEILLOT, Nouv. Dict. Hist. Nat., vol. 14, 1817, p. 103. (Paraguay.)

The boldly marked orange-billed saltator was found at Las Palmas, Chaco, on July 23 and 26, and was fairly common, though so shy that it came seldom in view. The species inhabited clumps of dense brush in pasture lands where safe cover was available. At intervals males came out on open perches on dead limbs, often near the tops of the trees in order to sing, but at the slightest alarm pitched down into heavy growth below. The manner of delivery of the song was quick and explosive, in a way suggesting the method of enunciation employed by the white-eyed vireo (*Vireo griseus*), so that on my first encounter with the species I was somewhat surprised to identify the singer as a member of the present family. The abrupt, but musical song may be rendered as *chu chu chu wheet a sur,* with the last syllable heavily accented. An adult pair was secured on July 26. The female has the coloration less intense and the black breastband restricted though still complete. The male when first killed had the bill zinc orange, marked obscurely with fuscous at the tip, especially along the distal third of the culmen, and in front of the nostrils; iris Rood's brown, tarsus and toes deep neutral gray. The duller coloration of the plumage in the female extended to the tint of her bill which was marked obscurely with honey yellow on the lower mandible, especially toward the tip and along the gonys, and elsewhere was dull black. The color of eyes and tarsi were similar to those of the male.

In Formosa, near the Riacho Pilaga, this subspecies was encountered on August 8, 13, and 14, when specimens were taken. One was preserved as a skeleton and the other two as skins. These last two are somewhat paler, especially on the undersurface, than those secured at Las Palmas. The Toba Indians named this bird *chi pi gih yo.*

In the extensive areas of dry scrub that covered the low hills near Tapia, Tucuman, the present form was common from April 6 to 13, 1921.   An adult bird, probably a female, in molt was taken April 6, and an immature male in process of change to the post-juvenal plumage on April 7.   Males sang occasionally during this period, in spite of their ragged feathers.

Measurements, in millimeters, of these specimens are appended in the following table.   It may be noted that the rectrices in this species often indicate considerable abrasion so that their measurement is variable.

Mu- seum No.	Sex	Locality	Date	Wing	Tail	Culmen	Depth [1] of bill	Tarsus
				Mm.	Mm.	Mm.	Mm.	Mm.
284100	Male ad___	Las Palmas, Chaco_____	July 26, 1920	91. 0	85. 0	17. 8	11. 9	27. 5
227427 [2]	_____do____	Kilometer 182, Formosa__	Aug. 8, 1920	87. 0	88. 7	18. 0	12. 3	26. 7
284098	_____do____	_____do_____	Aug. 13, 1920	88. 0	78. 5	18. 4	11. 6	25. 5
284099	_____do____	_____do_____	Aug. 14, 1920	92. 2	85. 5	18. 0	11. 5	25. 0
285214	Male im___	Tapia, Tucuman_____	Apr. 7, 1921	88. 0	81. 5	17. 5	11. 2	26. 6
284094	Female ad_	Las Palmas, Chaco_____	July 26, 1920	90. 0.	85. 6	17. 5	11. 6	26. 5
285213	Female ?__	Tapia, Tucuman_____	Apr. 6, 1921	86. 2	80. 5	17. 5	11. 4	26. 2

[1] Measured from posterior end of gonys to highest point on culmen.
[2] Preserved as a skeleton.

D. S. Bullock secured a male of this form (with culmen from base measuring 17.5 mm.) at Salta, on August 1, 1922.

SALTATOR AURANTIIROSTRIS NASICA Wetmore and Peters

Saltator aurantiirostris nasica WETMORE and PETERS, Proc. Biol. Soc. Washington, vol. 35, Mar. 20, 1922, p. 45.   (El Salto, alt. 1,800 meters, above Potrerillos, Province of Mendoza, Argentina.)

A specimen of this subspecies was collected near Victorica, Pampa, on December 27, 1920, and others were noted on December 29.   The birds frequented brushy areas in extensive forests of low, heavy-trunked trees, where it was difficult to gain an idea of their actual abundance.   Males were singing constantly from tree tops, and here rested among clumps of leaves rather than in the open.   At Potrerillos, Mendoza, this form was seen at intervals from March 15 to 21, and one was observed near Tunuyan, March 23.   In this Province they inhabited thickets, often of rather low brush, but had the retiring habit common to the species elsewhere.   The species seems to be a persistent singer, as from July to April I heard their songs frequently when in their haunts.

So far as known at present this subspecies ranges only in Mendoza (Alto Verde, Potrerillos, and Mendoza) and western Pampa (Victorica).   It is characterized by longer, larger, and heavier bill.   In males the culmen from base measures 19–21 mm.; depth of bill from posterior end of gonys to highest point on culmen, 13.5–14.5 mm.;

in females, culmen 19.3–20 mm.; depth of bill 13.5–14 mm. Berlepsch[95] has noted that skins from Mendoza differ from typical *aurantiirostris*, and also includes Cordoban specimens as similar. The typical subspecies is found to the north in Tucuman and Salta.

GUBERNATRIX CRISTATA (Vieillot)

Coccothraustes cristata VIEILLOT, Nouv. Dict. Hist. Nat., vol. 13, 1817, p. 531. (29° S. Lat., Argentina.)[96]

At Victorica, Pampa, this handsome cardinal was fairly common from December 27 to 29, 1920, and two males were prepared as skins. As this was the breeding season males were singing from perches among leaves in the tops of trees, a loudly whistled song, cheering in tone, and somewhat cardinallike, that may be represented as *wir-tu wir-tu tse kwa wir-tu*. Their usual call was a low *tsip*. They held the crest erect and presented a handsome, spirited appearance. The flight was undulating.

On January 25 and 31, 1921, several were seen near San Vicente, Uruguay, and on February 2 a male was taken at the Paso Alamo, on the Arroyo Sarandi. One was seen near Lazcano, Rocha, on February 7.

A male, taken December 27, had the base of the mandible dawn gray; rest of bill black; iris natal brown; tarsus and toes dull black.

PHEUCTICUS AUREO-VENTRIS (d'Orbigny and Lafresnaye)

Pitylus aureo-ventris d'ORBIGNY and LAFRESNAYE, Mag. Zool., 1837, cl. 2, p. 84. (Yungas, Sicasica, Bolivia.)

An adult male in full molt was brought to me by a boy at Tapia, Tucuman, on April 8, 1921. The maxilla was dull black, mandible dawn gray, shaded across gonys with gray number 6; iris carob brown; tarsus and toes dark plumbeus.

The species is in high repute as a cage bird under the name of *reina mora* or *rey de los pajaros cantadores*. It was observed in bird stores in the city of Mendoza, and on April 19 several young birds recently from the nest were offered for sale in the railroad station at Dean Funes, Cordoba.

CYANOCOMPSA CYANEA ARGENTINA (Sharpe)

Guiraca cyanea argentina SHARPE, Cat. Birds Brit. Mus., vol. 12, 1888, p. 73. (Fuerte de Andalgala, Catamarca.)

An adult male, taken July 23, 1920, at Las Palmas, Chaco, has a wing measurement of 83.5 mm., and so agrees with the present form.

---

95 Verb. V. Int. Ornith.-Kongr., 1911, p. 1146.

96 According to Azara, Apunt. Hist. Nat. Pax. Paraguay, etc., vol. 1, 1802, p. 464. The point mentioned, where Azara recorded three pairs, is assumed to be in the present Province of Corrientes.

Another (preserved as a skeleton) taken at the Riacho Pilaga, Formosa, August 18, was of equal size. Two immature birds, shot at Tapia, Tucuman, April 7 and 11, do not have the wing fully developed but are placed with this form on geographic grounds. According to a revision of this genus by Todd,[97] true *cyanea*, characterized by shorter wing (less than 80.5 mm.), ranges from Pernambuco and Rio Grande do Sul, west to Goyaz and Paraguay. These birds were found near the ground in heavy thickets where they were located with difficulty.

An adult male, taken July 23, had the bill black, becoming slate gray number 5 at base of mandible; iris Rood's brown; tarsus fuscous; toes dull black.

PAROARIA CRISTATA (Boddaert)

Fringilla Dominicana Cristata, BODDAERT, Tabl. Planch. Enl., 1783, p. 7, (Brazil.)

Mathews and Iredale [98] have indicated correctly that *Loxia cucullata* Latham 1790, long in vogue for the species known to aviculturists as the Brazilian cardinal, is preoccupied by *Loxia cucullata* Müller, so that the species must be titled as indicated above.

P. cristata was found as follows: Santa Fe, Santa Fe, July 4, 1920; Las Palmas, Chaco, July 21 and 30; Riacho Pilaga, Formosa, August 8 to 16; Formosa, Formosa, August 23 and 24; Kilometer 80, west of Puerto Pinasco, Paraguay, September 6 to 20 (three skins preserved); Lavalle, Buenos Aires, October 30 and November 2 (one taken); La Paloma, Uruguay, January 23, 1921; San Vicente, Uruguay, January 26 to 31 (one taken); Lazcano, Uruguay, February 5 to 9. Specimens from the Province of Buenos Aires have slightly larger bills than those from Paraguay. I have not seen skins with authentic localities from Brazil. The species was common in a number of localities visited in the Chaco and in eastern Uruguay, and was found occasionally on the pampas, where groves and thickets furnished cover. The birds fed often on the ground in wet localities, or on vegetation floating on little ponds where they walked about with ease as their large feet and long toes prevented their sinking deeply. At such times their bright colors contrasted handsomely with their green background. The species was common as a cage bird and was held in high esteem for its pleasant song.

A male, taken September 7, had the upper part of the maxilla and tip of the mandible deep mouse gray; sides of maxilla and rest of mandible whitish, tinged very slightly with gray; iris pecan brown; tarsus and toes fuscous black.

---

[97] Auk, 1923, pp. 58–69.
[98] Austr. Av. Rec., vol. 3, Nov. 19, 1915, p. 38.

COCCOPSIS CAPITATA (d'Orbigny and Lafresnaye)

Tachyphonus capitatus d'ORBIGNY and LAFRESNAYE, Mag. Zool., 1837, cl. 2, p. 29. (Corrientes, Argentina.)

From true *Paroaria* the group of species included in *Coccopsis* [99] is distinguished by more slender bill with more strongly curved culmen, less abrupt deflexure at gape, more exposed nostrils, and more strongly rounded tail. The present species is easily told from its congeners by the lack of dark markings on the bill.

The species was first recorded at Las Palmas, Chaco, July 17, 1920. A male was brought to me at the Riacho Pilaga, Formosa, on August 8, and at Formosa, Formosa, on August 23, several were seen. One was noted near Puerto Pinasco, Paraguay, September 3, and at Kilometer 80 west, where two were prepared as skins, the species was fairly common from September 6 to 20. To many this bird was known as *cardenilla* from the supposition that it was the female of *Paroaria cristata*.

Coccopsis capitata was found usually in pairs in or near thickets, and fed on the ground in pastures, wet meadows, or the borders of lagoons. The birds many times were wild and difficult to approach, but at Kilometer 80, in Paraguay, came familiarly about the ranch house. One was observed pulling bits of meat from a bone thrown out from the kitchen. One that I wounded was seized and carried away by a small hawk before I could get to it. The call of this species was a low *chew chew*.

An adult female, taken September 16, had the bill apricot orange, becoming slightly duller at the tip; iris English red; tarsus and toes vinaceous tawny; claws natal brown.

SPOROPHILA CAERULESCENS (Vieillot)

Pyrrhula Caerulescens VIEILLOT, Tabl. Encyc. Méth., vol. 3, 1823, p. 1023. (Brazil.)

This common seed-eater was reported at the following localities: Resistencia, Chaco, July 10, 1920 (adult male taken); Las Palmas, Chaco, July 17; La Paloma, Uruguay, January 23, 1921 (adult male taken); San Vicente, Uruguay, January 27 and 31; Lazcano, Uruguay, February 5 to 7 (a pair taken); Rio Negro, February 14 to 19 (immature female, February 19); Tunuyan, Mendoza, March 22 to 29. These little birds were found in brush grown arroyos, in open thickets, or in weed patches bordering cultivated fields, often in mixed parties with other small finches. When excited they came out on open perches and scolded with jerking tails. At other times they

---

[99] *Coccopsis* Reichenbach, Av. Syst. Nat., 1850, pl. 77; type, *Tanagra gularis* Linnaeus. Gray, Cat. Gen. Subgen. Birds, 1855, p. 74.

rested motionless as though made of wood. During summer males sang from the tops of low trees, an emphatic song in phrasing and pleasant sound like that of an indigo bunting. It was similar also to the notes of *Sicalis pelzelni*, but without the shrillness characteristic of that species. A female was seen feeding nearly grown young January 23.

An adult male, taken July 10, had the bill grape green, verging to dusky at tip, with a tinge of yellowish on tomia and toward base; iris bone brown; tarsus blackish brown number 1.

SPOROPHILA MELANOCEPHALA MELANOCEPHALA (Vieillot)

Coccothraustes melanocephala VIEILLOT, Nouv. Dict. Hist. Nat., vol. 13, p. 542. (Paraguay.)

At Las Palmas, Chaco, adult male and adult and immature females were taken July 22, 1920, and another adult male on July 23. Near Kilometer 80, west of Puerto Pinasco, Paraguay, an immature male was shot September 8 and an adult female on September 17. These are identified in accordance with Hellmayr's review of the species.[1] I have not seen specimens of *S. m. ochrascens* Hellmayr described therein (p. 534, type from "Rio Parana," taken by Natterer).

These little seed-eaters were found in heavy growths of weeds or other vegetation bordering lagoons, or occasionally in more open pastures. They flushed from heavy cover to fly with quickly tilting flight to safer distances, displaying the chestnut-marked rump prominently. Their call was a low *pree pree*. They share with *S. caerulescens* the Spanish name of *corbatita*.

CATAMENIA ANALIS (d'Orbigny and Lafresnaye)

Linaria analis d'ORBIGNY and LAFRESNAYE, Mag. Zool., 1837, cl. 2, p. 83. (Sicasica, Cochabamba, Bolivia.)

A male in molt was taken near Potrerillos, Mendoza, March 18, 1921, and others were seen the following day. Another was secured among low sand hills east of the Rio Tunuyan, near Tunuyan, Mendoza, March 27. Little flocks of these small birds frequented weed patches near irrigation ditches or small streams, or were found on the rock-strewn slopes above, sometimes among low bushes. The black and white markings of wings and tail were displayed prominently when on the wing. Their flight was undulating, like that of a siskin or goldfinch, while their soft notes suggested those birds rather than seed-eaters. Like many others of the small bird inhabitants of the hills, they were often very shy.

---

[1] Verb. K. K. Zool.-Bot. Ges. Wien, 1904. pp. 533–534.

SICALIS PELZELNI Sclater

Sycalis pelzelni SCLATER, Ibis, 1872, p. 42. (Buenos Aires.)

The present species, of wide distribution in Argentina, is represented by 12 skins from widely separated localities. It was recorded as follows: Berazategui, Buenos Aires, June 29, 1920; Resistencia, Chaco, July 10 to 31; Kilometer 182, Formosa, August 21; Formosa, Formosa, August 23 and 24; Kilometer 80, west of Puerto Pinasco, Paraguay, September 13 to 18; Lavalle, Buenos Aires, October 23 to November 13; Bahia Blanca, Buenos Aires, December 13; La Paloma, Uruguay, January 23, 1921; San Vicente, Uruguay, January 25 to 31; Lazcano, Uruguay, February 5 to 9; Rio Negro, Uruguay, February 14; Tapia, Tucuman, April 6 to 13.

A pair from Puerto Pinasco, Paraguay, are slightly smaller than specimens from farther south (wing in male, 64 mm.), so that possibly there are two forms in this species. Specimens from Sapucay, Paraguay, and Jujuy, and Resistencia, Chaco, in northern Argentina, do not differ in measurements from those from farther south (wing in males from southern localities, 67.8–71.4 mm.)

Immature males have the streaked sparrowlike plumage of females, and may not acquire the yellow of adult males until their second year, as a male shot at the Estancia Los Yngleses, near Lavalle, Buenos Aires, October 30, 1920, was breeding, though in immature dress.

These birds were found in pairs or little flocks that fed on the ground in bare spots among bushes, weeds, or clumps of saw grass. When startled they flew with undulating flight to a perch in bushes or low trees where they rested quietly. From October to the end of January males constantly sang an insistent song of pleasing character.

In eastern Uruguay the birds were common among the extensive palm groves in the lowlands.

SICALIS ARVENSIS ARVENSIS (Kittlitz)

Fringilla arvensis KITTLITZ, Mém. Acad. Imp. Sci. Saint-Pétersbourg, Div. Sav., vol. 2, 1835, p. 470, pl. 4. (Valley of Quillota, Chile.)

The *misto*, a widely distributed species, was encountered as follows: Lavalle, Buenos Aires, November 13, 1920 (adult male taken); General Roca, Rio Negro, November 23 to December 3 (a pair collected); Carhue, Buenos Aires, December 15 to 18 (three pairs); Carrasco (near Montevideo), Uruguay, January 9 and 16, 1921; La Paloma, Uruguay, January 23; San Vicente, Uruguay, January 27 to 31 (juvenile female); Lazcano, Uruguay, February 5 to 8; Tunuyan, Mendoza, March 22 and 23 (immature female). After comparison of these skins, with large series in the United States National

Museum, I agree with Hellmayr[2] that typical *arvensis* ranges from Matto Grosso and Rio Grande do Sul south to northern Patagonia (General Roca and Bariloche, Rio Negro) and Chile. Males from Chile are very slightly larger (wing, 75.6–78 mm.) than those from Brazil and central to·northern Argentina (wing, 71–76 mm.), but the difference is too slight to merit distinction. *S. a. luteola* Sparrman,[3] of more northern range, is smaller and brighter colored. *Sicalis chapmani* Ridgway, from Santarem on the lower Amazons, is apparently a distinct species and not a subspecies of *arvensis*, as it has a much larger, heavier bill with a more strongly curved culmen.

During the breeding season the *misto* is found about clumps of bunch grass or saw grass near streams·or marshes or, in irrigated sections, about alfalfa fields where it often gathers in little colonies. Males sing from perches in the grass or rise a few feet and scale down with long wings fully spread, while they utter their low, tremulous notes. Often before resting they rise to repeat the performance a second time, or sail about in erratic circles. The wings during this performance are fully spread and extended forward. Breeding colonies were observed at Lavalle, Buenos Aires, in November; at General Roca, Rio Negro, at the end of the same month; and at Carhue, Buenos Aires, the middle of December. A bird in juvenal plumage was taken at San Vicente, Uruguay, January 27. Two nests found near Carhue, December 18, contained four fresh eggs, and another on the same date held five young a week old. The nests, placed from 300 to 400 mm. from the ground, in clumps of stiff-stemmed grass, were cup-shaped structures made of fine grasses, lined with coarse hair from the manes and tails of horses. The eggs are white, tinged faintly with grayish green, with spots of Mars and chestnut brown, distributed over the egg, but concentrated heavily about the large end. The two sets taken differ considerably, as one is large and heavily blotched, while the other is smaller, rounder, and has much finer markings. The larger set measures 17.4 by 13.4, 17.5 by 13.2, 17.6 by 13.3, and 18.1 by 13.4 mm. The second set measures 16.1 by 13.7, 16.2 by 13.4, 16.3 by 13.6, and 16.5 by 13.6 mm. The eggs are fragile and thin shelled. (Pl. 20.)

SICALIS LUTEA (d'Orbigny and Lafresnaye)

Emberiza lutea, d'ORBIGNY and LAFRESNAYE, Mag. Zool., 1837, cl. 2, p. 74. (Summis Andibus, Bolivia.)

The statement of Todd[4] that the yellow finches of Middle and South America must be grouped in one genus, *Sicalis*, is substan-

[2] Nov. Zool., vol. 15, June, 1908, p. 34.
[3] According to Hartert, Vög. Pal. Fauna, Nachtr. 1, September, 1923, *Emberiza luteola* Sparrman, 1789, refers to the form known currently as *Sicalis a. minor* Cabanis.
[4] Ann. Carnegie Mus., vol. 14, October, 1922, p. 519.

tiated by careful examination of a large series that includes the majority of the species and subspecies. At first glance it appears that *Pseudosicalis* might be maintained as distinct from the fact that it has a much longer wing tip (one and one-half to two times as long as culmen), while in the species of true *Sicalis* the wing tip is much shorter (equal to or shorter than culmen). In *Sicalis arvensis*, however, both conditions hold, since in fresh fall plumage the wing tip may be no longer than the culmen, while in birds shot in the breeding season it may be twice as long. The two groups may be maintained merely as subgenera.

Under these circumstances *Pseudochloris* Sharpe may prove tenable as a subgeneric name and take precedence over *Pseudosicalis* Chubb[5] proposed to replace it on the grounds that the type of *Pseudochloris*, *Orospina pratensis* Cabanis, is considered a species of *Sicalis*. If the illustration given by Cabanis[6] is correct, however, in depicting the form of the wing tip *pratensis* would be subgenerically distinct from *Sicalis*. The species in question is not available at present for examination.

Near Potrerillos, Mendoza, *Sicalis lutea* was common from March 18 to 21, 1921. The birds were found in little flocks, family parties in most cases, that frequented dry, rock-strewn slopes or flats often high among the hills. Here they hopped about over the rough surface in search of grass seeds that had fallen to the ground or fluttered up to pull off achenes that still clung to the seed heads. Occasionally one rested for a short time on top of some bowlder. Attention was usually attracted to them by their musical call note, a pleasant *tweep tweep* that suggested familiar notes of other flocking finches of the *Carpodacus* or *Astragalinus* type. The birds, like many other finches that inhabit open mountain slopes, were shy and difficult to approach. Their flight was undulating and not strong. On the brushy flats near the foothills 15 miles west of Tunuyan, Mendoza, a dozen were seen March 24. A heavy storm in the mountains at the time may have driven them out on the plains.

Two adult males taken at Potrerillos, March 18 and 21, are in molt. Two other birds secured at the same time are in juvenal dress. The latter are plain brown, faintly streaked on the head, with only a hint of yellow on lower breast, abdomen, tail edgings, and under tail coverts. In this species there is a distinct knob that protrudes on the palate and apparently serves to aid in cracking seeds.

[5] Bull. Brit. Orn. Club, vol. 41, Feb. 24, 1921, p. 78.
[6] Journ. für Ornith., 1883, pl. 1, fig. 1.

PHRYGILUS FRUTICETI FRUTICETI (Kittlitz)

Fringilla fruticeti KITTLITZ, Kupf. Naturg. Vög., 1832, p. 18, pl. 23, fig. 1.
(Coast near Valparaiso, Chile.)

In a recent partial review of *Phrygilus*, Lowe [7] has proposed to divide the genus as ordinarily taken into four groups, basing his action apparently mainly on color pattern. (He does not include in his study *P. alaudinus* and its allies, which have been placed by some in a fifth segregation.) After a careful study of the entire group, and a consideration of the structural characters of the different species, I am led to follow a somewhat different course. The two species *melanodera* and *xanthogramma* differ from all others involved in the greatly exaggerated wing tip, where the distance from the longest secondaries to the tip of the longest primaries is one-third of the length of the wing, the ninth (outermost) primary is equal to or slightly longer than the eighth, and the other primaries decrease regularly in length to the first. (In true *Phrygilus* the wing tip is decidedly more rounded.) In addition the bill is more conical, more sharply pointed, a condition that reaches its maximum in *xanthogramma*. These two species may be separated as the genus *Melanodera* Bonaparte.[8] (I am uncertain as to the relationships of *Rowettia* to this group, as I have not seen specimens.)

All of the remaining species must be included in the genus *Phrygilus*, since there are no structural characters whereby groups of the species involved may be separated definitely from one another. The divisions that have been proposed may be considered one by one. The type of the supposed genus *Rhopospina*,[9] *Phrygilus fruticeti* (Kittlitz), has the wing formula, wing tip, tarsal and bill structure so similar to that of *Phrygilus gayi* (Eydoux and Gervais) that the two may not be separated except on the basis of color. *Phrygilus alaudinus* (Kittlitz), which has been segregated as the type of *Corydospiza* Sundevall,[10] has the inner secondaries almost as long as the primaries, a character shared by *Phrygilus carbonarius* (d'Orbigny and Lafresnaye), which from color is placed by Lowe in *Rhopospina*. The wing tip is gradually lengthened in *P. plebejus* Cabanis, and *P. ocularis* Sclater, until it approximates the length found in *P. gayi* and its allies. *Corydospiza* may be used as a subgenus to include the four species mentioned in which the wing tip is shorter than the culmen from base, as distinguished from a subgenus *Phrygilus* (*verus*), in which the wing tip is longer than the culmen. Transition between the two is so gradual

---

[7] Ibis, 1923, pp. 513–519.
[8] Consp. Gen. Av., vol. 1, 1850, p. 470. Type, *Emberiza melanodera* Quoy and Gaimard.
[9] Cabanis, Mus. Hein., pt. 1, April, 1851, p. 135.
[10] Av. Tent., 1872, p. 33.

that there is no sharp dividing line. *Phrygilus unicolor* (d'Orbigny and Lafresnaye), which has been proposed as type of a genus *Geospizopsis* Bonaparte,[11] is identical in form with *P. gayi* and must be considered a synonym of true *Phrygilus*.

Haplospiza Cabanis,[12] with *Haplospiza unicolor* Cabanis as its type, is closely similar to certain forms of *Phrygilus unicolor* in general appearance. It differs structurally from true *Phrygilus* in more rounded tail, more rounded wing tip, in having the ninth (outermost) primary shorter than the third, about equal to the second, more attenuate bill, and shorter lateral toes, with the fourth toe without claw reaching barely beyond the base of the last joint of the middle toe, and the second toe without claw barely extended to that point. *Haplospiza unicolor*, in addition to smaller size, is distinguished at a glance from any *Phrygilus* in lacking any edging of lighter color on the outer webs of the primaries.

Near General Roca, *Phrygilus fruticeti* was common from November 25 to December 2 (four prepared as skins), and others were recorded at Zapala, Neuquen (one taken). The birds were not breeding, but were gathered in little flocks that fed on the open ground under shelter of desert shrubs of various species. The birds were wary, and at any alarm rose and flew with a swiftly darting or undulating flight for long distances before again dropping to cover. In early morning they rested in low bushes in the sun. Their call was a low, mewing note or a sharp *plick*. On December 2 males were singing a song that sounded like the effort of some icterid.

On comparison of a good series that includes 10 skins from the highlands of Peru, 5 from central Chile, and 11 from Rio Negro and Neuquen, in Argentine Patagonia, it appears that *Phrygilus fruticeti peruvianus* Zimmer,[13] while not sharply differentiated, may be distinguished from true *fruticeti* in adult males by the grayer coloration above, with more sharply defined, blacker streaks, and in females by heavier streaking above, especially on the head. The white spots on middle and greater coverts are usually, but not always, larger in *peruvianus*. In size the two forms, from the series at hand, appear quite similar, except that in Peruvian birds the bill is slightly larger. In five males from La Raya and one from Arequipa, Peru, the wing ranges from 94 to 100.5 mm., the bill from 14.4 to 15.5 mm.; in three males from Santiago and central Chile the range is from 94 to 100.2 and 13.2 to 13.5 mm.; while in five from Rio Negro, Argentina (General Roca, Arroyo Cumallo, Arroyo Seco, Paja Alta, and Nahuel Niyeu), it is 91 to 98 and 13 to 14.7 mm. The length of wing, it will be noted, is practically identical.

[11] Compt. Rend., vol. 42, 1856, p. 955.
[12] Mus. Hein., pt. 1, April, 1851, p. 147.
[13] Field Mus. Nat. Hist., Zool. Ser., vol. 12, Apr. 19, 1924, p. 63.   (Matucana, Peru.)

PHRYGILUS CARBONARIUS (d'Orbigny and Lafresnaye)

Emberiza carbonaria d'ORBIGNY and LAFRESNAYE, Mag. Zool., 1837, cl. 2, p. 79. (Patagonia.)

The present species has the wing structure typical of the subgenus *Corydospiza*, with the inner secondaries elongated until they are nearly as long as the longest primaries.

Near Ingeniero White, the port of Bahia Blanca, Buenos Aires, an adult male was taken December 13, 1920, among greasewoods near the bay. Another was shot December 26 near Victorica, Pampa, in rather high bunch grass in an old pasture. Both specimens were very wild and were secured with difficulty. Their flight was undulating, and they alighted indifferently on the ground or on low bushes.

The male taken December 13, when first killed had the bill between mustard and primuline yellow; iris Vandyke brown; tarsus and toes chamois; nails deep neutral gray.

PHRYGILUS ALAUDINUS ALAUDINUS (Kittlitz)

Fringilla alaudina, KITTLITZ, Kupf. Naturg. Vög., 1832, p. 18, pl. 23, fig. 2. (Chile.)

Near Concon, Chile, April 28, 1921, these birds were fairly common amid scattered, scrubby bushes over a broad flattened hill top, where they fed on the ground in company with *Brachyspiza* and diuca finches. When flushed they flew away with undulating flight or perched in the tops of low bushes. Of three taken, a pair were preserved as skins and a single bird as a skeleton.

An adult male, when first killed, had the bill tipped with dark mouse gray, the remainder chamois; iris Vandyke brown; tarsus and toes honey yellow; claws dark neutral gray. In a female the base of the mandible except at the cutting edge was wood brown, and the remainder of the bill fuscous black; iris Vandyke brown; tarsus and toes honey yellow; claws dark neutral gray.

PHRYGILUS ALAUDINUS VENTURII Hartert

Phrygilus alaudinus venturii, HARTERT, Nov. Zool., December, 1909, p. 180. (Lagunita, Tucuman, 3,000 meters.)

An adult female and a juvenile male were taken at an altitude of 2,300 meters on the Sierra San Xavier, above Tafi Viejo, Tucuman, on April 17, 1921. The adult female has the following measurements: Wing (worn), 78; tail, 60.4; culmen from base, 13.2; tarsus, 24.5 mm. Four females of *alaudinus* from central Chile (Concon, Santiago, and two without locality) show the following range: Wing, 72.0–74.6; tail, 50.4–55.5; culmen from base, 12.1–13; tarsus, 20–21.8 mm. The bird from Tucuman is darker than those from Chile, and is duller white on the abdomen.

On the date mentioned three or four were found on open, rounded shoulders at the summit of the cumbre, where the ground was covered with tussock grass. In habits they resemble *P. a. alaudinus*.

DIUCA DIUCA (Molina)

Fringilla Diuca, MOLINA, Sagg. Stor. Nat. Chili, 1782, p. 249. (Chile.)

Handsome dinca finches were found everywhere in the valley of the Rio Aconcagua near Concon, Chile, from the open sand dunes of the coast inland to the brush-grown hillsides. Flocks were often seen on the ground, or single birds were noted perched on thistles, bushes, or the tops of trees, where they were readily recognized by their white throats when their stocky forms and heavy bills were not sufficient to identify them. In flight the white in the tail was prominent. Flocks were seen often feeding in growths of weeds or on open ground. They were a veritable pest in truck gardens as they destroyed the leaves and tender shoots of growing vegetables. In small fields it was common to see a 5-gallon oil tin suspended from a stick with a piece of tin hanging against one side and a wire leading to a hut, perhaps 70 or 80 meters away. As the diucas alighted to feed the wire was jerked causing the tin to rattle against the can, a noise that made the finches rise in alarm and pass on. On the whole the arrangement seemed very effective.

Two males and one female were preserved as skins. A male, shot April 25, had the tip of the bill dull black; base of maxilla dark mouse gray; a line of dark mouse gray along cutting edge of mandible, rest light Payne's gray; iris bone brown; tarsus dusky green-gray; toes sooty black.

DIUCA MINOR Bonaparte

Diuca minor BONAPARTE, Consp. Gen. Av., vol. 1, 1850, p. 476. (Patagonia.)

The lesser diuca finch was fairly common at General Roca, Rio Negro, from November 23 to December 3, 1920, where a male and four females were taken, and was found in equal numbers near Victorica, Pampa, from December 23 to 29, where two pairs were preserved as skins. One was seen in a warm north valley below Zapala, Neuquen, on December 9.

D. minor is treated here as specifically distinct from *D. diuca* as, though the two appear complementary in range and are of similar color pattern and color, in the series examined there is no apparent intergradation in size. Males of *diuca* have decidedly longer wings than males of *minor*, but the difference in females of the two are less pronounced. *D. diuca*, however, may always be told by the longer, heavier bill. The difference in bulk between well-made skins of the two is decided.

Near General Roca the lesser dinca finch was widely spread through the arid, gravel hills and the less somber brush-covered flats of the river-flood plain. During early morning they ran or hopped about on the ground in search for food; but later in the day were observed resting quietly in some low bush, perched in the sun. Their flight was strongly undulating. At this season males were singing a pleasant warbling song, in character like that of some grosbeak, but nesting had not begun so far as I could ascertain. At Victorica, Pampa, at the end of December, all were nesting, some having fresh eggs while others were feeding young a few days old. The small young have the down on the head dull gray, becoming lighter behind until it is white over the posterior part of the body. For nesting sites the finches chose the old stick nests of various tracheophones, *Anumbius*, and others, strongly made domiciles that withstand weathering for some time and with their thorny twigs offer armed protection against most enemies. The nests chosen ranged from 300 to 400 mm. in diameter. Eggs or young, as the case might be, were contained in a covered, concealed nest cavity lined warmly with plant downs and feathers, comfortable furnishings that had perhaps been inherited with the rest of the home from the original occupants. Finches hopped about on many of these nests, and from time to time I found one that they were occupying. One set of four fresh eggs was taken from a stick nest 7 feet from the ground in a low tree. The nest was in good repair with an entrance through an old tunnel at one side. The ground color of the eggs is very pale greenish white, while they are heavily marked with indefinite large and small spots of bone brown and pale dusky brown. They measure 19.6 by 16, 20.1 by 16.6, 20.4 by 16.2, and 21 by 16.2 mm. (Pl. 10.)

A second heavily incubated set contained one egg of the finch and one of *Molothrus b. bonariensis*. The finch's egg, similar in color to those described above, measures 20.3 by 16.7 mm.

An adult female of this species, taken November 27, 1920, when fresh, had the maxilla and tip of mandible blackish slate, rest of mandible dawn gray; iris, natal brown; tarsus, deep neutral gray.

BUARREMON CITRINELLUS Cabanis

Buarremon (Atlapetes) citrinellus CABANIS, Journ. für Ornith., 1883, p. 109. (Chaquevil and St. Xavier, Tucuman.)

An adult female was taken at an altitude of 1,800 meters on the Sierra San Xavier, above Tafi Viejo, Tucuman, April 17, 1921. The bird was found in a very dense growth of waist-high weeds at the border of a grove. It was quiet and slow in movement and silent except for a faint call, *tsip*.

In conformation of bill and wing this species is closely similar to the type of *Buarremon*, *B. torquatus*, and as it differs slightly in structural characters from the type of *Atlapetes*, where it has been placed in recent years, it is here located in *Buarremon*. Differences between *Buarremon* and *Atlapetes* appear intangible, and careful study of all the species may indicate that the two groups merge. A realignment of the species concerned seems necessary at least.

ARREMON POLIONOTUS POLIONOTUS Bonaparte

Arremon polionotus "Pucheran" BONAPARTE, Consp. Av., vol. 1, 1850, p. 488. (Corrientes, Argentina.).

A specimen of the present species was taken at Resistencia, Chaco, July 9, 1920, others were seen at Las Palmas, Chaco, July 14 and 31, and still others at the Riacho Pilaga, Formosa, on August 8 and 18 (when two were taken). The skin from Resistencia is from a short distance from the type locality. *A. p. devillei* Des Murs, from Corumba and eastern Bolivia to central Brazil, is marked by slightly paler dorsal surface and possibly by narrower pectoral band.[14]

Chubb[15] has placed *Arremon callistus* Oberholser, from Sapucay, Paraguay, in the synonomy of *A. polionotus*. The type of *callistus* differs, however, in that the wing coverts are like the back except for a faint wash of olive green on the lesser coverts in which it is distinct from all *polionotus* seen. The difference may be individual or the specimen may represent a true form.

These birds were found in dense brush, often near clearings, where they fed on or near the ground. Their call was a faint *tseet*. One of unknown sex, taken at Resistencia, when first killed had the greater part of the maxilla black; mandible and sides of maxilla for posterior three-fourths cadmium orange; iris very dark brown; tarsus benzo brown; toes, including claws on second to fourth, mouse gray; hind claw vinaceous buff. The pale color of the claw of the hallux is easily seen in fresh skins, but the distinction disappears when toes and tarsi bleach to yellowish brown as the specimen ages.

CORYPHOSPINGUS CUCULLATUS ARAGUIRA (Vieillot)

Fringilla araguira VIEILLOT, Hist. Nat. Ois. Chant., fasc. 5, 1808,[16] p. 52*. (not 52), pl. 28* (not 28). (Corrientes.[17])

Skins of *Coryphospingus cucullatus* from Argentina are distinctly grayer on the dorsal surface than those from Brazil and more north-

---

[14] See Hellmayr, Nov. Zool., vol. 13, 1906, p. 313.

[15] Ibis, 1910, p. 631.

[16] See Sherborn, Ind. Anim., pt. 3, 1923, p. 421.

[17] Though Vieillot mentions Guiana in connection with this bird he based his description largely on Azara's account of the *araguira* (Apunt. Hist. Nat. Pax. Paraguay, vol. 1, 1802, pp. 499–508). Azara states that he did not see the species south of 30° S. lat. The type-locality is hereby restricted to Corrientes, on the Rio Parana, which was included in Paraguay in the days of Azara.

ern localities, and may be distinguished under the subspecific name of
araguira. Skins from Tapia, Tucuman, and Salta are somewhat
brighter than those from Chaco and Formosa, while specimens from
Sapucay and Puerto Pinasco, Paraguay, are distinctly intermediate
between the typical form and its southern representative, but are
placed best with araguira. The range of C. c. araguira may be given
as from Santa Fe and Corrientes north through Paraguay, west
apparently into Bolivia (one specimen seen marked "La Paz").

At Resistencia, Chaco, this finch was observed commonly from
July 8 to 10, 1920, and a fine male was collected July 8. At Las
Palmas, Chaco, a few were noted from July 16 to August 1, and at
the Riacho Pilaga, Formosa, a few were observed from August 13
to 18, and male and female were taken August 14. The species was
noted near Formosa, Formosa, on August 24, and at Kilometer 80,
west of Puerto Pinasco, it was recorded occasionally from September
7 to 15. On September 30 one was taken from a little flock at the
base of the Cerro Lorito on the eastern bank of the Rio Paraguay
opposite Puerto Pinasco.

These handsome little birds were found in pairs or small bands in
low, heavy brush at the borders of fields or groves. They sometimes
fed like other finches on the ground among clumps of grasses, but
more frequently at an alarm appeared from the depths of thickets
to scold excitedly and perhaps to fly with an undulating flight for
a few yards to new cover. Sometimes they were encountered in
company with Poospiza melanoleuca. The feathers of the eyelids
are clear white, so that at a little distance the eye appears large and
white.

The Toba Indians in Formosa called this species koi yoh.

A male, taken July 8, had the maxilla dusky black; tip of mandible
dull slaty black; base pallid purplish gray; iris Army brown; tarsus
and feet dusky brown.

BRACHYSPIZA CAPENSIS CAPENSIS (P. L. S. Müller)

Fringilla Capensis P. L. S. MÜLLER, Vollst. Naturs., Suppl., 1776, p. 165.
(Cayenne.[18])

Although the *Brachyspiza* from the coast of Uruguay near
Montevideo belongs to the subspecies *argentina*, an adult male and
three specimens in juvenal plumage secured near Rio Negro in the
Department of Rio Negro, a little more than 240 kilometers north,
seem best placed under *B. c. capensis*. The adult male, taken February 17, 1921, in rather worn breeding plumage just beginning
the post-nuptial molt, when compared with *B. c. argentina*, is much
darker, with a narrower median crown stripe, broader, bolder black
markings on the wings and back, and deeper, more rufescent edgings

---

[18] See Berlepsch and Hartert, Nov. Zool., vol. 9, April, 1902. p. 28.

on the feathers of the entire dorsal surface. The sides and flanks are deeper in color and the wing and tail feathers blacker. Three juvenile birds, all fully grown (secured February 15, 17, and 19), compared with specimens of *argentina* in similar stage, have the dark markings blacker and average slightly more rufescent on the dorsal surface. The adult male has the following measurements: Wing, 73.3; tail, 62.5; culmen, 13.4; tarsus, 23.6 mm. Comparative material at hand representing *capensis* is slight, so that the specimens under discussion are allocated under that name because of their dark coloration, in spite of the fact that they seem a trifle large. They may be representatives of a distinct race not at present recognized.

From February 14 to 19 these birds were common in thickets near the Rio Negro, where they were found in little parties in which adults, young in molt, and those that had just attained juvenal plumage fed together in amity. The breeding season was about completed and adults and young were in many cases in ragged molting plumage.

BRACHYSPIZA CAPENSIS ARGENTINA Todd

Brachyspiza capensis argentina Todd, Proc. Biol. Soc. Washington, vol. 33, Dec. 30, 1920, p. 71. (Rio Santiago, near Buenos Aires, Argentina.)

Specimens assigned to this form were taken as follows: Buenos Aires (Province)—Berazategui (near Buenos Aires), June 29, 1920; Lavalle (formerly known as Ajo), October 23, 30, and November 1 (5 specimens); Carhue, December 15 and 17 (3 specimens); Guamini, March 4 and 7, 1921; Ingeniero White (the port of Bahia Blanca), December 13, 1920. Pampa—Victorica, December 23 and 27. Chaco—Resistencia, July 8 and 9; Las Palmas, July 16. Uruguay—San Vicente, January 25, 29, and 31 (4 specimens).

The series from the Province of Buenos Aires is uniform, with allowance for the changes due to seasonal wear. A single specimen from Berazategui in winter is from within a few miles of the type locality, while others from Lavalle are not far distant. Other specimens from Bahia Blanca, in the extreme south, are not noticeably different from those mentioned from near the type-locality. A breeding female from Victorica, Pampa, in appearance is also typical of this subspecies. Birds shot in winter near Resistencia and Las Palmas, Chaco, not far from the Paraguay River, have the measurements of *argentina*, but average somewhat darker, one female especially having an olive wash that deepens considerably the color of the dorsal surface. These seem to represent an approach to what is now considered *capensis*, but lack the distinctly rufescent markings of that bird. A small series of skins from San Vicente,

Uruguay, are distinctly intermediate, as they show a tendency toward a narrower median crown stripe and darker, more rufescent dorsal surface. They may, however, be placed without difficulty with *argentina*. Specimens that I have examined from Montevideo are similar to *argentina* from northern Buenos Aires, while one from Quinta, a few miles from the town of Rio Grande do Sul, Brazil (collected by E. G. Holt), is representative of *capensis*. The skins from San Vicente thus are connecting intermediates between the two forms. -

In the following table are given pertinent measurements, in millimeters, of the series taken of the present form.

Catalogue No.	Sex	Locality	Date	Wing	Tail	Culmen	Tarsus
				Mm.	*Mm.*	*Mm.*	*Mm.*
284123	Male	Berazategui, Buenos Aires	June 29, 1920	70.0	59.5	11.5	22.1
284135	...do	LaValle, Buenos Aires	Oct. 23, 1920	72.2	60.5	12.5	23.0
284122	...do	....do	....do	68.5	57.5	11.3	21.5
284138	...do	Carhue, Buenos Aires	Dec. 15, 1920	69.5	56.7	12.5	20.2
284133	...do	....do	....do	71.2	60.1	11.7	21.0
284129	...do	....do	Dec. 17, 1920	69.0	56.5	11.5	22.2
284144	...do	Ingeniero White, Buenos Aires	Dec. 13, 1920	73.5	59.2	11.7	20.5
284131	...do	....do	....do	71.0	60.5	11.7	20.0
284134	...do	....do	....do	69.0	57.0	12.5	23.0
284145	...do	Resistencia, Chaco	July 8, 1920	71.5	61.5	11.7	20.7
284145	...do	Las Palmas, Chaco	July 16, 1920	70.0	61.0	12.5	22.1
284127	...do	San Vicente, Uruguay	Jan. 25, 1920	73.0	64.0	12.4	22.5
284139	...do	....do	Jan. 31, 1920	72.5	60.2	12.7	23.0
		Average for 13 specimens		70.8	59.6	12.0	21.7
284137	Female	LaValle, Buenos Aires	Oct. 23, 1920	66.2	57.5	12.0	22.0
284140	...do	....do	Oct. 30, 1920	66.0	53.5	13.0	22.0
284126	...do	....do	Nov. 1, 1920	65.0	53.0	12.6	22.1
284134	...do	Victorica, Pampa	Dec. 23, 1920	70.0	57.6	11.7	20.5
284132	...do	Resistencia, Chaco	July 9, 1920	68.5	55.5	13.1	21.0
284142	...do	San Vicente, Uruguay	Jan. 29, 1920	67.0	59.5	12.6	20.6
284147	...do	....do	....do	69.1	57.5	12.4	21.0
		Average for 7 specimens		67.4	56.3	12.6	21.3

Birds of this subspecies were recorded in my notes as follows (where not represented by specimens these records are allocated on the basis of probability): Berazategui, Buenos Aires, June 29, 1920; Resistencia, Chaco, July 9 to 10; Las Palmas, Chaco, July 13 to August 1; Formosa, Formosa, August 23 and 24 (possibly *B. c. mellea*, as no specimens were taken); Dolores, Buenos Aires, October 21; Lavalle, Buenos Aires, October 23 to November 15; Ingeniero White, Buenos Aires, December 13; Bahia Blanca, Buenos Aires, December 14; Carhue, Buenos Aires, December 15 to 19; Victorica, Pampa, December 23 to 30; Montevideo, Uruguay, January 9 and 14, 1921; Carrasco, Uruguay, January 16; La Paloma, Rocha, Uruguay, January 23; San Vicente, Uruguay, January 25 to February 2; Lazcano, Uruguay, February 3 to 9; Guamini, Buenos Aires, March 3 to 8.

Specimens in full juvenal plumage were secured at Victorica, Pampa, December 27 and Guamini, Buenos Aires, March 7. Birds were in molt from the latter part of January until the middle of April.

In a region where finches in variety of species are far from common *Brachyspiza* is a welcome sight to the traveler from other lands, though at times its familiar form may seem somewhat out of place when in company with tracheophones or others of the more typical South American birds. In their adaptability to diverse faunal areas, their abundance, and their trustful acceptance of man and the changes that he has wrought in the face of the earth, these sweet-voiced finches have gained a place in the esteem of the countrymen held by few in lands where anything bearing feathers is regarded with interest mainly as a potential source of food. Though found usually in little flocks or at times alone about small openings in brush or scrub, in the pampas, where such cover is scant, they come to frequent tracts of weeds or even low grasses or other scant shelter that may be available. In towns, where they are scattered through gardens and plazas, and even enter the small patios where shrubbery may offer shelter, their friendly traits as they come about in search of crumbs and their restful songs endear them to the hearts of all.

They feed almost invariably on the ground in a manner usual among small sparrows, searching for food in open spaces or scratching with both feet at once where dead vegetation may conceal possible tidbits. The flight is tilting, and when flushed from the open they may take refuge in heavy cover or may fly up to rest in a weed, shrub, or low tree. Their usual call note is a low *tsip*, that with excitement or fear becomes more emphatic and insistent. The song is a clear, modulated whistle low in tone, but still with sufficient volume to make it audible for some distance, while not harsh or unpleasant to the ear when heard near at hand. In some of its inflections it suggests the notes of an eastern meadow lark (*Sturnella magna*), but on the whole has a closer similarity to the utterances of *Zonotrichia leucophrys*. Though heard at its best in the breeding season from October to December, it is given more or less constantly throughout the year, even during the period of molt.

Near Dolores, Buenos Aires, a *chingolo*, as this bird is known universally, was seen carrying nesting material on October 21, and near Lavalle I found the birds in pairs on October 23. In this region of low elevations they frequented slightly elevated ground about marshes. At the Estancia Los Yngleses a nest collected October 30, was placed under a little shrub in an open space surrounded by small trees that grew among low sand dunes. The nest, entirely concealed from above, was a very slight structure.

as the sandy soil in which it was placed was soft and yielding, so that little padding was required to protect the eggs from injury. In fact, the nest material was formed into a mere rim that main-tained the form of the depression and prevented sand from sifting down on the eggs. This nest contained three eggs with slight in-cubation. The color of these is white tinged faintly with bluish green, with diffuse markings of vinaceous russet heavier toward the larger end, where there is an occasional dot of black, but distributed in fine irregular dots over the entire surface. The three measure 21.5 by 16.1, 21.8 by 16, and 21.6 by 15.5 mm. Another nest dis-covered November 1, on the ground in a growth of weeds in a garden, was a strong, firmly built structure of grasses with a few horsehairs in the lining. It contained three eggs with incubation begun, similar in color to the set described above, but with less diffusion of the brown markings so that the spotting appears bolder and heavier. These eggs measure 22 by 15.8, 20.4 by 15.6, and 19.7 by 15.4 mm. The larger volume of the eggs in these two sets of *B. c. argentina*, when compared with the single sets of *B. c. choraules* and *B. c. canicapilla* secured, is noticeable, though the series ex-amined is too small to permit much generalization.

During the first week in November, 1920, when I pitched tem-porary camp in an unoccupied hut far from other habitation in the sand dunes below Cape San Antonio, eastern Buenos Aires, a pair of chingolos came to the threshold as soon as my outfit had been placed under cover. And during a three days' gale of violent wind and rain that followed these sparrows, with the *toldero* (*Em-bernagra platensis*), came invariably at meal time to secure bits of hard bread that I tossed out to them, the only birds undismayed by the force of the elements.

On the whole, *B. c. argentina* would seem to be more or less sedentary as its range does not include areas of rigorous climate. In its typical form it is essentially a subspecies of the pampas region, where it occurs universally from the drier interior, to the salt marshes of the coast.

BRACHYSPIZA CAPENSIS HYPOLEUCA Todd

Brachyspiza capensis hypoleuca TODD, Proc. Biol. Soc. Washington, vol. 28, Apr. 13, 1915, p. 79. (Rio Bermejo, between Oran and Embarca-cion, Province of Salta, Argentina, altitude 400 meters.[19])

The present subspecies, recorded previously in literature only from the type locality, was secured at Tapia, Tucuman, a consider-

---

[19] In the original description the type was said to have come from the "Rio Bermejo, Argentina." Mr. Todd informs me that the bird was secured by Steinbach, probably between Oran and Embarcacion; the type specimen is marked "Rio Bermejo, Prov. Salta, Argentina, 400 m."

able southward extension in its known range, and I have seen a male taken by D. S. Bullock, July 29, 1922, at Jujuy. Through W. E. Clyde Todd, of the Carnegie Museum, I have been permitted to examine the type of *B. c. hypoleuca*, two additional specimens from the type locality, and two others from Guanacos, Province of Cordillera, Bolivia. The present form differs from *B. c. argentina* in more rufescent dorsal surface, with the dark longitudinal stripes extended forward practically to the brown neck collar. Below *hypoleuca* is whiter with little or no gray on the breast, and with the lateral underparts brighter, more buffy in color.

At Tapia, Tucuman, from April 6 to 13, 1921, these birds were common in weed-grown areas scattered through the dry, scrubby forest, or in weed patches that bordered fields and low thickets. It is possible that the number of residents in this region had been augmented by the arrival of migrants from the higher levels of the mountains a few kilometers westward, as it was usual to find them in flocks that contained nearly 100 individuals. Though a part were in fresh fall plumage, others were in ragged molt. Their flocks often were found in company with *Coryphospingus* and *Saltatricula*. In spite of their disreputable feathers, males sang regularly, and I saw one, more exuberant than others, rise to give a flight song. One male and three females collected on April 6, 8, and 9 offer the following measurements:

Catalogue No.	Sex	Locality	Date	Wing	Tail	Culmen	Tarsus
				Mm.	*Mm.*	*Mm.*	*Mm.*
285254	Male	Tapia, Tucuman	Apr. 9, 1921	73. 7	65. 2	12. 5	21. 0
285248	Female	____do____	Apr. 6, 1921	68. 2	58. 7	12. 0	21. 8
285257	___do___	____do____	Apr. 8, 1921	67. 0	57. 5	12. 0	22. 7
285252	___do___	____do____	Apr. 9, 1921	67. 5	58. 5	12. 0	20. 0

In the city of Tucuman, *Brachyspiza* were recorded in song on April 1. Above Tafi Viejo, on April 17, they were found on the higher, more open slopes of the Sierra San Xavier, in little flocks that fed on the ground among growths of composites, or rested in the pleasant warmth of the sun at the borders of thickets of alder.

BRACHYSPIZA CAPENSIS MELLEA Wetmore

Brachyspiza capensis mellea WETMORE, Proc. Biol. Soc. Washington, vol. 35, Mar. 20, 1922, p. 39. (80 kilometers west of Puerto Pinasco, Paraguay.)

This pallid form, seemingly the *Brachyspiza* of the interior of the central and northern Chaco, was found at the Riacho Pilaga, Formosa, and near the ranch at Kilometer 80, west of Puerto Pinasco,

Paraguay. *B. c. mellea* is a well-marked race, easily distinguished from others that adjoin by its much paler coloration both above and below. The following measurements were taken from two adult males: Wing, 69.9–70; tail, 56.2–57.8; culmen, 11.8–12.2; tail, 20–21.2 mm. A female measures as follows: Wing, 64; tail, 55.2; culmen, 12; tarsus, 20.5 mm. The specimen from Formosa is slightly darker than those from Paraguay and seems to represent a somewhat intermediate condition verging toward *B. c. argentina.*

Individuals were observed at the Riacho Pilaga, Formosa, on August 13 and 14, 1920, and one was taken on the former date. At Kilometer 80, west of Puerto Pinasco, Paraguay, they were recorded from September 6 to 18, and birds were shot on September 16. Little flocks or single individuals came familiarly about the ranch house under the dirt-floored porches or in the bare patio, skipping nimbly away as people passed, their gray backs blending in such a manner with the color of the earth that by most they passed unnoticed.

The status of chingolos, as these birds are commonly known, recorded along the Rio Paraguay, at Puerto Pinasco, on September 3 and 4, is uncertain, as no specimens were taken.

BRACHYSPIZA CAPENSIS CHORAULES Wetmore and Peters

Brachyspiza capensis choraules WETMORE and PETERS, Proc. Biol. Soc. Washington, vol. 35, Mar. 20, 1922, p. 44. (General Roca, Gobernacion de Rio Negro, Argentina.)

The present form, common at General Roca, Rio Negro, the type locality, may be supposed to range in the valley of the Rio Negro from Neuquen eastward. Adult males were taken at Roca on November 23 and 30, and adult females on November 24 and 30, all in breeding condition. Specimens shot farther north, at Tunuyan, Mendoza, vary slightly toward *B. c. chilensis,* but more nearly resemble *choraules,* with which they are identified. Adult females taken on March 25 and 27 have just completed the post-nuptial molt. One has the black lateral crown stripes restricted to narrow lines and may be a migrant from the south, as it represents an approach to *canicapilla.* One immature female shot March 24 is in full fall plumage; another taken on the same date is just beginning the molt from the juvenal stage. This latter specimen is paler and browner (less blackish) above than *B. c. argentina* of the same age, and differs from *B. c. chilensis* (from Potrerillos, Mendoza, in similar condition) much as do adults of *choraules,* as it is paler above, more especially on the sides of the neck, and has a more extensively white superciliary stripe. Measurements of fully grown individuals are given below.

Catalogue No.	Sex	Locality	Date	Wing	Tail	Culmen	Tarsus
				Mm.	Mm.	Mm.	Mm.
284130	Male ad____	General Roca, Rio Negro_____	Nov. 23, 1920	77.2	59.2	12.2	22.5
284125 [1]	____do_____	____do_____	Nov. 30, 1920	78.2	60.5	12.0	21.3
284128	Female ad__	____do_____	Nov. 24, 1920	73.3	58.2	11.4	21.5
284136	____do_____	____do_____	Nov. 30, 1920	75.7	63.0	11.5	21.3
285247	Female im__	Tunuyan, Mendoza_____	Mar. 24, 1921	75.0	61.2	11.2	21.2
285261	Female ad__	____do_____	Nov. 25, 1921	76.7	60.7	11.7	21.5
285246	____do_____	____do_____	Nov. 27, 1921	79.8	67.5	11.3	21.0

[1] Type specimen.

At General Roca these finches were common from November 23 to December 3, where they found conditions suitable for their habitation. They were most abundant in the brush-grown flats toward the Rio Negro, and seemed on the whole restricted in their distribution by the availability of water, as toward the gravel hills that bordered the flood plain of the stream they were found only where irrigation ditches furnished the means to satisfy thirst.

As a whole, *Brachyspiza* is adaptable and will increase in abundance as extended irrigation projects offer it new range in regions at present desolate and arid.

The period of my visit to Roca corresponded with the height of the breeding season, and males were in full song, while their mates were busy with their nests or with the quieter duties of incubation. Occasionally males rose a few meters in the air to give a flight song, louder and clearer than their ordinary efforts, and then pitched down to cover. On November 24 I flushed a female from a nest placed under a tussock of grass in a little opening in the brush. The nest, a slight cup made of grass sunk in the sandy soil until it was level with the surface, contained three fresh eggs. The eggs are similar in color to those described under *B. c. canicapilla;* one has bold, somewhat restricted markings, the other two a more suffused wash that in one nearly conceals the ground color. Measurements of this set are appended: 19.4 by 14.2, 18.9 by 14.3, 18.6 by 14.1 mm. A pair that nested in the patio at my hotel brought out two young on November 25. Cats killed one of the young birds in a short time and the other was placed in a cage hung on a post where it was safe. The parents continued to attend the wants of their offspring and scolded sharply whenever possible enemies appeared. The male became so reconciled to the new arrangement that frequently he sang his pleasant song from a perch on the cage. Like many other birds of this region I found *Brachyspiza*, elsewhere tame and confiding, unusually wild.

At Tunuyan, Mendoza, from March 22 to 29, 1921, these birds were found in scattered flocks spread through extensive weed patches. A part of them seemed to be migrants, or so it appeared from their actions. Though the breeding season was past, males sang pleasantly at frequent intervals or even on rarer occasions rose in the air for a flight song.

BRACHYSPIZA CAPENSIS CANICAPILLA (Gould)

Zonotrichia canicapilla GOULD, Zool. Voy. Beagle, pt. 3, Birds, 1841, p. 91.
(Port Desire, Patagonia, and Tierra del Fuego.)

The present form of *Brachyspiza* was the breeding species at Zapala, Neuquen, though at General Roca, a few meters lower, I found the more familiar type with a black streak on either side of the crown. *B. c. canicapilla* was fairly common through the brush that grew in scattered clumps over the gravelly hills above the town of Zapala on December 7, 8, and 9, and one specimen was taken on each of the two dates last mentioned. On December 7 a female flushed from a nest placed in a depression in the sand under a little bush and ran rapidly away with lifted wings. The nest was a cup of grasses and weed stems, lined warmly with small rhea feathers. The two heavily incubated eggs have a ground color paler than pale Niagara green, finely dotted with Rood's brown, the dots more or less confluent in small irregular blotches, heavier about the larger end, where they are accompanied by minute, scattered spots of black. The markings in one egg are soft and suffused, in the other bolder and more restricted. These eggs measure, respectively, 18.5 by 14.4, and 18 by 14.5 mm.

At Potrerillos, Mendoza, on March 20, 1921, I killed an immature female of the present race, in company with *B. c. chilensis*. I supposed that it was a migrant from some more southern region. These records probably mark near the northwestern extension of the range of *canicapilla*, both during the breeding season and in migration.

BRACHYSPIZA CAPENSIS CHILENSIS (Meyen)

Fringilla chilensis MEYEN, Nov. Act. Acad. Caes. Leop.-Carol. Nat. Curios.,
vol. 16, Suppl., 1834, p. 88. (Santiago, Chile.[20])

At Concon, Chile, this form was common from April 24 to 28, 1921, and five specimens were taken on April 24, 25, 26, and 27. These include adult and immature birds all in full fall plumage. Five additional specimens secured at Potrerillos, Mendoza, in the Andean foothills, while not wholly typical, are so near *chilensis* from Chile as to forbid their separation. The bird was common in this vicinity, in brushy areas near streams, in the valleys, and along the canyon walls from March 15 to 21, 1921. Two adult males (taken March 19 and 20) have not quite completed the post-breeding molt. An adult

[20] For this reference I am indebted to notes from this publication made by Dr. C. W. Richmond.

female (March 15) is still in worn breeding dress, and a male and a female (March 17 and 19), the former with tail not fully grown, are in juvenal plumage. These specimens were secured at altitudes ranging from 1,500 to 1,800 meters, the latter the elevation at a ranch known as El Salto at the base of the main Cordillera. These birds measure about as in *B. c. choraules* found on the plains beyond the foothills in the same Province, but are distinctly darker. *B. c. chilensis* seems to range across the Andes in this latitude as far as the foothills on the eastern slope. Measurements of specimens of this subspecies are tabulated below.

Catalogue No.	Sex	Locality	Date	Wing	Tail	Culmen	Tarsus
				Mm.	*Mm.*	*Mm.*	*Mm.*
284952	Male ad	Concon, Chile	Apr. 25, 1921	76.7	59.0	12.2	22.5
284951	----do-----	----do-----	Apr. 27, 1921	74.2	59.2	12.2	22.0
285256	----do-----	Potrerillos, Mendoza	Mar. 19, 1921	78.0	56.5	12.2	22.0
285249	----do-----	----do-----	Mar. 20, 1921	80.5	65.5	12.1	21.2
284953	Female im	Concon, Chile	Apr. 24, 1921	77.5	62.5	12.5	21.6
284950	Female ad	----do-----	Apr. 26, 1921	72.3	57.5	12.0	21.0
284949	Female im	----do-----	Apr. 27, 1921	72.5	60.7	12.1	22.0
285258	Female ad	Potrerillos, Mendoza	Mar. 15, 1921	77.2	62.0	12.0	21.5

POOSPIZA MELANOLEUCA (d'Orbigny and Lafresnaye)

Emberiza melanoleuca d'ORBIGNY and LAFRESNAYE, Mag. Zool., 1837, cl. 2, p. 82. (Chiquitos, Bolivia.)

Specimens taken are similar to a single skin seen from Sumaipata, Bolivia, though the latter appears to have a slightly longer tail than birds from the Argentine. The species was recorded as follows: Resistencia, Chaco, July 10 (immature male taken); Las Palmas, Chaco, July 13 to 31 (adult male; July 13, adult female, July 20); Riacho Pilaga, Formosa, August 8 to 18 (three females, August 9 and 11); Formosa, Formosa, August 24; Kilometer 80, west of Puerto Pinasco, Paraguay, September 11; Kilometer 200, west of Puerto Pinasco, September 25; Tapia, Tucuman, April 7 to 13 (adult female, April 7). The female has the crown gray, with only an indication of the black found in the male. The bird taken April 7 is in molt. One fully adult male has the flanks distinctly cinnamon, a marking seen in no other specimen. Hellmayr [21] notes that *Poospiza cinerea* Bonaparte of Brazil is specifically distinct from *P. melanoleuca* because of its longer wings, larger bill, and lack of black on the crown in the adult male. These attractive little finches were encountered in bands of a dozen or less that frequented the ground or low coverts amid thickets, scattered bushes or growths of saw grass. At an alarm they came out into the open for a few seconds, chipping excitedly and flashing the white in the tail, but

---

[21] Nov. Zool., vol. 15, 1908, p. 35.

disappeared at once, slipping away one at a time until almost before I was aware of it all had gone.

A male had the bill black; tarsus and toes blackish brown number 7; iris Prussian red.

POOSPIZA NIGRO-RUFA (d'Orbigny and Lafresnaye)

Emberiza nigro-rufa d'ORBIGNY and LAFRESNAYE, Mag. Zool., 1837, cl. 2, p. 81. (Santa Fe, Argentina.)

A common species in central and northern Argentina that I found at Berazategui, Buenos Aires, June 29, 1920 (adult female taken); Las Palmas, Chaco, July 22 and 28 (adult and immature female and immature male shot); Lavalle, Buenos Aires, November 9 (immature female taken); Carrasco, Uruguay, January 9, 1921; La Paloma, Uruguay, January 23; San Vicente, Uruguay, January 27 and February 2 (adult male taken); Lazcano, Uruguay, February 5 to 9 (female in immature dress shot); Rio Negro, Uruguay, February 14 (adult male and immature female killed).

Plumage in this species is highly variable from the heavily streaked juvenal dress to the rich color of the adult, so that almost every shade between the two may be found. Chubb [22] has remarked that northern birds have longer wings than those from the south and west, a distinction that is not evident in the series before me. This, however, does not include skins from Paraguay.

A male taken February 14 is in full molt.

This finch frequented saw-grass swamps, reeds bordering lagoons and channels, thickets growing in low, wet grounds, or growths of weeds bordering canefields in low situations, where it sought heavy cover and only appeared when alarmed to look about for an instant or to fly away with quickly tilting flight. Only occasionally was it observed resting in the open.

POOSPIZA WHITII Sclater

Poospiza whitii SCLATER, Proc. Zool. Soc. London, 1883, p. 43, pl. 9. (Córdoba.)

An immature female was taken April 10, 1921, at Tapia, Tucuman, from a perch in the top of a bush near a growth of weeds. It differs from the female of *P. nigro-rufa* in having nearly half of the distal end of the outer rectrix white, in paler flanks, paler dorsal surface, and more white on the breast.

A male collected at Salta, July 30, 1921, by D. S. Bullock, differs in general from *nigro-rufa* as does the female mentioned above, but has the white in the tail slightly less extensive.

---

[22] Ibis, 1910, p. 640.

POOSPIZA ORNATA (Landböck)

Phrygilus ornatus LANDBÖCK, Journ. für Ornith., 1865, p. 405. (Between the guard house at the Portillo Pass and Melocoton, Mendoza.)

Near Victorica, Pampa, this handsome finch was common from December 23 to 29, 1920; four specimens taken include a male and two adult and one juvenile females. The species has been recorded previously from Mendoza, Rioja, Tucuman, and Cordoba. The female, in juvenal plumage, is dark mouse gray above with dull avellaneous margins on the feathers, the margins somewhat brighter on the rump; a faint superciliary of dull cream buff; remiges and wing coverts blackish mouse gray, margined with dull avellaneous; rectrices blackish mouse gray, outer web and half of inner web of outermost white; three succeeding pairs tipped on inner web with white; cheeks chaetura drab, faintly mottled with dull buff; below whitish, washed with pinkish buff streaked everywhere with dark mouse gray; undertail coverts pinkish buff.

These finches ranged among bushes or low trees in pasture lands, where they fed quietly on the ground amid scattered bunch grass or rested on small limbs sometimes at the summits of little trees. At this season adults were accompanied by grown young. Males still sang at intervals a low song, but one sharp and emphatic in its inflection. Their tilting flight was accompanied by a prominent display of white in the tail. They were quiet and undemonstrative and uttered no call notes.

POOSPIZA ASSIMILIS Cabanis

Poospiza assimilis CABANIS, Mus. Hein., pt. 1, 1851, p. 137. (Southern Brazil and Paraguay.)

Fairly common at San Vicente, Uruguay, from January 28 to 30, 1921, when two males, a female, and a juvenile bird were taken; seen at Lazcano, Uruguay, February 7 and 8, along the Rio Cebollati; and taken at Rio Negro, Uruguay, February 17 and 18 (two adult males). The bird in juvenal plumage has breast, throat, sides of head, and forepart of crown washed heavily with olive yellow. The greater coverts are margined heavily with white, a marking that may be present or absent in adults. Adult birds taken are in worn plumage or are molting.

Poospiza assimilis was found in dense thickets, where it moved about in a slow and somewhat sluggish manner. It was decoyed out into view readily, but after a few seconds dodged back out of sight. Adult males sang, in spite of rain and wind, from perches among leaves near the tops of bushes, uttering a shrill song, given earnestly but with little carrying power. Their song may be written *wheet see a wheet wheet;* the call note was a low *chwit.* The young, with greenish heads, were very tame and perched unafraid almost within reach.

An adult male had the maxilla dull black; base of gonys dark olive buff; rest of mandible somewhat lighter than andover green; iris russet; tarsus and toes chaetura drab.

POOSPIZA TORQUATA (d'Orbigny and Lafresnaye)

Emberiza torquata d'ORBIGNY and LAFRESNAYE, Mag. Zool., 1837, p. 82. (Sicasica, Bolivia.)

At Kilometer 200, west of Puerto Pinasco, Paraguay, an adult female was taken September 25, 1920, from a little flock found working among small branches in heavy brush. Subsequently others were noted at Victorica, Pampa, on December 24 and 29, and two more adult females were taken. They frequented low, scrubby growth near the ground. The flight was tilting, accompanied by a prominent display of white in the tail.

LOPHOSPINGUS PUSILLUS (Burmeister)

Gubernatrix pusilla BURMEISTER, Journ. für Ornith., 1860, p. 254. (Tucuman.)

At Laguna Wall, 200 kilometers west of Puerto Pinasco, Paraguay, on September 25, 1920, I found a pair of these little, crested finches feeding on the ground at the border of a thicket of *vinal*; and secured a female. The birds flew with quickly tilting flight, with a prominent display of the white tail markings. Near Tapia, Tucuman, on April 11, 1921, I secured another near a large barranca in a tract of low brush grown with weeds, where it was on the ground with a mixed flock of *Brachyspiza* and *Saltatricula*. The species has been recorded from Cordoba, Tucuman, Salta, and Jujny, so that its occurrence in western Paraguay is not astonishing. The male is in molt. The female shows a black throat patch of irregular form, and smaller size than in the male. In the Ibis for 1880 (pl. 9) a female taken in Tucuman by Durnford is figured with the throat white.

SALTATRICULA MULTICOLOR (Burmeister)

Saltator multicolor BURMEISTER, Journ. für Ornith., 1860, p. 254. (Parana.)

This handsome finch was first encountered at the Laguna Wall, 200 kilometers west of Puerto Pinasco, Paraguay, where a male was taken on September 25, 1920, a considerable extension of range as it has not been recorded in literature outside of Argentina, where it is found from Cordoba and Mendoza north to Salta. Near Tapia, Tucuman, the species was common from April 7 to 13, 1921, and four were skinned. The single bird from Paraguay is paler brown on back, wing edgings, and sides than others seen from Tucuman and Salta, and may represent a geographic race not yet described. Specimens taken at Tapia include adults in molt and a female in juvenal plumage which differs from the adult in being paler brown, and much paler on sides and flanks.

These birds ranged in pairs or parties of five or six on the ground among low bushes, in burns, or at the borders of fields or thickets. When alarmed they flew up, with a flash of black and white from the tail and a tilting flight, to perch in thickets sometimes 3 or 5 meters from the ground. Their light coloration made them conspicuous, but they rested quietly without apparent fear to drop back after a minute or two out of sight. Their call is an insistent chipping note like that of a junco.

A male, taken April 7, had the maxilla dark mouse gray; cutting edge of maxilla and mandible mustard yellow; iris natal brown; tarsus and toes neutral gray. A juvenile individual had the bill dull black with the cutting edges faintly marked with dull yellow.

AIMOPHILA STRIGICEPS DABBENEI (Hellmayr)

Zonotrichia strigiceps dabbenei HELLMAYR, Verh. Ornith. Ges. Bayern, vol. 11, 1912, p. 190. (Tapia, Tucuman.)

Two specimens were taken at Tapia, Tucuman, April 10, 1921, one of which was prepared as a skeleton. Doctor Hellmayr in describing the present form differentiated it from *strigiceps* (type-locality Santa Fe) on larger size, stronger bill, and darker brown stripes on the side of the head. The specimens taken at Tapia are the only ones that I have seen. The skin, an adult male, has the following measurements: Wing, 72.6; tail, 78.3; culmen from base, 14; tarsus, 22 mm.

The generic relationships of this species have been in doubt. The bird belongs without question in the supergeneric group entitled the Zonotrichiae by Mr. Ridgway,[23] in which genera, in most cases, are closely related. It is certainly not a *Zonotrichia*, in which Hellmayr placed it provisionally, as it has the tail longer than the wing, a different wing formula and a heavier bill, reasons that likewise prevent its allocation in *Brachyspiza*. In general appearance, wing, tail, and bill, it is closely similar to *Aimophila rufescens*, the type of *Aimophila* Swainson, differing from that bird structurally mainly in its smaller more delicate feet. Although, as Mr. Ridgway has indicated, *Aimophila* is a somewhat heterogeneous group, still it may not be divided, and *strigiceps* may be placed in it without difficulty, as has been done by Salvadori and Dabbene and others. As proof of this the species runs readily to the genus *Aimophila* in Mr. Ridgway's key to the finches of North and Middle America.

---

[23] Birds North and Middle America, vol. 1, 1901, p. 28.

These finches were found associated with chingolos (*Brachyspiza*) in growths of more or less open brush and weeds. Three seen rested quietly on low perches, easily distinguished from *Brachyspiza* by their larger size and grayer coloration. In color and bearing they are strikingly similar to North American song sparrows. Those taken uttered a sharp, chipping note. They were breeding at this season.

EMBERNAGRA PLATENSIS (Gmelin)

Emberiza platensis GMELIN, Syst. Nat., vol. 1, pt. 2, 1789, p. 886. (Buenos Aires.)

In a review of the genus *Embernagra*, Chubb[24] has recognized three forms of *E. platensis*, the typical subspecies from "eastern Argentina," *E. p. poliocephala* Gray from southern Uruguay, and *E. p. paraguayensis* described as new from Paraguay, Rio Parana, and northeastern Argentina. Of the latter no type is cited, but from the context it appears that the type specimen may have been one taken by Foster at Sapucay. *E. p. poliocephala* is said to differ from *platensis* in possessing a whitish abdomen which contrasts with the buffy flanks. In material at hand, which includes skins from San Vicente, Uruguay, and Taquara do Mundo Novo and Quinta, Rio Grande do Sul, I can see no distinction between birds from southern Uruguay and those from the opposite shore of the Rio de la Plata in Buenos Aires.

E. p. paraguayensis is said to have the upper surface darker green and the abdomen whiter than *platensis*. A skin from Sapucay, Paraguay, and one from Resistencia, Chaco, should represent this race but offer no evident differences from a fair series from Conchitas and Lavalle, Buenos Aires.

E. platensis was recorded as follows: Resistencia, Chaco, July 9 (male taken) to July 10, 1920; Las Palmas, Chaco, July 15 to July 31; Riacho Pilaga, Formosa, August 7 to 21; Formosa, Formosa, August 23 and 24; Kilometer 25, west of Puerto Pinasco, Paraguay, September 1; Dolores, Buenos Aires, October 21; Lavalle, Buenos Aires, October 23 to November 15 (one male and two females taken); General Roca, Rio Negro (?), December 3; Carhue, December 15 to 18; Carrasco, Uruguay, January 16, 1921; San Vicente, Uruguay, January 26 and 27 (male taken); Lazcano, Uruguay, February 5 to 9; Rio Negro, Uruguay, February 15; Guamini, Buenos Aires, March 3 and 4 (two taken).

A bird of this genus seen near the Rio Negro, below General Roca, is placed here tentatively, as it was not collected. It is possible that it was *E. o. gossei*.

The present species inhabited marshy localities, where it ranged in saw grass or low bushes, or on the pampas at times frequented

---

[24] Ibis, 1918, pp. 1–10, 1 pl.

clumps of *cola de zorro*, a plumed grass, standing amid thickets of thistles. The birds were usually in pairs, though in the Chaco, where they were common, five or six frequently congregated. They fed in little open places on the ground where they walked about sedately, or when in a hurry progressed with long hops. Their flight was rapidly tilting. When excited they perched in low bushes or clumps of grass, appearing big and bluff as they are strong and robust of body. During winter, males sang an absurd little whispered song that was barely audible at 50 meters, but with the opening of the nesting season they became more noisy and sang persistently a curious accented song. The strangely phrased notes may be represented as *oo ro fa la tsuf la tsee*. Their call is a faint *tsip*.

Occasionally one came hopping about camp in search of bread crumbs, and they often pecked avidly at bits of cooked meat or at freshly skinned bird bodies. Their tameness under such circumstances has given them the name of *toldero* in eastern Buenos Aires. In the Chaco they were known as *Juan che vido*, while elsewhere they were called *cotorrón* or *cotorrito*.

Young in streaked juvenal plumage were taken at Guamini, Buenos Aires, March 3. The plumage of adults is subject to much wear and abrasion, perhaps from the saw-edged grasses amid which they live, so that the green shade of the upper surface changes to gray; in some the prominent dark streaks of the back entirely disappear.

An adult male, taken July 9, had the upper part of the maxilla blackish slate; sides of maxilla and mandible zinc orange; iris dark brown; tarsus sayal brown; toes and claws hair brown; underside of toes yellowish. A juvenile male, taken March 3, had the maxilla (except at sides adjacent to the feathers) and a stripe along sides of mandible dull black; base of maxilla and lower margin of mandibular rami chamois, shading at margin to antimony yellow; anterior half of cutting edge of mandible and region over gonys slightly paler than avellaneous; iris natal brown; tarsus between benzo brown and fuscous; nails blackish.

EMBERNAGRA OLIVASCENS OLIVASCENS d'Orbigny

Embernagra olivascens d'ORBIGNY, Voy. Amér. Mérid., Ois., 1835–1844, p. 285. (Enquisivi, Province of Sicasica, Palca, Province of Ayupapa, and valley of Cochabamba.)

The present species is distinguished from *E. platensis* by the plain, unstreaked back. The typical subspecies is represented by a bird molting from juvenal to first fall plumage secured at Tapia, Tucuman, on April 6, 1921. This is distinctly greener above than skins of *E. o. gossei* of similar age. An adult male taken August 7, 1922, at Tucuman, by D. S. Bullock, also is distinctly greener on the back than a small series of *gossei*.

EMBERNAGRA OLIVASCENS GOSSEI Chubb

Embernagra gossei CHUBB, Ibis, 1918, p. 3, pl. 1, fig. 2. (Lujan, Mendoza.)

The southern subspecies of *olivascens*, described by Chubb as *gossei*, is characterized by the gray dorsal surface. At Tunuyan, Mendoza, it was fairly common from March 22 to 29 near wet, swampy localities. Specimens taken include an adult female and two young in juvenal plumage. Adults at this season were in worn dress and were molting, while juveniles were beginning to assume first fall plumage. In the juvenal stage this species is duller in color than *E. platensis* and has the dark streaks narrower and less sharply defined, particularly on the upper surface.

The birds at this season frequently found congenial haunts in cornfields bordering the saw-grass tracts of the extensive cienagas that formed their usual homes.

DONACOSPIZA ALBIFRONS (Vieillot)

Sylvia albifrons VIEILLOT, Nouv. Dict. Hist. Nat., vol. 11, 1817, p. 276. (Paraguay.)

Strange little long-tailed finches of this species were first seen in dense grass bordering a marsh at Las Palmas, Chaco, July 28, 1920 (adult male secured). Later, on October 25 and November 15 they were found breeding in growths of rushes near the mouth of the Rio Ajo, below Lavalle, Buenos Aires, where three were taken. They perched on the rush stalks with long tails blowing in the wind, or flew to secure coverts with rapid, tilting flight. Their call was a low *chip* or *zit*. The song was a pleasant buzzing warble that may be written as *tsef tsef tsef wee tsu wee tsu wee tsu*, barely audible above the rustling of the wind in the grass.

The species in general suggests a *Poospiza* of the *nigro-rufa* type with a greatly elongated tail.

MYOSPIZA HUMERALIS HUMERALIS (Bosc)

Tanagra humeralis Bosc, Journ. Hist. Nat., vol. 2, 1792, p. 179, pl. 34, fig. 4. (Cayenne.)

As indicated by Chubb,[25] *Tanagra humeralis* of Bosc from Cayenne has priority over *Fringilla manimbe* Lichtenstein,[26] as birds from Bahia and the Guianas appear identical. In recent years several forms have been distinguished in this widely distributed

---

[25] Bull. Brit. Orn. Club, vol. 31, Jan. 25, 1913, p. 39.
[26] *Fringilla Manimbe* Lichtenstein, Verz. Doubl. Zool. Mus. Berlin, 1823, p. 25. (Bahia.)

species, and as there is considerable variation in individual series identification is somewhat complicated. For comparison with my skins from Paraguay, Uruguay, and Argentina, I have assembled in all a series of 80 specimens from the United States National Museum, the American Museum of Natural History, the Museum of Comparative Zoölogy, and the Carnegie Museum. On first review it appears that the series may be easily separated into two groups, a northern one ranging from Paraguay northward, in which the dorsal surface is distinctly reddish brown, and a southern one, spread from northern Paraguay and Rio Grande do Sul southward, in which the prevailing coloration of the dorsal surface is blackish and gray with little or no brown except in the subspecies *tucumanensis*, which, however, is distinctly gray. Skins are subject to great seasonal wear, a fact that needs attention as northern specimens in worn breeding dress appear much blacker above than normal. Immature individuals in juvenal plumage are also darker above than normal. There is no appreciable difference apparent in measurements, though a few skins from the Argentine are larger than any examined from farther north. Others, however, have the size identical with skins from elsewhere. The tail is subject to wear and is highly variable in length.

In this study, as I have been intent on identifying southern skins, I have not attempted to assemble all available specimens from Venezuela and Colombia, and have accepted two northern subspecies in addition to the typical race somewhat on faith. The few skins seen indicate that the forms in question are distinct. Five current subspecies may be recognized as follows:

1. MYOSPIZA HUMERALIS HUMERALIS (Bosc).

Tanagra humeralis Bosc, Journ. Hist. Nat., vol. 2, 1792, p. 179, pl. 34, fig. 4. (Cayenne.)

Decidedly brown above, with well-defined black streaks.

Specimens seen from Venezuela (Maripa, Rio Caura; Suapore); British Guiana; Surinam (District of Para); Brazil (Santarem, Para; Ceara; Bahia; Chapada and Arapua, Matto Grosso); Bolivia (Buenavista, Prov. del Lara; Santa Cruz de la Sierra; Rio San Julian and Rio Quiser, Chiquitos); and Paraguay (Puerto Pinasco and Sapucay).

Specimens from Bahia seem closely similar to those from British and Dutch Guiana, here assumed to be typical since no birds from Cayenne are available, so that *manimbe* Lichtenstein (1823) for the present at least is considered a synonym of *humeralis*. Three specimens from Santarem on the Amazons are very gray above, with the black streaks considerably reduced. Mrs. E. M. B. Naumburg informs me that she has seen gray birds from the Islands of Marajos.

Skins from Bolivia and Matto Grosso are often paler than typical *humeralis*, but on the whole agree very well. With more material from Brazil it is possible that *manimbe* may be recognized as distinct, though I consider this doubtful.

2. MYOSPIZA HUMERALIS MERIDANA Todd.

Myospiza humeralis meridanus TODD, Proc. Biol. Soc. Washington, vol. 30, July 27, 1917, p. 127. (Guarico, Lara, Venezuela.)

Similar to *humeralis* but general coloration darker, pileum more heavily streaked with black, especially in front; gray edging of feathers of back less prominent, with brown more conspicuous; prevailing tone of upperparts brown, not black as in *columbiana*.

Specimens seen from Colombia (Palmar, Boyaca, and Paramo de Macatama, 2,800 meters) and British Guiana? (Mount Roraima?).

The present subspecies is intermediate between *columbiana* and *humeralis*. Two specimens from Mount Roraima, British Guiana, with a distinct wash of brown on sides and upper surface, are placed here tentatively but they do not fit well in any series. They are much browner than any others seen.

3. MYOSPIZA HUMERALIS COLUMBIANA (Chapman).

Myospiza manimbe columbiana CHAPMAN, Bull. Amer. Mus. Nat. Hist., vol. 31, July 23, 1912, p. 162. (Cali, Cauca, Colombia.)

Similar to *humeralis* but averaging darker, much more heavily streaked with blackish above. Darker than *meridana*.

Specimens seen from Colombia (Cali, Cauca; Yumbo, Valle.)

4. MYOSPIZA HUMERALIS DORSALIS (Ridgway).

Coturniculus manimbe, var. *dorsalis* RIDGWAY, in Baird, Brewer, and Ridgway, Hist. North American Birds, vol. 1, 1874, p. 549. (Buenos Aires.)

Without distinct rufescent or brownish markings above, prevailing tone of upperparts gray and blackish, very distinct from the reddish northern forms. Specimens seen from Paraguay (Puerto Pinasco); Argentina (Riacho Pilaga, Formosa; Las Palmas, Avia Terai, and General Pinedo, Chaco; Concepcion del Uruguay, Entre Rios; Buenos Aires, Guamini, and Carhue, Buenos Aires); Uruguay (Montevideo and Lazcano); and Rio Grande do Sul (Santa Maria).

This form was indicated originally as from Buenos Aires and Uruguay. The type-specimen in worn breeding dress, collected by J. K. Townsend, was taken at Buenos Aires. *Myospiza manimbe nigrostriata* Cherrie [27] is identical with the present form as is shown by examination of the type specimen.

---

[27] Bull. Amer. Mus. Nat. Hist., vol. 35, May 20, 1916, p. 189. (Rio Negro, a small tributary of the Rio Pilcomayo, entering 35 or 40 miles from its mouth, Paraguayan Chaco.)

The present subspecies is characterized by the prevalence of black and gray above, and in worn breeding dress appears very black in-deed, especially on the rump. Skins from Uruguay are similar to those from Buenos Aires, but those from Rio Grande do Sul verge toward *humeralis* as they are more brownish above. A good series from Avia Terai and General Pinedo, in central and western Chaco, are intermediate between *dorsalis* and *tucumanensis* as is to be ex-pected. Some, in fact, appear identical with *tucumanensis*, others resemble *dorsalis*, and others are halfway between.

A puzzling circumstance is that a specimen labelled Puerto Pinasco, September 4, 1916, taken by Cherrie, is typical of *dorsalis*, while another that I secured at the same point September 3, 1920, is as typical of the brown-backed *humeralis* from Brazil and north-ward. Further specimens alone can solve the riddle presented, espe-cially since *humeralis* is the form also from Sapucay, not far from Asuncion, Paraguay.

M. dorsalis is the best marked of the forms of the species.

5. MYOSPIZA HUMERALIS TUCUMANENSIS Bangs and Penard.

Myospiza humeralis tucumanensis BANGS and PENARD, Bull. Mus. Comp. Zoöl., vol. 62, April, 1918, p. 92. (Tapia, Tucuman.)

Similar to *dorsalis* but paler above with dark central markings of feathers narrow, with a buffy-brown cast to upper surface. De-cidedly paler and grayer above than *humeralis*.

Specimens seen from Tapia, Tucuman (the type specimen, col-lected by Dinelli) and Victorica, Pampa. A few skins taken by Miller and Boyle at the end of April and the beginning of May at Avia Terai and General Pinedo, Chaco, are closely similar to the type and may possibly be migrant individuals of this form, though taken at the same time as birds representative of *dorsalis*.

The only specimen of *M. h. humeralis* that I secured was an adult male taken September 3, 1920, at Puerto Pinasco, Paraguay, near the Rio Paraguay. On this date the birds were common in open, grassy localities, and one little colony, established in an area of bunch grass near the river, was about ready to breed. Males sang, from weeds or fence posts, a low ditty that may be represented as *chip-p-p chee-ee-ee-ee chee chee* in the cadence and tone of a song sparrow (*Melospiza melodia*). Their call note was a faint *tsip*. On the ground, they ran rapidly ahead of me from cover to cover, or flushed with an undulating flight. As no others were taken those recorded at Kilometer 80, September 9 to 20, and Kilometer 200, September 25, west of Puerto Pinasco, are assigned here with reser-vation.

The single male preserved is absolutely typical of the reddish brown northern form.

MYOSPIZA HUMERALIS DORSALIS (Ridgway)

Coturniculus manimbe, var. *dorsalis* RIDGWAY, in Baird, Brewer, and Ridgway, Hist. North American Birds, vol. 1, 1874, p. 549. (Buenos Aires.)

This common form was encountered at Riacho Pilaga, Formosa, August 8 to 21, 1920 (adult male, August 11); Formosa, Formosa, August 23 and 24; Las Palmas, Chaco, July 15 to 31 (4 adult males); Carhue, Buenos Aires, December 17 (adult male); Guamini, Buenos Aires, March 6, 1921 (adult male); Carrasco, Uruguay, January 16; La Paloma, Uruguay, January 23; Lazcano, Uruguay, February 2 to 9 (adult male taken); and Rio Negro, Uruguay, February 14. The birds were found in weed or grass grown fields usually near but not in marshy localities. They have the habits and mannerisms of the grasshopper sparrows of the United States, but appear decidedly darker in color. On the southern pampas they were found at times in pastures with very little cover, where they crept about on the ground as inconspicuously as possible. During the winter season they were entirely silent, but were in song the middle of February. One taken at Guamini, March 6, had begun to molt.

The Toba Indians called this bird *po ko likh*.

MYOSPIZA HUMERALIS TUCUMANENSIS Bangs and Penard

Myospiza humeralis tucumanensis BANGS and PENARD, Bull. Mus. Comp. Zoöl., vol. 62, April, 1918, p. 92. (Tapia, Tucumán.)

An adult male, shot December 26, 1920, at Victorica, Pampa, agrees in coloration with the type-specimen. At Victorica the birds were fairly common on rolling hills covered with bunch grass. They were breeding and were in song at this time.

PASSER DOMESTICUS (Linnaeus)

Fringilla domestica, LINNAEUS, Syst. Nat., ed. 10, vol. 1, 1758, p. 183. (Sweden.)

The familiar house sparrow, or gorrión, is well established now throughout the Argentine, where, according to Berg,[28] it was first introduced in Buenos Aires by E. Bieckert in 1872 or 1873, for the purpose of destroying a Psychid *Oiketicus platensis* Berg. Several importations may have been made, however, as Doctor Holmberg[29] reports that they were brought in by one Peluffo, and Gibson[30] cites a rumor that they were introduced by a German brewer. Sparrows first attracted attention in the nineties, as Gibson mentions them on the Calle Florida, in Buenos Aires, in 1890, and E. L. Holm-

---

[28] Com. Mus. Nac. Buenos Aires, vol. 1, 1901, p. 284. See also F. Lahille, El. Hornero, vol. 2, 1921, p. 216.
[29] Quotation from Rev. Jardin Zool., June 15, 1893, in El Hornero, vol. 2, 1920, p. 71.
[30] Ibis, 1918, pp. 386–387.

berg,[31] in 1898, says they had spread over a radius of 50 leagues from that city. At the present time they are found throughout the settled central Provinces and are extending the area of occupation rapidly. Mr. E. Lynch Arribalzaga [32] reported in 1920 that they had appeared about the church in Resistencia, Chaco, about 11 years previous, and I found them fairly common at Las Palmas, Chaco, where they were said to have arrived about 1917. In traveling north from Santa Fe I recorded them from the train as far as Vera, Santa Fe, but did not see them again until I reached Resistencia. J. H. Reboratti [33] notes that they arrived at Concepcion, Corrientes, about 1916.

In Formosa they were fairly common, and I was astonished, on August 21, 1920, to note a few about the railroad station at Kilometer 182, the farthest point inland to which I penetrated along the government railroad. It was an even greater surprise to note a pair, on October 6, in the little plaza opposite the railroad station in Asuncion, Paraguay, where they were nesting in a hollow in a tree. At the time I supposed this to be the first record of them for Paraguay, but since have seen a statement by Bertoni [34] that the species has been seen in some numbers in the streets of Ascuncion, since 1920. Bertoni cites a rumor that the sparrow had been imported from Buenos Aires on two occasions, but considers it probable that it has invaded Paraguay of its own volition. The latter is credible, as *Passer* at the time in question had established itself near the Argentine frontier.

English sparrows were common at Lavalle, Buenos Aires, and one even appeared, on November 7, at a remote cabin hidden among the sand dunes of the coast, 25 kilometers below Cape San Antonio. To the south they were common in Bahia Blanca, and extended along the line of railroad to the west as far as Zapala, Neuquen, at the base of the Andes. Peters [35] noted them in Rio Negro during 1920 and 1921 at Puesto Horno, Maquinchao, and Huanuluan, and I heard rumor that they had spread as far as 16 de Octubre, in northwestern Chubut. Bennett [36] has noted their arrival in Port Stanley, in the Falkland Islands, in November, 1919, as stowaways aboard four sailing vessels from Montevideo.

I found the sparrow common at Victorica, Pampa, December 23, and saw it at Potrerillos, Mendoza, March 15, Tunuyan, Mendoza,

[31] Fauna Arg., 1898, p. 545.
[32] El Hornero, vol. 2, 1920, pp. 97–98.
[33] El Hornero, vol. 7, 1919, p. 194.
[34] Rev. Soc. Cient. Paraguay, vol. 1, July, 1923, pp. 73–74.
[35] Bull. Mus. Comp. Zoöl., vol. 65, May, 1923, p. 331.
[36] El Hornero, vol. 2, 1921, p. 225. See also p. 204.

March 23 to 29, and Tapia, Tucuman, April 6 to 14. Serié[37] has noted its recent arrival at Amaicha, Tucuman.

At the Museo de Historia Natural in Montevideo I was told that the sparrows of Uruguay were supposed to have come from an importation made at Colonia about 30 years ago. Alvarez,[38] in 1913, speaks of them as common throughout Uruguay, in corroboration of which I noted them in fair numbers at San Vicente, Lazcano, Corrales, and Rio Negro. E. G. Holt informs me that in 1919 they were established in the city of Rio Grande do Sul in southeastern Brazil. (I saw large numbers in Rio de Janeiro, in June, 1920.) Serié also reports them as abundant in Rio Grande do Sul.

The three preserved as skins were taken at Lavalle, Buenos Aires, November 1, 1920; Guamini, Buenos Aires, March 5, 1921; and Tunuyan, Mendoza, March 23, 1921.

SPINUS ICTERICUS (Lichtenstein)

Fringilla icterica LICHTENSTEIN, Verz. Doubl. Zool. Mus. Berlin, 1823, p. 26. (Sao Paulo, Brazil.)

Seven specimens are referred to the present species: A male from Dolores, Buenos Aires, October 21, 1920; two males from Lavalle, Buenos Aires, October 27 and November 1; a pair from San Vicente, Uruguay, January 26, 1921; and two immature males taken at El Salto, above Potrerillos, Mendoza, March 19. The last two are in juvenal dress and are identified tentatively as comparative material of Neotropical *spinus* is not at this moment available. These two are dull in color and have the under tail coverts striped with dusky which suggests that they may be *barbatus*.

On October 21, 1920, three were observed and one taken on low ground bordering a marsh near Dolores. In the vicinity of Lavalle, Buenos Aires, the species was common from October 27 to November 13. On October 30, at Los Yngleses, a pair were busy with the construction of a cup-shaped nest in a climbing rose near a door. And on the following day another nest was begun 7 meters from the ground in a pine tree. During the first week in November siskins were noisy and demonstrative, and were continually flying about with great display of the yellow markings in wings and tail, but by November 10, with incubation begun, they were quieter and less frequently in view, though the song of the male, a pleasant chattering warble, uttered brokenly and rapidly like that of a pine siskin, was still given constantly. They delighted in feeding on the ground on recently cut lawns.

Near San Vicente, Uruguay, the species was breeding commonly in the great palm groves that grew in the swampy lowlands, and at

---

[37] El Hornero, vol. 3, 1923, p. 190.
[38] T. Alvarez, Exterior de las aves Uruguayas, etc., Montevideo, 1913, pp. 59–60.

Lazcano, Uruguay, they were fairly common from February 5 to 8. At Potrerillos, Mendoza, a little flock was found at an elevation of 6,000 feet on March 19, and others were noted lower, near the hotel, on March 21. They were seen at Tunuyan, Mendoza, March 27, and at Tapia, Tucuman, April 7 to 13.

SPINUS BARBATUS (Molina)

Fringilla barbata MOLINA, Sagg. Stor. Nat. Chili, 1782, p. 247. (Chile.)

A male from Concon, Chile, taken April 26, 1921, is supposedly typical of this species. Another shot at General Roca, Rio Negro, November 24, 1920, is in a peculiar immature stage of plumage, though in breeding condition.

Near General Roca, Rio Negro, these birds were seen in small numbers along irrigation ditches, where they were breeding. At Concon, Chile, small flocks of 25 or 30 were found singing sweet medleys from the tops of trees standing in the open, or passing with undulating flight to feed among bushes and weeds.

INDEX

435

CPSIA information can be obtained
at www.ICGtesting.com
Printed in the USA
BVOW06s1427090617
486278BV00024B/313/P